BASIC PLUMBING SKILLS

TAFE NSW

PLUMBING SKILLS Series

BASIC PLUMBING SKILLS

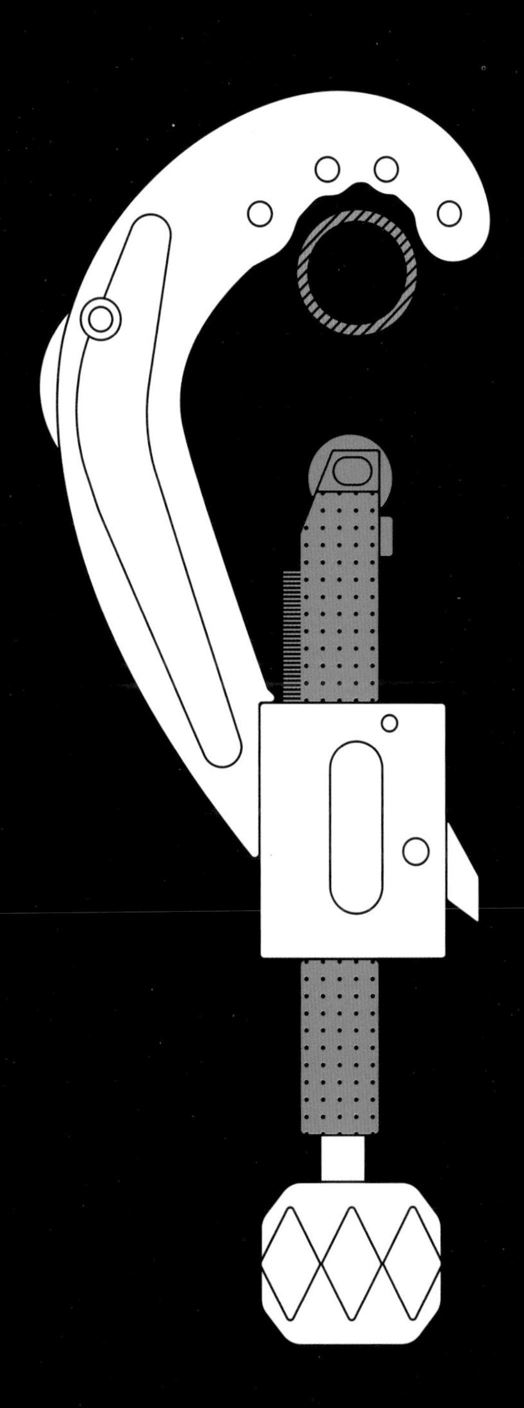

Basic Plumbing Skills
4th Edition
Dean Carter

Head of content management: Dorothy Chiu
Content manager: Sandy Jayadev/Chee Ng
Content developer: Christin Quirk/James Cole
Project editor: Raymond Williams
Editor: Peter Cruttenden
Proofreader: James Anderson
Indexer: Max McMaster
Permissions/Photo researcher: Wendy Duncan
Text designer: Linda Davidson
Cover design: Chris Starr (MakeWork)
Typeset by KnowledgeWorks Global Ltd

Third edition published 2017

For product information and technology assistance,
in Australia call 1300 790 853;
in New Zealand call 0800 449 725

For permission to use material from this text or product, please email
aust.permissions@cengage.com

National Library of Australia Cataloguing-in-Publication Data
ISBN: 9780170424691
A catalogue record for this book is available from the National Library of Australia.

Cengage Learning Australia
Level 7, 80 Dorcas Street
South Melbourne, Victoria Australia 3205

Cengage Learning New Zealand
Unit 4B Rosedale Office Park
331 Rosedale Road, Albany, North Shore 0632, NZ

For learning solutions, visit cengage.com.au

Printed in China by 1010 Printing International Limited.
6 7 25 24

BRIEF CONTENTS

CONTENTS

Guide to the text

As you read this text you will find a number of features in every chapter to enhance your study of plumbing and help you understand how the theory is applied in the real world.

PART-OPENING FEATURES

The **Part overview** helps students understand the relationships between the units and topics.

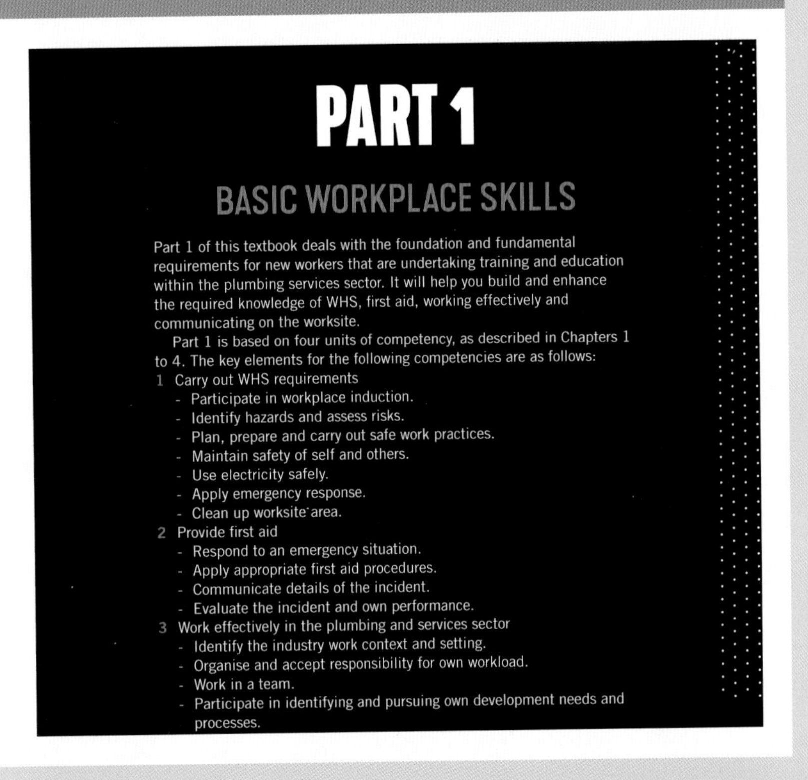

PART 1

BASIC WORKPLACE SKILLS

Part 1 of this textbook deals with the foundation and fundamental requirements for new workers that are undertaking training and education within the plumbing services sector. It will help you build and enhance the required knowledge of WHS, first aid, working effectively and communicating on the worksite.

Part 1 is based on four units of competency, as described in Chapters 1 to 4. The key elements for the following competencies are as follows:

1 Carry out WHS requirements
   - Participate in workplace induction.
   - Identify hazards and assess risks.
   - Plan, prepare and carry out safe work practices.
   - Maintain safety of self and others.
   - Use electricity safely.
   - Apply emergency response.
   - Clean up worksite area.
2 Provide first aid
   - Respond to an emergency situation.
   - Apply appropriate first aid procedures.
   - Communicate details of the incident.
   - Evaluate the incident and own performance.
3 Work effectively in the plumbing and services sector
   - Identify the industry work context and setting.
   - Organise and accept responsibility for own workload.
   - Work in a team.
   - Participate in identifying and pursuing own development needs and processes.

CHAPTER OPENING FEATURES

The **Chapter overview** lists the topics that are covered in the chapter.

Identify the key concepts you will engage with through the **Learning objectives** at the start of each chapter.

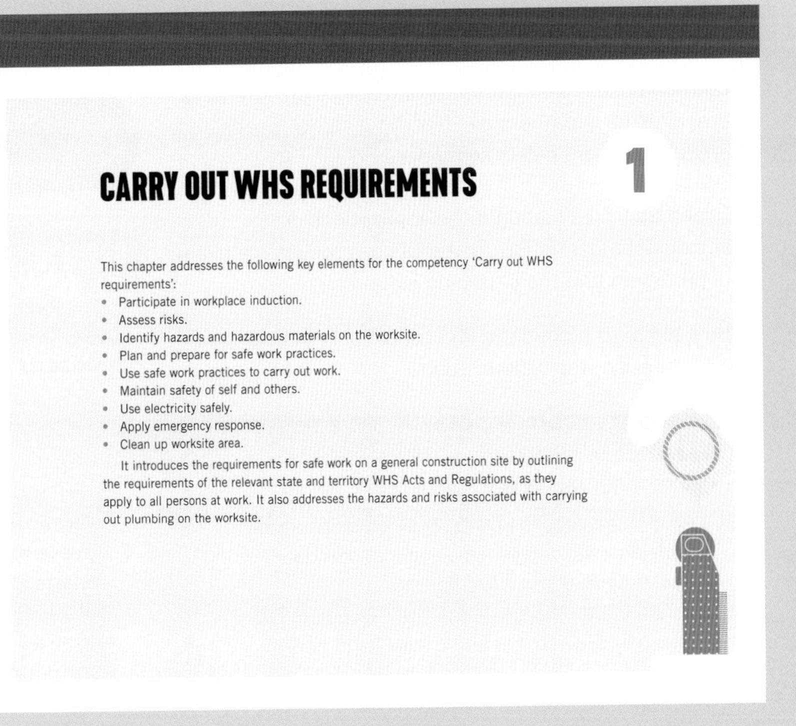

CARRY OUT WHS REQUIREMENTS

1

This chapter addresses the following key elements for the competency 'Carry out WHS requirements':

- Participate in workplace induction.
- Assess risks.
- Identify hazards and hazardous materials on the worksite.
- Plan and prepare for safe work practices.
- Use safe work practices to carry out work.
- Maintain safety of self and others.
- Use electricity safely.
- Apply emergency response.
- Clean up worksite area.

It introduces the requirements for safe work on a general construction site by outlining the requirements of the relevant state and territory WHS Acts and Regulations, as they apply to all persons at work. It also addresses the hazards and risks associated with carrying out plumbing on the worksite.

FEATURES WITHIN CHAPTERS

Engage actively with the learning by completing the practical activities in the **Learning task** boxes.

LEARNING TASK 1.1

PARTICIPATE IN WORKPLACE INDUCTION

NEW

Choose a simulated workplace such as a practical workshop and discuss the WHS requirements for a workplace induction.

Provide short answers to the following questions:

1 What are the risks and hazards specific to the construction area?
2 How are hazards reported and who are they reported to?
3 Identify the location and type of emergency equipment currently located in the workshop.
4 What are the emergency evacuation procedures for the site?

The **Standards** box highlight where Plumbing Standards are addressed, to strengthen knowledge and hone research skills.

1.1 STANDARDS

- AS/NZS 1801 Occupational protective helmets
- AS/NZS 1800 Occupational protective helmets – Selection, care and use

From experience boxes explain the responsibilities of employees, including skills they need to acquire and real-life challenges they may face at work to enhance employability skills on the job site.

FROM EXPERIENCE

Effective communication skills and working as a team to achieve a common outcome will reduce the risk of injury when performing manual handling tasks.

Coordinating team lifting uses the following steps:

- One person should be nominated and give the orders and signals; this person must be able to see what is happening.
- The movements of the team members should be performed simultaneously (i.e. all lifting together).
- All people involved in the lift should be able to see or hear the one giving the orders.

Caution boxes highlight important advice on safe work practices for plumbers by identifying safety issues and providing urgent safety reminders.

Exposure to carbon dioxide in a confined area could cause asphyxiation. Move clear of the area immediately after use and ventilate the area once the fire is out.

Green tip boxes highlight the applications of sustainable technology, materials or products relevant to plumbers and the plumbing industry.

GREEN TIP

As plumbers, it is essential we commit to engaging in sustainable and innovative practices to ensure we continue to secure our water supply for many years to come.

How to boxes highlight a theoretical or practical task with step-by-step walkthroughs.

HOW TO

MAKE A PEINED-DOWN JOINT

1 From the supplied plans, mark out and cut the material into shape, ensuring that adequate material is allowed for the laps.
2 Turn out the flange on the body section of the job using a hammer and a steel hand dolly (see Figure 8.27 (c), below) or a burring machine to achieve an angle of 90° (see Figure 8.20).
3 Dress and round up the cylinder. Measure the outside diameter in several places and average the measurement to ensure the cylinder is round (see Figure 8.21).

END-OF-CHAPTER FEATURES

At the end of each chapter you will find several tools to help you to review, practise and extend your knowledge of the key learning objectives.

References and **Further reading** sections provide you with a list of each chapter's references, as well as links to important text and web-based resources.

REFERENCES AND FURTHER READING

Acknowledgements
Reproduction of the following resource list references from DET, TAFE NSW C&T Division (Karl Dunkel, Program Manager, Housing and Furniture) and the Product Advisory Committee is acknowledged and appreciated.

Texts
Graff, D.M. & Molloy, C.J.S. (1986), *Tapping group power: A practical guide to working with groups in commerce and industry*, Synergy Systems, Dromana, Victoria.

Web-based resources
Regulations/Codes/Laws
AustLII legislation database: **http://www.austlii.edu.au/databases.html**
Australian Codes of Practice: **http://www.safeworkaustralia.gov.au**
SafeWork NSW (1999), Hazard profile for demolition: **http://www.workcover.nsw.gov.au/Publications/Industry/Construction/demolition.htm**
WA Government OHS site – manual handling: **http://www.wa.gov.au**

Audiovisual resources
Short videos covering topics such as safe manual handling, workplace housekeeping, hazardous substances and accident investigation are available from the following organisations:
Safetycare: **http://www.safetycare.com.au**
TAFE NSW: **https://www.tafensw.edu.au**.

SafeWork NSW publications
'Applying the new safety regulations', Cat. no. 229 (also 100, 110, 1008, 2001)

After you have worked through the chapter, reinforce the practical component of your training with the **Get it right** section.

NEW

GET IT RIGHT

1

The photo below shows an incorrect practice that can be performed when installing pipe work near asbestos sheeting.

Identify the incorrect method and provide reasoning for your answer

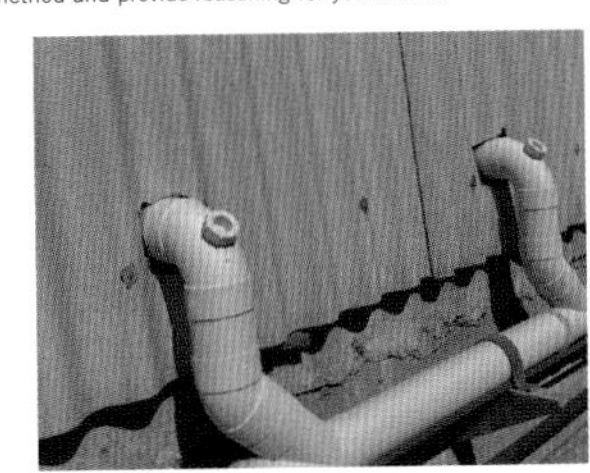

Worksheets give you the opportunity to test your knowledge and consolidate your understanding of the chapter competencies.

WORKSHEET 1

WORKSHEETS 1

To be completed by teachers	
Satisfactory	☐
Not satisfactory	☐

Student name:

Enrolment year:

Class code:

Competency name/Number:

Task: Review the section 'Participate in workplace induction' in this chapter and answer the following questions.

1. Describe what the term PCBU means.
2. List three categories of workers that an employer is responsible for on site.
3. What measures must a worker take regarding WHS Regulations?

Worksheet icons indicate in the text when a student should complete an end-of-chapter worksheet.

Guide to the online resources

FOR THE INSTRUCTOR

Cengage is pleased to provide you with a selection of resources that will help you prepare your lectures and assessments. These teaching tools are accessible via cengage.com.au/instructors for Australia or cengage.co.nz/instructors for New Zealand.

MINDTAP

Premium online teaching and learning tools are available on the *MindTap* platform – the personalised eLearning solution.

MindTap is a flexible and easy-to-use platform that helps build student confidence and gives you a clear picture of their progress. We partner with you to ease the transition to digital – we're with you every step of the way.

The *Cengage Mobile App* puts your course directly into students' hands with course materials available on their smartphone or tablet. Students can read on the go, complete practice quizzes or participate in interactive real-time activities. *MindTap* is full of innovative resources to support critical thinking, and help your students move from memorisation to mastery!

The *Series MindTap for Plumbing* is a premium purchasable eLearning tool. Contact your Cengage learning consultant to find out how MindTap can transform your course.

SOLUTIONS MANUAL

The **Solutions manual** provides detailed answers to every question in the text.

MAPPING GRID

The **Mapping grid** is a simple grid that shows how the content of this book relates to the units of competency needed to complete the Certificate III in Plumbing.

INSTRUCTORS' CHOICE RESOURCE PACK

This optional, purchasable pack of premium resources provides additional teaching support, saving time and adding more depth to your classes. These resources cover additional content with an exclusive selection of engaging features aligned with the text. Contact your Cengage learning consultant to find out more.

COGNERO® TEST BANK

A **bank of questions** has been developed in conjunction with the text for creating quizzes, tests and exams for your students. Create multiple test versions in an instant and deliver tests from your LMS, your classroom, or wherever you want using Cognero. Cognero test generator is a flexible online system that allows you to import, edit, and manipulate content from the text's test bank or elsewhere, including your own favourite test questions.

POWERPOINT™ PRESENTATIONS

Cengage **PowerPoint lecture slides** are a convenient way to add more depth to your lectures, covering additional content and offering an exclusive selection of engaging features aligned with the textbook, including teaching notes with mapping, activities, tables, photos, and artwork.

ARTWORK FROM THE TEXT

Add the **digital files** of graphs, tables, pictures and flow charts into your learning management system, use them in student handouts, or copy them into your lecture presentations.

FOR THE STUDENT

MINDTAP

MindTap is the next-level online learning tool that helps you get better grades!

MindTap gives you the resources you need to study – all in one place and available when you need them. In the *MindTap Reader*, you can make notes, highlight text and even find a definition directly from the page.

If your instructor has chosen *MindTap* for your subject this semester, log in to *MindTap* to:

- Get better grades
- Save time and get organised
- Connect with your instructor and peers
- Study when and where you want, online and mobile
- Complete assessment tasks as set by your instructor

When your instructor creates a course using *MindTap*, they will let you know your course link so you can access the content. Please purchase *MindTap* only when directed by your instructor. Course length is set by your instructor.

FOREWORD

In Australia, the plumbing industry provides employment in a range of service areas, including water supply, fire and sprinkler systems, sanitary plumbing, drainage, gas installation, roof plumbing and mechanical services. The industry is one of the biggest employers of tradespeople in the country and they provide important services to consumers, business and other industries.

In order for plumbing enterprises to keep pace with global change and sustainable practices, the vocational and training (VET) sector must respond quickly and efficiently to meet industry needs. This text is designed to meet the requirements of the latest national Training Package (CPC08) by providing information and activities that reflect the ever-changing skills required to undertake safe and effective activities in the water supply services area. The knowledge and skills derived from this text will provide learners with the tools for future learning and will prepare new and existing workers for a long and rewarding career in the industry.

I thank all the teachers of plumbing who have contributed their time and expertise to ensure that this text meets the outcomes of the training package. I especially thank Dean Carter for his recent contribution to review and update this text. His knowledge and attention to detail will ensure that the text continues to be the preferred resource for training in the plumbing workforce of Australia.

Shayne Fagan
Head of Skills Team | Innovative Manufacturing, Robotics and Science
TAFE NSW – South Western Sydney Institute | Granville College

PREFACE

Plumbing as a whole is an integral part of our everyday lives. It is imperative for the success and economic growth of our nation that the health and hygiene of all is maintained within our communities to ensure a long and prosperous life. Plumbers play an important role in providing the basic services that many take for granted, from the installation of clean drinking water to maintaining sanitary drainage. Although these services are vital, there are a great deal of other important services that plumbers can provide. In recent years, the conservation of energy and providing the installation and maintenance of products that are sustainable and reduce the impact on the environment have been contributing factors for people and how they live. Due to increased demand in products that are sustainable for the environment, comes a continual change in how materials are manufactured and installed. Plumbers are continually having to keep abreast of not only the materials that are changing but also the techniques and installation methods of these new materials, while maintaining essential hand skills and techniques of existing materials. These skills are continually being passed down from tradespersons to apprentices as it has been done for many hundreds of years.

In addition to the maintaining of skills, plumbers are required to adhere to various Australian Standards, Codes of Practice and relevant legislation. Therefore, it is a requirement that a plumber remains current in all facets of the trade and adopts a best practice approach to ensure that at the completion of the job, all work that is undertaken meets all the requirements both legally and professionally.

The *Basic Plumbing Services Skills* addresses a number of competencies that are undertaken during the early stages of an apprenticeship. The chapters within this text cover the introductory topics that are essential in providing apprentices with the foundational underpinning skills necessary throughout their plumbing career.

Starting a new plumbing apprenticeship is an exciting yet at times a daunting, experience for new apprentices. This book provides learners with information on current workplace practices within the building and construction industry and will assist in providing guidance for a path of lifelong learning. It has been written to allow apprentices to understand what to expect in the trade and what expectations may be required of themselves, so they can start their plumbing career in the most informed and best possible way.

Dean Carter
Western Sydney Institute

ABOUT THE REVISING AUTHOR

Dean Carter, Graduate Diploma of Adult Education, Cert IV TAE, Hydraulic Diploma Plumbing, Cert IV Plumbing, Cert III Plumbing, is a Plumbing Teacher in the Western Sydney Institute of TAFE NSW. He has taught all facets of the Cert III & IV Plumbing courses for over thirteen years. He originally started his plumbing career through enrolling in the Joint Schools Secondary TAFE program, which is now commonly known as TVET while completing his higher school certificate. He has worked in a number of different areas of plumbing which include industrial, commercial and residential plumbing, as well as hydraulic consulting and has worked in the UK in a number of plumbing roles. He is still engaged in industry and continues to conduct his own plumbing business to ensure industry currency. Dean is very committed to his trade and appreciates the opportunities that plumbing has provided to him. He is dedicated to his students and thoroughly enjoys passing his knowledge on to the new group of future plumbers.

ACKNOWLEDGEMENTS

This book would not have been possible without the efforts from industry reviewers and their valued feedback and colleagues within the plumbing industry who have written and updated chapters in previous editions. Many thanks to Peter Smith, Anthony Backhouse, Bob Bulkeley, Scott Bullow, Shayne Fagan, Richard Hickey, John Humphrey, Jon McEwan, Robert Neeson, Peter Towell and Rob Young for their significant contributions.

I would like to thank my colleagues at Miller TAFE for their continual support and expertise and David George for his valuable contributions toward the Asbestos section.

Also, many thanks to Cengage for the opportunity to revise this book and for the constant support from Sandy Jayadev (Content manager), Chee Ng (Content manager), Christin Quirk (Content development) and James Cole (Content development).

Finally I would like to thank my wife and children for their continual support and understanding during the revision of this text.

The authors and Cengage would like to thank the following reviewers for their incisive and helpful feedback:

Aliceson Parker – TAFE NSW, Wayne Diffey – Holmesglen TAFE, Phil Skinner – Holmesglen TAFE.

Finally, a special thank you to Shaun Tinnion for writing chapter 2: Provide First Aid, along with the supporting materials.

UNIT CONVERSION TABLES

TABLE 1 Length units

Millimetres *mm*	Metres *m*	Inches *in*	Feet *ft*	Yards *yd*
1	0.001	0.03937	0.003281	0.001094
1000	1	39.37008	3.28084	1.093613
25.4	0.0254	1	0.083333	0.027778
304.8	0.3048	12	1	0.333333
914.4	0.9144	36	3	1

TABLE 2 Area units

Millimetre square mm^2	Metre square m^2	Inch square in^2	Yard square yd^2
1	0.000001	0.00155	0.000001
1000000	1	1550.003	1.19599
645.16	0.000645	1	0.000772
836127	0.836127	1296	1

TABLE 3 Volume units

Metre cube m^3	Litre *L*	Inch cube in^3	Foot cube ft^3
1	1000	61024	35
0.001	1	61	0.035
0.000016	0.016387	1	0.000579
0.028317	28.31685	1728	1

TABLE 4 Mass units

Grams *g*	Kilograms *kg*	Pounds *lb*	Ounces *oz*
1	0.001	0.002205	0.035273
1000	1	2.204586	35.27337
453.6	0.4536	1	16
28	0.02835	0.0625	1

TABLE 5 Volumetric liquid flow units

Litre/second *L/sec*	Litre/minute *L/min*	Metre cube/hour *m³/hr*	Foot cube/minute *ft³/min*	Foot cube/hour *ft³/hr*
1	60	3.6	2.119093	127.1197
0.016666	1	0.06	0.035317	2.118577
0.277778	16.6667	1	0.588637	35.31102
0.4719	28.31513	1.69884	1	60
0.007867	0.472015	0.02832	0.01667	1
0.06309	3.785551	0.227124	0.133694	8.019983

TABLE 6 High pressure units

Bar bar	Pound/ square inch *psi*	Kilopascal *kPa*	Megapascal *mPa*	Kilogram force/ centimetre square *kgf/cm²*	Millimetre of mercury *mm Hg*	Atmospheres *atm*
1	14.50326	100	0.1	1.01968	750.0188	0.987167
0.06895	1	6.895	0.006895	0.070307	51.71379	0.068065
0.01	0.1450	1	0.001	0.01020	7.5002	0.00987
10	145.03	1000	1	10.197	7500.2	9.8717
0.9807	14.22335	98.07	0.09807	1	735.5434	0.968115
0.001333	0.019337	0.13333	0.000133	0.00136	1	0.001316
1.013	14.69181	101.3	0.1013	1.032936	759.769	1

TABLE 7 Temperature conversion formulas

Degree Celsius (°C)	(°F – 32) × 0.56
Degree Fahrenheit (°F)	(°C × 1.8) + 32

TABLE 8 Low pressure units

Metre of water mH_2O	Foot of water ftH_2O	Centimetre of mercury *cmHg*	Inches of mercury *inHg*	Inches of water inH_2O	Pascal *Pa*
1	3.280696	7.356339	2.896043	39.36572	9806
0.304813	1	2.242311	0.882753	11.9992	2989
0.135937	0.445969	1	0.39368	5.351265	1333
0.345299	1.13282	2.540135	1	13.59293	3386
0.025403	0.083339	0.186872	0.073568	1	249.1
0.000102	0.000335	0.00075	0.000295	0.004014	1

PART 1

BASIC WORKPLACE SKILLS

Part 1 of this textbook deals with the foundation and fundamental requirements for new workers that are undertaking training and education within the plumbing services sector. It will help you build and enhance the required knowledge of WHS, first aid, working effectively and communicating on the worksite.

Part 1 is based on four units of competency, as described in Chapters 1 to 4. The key elements for the following competencies are as follows:

1. Carry out WHS requirements
   - Participate in workplace induction.
   - Identify hazards and assess risks.
   - Plan, prepare and carry out safe work practices.
   - Maintain safety of self and others.
   - Use electricity safely.
   - Apply emergency response.
   - Clean up worksite area.
2. Provide first aid
   - Respond to an emergency situation.
   - Apply appropriate first aid procedures.
   - Communicate details of the incident.
   - Evaluate the incident and own performance.
3. Work effectively in the plumbing and services sector
   - Identify the industry work context and setting.
   - Organise and accept responsibility for own workload.
   - Work in a team.
   - Participate in identifying and pursuing own development needs and processes.
   - Participate in workplace meetings.
   - Observe sustainability principles when preparing for and undertaking work processes.

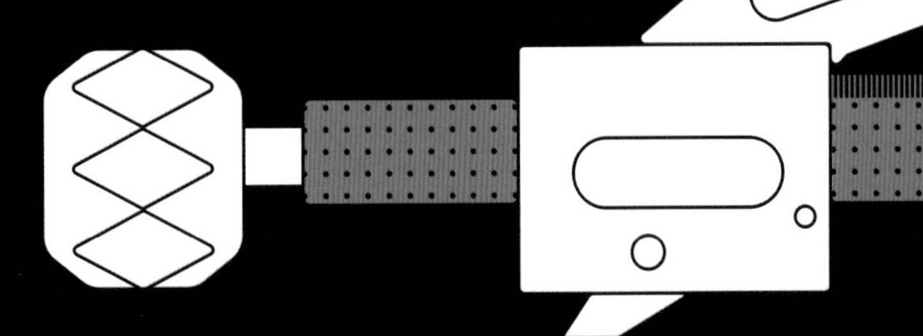

4 Carry out interactive workplace communication
   - Understand good communication.
   - Apply oral communication.
   - Apply visual communication.
   - Apply written communication and signage.
   - Understand alternative forms of communication.

Work through Part 1 and engage with your teacher and peers to prepare for the basic workplace skills that are vital for keeping yourself and those around you safe and for effectively communicating on the worksite.

The learning outcomes for each chapter are a good indicator of what you will be required to know and perform, what you will need to understand, and how you will apply the knowledge gained on completion of each chapter. Teachers and students should discuss the knowledge and evidence requirements for the practical components of each unit of competency before undertaking any activities.

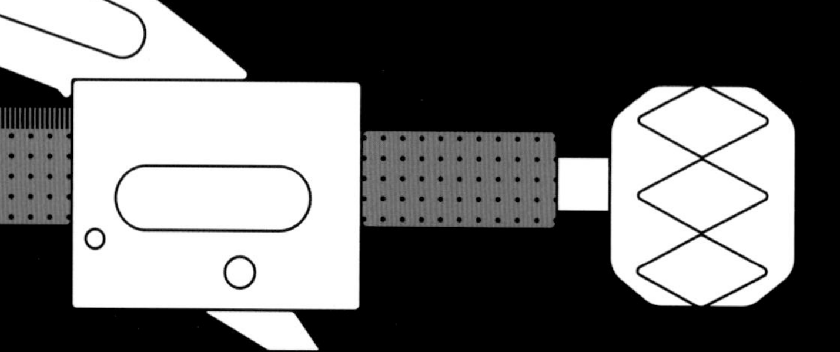

CARRY OUT WHS REQUIREMENTS

1

This chapter addresses the following key elements for the competency 'Carry out WHS requirements':

- Participate in workplace induction.
- Assess risks.
- Identify hazards and hazardous materials on the worksite.
- Plan and prepare for safe work practices.
- Use safe work practices to carry out work.
- Maintain safety of self and others.
- Use electricity safely.
- Apply emergency response.
- Clean up worksite area.

It introduces the requirements for safe work on a general construction site by outlining the requirements of the relevant state and territory WHS Acts and Regulations, as they apply to all persons at work. It also addresses the hazards and risks associated with carrying out plumbing on the worksite.

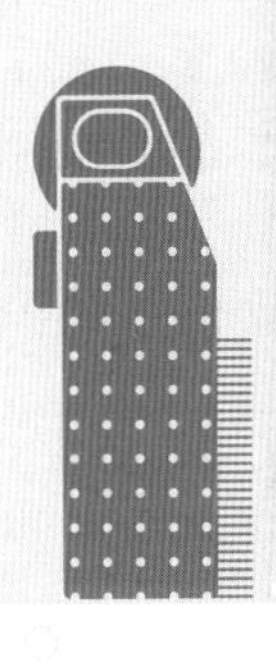

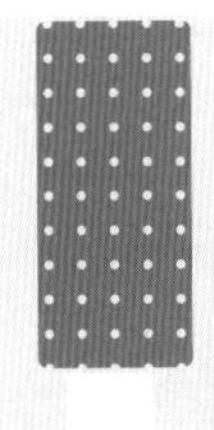

Participate in workplace induction

The contents of this chapter are designed to increase your awareness of work health and safety (WHS) principles. These WHS principles are 'embedded' in nearly every work activity in which you are involved.

Your health, welfare and safety, as well as that of others, is of absolute importance in your everyday work activities. You should plan, practise and maintain your awareness of WHS principles during the planning, ongoing and completion stages of every activity.

You will develop a greater awareness of WHS principles as you gain more experience in your trade. Never relax in your approach to practising and developing that awareness.

If you are unsure of the safety in any situation, always seek advice from a responsible person on site. If the situation looks or feels unsafe, do not put yourself or anyone else in a position of potential or real danger. Take all measures to prevent an accident from occurring.

Always maintain safety awareness on site and never become complacent with activities that you may be proficient in.

The term 'work health and safety' is used to cover a broad range of workplace practices that, together with federal and state laws, aim to improve the standards of workplace health and safety. The aim is to reduce work-related injuries, making a healthier and safer working environment for all concerned.

In 2011 Safe Work Australia developed a set of *model WHS laws* to be implemented across Australia. The Commonwealth, states and territories are required to implement their own laws based on this legislation, and they include model WHS Acts, model WHS Regulations and model Codes of Practice. The Safety Act that most states and territories within Australia comply with is the current *Work, Health and Safety Act 2011 (WHS Act 2011)*, which includes substantial amendments since its publication and is current as of 2020.

Previously, WHS was known as 'occupational, health and safety' (OHS). Although some states still comply with their own individual state and territory Acts and still commonly use the term OHS, for the purpose of this text all reference will be towards the current *WHS Act 2011* and the acronym WHS will be used. For further reference of individual state and territory Acts refer to Table 1.1.

What is a PCBU?

A PCBU (Person Conducting a Business or Undertaking) can range from a sole trader or small partnership through to a large company. It can also consist of a corporation, an association or even a volunteer organisation that engages or employs someone to carry out some form of work for them, whether it is for profit or non-profit. PCBU is a broad term used to describe all forms of working arrangements between the PCBU and its workers, and is essentially the person or group that employs you to undertake plumbing work on their behalf. For further clarification refer to the *WHS Act 2011*.

Note: Due to new plumbers being employed as apprentices by a PCBU, the text will use the term 'employer' as opposed to PCBU for simplicity. The term 'PCBU' may be used in some contexts where required.

What is a worker?

As the name implies, a worker is someone who is assigned to work by another party (the PCBU), whether they are in paid employment or working as a volunteer. Examples of a worker include an apprentice or trainee, an employee, someone doing work experience, a contractor or subcontractor, and a volunteer.

Everyone involved in a work-related activity has a duty of care to consider their own safety and that of others who may be affected by either acts or omissions, and this means both the employer and the worker. Workers must also take all necessary measures to comply with any efforts made by the employer to comply with current WHS Regulations.

Australian worker fatalities and injuries

Safe Work Australia statistics show that in 2019 174 Australian workers were fatally injured in work-related accidents, in comparison to 144 workers in 2018, with many more associated deaths from hazards such as asbestosis and long-term exposure to dangerous substances. However, the number of work-related deaths has fallen over the years, quite possibly due to increased safety awareness and legislative change. In addition, many workers suffer from work-related injuries or illness, with the most common injuries relating to musculoskeletal disorders such as joint/ligament and muscle/tendon injuries. The importance of WHS is evident in these statistics. Historically, the construction industry is one of the most dangerous for workers.

The prevention of incidents in industry is a concern for all workers and they must learn how to work without hurting themselves or endangering fellow workers. Every worker's own efforts in keeping the workplace safe, and reporting possible causes of injury and illness, are most important.

General construction induction training

Outlined within the relevant WHS legislation are the duties and obligations of the worker and the PCBU (employer). The PCBU must ensure general construction induction training (white card training) is undertaken by the person who is to carry out construction work. Non-compliance could invoke a breach of the relevant WHS Act and subsequent fines may apply.

It is a requirement under the relevant WHS legislation, enforced by the WHS regulatory authorities from each state and territory, that all workers carry out general construction induction training to familiarise themselves with:

- identification of health and safety legislation requirements
- the rights and responsibilities of employers and workers in relation to the Act
- duty of care requirements
- safe work practices
- identification of construction hazards and risk control measures
- identification and use of essential PPE (personal protective equipment)
- procedures for reporting hazards
- identification of incident and emergency response procedures
- procedures for first aid
- identification of fire safety equipment.

Statement and proof of induction training

On successful completion of and attendance at a general construction induction training session, a worker will be issued with a statement that outlines the training, identifies the training body and assessor, and states the date of the assessment.

In states and territories where there is compulsory general construction induction training, after the course each person is provided with a general construction induction card (white card), which confirms that they have completed the necessary induction training.

Examples of general construction induction cards that you may come across are shown in Figures 1.1–1.4. They show the person's name, the date the training was completed and the name of the body carrying out the training.

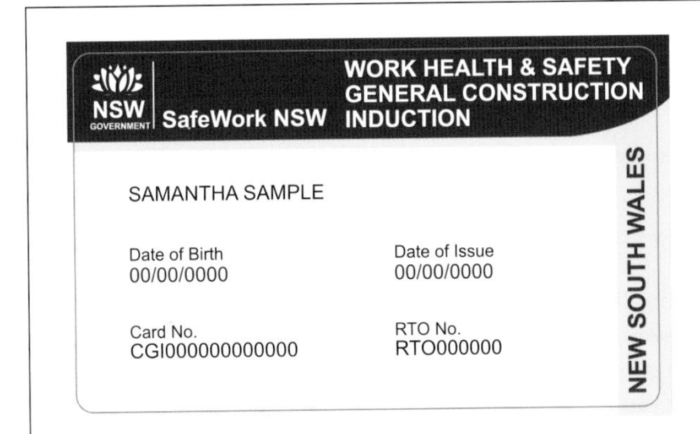

FIGURE 1.1 Construction induction card sample as issued in NSW

Source: With permission SafeWork NSW, https://www.safework.nsw.gov.au/resource-library/licence-and-registrations/recognition-of-general-construction-induction-training-cards-fact-sheet.

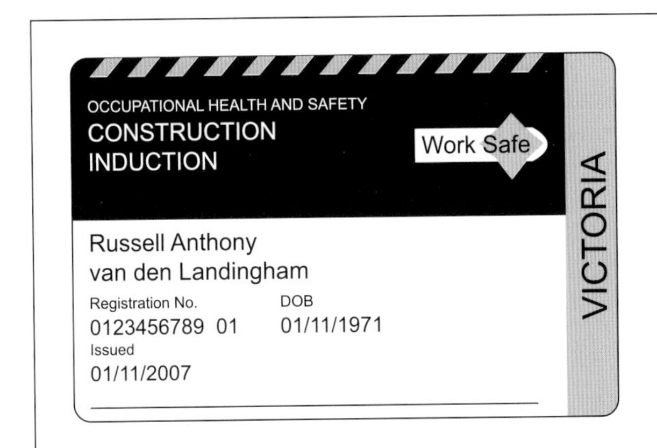

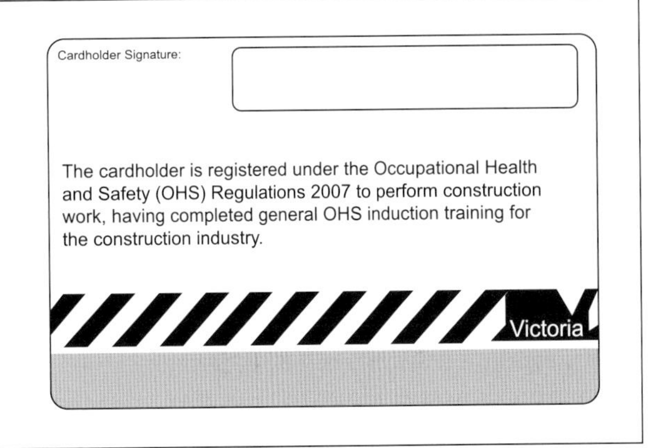

FIGURE 1.2 Construction induction card sample as issued in Victoria (post 1 July 2008)

Source: With permission SafeWork NSW, https://www.safework.nsw.gov.au/resource-library/licence-and-registrations/recognition-of-general-construction-induction-training-cards-fact-sheet.

Queensland Government

WORKPLACE HEALTH AND SAFETY QUEENSLAND
CONSTRUCTION INDUCTION

QUEENSLAND

Cardholder's name | Date of birth / /

RTO No. | Issue date / /

Card No.

WORKING ACROSS BORDERS

The cardholder has met the requirements of National OHS General Induction Training.

Cardholder's signature

If you lose this card, contact your training provider.
If you find this card, please return to:
Workplace Health and Safety Queensland
Reply Paid 69, BRISBANE Qld 4001.

FIGURE 1.3 White card sample as issued in Queensland

Source: Courtesy of Queensland Government.

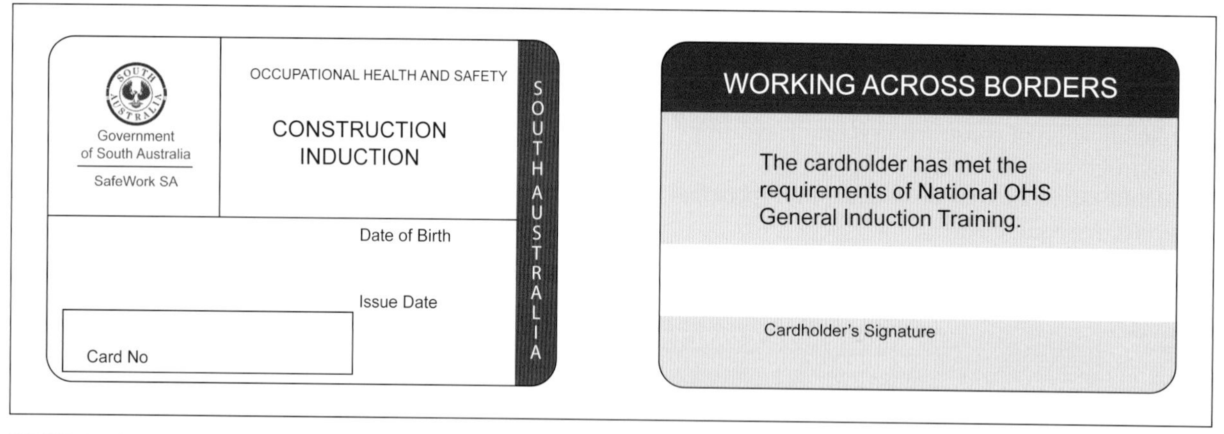
SOUTH AUSTRALIA
Government of South Australia
SafeWork SA

OCCUPATIONAL HEALTH AND SAFETY
CONSTRUCTION INDUCTION

SOUTH AUSTRALIA

Date of Birth

Issue Date

Card No

WORKING ACROSS BORDERS

The cardholder has met the requirements of National OHS General Induction Training.

Cardholder's Signature

FIGURE 1.4 White card sample as issued in South Australia

Source: With permission SafeWork NSW, https://www.safework.nsw.gov.au/resource-library/licence-and-registrations/recognition-of-general-construction-induction-training-cards-fact-sheet.

There is a mutual agreement between the applicable state and territory WHS regulatory authorities, known as 'working across borders', to acknowledge acceptance for these general construction induction cards. As long as the training meets existing standards for currency, as per the relevant state or territory requirements, and the construction worker can provide sufficient evidence that they hold one of the general construction induction cards as shown in Figures 1.1–1.4, or similar, then they will be permitted to carry out work on a construction site without having to undertake the general construction induction course relevant to the state or territory in which they are seeking employment. The relevant current card should be carried on site at all times and produced on demand for inspection.

Note: In some states and territories, the general construction induction card may become void if you are not working in the construction industry. SafeWork NSW states that if a worker in NSW has not carried out construction work for two consecutive years, the general construction induction card (white card) will become void. This means that if you have not carried out construction work for the period stipulated by the applicable state or territory WHS regulatory authority, you must undergo general construction induction training again. It is recommended that readers become familiar with their appropriate state or territory induction training programs, by either getting in contact with their respective WHS regulatory authority or viewing its website, as detailed in Table 1.1.

Workplace induction

Workplace-specific induction, which is often referred to as site-specific induction, must be carried out for every worker or visitor before work starts or before they enter a construction zone. The induction is to inform people of the hazards and risks that are specific to the construction workplace and how to report them. This will also include identifying the location of emergency equipment, emergency and evacuation procedures, site layout and PPE requirements. This induction training is to be provided for all workers at no charge.

Origin of modern WHS legislation

To understand how safe working conditions are determined, it is important identify how and when they began. To comprehend any piece of legislation (law), it is important to appreciate its origin and the reasons for the law first being introduced.

The industrial workplace in Australia is governed by either federal or state legislation, which varies in different states, territories and industries. Legislation and Regulations provide a set of minimum standards of protection for the health and safety of workers.

In Australia, prior to the concept behind the implementation of the model *Work Health and Safety Act 2011*, each state and territory had individual responsibility for making laws covering WHS and for enforcing those laws. Today, each individual state and territory WHS Act sets out the requirements for ensuring that workplaces are safe and healthy. These requirements outline the duties of different groups of people who play a role in WHS (see Table 1.1).

The main objective of the WHS legislation was to provide a balanced and nationally consistent framework for health and safety of workers by harmonising WHS legislation. Prior to the model WHS legislative framework, only one-third of the workforce was covered by any WHS legislation and there were up to 26 different Acts in one state alone relating to WHS. Enforcement procedures for these Acts complicated legal requirements. Allowing organisations to regulate their own WHS programs was not working and there was a need for a nationally consistent approach for workers' compensation to bring it into line with the modern working environment.

On a federal level, Safe Work Australia was established by the *Safe Work Act 2008* with the responsibility to lead the development of policy to improve WHS and workers' compensation arrangements across Australia. Safe Work Australia seeks to build cooperation between the three groups involved – governments, employers and employees – with a view of bringing them together to forge solutions and decide on policy. From this, the states and territories develop their own legislation and policies.

South Australia was the first state to introduce WHS legislation in 1972, with the introduction of the *Industrial Safety and Welfare Act*. In 1986, it adopted the *Occupational Health, Safety and Welfare Act*, following the general form of Victorian state legislation.

TABLE 1.1 Relevant Australian state and territory OHS/WHS Acts and Regulations as of May 2020

State or territory	Current OHS/ WHS Act	Current OHS/WHS Regulation	OHS/WHS regulatory authority	Website and contact numbers
ACT	*Work Health and Safety Act 2011*	Work Health and Safety Regulation 2011	WorkSafe ACT	Website: https://www.accesscanberra.act.gov.au Phone: 13 22 81
NSW	*Work Health and Safety Act 2011*	Work Health and Safety Regulation 2017	SafeWork NSW	Website: https://www.safework.nsw.gov.au Phone: 13 10 50
NT	*Work Health and Safety (National Uniform Legislation) Act 2011*	Work Health and Safety *(National Uniform Legislation)* Regulations 2011	NT WorkSafe	Website: http://www.worksafe.nt.gov.au Phone: 1800 019 115
Qld	*Work Health and Safety Act 2011*	Work Health and Safety Regulation 2011	WorkCover Qld	Website: https://www.worksafe.qld.gov.au Phone: 1300 362 128
SA	*Work Health and Safety Act 2012 (SA)*	Work Health and Safety Regulations 2012 (SA)	SafeWork SA	Website: https://www.safework.sa.gov.au Phone: 1300 365 255
Tas	*Work Health and Safety Act 2012 (Tas)*	Work Health and Safety Regulations 2012 (Tas)	WorkSafe Tasmania	Website: https://worksafe.tas.gov.au Phone: 1300 366 322 (inside Tas) or 03 6166 4600 (outside Tas)
Vic	*Occupational Health and Safety Act 2004*	Occupational Health and Safety Regulations 2017	WorkSafe Victoria	Website: http://www.worksafe.vic.gov.au Phone: 1800 136 089
WA	*Occupational Safety and Health Act 1984*	Occupational Safety and Health Regulations 1996	WorkSafe WA	Website: https://www.commerce.wa.gov.au/worksafe Phone: 1300 307 877 or 08 9388 5555

In NSW, the *Occupational Health and Safety Act* was enacted in 1983, following the Williams Inquiry into health and safety practices in the workplace. The Inquiry was commissioned in 1979 as a result of pressure from trade unions and community groups over serious hazards faced by workers in NSW. The Act amended and complemented other legislation that previously covered WHS in NSW under seven separate Acts. In 1987, major changes were made to the Act.

Victoria introduced legislation in 1985 that became the model for the rest of the country. It had far-reaching social and industrial concepts incorporated into the legislation.

WHS Regulations and Codes of Practice

Some workplaces have specific hazards and risks that have the potential to cause injury or disease such that specific Regulations or Codes of Practice are necessary. These Regulations and codes adopted under state and territory WHS Acts explain the duties of particular groups of people in identifying and controlling the risks associated with specific hazards.

What is the difference between the WHS Act, Regulations and Codes of Practice?

The model WHS Act sets out WHS responsibilities. The WHS Regulations expand on the requirements of the Act, with details of how certain sections of the Act are implemented and specific direction on how to meet those obligations.

The various Codes of Practice provide guidance on achieving the standard of health and safety that can apply to a profession, trade or industry. They provide detailed information on particular areas of an Act or Regulation, and outline activities, actions, technical requirements, responsibilities, and responses to events or conditions within a workplace.

The basic purpose of Codes of Practice is to provide workers in the building industry with practical, common sense, industry-acceptable ways of dealing with the WHS legislation and working safely. They are compiled and published by Safe Work Australia and the WHS regulatory authority from each state and territory, and cover such areas as excavation, electrical safety, handling of asbestos, PPE and managing the risk of falls.

Note the following:

- Regulations are legally enforceable.
- Codes of Practice provide advice on how to meet regulatory requirements. As such, Codes of Practice are not legally enforceable, but they can be used in court as evidence that legal requirements have or have not been met.

Rights and responsibilities of employers and workers

In each state and territory there are specific rights and responsibilities for employers and workers under the state or territory workplace WHS legislation.

Requirements of state and territory regulations

Each state has its own specific requirements, which may include any of the following.

The employer

Employers must provide for the health, safety and welfare of their workers at work. To do this, employers must:

- provide and maintain equipment and systems of work that are safe and without risks to health
- make arrangements to ensure the safe use, handling, storage and transport of equipment and substances
- provide the information, instruction, training and supervision necessary to ensure the health and safety of workers at work
- maintain places of work under their control in a safe condition and provide and maintain safe entrances and exits
- make available adequate information about research and relevant tests of substances used at the place of work.

Employers must not require workers to pay for anything that is to be provided or completed to meet specific requirements made under the Acts or associated legislation. They must also ensure the health and safety of people visiting their places of work who are not workers, such as plumbing supplier deliveries and people undertaking work experience.

The worker

Workers must take reasonable care of the health and safety of others and must cooperate with employers in their efforts to comply with WHS requirements.

Workers must not:

- interfere with or misuse any item provided for the health, safety or welfare of persons at work
- obstruct attempts to give aid or attempts to prevent a serious risk to the health and safety of a person at work
- refuse a reasonable request to assist in giving aid or preventing a risk to health and safety.

Offences and penalties

Nationally, under the relevant WHS Acts and Regulations of each state and territory, there are offences, penalties and infringement systems in place. States and territories have adopted a system that applies a varying number of penalty units according to the severity of the offence, with each unit having a set monetary value. While the infringement value of offences varies, the underlying principles behind the infringement system of each state and territory are the same.

The harshness of fines issued is also influenced by whether an individual or a corporation is guilty of the offence and whether or not they are previous offenders. It should also be noted that, as well as imposing fines, courts may choose to impose a sentence of imprisonment. They also have the option to order offenders to undertake any or all of the following:

- Take steps to remedy or restore any matter caused by the offence.
- Pay state or territory safety regulatory authorities for the costs of the investigation.
- Publicise or notify other persons of the offence.
- Carry out a project for the general improvement of health and safety.

It is recommended that readers become familiar with their relevant state or territory infringement systems by either contacting their respective WHS regulatory authority or accessing their website, as detailed in Table 1.1.

LEARNING TASK 1.1

PARTICIPATE IN WORKPLACE INDUCTION

Choose a simulated workplace such as a practical workshop and discuss the WHS requirements for a workplace induction.

Provide short answers to the following questions:

1. What are the risks and hazards specific to the construction area?
2. How are hazards reported and who are they reported to?
3. Identify the location and type of emergency equipment currently located in the workshop.
4. What are the emergency evacuation procedures for the site?
5. Where are the emergency assembly points?
6. What is the practical workshop layout? Are there amenities?
7. What PPE is required to start work?

COMPLETE WORKSHEET 1

Assess risks

In the building industry, electricity, falls, collapsing trenches and melanoma can be fatal. Chemicals, corrosives, noise and dust inhalation can result in blindness, deafness, burns and injuries to lungs. Back problems or other serious strains or sprains can slow workers down and prevent them from working for weeks or even permanently. Therefore, it is important to identify the hazards on the worksite and assess the risks and the likelihood that they may occur.

Identify hazards

A hazard is a potential source of harm where a person is exposed to a dangerous or harmful situation, physical or otherwise, that may affect the health and safety of that person.

Acute hazards

An acute hazard is one where short-term exposure to the hazard will cause an injury or sickness (e.g. being burnt in an explosive fire).

Chronic hazards

A chronic hazard is one where long-term exposure to the hazard will cause an injury or sickness; for example, melanoma from extended exposure to the sun, slow poisoning from chemicals building up in the body's system over a long period of time, or lung disease (mesothelioma) from exposure to asbestos or asbestos-containing material.

If every person is aware of the hazards to health and safety to which they may be exposed, and follow the safety rules with a common-sense approach, both employers and workers will benefit.

Workplace hazards

To be aware of potential dangers at work we must be able to identify workplace hazards. We come into contact with these hazards every day. Some of the more common hazards that influence health and safety in the workplace are:

- lifting and handling materials
- falls by objects and people
- machinery, both power and hand tools
- chemicals and airborne dust
- noise
- vibration
- thermal discomfort
- illumination (visibility)
- potential fire hazards.

Hazard groups

Hazards can be described as being:

- safety hazards
- health hazards.

Health hazards can be further subdivided into:

- physical
- chemical
- biological
- stress.

This makes five major *groupings* under which to define the various *types* of hazards found in the workplace (see Figure 1.5).

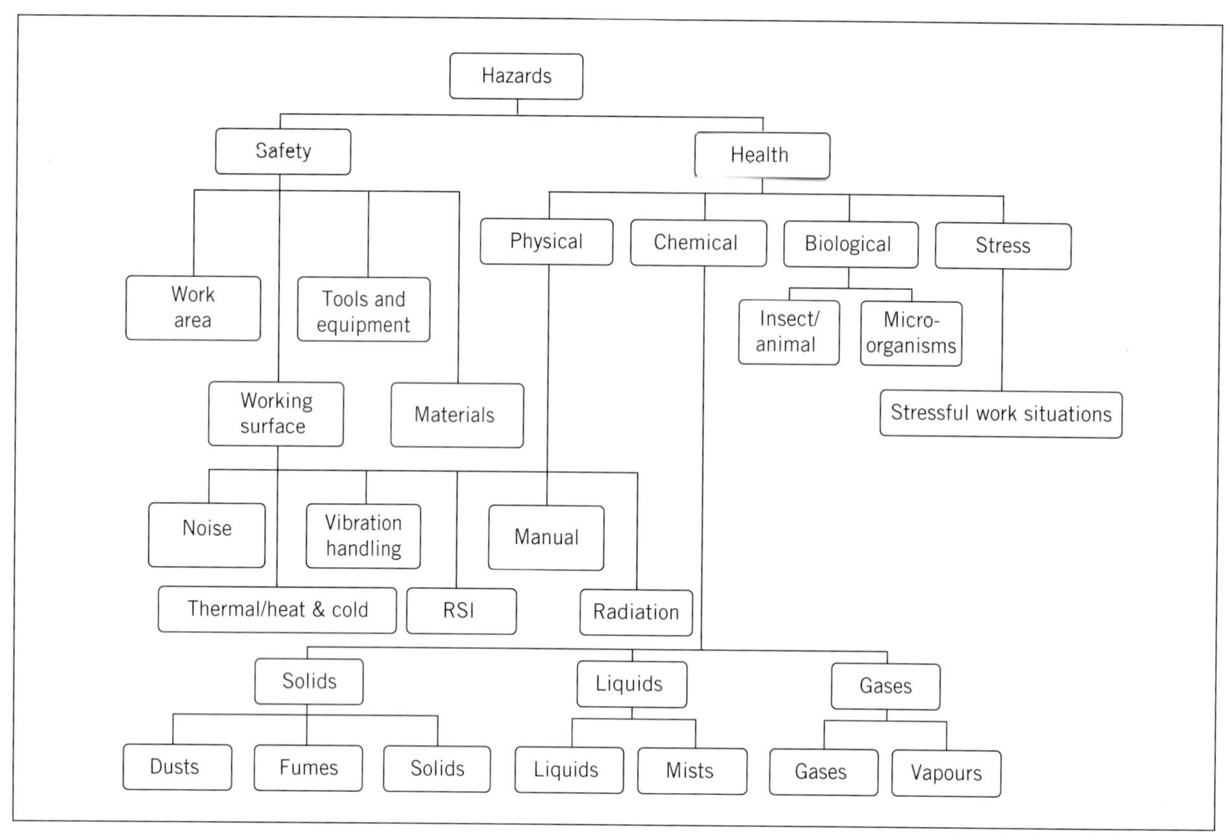

FIGURE 1.5 Hazard groups

Common types of workplace hazards

The following is a list of common hazards found on building and construction sites.

Safety hazards

Safety hazards include the following:

- *poor housekeeping* – untidy sites, lack of guardrails, inadequate walkways and rubbish left in work areas.
- *electricity* – underground and overhead supply cables, site wiring and power leads
- *water* – presence of water in a work area, which can cause slips and falls, electrocution, drowning and/or trench excavation collapse
- *ladders and scaffolding* – slips and falls from ladders and scaffolding
- *hand tools* – that can cause serious injuries if not used and maintained correctly
- *electrical tools* – that can cause serious cuts through contact with unguarded moving parts, electrocution, or falls and burns from electrical shock
- *heavy equipment* – heavy motorised equipment moving around construction sites
- *sharp materials* – materials with sharp edges that can cause cuts and lacerations
- *airborne materials (projectiles)* – falling or flying materials, which are one of the most common causes of injury in the industry.

Physical hazards

Physical hazards include the following:

- *noise* – exposure to excessive noise that can cause temporary and permanent hearing loss, stress and annoyance
- *temperature, both hot and cold* – that can cause a reduction in concentration and heat-related medical conditions
- *vibration* – whole-body vibration injuries from working with heavy equipment; and hand and arm vibration injuries caused by the use of vibrating tools
- *manual handling* – injuries from lifting, carrying, pushing and pulling, which are very common injuries in the building industry.

Chemical hazards

There are different hazardous chemicals that may be harmful to health and are used in the construction industry in the forms of:

- *solids* – dusts, fumes and solid materials
- *liquids* – liquid materials and mists
- *gases* – gases and vapours.

Some of these chemicals can cause acute or chronic injuries and severe medical conditions.

Toxic chemicals

A large number of chemicals are toxic or poisonous. The effects of chemicals on your health are difficult to identify at the source of exposure. Often the symptoms appear in parts of the body far removed from the point of absorption or perhaps do not appear until some years later.

Always read the SDS when using hazardous chemicals to ensure they are being handled and stored correctly.

Biological hazards

Biological hazards include the following:

- *insect/animal* – poor sanitation in dining, food prep and ablution areas, and poor practices of rubbish disposal, which can increase the chances of the spread of disease.
- *micro-organisms* – health and welfare facilities not being cleaned and maintained, leading to bacteria and viruses growing and spreading, causing disease; some work areas may also harbour dangerous viruses and bacteria.

Stress hazards

Stress is normally experienced as fatigue, anxiety and depression. Factors that may lead to stress disorders include:

- workload
- work pressure – excessive pace
- excessive overtime
- repetitive work
- uncertainty of responsibilities
- poor physical environment
- family matters
- substance abuse.

Once the hazards on the construction site have been determined, a risk assessment can now be used to ascertain the likelihood and the severity of risk.

Risk management

Identifying, assessing and controlling risks is an important process, and workers have a duty to take reasonable care of themselves and others regarding health and safety on the construction site. When hazards are to be identified in the workplace, the following four steps must be followed for managing health and safety risks:

- Identify the hazard.
- Assess the risk.
- Control the risk.
- Review control measures.

Identify the hazard

Hazards can be identified through site inspections, equipment to be used, materials required and how people work. Hazards may not always be obvious and may change over time on the job, so it is important to keep a list of the hazards found and consult with others.

Assess the risk

A risk assessment looks at what could happen if a person is exposed to a hazard and how the hazard may cause harm. A risk register is used as part of the risk assessment to determine the likelihood and the severity of the hazard (see Table 1.2). Risk assessments are mandatory for high risk activities.

TABLE 1.2 Plumbing risk register example

Plumbing risk register example			
Hazard	Worker falling from roof at height of over 2 m		
What is the harm that the hazard could cause?	Could result in serious injury or death		
What is the likelihood that the harm would occur?	High	What is the level of risk?	High
Wha controls are currently in place?	Correct use of safety harness Perimeter edge protection Safety mesh installed on roof Safe access to and from roof area through scaffolding		
Are further controls required?	Proof of general construction induction training required before entry onto site Workplace specific induction to be conducted prior to entry onto site		
Actioned by	John Smith	Date Due	16/11/2020
		Date Complete	16/11/2020
Maintenance and review	23/11/2020		

Source: https://www.safework.nsw.gov.au/__data/assets/pdf_file/0012/50070/How-to-manage-work-health-and-safety-risks-COP.pdf

Hierarchy of control measures

The order of preference for controlling hazards is known as the 'hierarchy of control' (see Figure 6.2 in Chapter 6). Removing or reducing the hazard is achieved by ranking control measures from the highest level of protection to the lowest:

- *Elimination* – eliminating risks or hazards is the most effective control measure.
- *Substitution* – substitute less hazardous materials, equipment or substances to reduce the risk.
- *Isolation* – isolate the hazard from people.
- *Engineering controls* – employ physical control measures by use of mechanical aids.
- *Administrative controls* – use job rotation to reduce exposure, and enforce policies and safe work methods.
- *Personal protective equipment (PPE)* – provide training in its use, and *ensure the equipment is used*. PPE should not be uncomfortable or disrupt the wearer in their duties.

APPENDIX A - HIGH RISK CONSTRUCTION WORK SAFE WORK METHOD STATEMENT TEMPLATE

NOTE: Work must be performed in accordance with this SWMS.

This SWMS must be kept and be available for inspection until the high risk construction work to which this SWMS relates is completed. If the SWMS is revised, all versions should be kept.

If a notifiable incident occurs in relation to the high risk construction work in this SWMS, the SWMS must be kept for at least 2 years from the date of the notifiable incident.

[PCBU Name, contact details]		**Principal Contractor (PC)**
Works Manager: Contact phone:		**Date SWMS provided to PC:**
Work activity:	[Job description]	**Workplace location:**
High risk construction work:	• Risk of a person falling more than 2 metres (*Note*: in some jurisdictions this is 3 metres)	• Work on a telecommunication tower
	• Likely to involve disturbing asbestos	• Temporary load-bearing support for structural alterations or repairs
	• Work in or near a shaft or trench deeper than 1.5 m or a tunnel	• Use of explosives
	• Work on or near chemical, fuel or refrigerant lines	• Work on or near energised electrical installations or services
	• Tilt-up or precast concrete elements	• Work on, in or adjacent to a road, railway, shipping lane or other traffic corridor in use by traffic other than pedestrians
	• Work in areas with artificial extremes of temperature	• Work in or near water or other liquid that involves a risk of drowning

FIGURE 1.6 A typical SWMS format (in part)

In many cases control methods can be used to manage risks and control hazards. The emphasis is on controlling the hazard or risk at its source by completely removing it from the situation; however, this may not always be possible, so choosing the most effective controls for minimising the hazard or risk is required to provide the highest level of protection possible.

Review of control measures

Control measures that are put in place must be regularly reviewed to ensure that they are effective as the construction site and hazards will continually change.

Workplace safety inspections

These are regular inspections of the workplace to determine what hazards exist by observation. They are conducted by management and representatives of the workforce. Reports and recommendations from these inspections will allow hazard control measures to be undertaken.

At workplaces with more than five workers, or as per WHS Regulations according to individual states and territories, permanent safety committees are normally established. Safety committees, which are made up of representatives from management and the workforce, are then responsible for carrying out the inspections.

In addition to workplace safety inspections that are conducted by management and workplace representatives, WHS regulatory state and territory inspectors can inspect the workplace in response to a complaint, incident or request for advice.

Responsibility and duty of care

As discussed earlier in this chapter, employers have a duty of care in relation to the health, safety and welfare of their workers. An employer must identify any foreseeable hazard that may arise from the work carried out by the company or partnerships.

To identify these risks and state how the employer will safeguard workers and all other persons at work, the employer is required to prepare a WHS site management plan, in accordance with the relevant state or territory WHS Regulations (see Table 1.1).

This management plan is site-specific and relates to individual sites being worked on at one location only and must be maintained and kept up to date during the course of the construction work. Non-compliance could invoke a breach of relevant WHS Acts and subsequent fines may apply.

Safe Work Method Statements (SWMS)

Like employers, workers have a duty of care to ensure that they work in a safe manner, obey site rules and do not interfere with any safety item or system. Therefore, it is a requirement under the relevant state or territory WHS Act that all people on a construction site adhere to safety requirements and take part in the formation and application of safety plans, which include SWMS for all work undertaken.

A Safe Work Method Statement is a document that: describes how work is to be carried out

- identifies and assesses the safety risks of work activities
- describes the risk control measures that will be applied to those work activities
- includes a description of the equipment used in the work
- states the standards or codes to be complied with
- outlines the required qualifications of the personnel doing the work
- reviews the control measures as required.

Workers' responsibilities

Each worker, whether they are a subcontractor or employed, should be involved in the development of a SWMS for any work to be carried out. This should include an assessment associated with that work and should also involve the employer.

Format of the Safe Work Method Statements

A SWMS should be kept as simple as possible and show the greatest amount of information (see Figure 1.6). Workers do not want to spend an excessive amount of time on the preparation and maintenance of these statements, so a standard format should be developed and used by individuals to suit the type of work being carried out.

Workers may reuse the same statement with modifications to suit new sites and/or conditions, as companies may specialise in one particular area and carry out the same type of work on most jobs.

LEARNING TASK 1.2

ASSESS RISKS

Choose a practical task that is to be undertaken in the practical workshop and discuss the requirements for developing a SWMS. Refer to Table 1.1 and visit the website for your relevant state and territory WHS regulator to download any templates or information that would assist in developing a SWMS.

COMPLETE WORKSHEET 2

Identify hazards and hazardous materials on the worksite

Construction sites contain numerous hazards and hazardous materials; therefore, it is important to identify the hazards before work begins in accordance with workplace procedures and legislative requirements. Codes of Practice that relate to specific hazards should be sought and can be obtained through Safe Work Australia and state and territory WHS regulators. These can provide guidance in the production of a SWMS. Specific hazards that plumbers may be exposed to include those associated with:

- asbestos
- crystalline silica
- excavation
- confined spaces
- working on roofs and at heights.

These hazards are discussed in turn, along with guidance on how to control risks associated with them.

Asbestos

One extremely important hazard that all employers and workers need to be aware of is *asbestos*, or, more specifically, what is termed 'asbestos-containing material' (ACM) (see Figure 1.7). In addition to ACM, asbestos-contaminated dust or debris (ACD) may also be found in the workplace and can be left after removal has taken place. The following list provides definitions of asbestos materials that you may come into contact with.

- *ACM* – this is any material that contains asbestos and is categorised in two forms:
  - friable material, which can be crumbled into a powder form by hand pressure when dry, and contains asbestos (see Figure 1.8)
  - non-friable material, which contains asbestos fibres reinforced with a bonding compound (see Figure 1.9).

FIGURE 1.7 Asbestos containing material (ACM)

FIGURE 1.8 Friable asbestos

FIGURE 1.9 Non-friable asbestos

- *ACD* – this is dust or debris that has settled within a workplace and is (or assumed to be) contaminated with asbestos.

Note: Non-friable materials may become friable due to deterioration from general exposure to the elements and damage sustained from extreme weather events such as hail.

All workers should be made aware of the availability of the *Model Code of Practice: How to manage and control asbestos in the workplace* (see Safe Work Australia: http://www.safeworkaustralia.gov.au) to cover situations where work involves asbestos. This is available through relevant state and territory WHS regulators' websites.

An employer must not carry out, direct or allow a worker to carry out work involving asbestos if that work involves manufacturing, supplying, transporting, storing, removing, using, installing, handling, treating, disposing of or disturbing asbestos or ACM except in prescribed circumstances.

The Model Code of Practice has been developed to provide practical guidance on how to manage risks associated with ACM and ACD at the workplace and to minimise the incidence of asbestos-related diseases such as mesothelioma, asbestosis and lung cancer.

What is asbestos?

Asbestos is a naturally occurring fibre that is invisible to the naked eye and is found all over the world. It was mined for its insulation properties. There are six types of asbestos, with the three most commonly mined and previously used in Australia being crocidolite (blue) asbestos, amosite (brown) asbestos and chrysotile (white) asbestos.

Where is asbestos found on the worksite?

Asbestos is commonly found in existing properties ranging from residential houses to commercial and industrial buildings. It was first manufactured in Australia in the 1920s and was used extensively in the 1950s to the 1970s, and, to a lesser extent, up to the 1980s when it began to be phased out. A total ban was enforced in Australia on the manufacture and use of asbestos from 31 December 2003. Most homes that were built before the mid-1980s have a high chance of containing ACM. Houses built after this period have a reduced likelihood of this, although some homes built in the 1990s and early 2000s may contain ACM, and those who are undertaking maintenance and renovations on buildings of all these time periods need to exercise a high degree of caution.

FIGURE 1.10 Asbestos roof sheeting

ACM products that may be found in the workplace include:

- roof sheeting (see Figure 1.10), including eaves, downpipes and gutters
- asbestos used for exterior sheeting (see Figure 1.11) on buildings and interior sheeting used in wet areas
- insulation on hot water pipes (see Figure 1.12) and mechanical services (see Figure 1.13)
- floor tiles and vinyl flooring
- asbestos piping (see Figure 1.14)
- asbestos unearthed during excavation (see Figure 1.15)
- asbestos cement compressed floor sheeting
- asbestos roof insulation (see Figure 1.16)

FIGURE 1.11 Exterior sheeting on a building

FIGURE 1.12 Asbestos insulation on hot water service in a domestic property

FIGURE 1.13 Asbestos insulation on commercial mechanical services

FIGURE 1.14 Asbestos piping

FIGURE 1.15 Asbestos unearthed during excavation

- spray-on insulation and ceiling linings (see Figure 1.17)
- asbestos waterproofing membrane (see Figure 1.18).

How can asbestos harm you?

Asbestos exposure can cause cancers that include lung cancer and mesothelioma, and lung diseases such as asbestosis. People who come into contact with airborne fibres are at risk, but when asbestos is left alone and not disturbed it poses no risk. Asbestos should never be cut, drilled, ground or subjected to any other process where particles of asbestos fibre may become airborne.

FIGURE 1.16 Roof insulation containing asbestos

FIGURE 1.17 Ceiling linings containing asbestos

What if asbestos is found?

Management must assume asbestos is present on the workplace if a 'competent person' reasonably believes that a material present is an ACM or an inaccessible part of the workplace is likely to contain an ACM.

It is not necessary to engage a competent person if management assumes asbestos is present. If ACM is found to be in good condition and left undisturbed, it is unlikely that airborne asbestos will be released in the air. If the ACM has deteriorated and ACD is found, then there is a higher risk of airborne asbestos. Identification of asbestos is the first step of managing the risk of exposure to asbestos. If asbestos is identified, then management must have all ACM clearly indicated – with labels where reasonably practicable. An asbestos register may be required for buildings constructed prior to 31 December 2003.

FIGURE 1.18 Waterproofing membrane containing asbestos

Note: Refer to the *Model Code of Practice: How to manage and control asbestos in the workplace* (see http://www.safeworkaustralia.gov.au) for the definition of a competent person.

Asbestos register

An asbestos register is a document that is used to identify and record asbestos in the workplace. Information that should appear on the register includes the date the asbestos was identified and the location, type and condition of the asbestos. If an asbestos register is not available for the workplace, work must not be carried out until a competent person determines if asbestos is present.

Asbestos management plan

An asbestos management plan sets out how the identified ACM will be managed, which may include procedures, risks and control measures. Management must ensure plans are reviewed at least once every five years.

Who can remove asbestos?

It is ideal to have a workplace free of asbestos and removal may be necessary. Where possible and if the ACM is in good condition, look for alternative routes or locations for installations where asbestos is not located to prevent it from being disturbed.

It is recommended that you refer to relevant state and territory WHS regulators' requirements and relevant Codes of Practice to determine the requirements of asbestos removal and disposal. Do not attempt to remove asbestos as specialist PPE (see Figure 1.19) and equipment is required, and exposure can cause cancer and lung disease.

FIGURE 1.19 PPE required for asbestos removal

Crystalline silica

Crystalline silica, like asbestos, is dangerous when inhaled and exposure can cause scarring of the lungs. Crystalline silica is found in the workplace in natural stone, bricks, mortar, concrete and sand. It is also found in manufactured composite stone, which is a popular choice amongst consumers for kitchen and bathroom benchtops. When you cut, saw or grind products that contain crystalline silica, airborne dust particles are generated and can embed themselves into the lungs, causing illness or disease.

If a worker is exposed to crystalline silica dust, they could develop chronic bronchitis, emphysema, silicosis of varying degrees, lung cancer and/or kidney damage. Health effects can vary, and it will depend on whether the person is exposed on a short- or long-term basis.

Excavation

Excavation can be potentially hazardous if adequate planning is not undertaken. When working in trenches it is important to identify the following hazards.

Hidden services

Plumbers are involved in excavation activities for a number of services. Many trenches are shallow and may not seem to present a serious danger. However, existing electrical cables or gas services, if damaged and exposed, are significant risks. Other services carry less risk but can be very expensive to repair. Locating all existing services before digging is strongly advised. This can be done with the assistance of site plans, working out likely service locations, non-destructive excavation (potholing), observation of the ground surface and exploratory digging.

Note: Plans of public services should be obtained from the Dial Before You Dig service by calling 1100 or going online at http://www.1100.com.au.

Collapse

Depending on the type of soil and location of an excavation as it gets deeper, the hazards increase significantly, with the possibility of collapse. At 1.5 m, measures need to be taken to support trench walls, but shallower depths may also require control measures depending on the composition of the soil. The weight of soil or rock can crush a person or prevent them from breathing (see Figure 1.20).

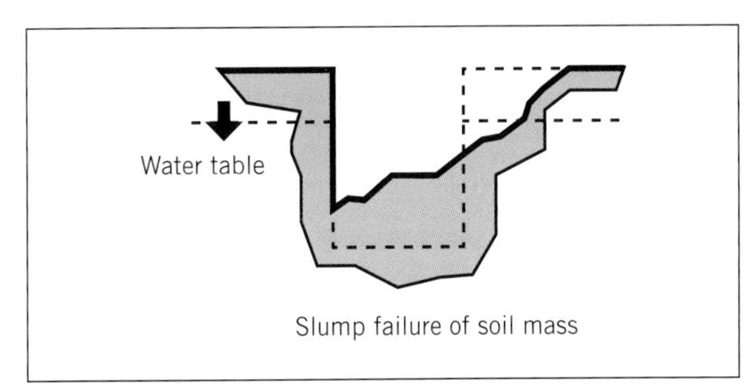

FIGURE 1.20 Trench collapse

The factors that contribute to instability include:

- soil type
- soil moisture content
- loads due to existing structures
- loads due to spoil placement
- the operation of nearby plant and equipment
- the amount of time the excavation is open
- water flooding the excavation
- objects falling into the excavation
- people falling into the excavation.

Control measures

Trench support is the main way of providing excavation stability (see Figure 1.21). There are many different methods, and selection in some cases might best be handled by an engineer. On large sites where excavation machinery is required, benching and battering may be used. Benching consists of creating a series of steps or horizontal and vertical sections in the excavation wall to reduce the wall height. Battering is where the wall of the trench is sloped back from the base of the trench to a predetermined angle. A combination of the two may also be used to minimise the risk of trench collapse.

FIGURE 1.21 Trench support: (a) Drag box being prepared to be installed; (b) Drag box positioned in trench; (c) Drag box in trench (note protective fencing around excavation and the ladder for access to the trench); (d) Drag box in trench (note safety signage)

In addition to benching or battering, support strategies are assisted by:

- restricting access around the excavation area to machinery, unauthorised personnel and surface water
- providing access to and from the excavation and appropriate PPE
- training people involved in the activity and regularly supervising and monitoring the conditions.

Refer to the *Model Code of Practice: Excavation work* (see http://www.safeworkaustralia.gov.au) for further information and guidance on managing risks associated with excavation.

Confined spaces

A confined space is an enclosed or partially enclosed area that can be hazardous due to unsafe breathing conditions, entrapment or potentially explosive vapours. Access to or from the confined space may also be restricted. Confined spaces that plumbers may be exposed to include:

- tanks
- pressure vessels
- access chambers
- sewers
- large pipes
- ducts
- shafts
- ceiling spaces
- under-floor spaces
- excavations.

Asphyxiation

Asphyxiation is extremely dangerous, as you may not realise there is a lack of oxygen until it is too late. Most people don't realise how fine a balance there is to having enough oxygen in order to be able to breathe. The air we are used to breathing contains approximately 21% oxygen. If this level drops below 19%, dizziness, then slipping into unconsciousness, can occur within a minute, and if oxygen is not quickly restored, asphyxiation soon follows. Confined spaces with restricted airflow can allow other gases to build up to the point where the oxygen level is too low to be safe. (The other gases may not be detectable by smell.) If someone in a confined space slips into unconsciousness, always alert the appropriate emergency personnel. Avoid attempting to rescue without the appropriate retrieval equipment and oxygen monitor as you may also slip into unconsciousness and risk death.

Poisoning

Some gases in confined spaces may also be toxic. These gases may be a result of industrial processes or the biological breakdown of waste materials. They are a hazard even if there is enough oxygen. The main toxic gas is carbon monoxide (CO), which is a by-product of combustion engines. It is particularly dangerous to allow the exhaust fumes from these engines to enter and/or settle in or near confined spaces.

Volatile gases

Other gases can be explosive, with obvious consequences. It is necessary that confined spaces are purged and ventilated thoroughly before entry and that any gas equipment taken into the space be checked for leaks.

Plumbers tend to transport gas in their vehicles (e.g. oxyacetylene and LPG) and are therefore subject to hazards, particularly if they are transported or stored incorrectly. Gases have a tendency to build up in a confined space and can easily be ignited.

Oxygen and acetylene bottles, when stored in a work vehicle, should be stored upright in a purpose-built container that is vented to the atmosphere. This can prevent gases from building up inside the vehicle and exploding.

Entrapment

In addition to hazards associated with ventilation, confined spaces may also need to be accessed and moved through in awkward ways. This means that if something does go wrong it is more difficult to remove yourself from the situation. People may experience claustrophobia, especially in emergencies. It is a sad fact that people die in confined spaces after entering to assist an injured worker.

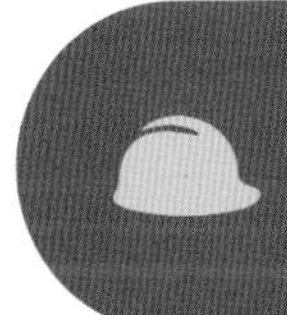

If entering a confined space to aid an unconscious worker, establish if it is safe to do so and use oxygen- and gas-monitoring equipment to establish if sufficient oxygen is present; otherwise wait for emergency personnel to arrive.

Control measures

Gas detection and monitoring of confined spaces must be carried out, along with ensuring that the space is structurally stable. Other measures include:

- providing confined space training
- restricting access to the space
- providing adequate ventilation
- having standby personnel and equipment
- understanding emergency procedures
- providing PPE.

Refer to the *Model Code of Practice: Confined spaces* (see http://www.safeworkaustralia.gov.au) for further information and guidance on managing risks associated with confined spaces.

Working on roofs and at heights

Plumbers are often required to work on a roof or at heights and the hazards associated with roofs are far more obvious, and the consequences quite serious.
A variety of activities can be carried out on the roof or at heights, including new installations, refurbishments, heritage work and maintenance. When planning to work on the roof, the hazards must first be identified, and then risks assessed to ensure compliance with relevant state and territory WHS requirements.
There are many factors that must be considered. The following Codes of Practice from Safe Work Australia (http://www.safeworkaustralia.gov.au) should be consulted to provide guidance on meeting the requirements of relevant state and territory Acts and legislation:

- *Model Code of Practice: Managing the risk of falls in housing construction*
- *Model Code of Practice: Managing the risks of falls at the workplace.*

Refer to Table 1.1 for your local authority.

Working on the roof

When working on the roof or at heights, workers are at a high risk of falls from unprotected edges of the roof, unguarded openings, deteriorated skylights and fragile roofs, such as those made from asbestos. WHS Regulations state that where there is risk of a person falling more than 2 m it is classified as high-risk construction work and a SWMS must be developed to provide adequate protection for the worker. This can be implemented through a number of control measures, and plans and site visits may be required to determine the requirements for the worksite.

Fall protection

Measures for fall protection must always be put in place. Equipment and PPE can include a full body harness with a double lanyard (see Figure 1.22), static anchorage lines (see Figure 1.23), inertia reels (see Figure 1.24), temporary work platforms, safety nets, safety mesh and anchorage points (see Figure 1.25). A suitable harness with a use-by date that has not passed must be used, along with a double lanyard attached to the worker at the chest position or dorsal attachment point, which then must be attached to suitable anchorage points. Connection to an anchorage point must be done as high as possible and workers should avoid working above the anchorage point to avoid long free falls. Consideration must also be given to the potential fall distance and the distance that workers are away from the anchorage point.

FIGURE 1.22 Correctly fitted harness: (a) Front; (b) Back

FIGURE 1.23 Static line installed on the roof

Workers should never work alone when wearing a harness, because if they were to fall from their position, suspension intolerance can occur. If the worker has an arrested fall and is in suspension, the lower legs can accumulate and store large amounts of blood, which reduces the return of blood to the heart, therefore slowing the heart rate and increasing the likelihood of fainting. Renal failure and death can occur based on the worker's condition and rescue attempts should be prompt. Workers that are accessing roofs should be trained in the rescue of a fallen worker. If a fall were to occur, workers should raise their legs in a horizontal position or carry straps to provide footholds to do so where possible.

Always wear a harness or an approved safety device when accessing and working on a roof.

Access to and exit from the roof

Where possible all work should be performed on the ground, which may involve prefabrication of components before they are required on the roof. Safe access to the roof can be gained in a number of ways, with the most common being ladders and scaffolding. Ladders should only be used as a means of access to or exit from a work area or for light work for a short duration. A mobile scaffold under 4 m can be erected by a competent person, and workers should be trained in its use; however, scaffolds that are over 4 m in height must only be erected or altered by a person with a high-risk scaffolding licence. Do not attempt to access incomplete or defective scaffolds. A harness should

FIGURE 1.24 An inertia reel

FIGURE 1.25 An anchorage point installed on the roof

always be worn with at least one point of contact connected to a suitable anchorage point. Consideration should be given to a hoist, which may be required for tools or bulky items to be brought to the work area.

Working at heights

You may not be required to work on a roof but still be working at a height over 2 m. Because ladders should only be used for light work, temporary work platforms such as scaffolding, work boxes and elevated work platforms (EWP) can be used to access the work area. An EWP may include a scissor lift (see Figure 1.26) or a boom type lift (see Figure 1.27). Some EWPs are designed for specific terrains or indoor purposes only; therefore it is important to check the landscape and area prior to obtaining an EWP for the work area. Pinch and crush hazards must be considered when working near or on EWPs. Never attempt to operate an EWP without the specific training and licensing required and always wear a suitable full body harness that is attached to a suitable anchorage point.

Edge protection

Edge protection on roofs may come in the form of parapet walls, temporary scaffolding (see Figure 1.28) and permanent or temporary guard rails. Due to a large number of plumbers working in construction, most

FIGURE 1.26 Scissor lift

FIGURE 1.27 Telescopic boom lift

FIGURE 1.29 Temporary perimeter guard rails on a new home

forms of edge protection that a plumber will encounter are temporary perimeter guard rails (see Figure 1.29) on new homes or barricade mesh fencing on large building sites. Toe boards and mid rails may be required to control falling hazards such as tools and equipment.

Fragile roof

FIGURE 1.28 Temporary scaffolding on a new home

Most working areas are stable and structurally sound; however, when working on an asbestos roof or near skylights, caution should always be exercised. Asbestos roofing (see Figure 1.30) is quite hard and brittle, and you could fall through the roof. Always walk along the screw line as it has support beneath. Skylights can become brittle over time due to UV exposure, and although in a lot of circumstances safety mesh is

FIGURE 1.30 Asbestos roof and warning sign due to brittle roof

installed, it is not always guaranteed and there is still the possibility of falling through. Always observe where skylights are located when walking on the roof.

Exposure to elements

Exposure to the sun, wind and rain are always concerns when working on a roof. The weather should always be checked prior to accessing the roof and can be done days ahead through checking weather forecasting. Working in adverse conditions such as wind and rain is extremely dangerous and increases the risk of illness and injury dramatically. When working in the sun, an SPF 50+ sunscreen and PPE that protects the skin from UV exposure should be worn, along with an appropriate pair of sunglasses to protect the eyes from excessive glare.

Electrical hazards

Electrical hazards on the roof work area are due to working near overhead powerlines and solar panels (see Figure 1.31). A safe distance must be kept from these electrical sources and this must be determined prior to accessing the roof. In some circumstances the local electrical authority may be required to disconnect or insulate power lines (see Figure 1.32) to prevent

FIGURE 1.31 Solar panels on a roof

FIGURE 1.32 Insulated power lines

electrocution. In addition to these two hazards, control measures for the use of power tools through proper use must be considered.

Codes of Practice give guidance on how the hazards associated with roof work can be managed, and can be accessed through the Safe Work Australia website (http://www.safeworkaustralia.gov.au). As with all hazards that may be present in the workplace, it is important to identify the hazard, assess the risk, control the hazard using the hierarchy of control and review control measures as the work progresses.

LEARNING TASK 1.3

IDENTIFY HAZARDS AND HAZARDOUS MATERIALS ON THE WORKSITE

Go to the Safe Work Australia website (http://www.safeworkaustralia.gov.au) and download the *Model Code of Practice: Managing the risk of falls in housing construction* and answer the following questions.

1 When inspecting the workplace, what are key hazards to look for?
2 What is the definition of a temporary work platform?
3 At or above what height is a SWMS required to be developed?
4 When using a mobile scaffold, workers should be trained in their use. What are the key points to ensure safe use when using a mobile scaffold?
5 At what distance should perimeter guard rails incorporate a top rail above the working surface?
6 When using a ladder to access to and exit from a roof, at what ratio should the ladder be set up?

COMPLETE WORKSHEET 3

Plan and prepare for safe work practices

Planning and preparation of the job is essential to ensure the job runs smoothly and to prevent unnecessary delays, injuries and illness. Once the WHS requirements have been addressed, the quality assurance requirements for company procedures and safe work practices must be identified and followed. As most tasks require tools, equipment and PPE, it is important to determine and select what is required.

Personal protective equipment (PPE)

Personal protective equipment (PPE) is the last line of defence to protect your health and safety from workplace hazards and should be worn by all workers when undertaking plumbing tasks to assist in carrying out work safely. It is the responsibility of the employer to protect the worker by providing PPE and clothing for tasks to be carried out. Appropriate instruction and training on the correct use and fitting of PPE should always be given by the employer. It is the worker's responsibility to wear and look after the equipment provided. PPE that does not fit or has not been fitted correctly may result in illness or injury and defeat the purpose of wearing the PPE in the first place.

PPE must be appropriate and designed and manufactured to provide protection from a specific hazard to a particular part of the body. No single design can be expected to provide protection from all types of hazards in the workplace. To decide what PPE and clothing is required, you must first be able to identify the hazards involved. Once the hazards have been identified, suitable equipment and clothing must be selected to give the maximum protection.

PPE can be grouped according to the part of the body it will protect, including the head (face and eyes), ears, respiratory system, hands, feet and body.

Head protection

Wearing head protection on site can reduce the risk of head injuries, and in some circumstances is mandatory. Listed below are common ways of protecting the head and face while working on site.

Safety helmets

Wearing safety helmets on construction sites may prevent or lessen a head injury from falling or swinging objects, or through striking a stationary object.

Safety helmets must be worn on construction sites when:

- it is possible that a person may be struck on the head by a falling object
- a person may strike their head against a fixed or protruding object
- accidental head contact may be made with electrical hazards
- carrying out demolition work
- instructed by the person in control of the workplace.

Safety helmets must comply with *AS/NZS 1801 Occupational protective helmets*. They must carry the AS or AS/NZS label, and must be used in accordance with *AS/NZS 1800 Occupational protective helmets – Selection, care and use* (see Figures 1.33 and 1.34).

1.1 STANDARDS

- AS/NZS 1801 Occupational protective helmets
- AS/NZS 1800 Occupational protective helmets – Selection, care and use

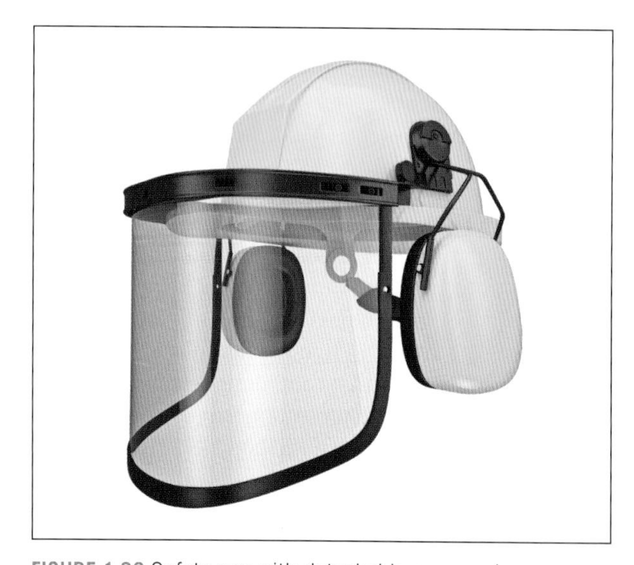

FIGURE 1.33 Safety cap with detachable accessories

Source: Shutterstock.com/ARTYuSTUDIO.

When wearing a helmet, the harness – which is located on the inside of the helmet – should be adjusted to allow for stretch on impact. There should be no potential for contact between the skull and the shell of the helmet when it's subjected to impact.

Sun shade

The risk of skin cancer for workers is increasing, with the neck, ears and face being particularly exposed.

FIGURE 1.34 Full welding helmets

Workers should always wear sun protection when working outdoors, even during wintertime.

Sun shades include wide-brimmed hats and foreign legion-style sun shields fixed to the inner liner of safety helmets (see Figure 1.35).

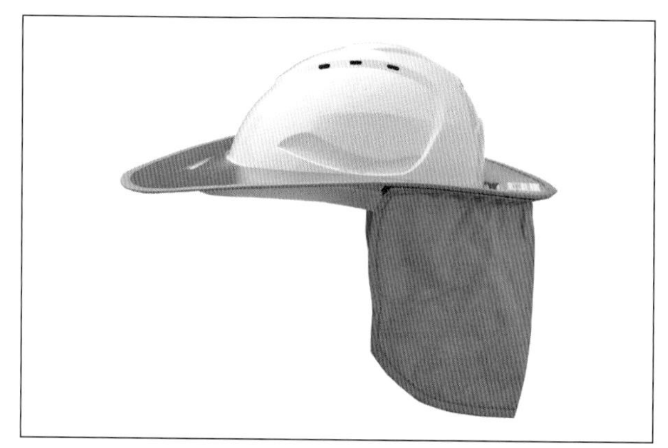

FIGURE 1.35 Fabric sun brim accessory for a safety cap

Source: Big Safety, www.bigsafety.com.au

Eye/face protection

The design of eye and face protection is specific to the application. It must conform to *AS/NZS 1337.1 Personal eye protection, Part 1: Eye and face protectors for occupational applications*.

1.2 STANDARDS

- AS/NZS 1337.1 Personal eye protection, Part 1: Eye and face protectors for occupational applications

There are three hazard categories for the eyes:

- *physical* – dust, flying particles or objects, and molten metals
- *chemicals* – liquid splashes, gases and vapours, and dusts
- *radiation* – sun, laser and welding flash.

The selection of the correct eye protection to protect against multiple hazards on the job is important. Most eyewear is available with a tint for protection against the sun's UV rays, or may have radiation protection included (see Figures 1.36 and 1.37).

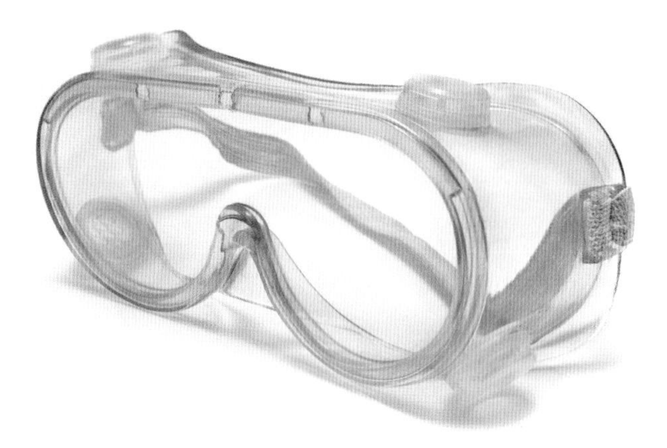

FIGURE 1.36 Clear wide-vision goggles

Source: Shutterstock.com/Gavran333

FIGURE 1.37 Clear-framed spectacles

Source: Alamy Stock Photo/Andrew Paterson

Face shields

Face shields give full face protection, as well as eye protection. They are worn when carrying out grinding and chipping operations, and when using power tools on timber. The shield may come complete with a head harness or be supplied for fitting to a safety helmet (see Figure 1.38).

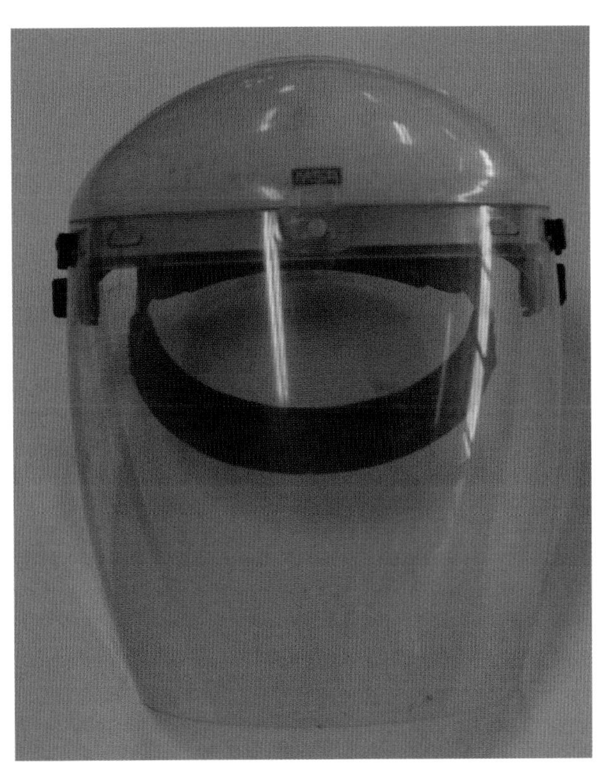

FIGURE 1.38 Face shield

Source: Shutterstock.com/PRILL.

Ear protection

You should always wear ear protection in areas where loud or high-frequency noise operations are being carried out, or where there is continuous noise. Always wear protection where you see a 'Mandatory hearing protection' sign, and when you are using or are near noisy power tools.

It is good etiquette to let other tradespeople in the vicinity know before you start up power tools – that they may require hearing or eye protection to be worn.

The two main types of protection available for ears (see Figure 1.39) are:

- *ear plugs* – semi- and fully disposable
- *ear muffs* – also available to fit on hard hats where required.

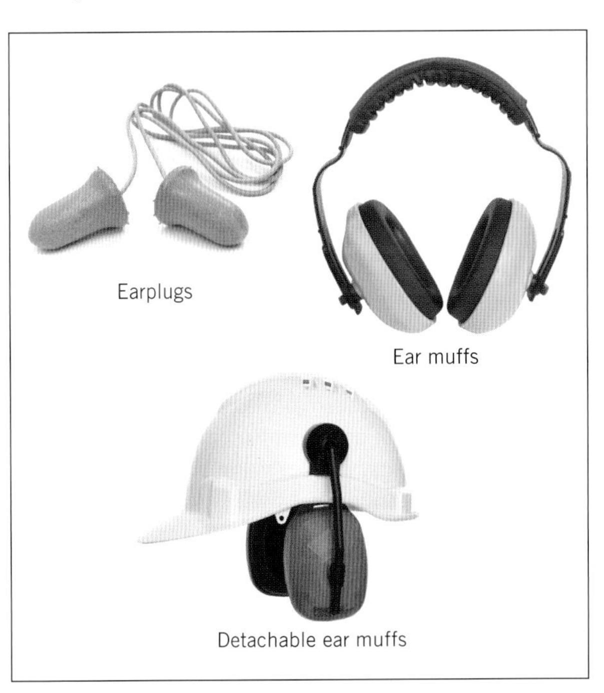

FIGURE 1.39 Hearing protection

Source: Shutterstock.com/Joe Belanger; Shutterstock.com/79mtk; Big Safety, www.bigsafety.com.au

Choose the one that best suits you and conforms to *AS/NZS 1270 Acoustics – hearing protectors*.

- AS/NZS 1270 Acoustics – hearing protectors

Respiratory protection

The increased use of mechanical equipment and chemicals for building construction work has increased the need for personal respiratory protection. Breathing contaminated or oxygen-deficient air creates a health hazard that can range from mild discomfort to chronic or acute poisoning, and even death.

Dust masks

Dust masks are generally used to filter out nuisance dusts, such as when undertaking general housekeeping duties (see Figure 1.40), but can also be used to filter out mist, smoke and aerosols, depending on the type of mask. They come in a variety of styles and can be purchased with exhalation valves. Most dusk masks available are rated for a P2 application and they should not be used where toxic dusts are likely to be found. Always refer to manufacturer's specifications before using a mask for a specific application.

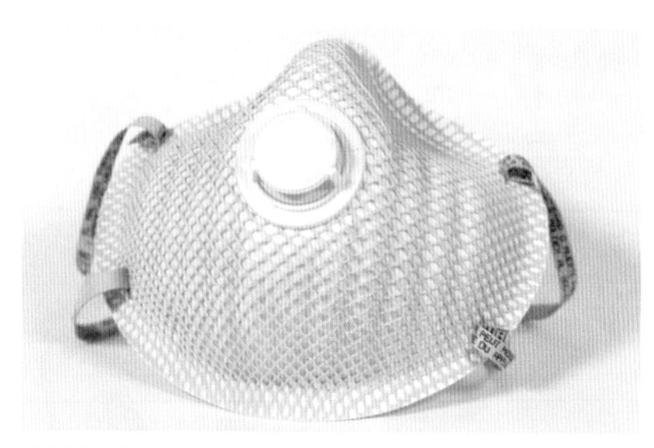

FIGURE 1.40 Dust mask for nuisance dust

Source: iStock.com/sethuphoto.

Respirators

Half-face and full-face respirators can have filters designed to keep out dusts, smoke, metal fumes, mists, fogs, organic vapours, solvent vapours, gases and acids.

The type of cartridge in a filter will define the type of protection it will give. Cartridge types can be identified by the classification ratings from *AS/NZS 1716 Respiratory protective devices*. It is important to learn about the wide range of face masks and respirators for various conditions.

- AS/NZS 1716 Respiratory protective devices

Respirators fitted with P2 class dust filters (see Figure 1.41) are suitable for use with the general low-toxic dusts and welding fumes that are commonly found on construction sites. Particulate, gas and vapour filters can be obtained to suit the desired application.

FIGURE 1.41 Half-face respirator with P2 class dust filters fitted

Source: Shutterstock.com/science photo.

Further information on respirators and dust masks should be obtained from suppliers and manufacturers. It is very important to be trained in the correct methods of selecting, fitting, wearing and cleaning the equipment in accordance with *AS/NZS 1715 Selection, use and maintenance of respiratory protective equipment*. Respirators and masks must be close-fitting to ensure that all air entering your respiratory passages has been fully filtered.

1.5 STANDARDS

- AS/NZS 1715 Selection, use and maintenance of respiratory protective equipment

It is advisable to be clean shaven – or use a hood, helmet or visor-type respirator for those with a beard – to minimise inhalation of potential contaminants.

Hand protection

Hands require protection from both physical and chemical hazards.

Gloves

Gloves can be used to give protection from both physical and chemical hazards. Stout gloves are required when handling sharp or hot materials. Chemical-resistant gloves are used when handling hazardous chemical substances. Gloves should conform to *AS/NZS 2161.1 Occupational protective gloves, Part 1: Selection, use and maintenance* (see Figure 1.42).

1.6 STANDARDS

- AS/NZS 2161.1 Occupational protective gloves, Part 1: Selection, use and maintenance

FIGURE 1.42 Gloves

Creams

Barrier cream is applied to the hands in industrial applications and places a physical barrier between the skin and contaminants. It may be used, when gloves are too restrictive, to protect the hands from the effects of cement, manual handling of copper and metals, and similar hazards (see Figure 1.43).

FIGURE 1.43 Barrier cream

Source: Lightning Products.

Foot protection

It is mandatory to wear protective footwear at the workplace at all times. *Open footwear is not permitted at any time.* Footwear should conform to *AS/NZS 2210.1 Safety, protective and occupational footwear, Part 1: Guide to selection, care and use.*

1.7 STANDARDS

- AS/NZS 2210.1 Safety, protective and occupational footwear, Part 1: Guide to selection, care and use

All safety footwear must have:

- stout soles or steel midsoles to protect against sharp objects and protruding nails
- good uppers to protect against sharp tools and materials
- reinforced toecaps to protect against heavy falling objects.

Safety boots should be worn in preference to safety shoes on construction sites to give ankle support over the rough terrain.

Safety shoes may be required when carrying out roof work or scaffold work and must have reinforced toecaps and rubber soles for durability and maximum grip.

Rubber boots should be worn when working in wet conditions, in wet concrete, or when working with corrosive chemicals. They must have reinforced toecaps.

See Figure 1.44 for examples of foot protection.

FIGURE 1.44 From left: Low-cut safety shoes, gumboots and steel-cap safety boots

Source: Shutterstock.com/Nikitin Victor; Shutterstock.com/canonzoom; Shutterstock.com/homydesign.

Body protection

Protecting your body through appropriate clothing can help to reduce the risk of injury and UV radiation. Listed below are common ways to protect your body while working on site.

Good-quality, durable work clothing is appropriate for construction work. It should be kept in good repair and cleaned regularly. If the clothing has been worn when working with hazardous substances it should not be taken home to launder but sent to a commercial cleaning company, as this will prevent the hazards from contaminating the home and the environment.

A good fit is important, as loose-fitting clothing is easily caught in machine parts or on protruding objects. Work pants should not have cuffs or patch pockets, as hot materials can lodge in these when worn near welding or cutting operations.

Rings, bracelets and neck chains should not be worn at work. It is recommended to cover or remove exposed body piercings in some situations.

Clothing should give protection from the sun's UV rays, cuts, abrasions and burns. Lightweight, collared, long-sleeved shirts provide good neck and arm protection from UV rays.

Industrial clothing for use in hazardous situations should conform to *AS/NZS 4501.2 Occupational protective clothing – General requirements*.

- AS/NZS 4501.2 Occupational protective clothing – General requirements

Sun protection

Sun protection, in the form of a SPF 30+ minimum and preferably a SPF 50+ sunscreen, should be provided for workers on construction sites. It should be applied on a regular basis – with most experts recommending application every two hours – to all areas of the body exposed to the sun.

Cleaning and maintenance

It is crucial that cleaning and maintenance be carried out on all PPE on a regular basis. This must be done by someone who is competent in inspecting and maintaining the equipment and be in accordance with manufacturers' guidelines.

Tools and equipment

Tools and equipment play a major role in how tasks are undertaken as they can be used to make the job easier and allow tasks to be undertaken more efficiently. It is essential that all tools and equipment are used in line with their design and purpose in order to avoid the risk of illness and injury.

Guards for tools and equipment

A guard on a power tool, static machine or any equipment that has a moving blade is another form of protection against injury. Its purpose is to prevent material and waste offcuts from being projected towards the operator, as well as preventing the fingers or hands from being drawn into moving parts or blades. Guards are also used to prevent pieces of shattered blade or abrasive disc from striking the operator when they are faulty or disintegrate due to being jammed.

A guard is fitted as the first line of protection for an operator, and therefore should *never* be removed or tied back while the tool or machine is in use. The only time the guard should be allowed to move from its safety position is when the tool is in use and it retracts as it is fed into the material or when the power has been disconnected from the tool to allow the blade or disc to be removed for replacement or cleaning.

Portable hand-held power tools and static machines that have guards fitted include:

- circular power saws (see Figure 1.45)
- angle grinders
- jig saws
- polyethylene butt welders
- drop saws
- wall chasing saws.

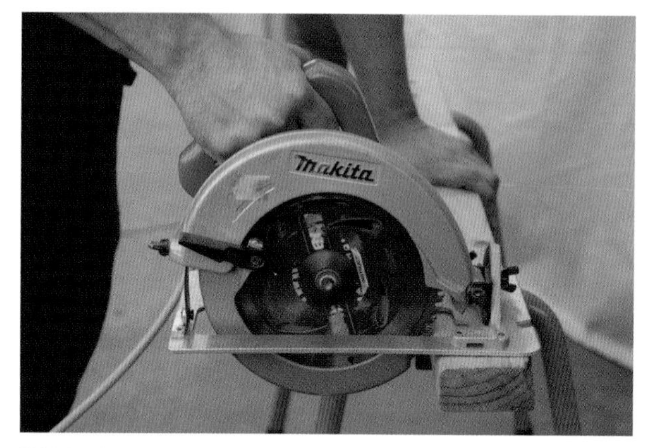

FIGURE 1.45 Circular saw fitted with a guard

Mechanical aids

Mechanical aids reduce the amount of manual handling and effort required for lifting and carrying tasks undertaken in the workplace. Some aids will give a mechanical advantage to the user to reduce the amount of effort required to carry out the task; for example, the use of a manhole cover lifter (see Figure 1.46). Some of the mechanical aids available are as follows.

- *Crowbars and levers* give a mechanical advantage when lifting or moving an object and can be used to set a heavy object in motion, as when using rollers or a crowbar to slide a heavy load forward.
- *Rollers* are placed under heavy loads to move them into position. The rollers may be pieces of water

FIGURE 1.46 Manhole cover lifter

pipe or round rod, or may be more sophisticated, such as air bags, for use over rough terrain, or multi-wheeled skates to move very heavy loads. The larger the diameter of the roller, the easier the object is to move. The path of travel of the load must be cleared of all obstacles before commencing the move.

- *Wheelbarrows* are the most common carrying aid used on building sites to cart concrete, bricks, pipe fittings and tools over all types of site conditions. They make it easy to negotiate tight situations due to the single wheel (see Figure 1.47).

FIGURE 1.47 A typical wheelbarrow

- *Hand trucks and trolleys* are carrying aids that take most of the weight of the load (see Figure 1.48).
- *Cranes and hoists* lift heavy loads without the use of manual handling. They may be hand or power operated (see Figure 1.49).
- *Jacks and lifting tackles* lift heavy loads. Jacks may be hydraulic or mechanical. Lifting tackle may be pulley blocks with rope tackle, chain blocks (see Figure 1.50) or wire rope tackle.
- *Forklift trucks and pallet trucks* are hand or power driven. These move large quantities of materials fast and safely. Materials are normally stacked on pallets for ease of handling (see Figure 1.51).
- *Lifting grips* are used to allow safe lifting and carrying of awkward materials; for example, suction grips for handling glass, and carry grips for lifting and carrying sheet material (see Figure 1.52).

Barricades

When moving tools and equipment on site, barricades and hoardings may be required to prevent a hazardous or dangerous situation occurring. Where required, barricade off the area to warn others or prevent the public from gaining access to the work area and erect hazard warning signs. Report any unsafe situations to your immediate supervisor or site safety officer.

FIGURE 1.48 Hand trolley

FIGURE 1.49 Hand-operated mobile crane

FIGURE 1.50 Chain block pulley system

Source: Alamy Stock Photo/Sarah Lack

FIGURE 1.51 Forklift

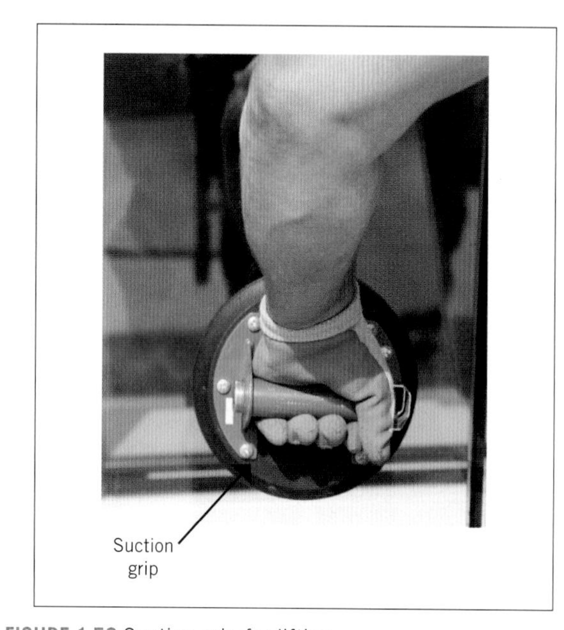

FIGURE 1.52 Suction grip for lifting

Source: iStock.com/BanksPhotos.

Safety Data Sheets (SDS)

In preparation for the job, a material list will be written and then either delivered or picked up. Plumbers use multiple products that all have different compositions, ingredients and chemicals. Different chemicals can cause different health problems. They also have different safe use and storage requirements. It is important to know about the chemical products used at your workplace.

A **Safety Data Sheet (SDS)** is a document that provides critical information on hazardous chemicals and is prepared by the manufacturer of products that are used in the workplace (see Figure 1.53). Safety Data Sheets, which were previously referred to as Material Safety Data Sheets (MSDS), have had their title changed to reflect the adoption of the Globally Harmonised System of Classification and Labelling of Chemicals (GHS) in 2012 and to meet the requirements for managing risks under the WHS Regulations. They are available from the manufacturer or supplier.

Any hazardous material being delivered to a construction site should have an SDS for the product provided at the site before it is delivered. The SDS should be kept on site and should be reviewed as required by WHS Regulations. If you have not read the SDS for a substance you are going to use, or will be exposed to when someone else uses it, make sure you obtain a copy and read it first. If you are unsure of any specific items mentioned in the SDS, have your employer or a competent person explain it to you. If any special training is required, you should complete the training before the material is handled.

Manufacturers and suppliers are required to update the SDS for a product every five years and to provide information about any conditions necessary to ensure that the substances will be safe and without risk to health when properly used, and that this information is included on the SDS for that product.

Employers are required to provide information, instruction, training and supervision as necessary to ensure the health and safety of workers.

An SDS gives advice on:

- the materials that make up the product
- the health effects of a product and first aid instructions
- precautions for use
- safe handling and storage information.

An SDS will also help you to:

- be aware of any health hazards of a product
- check that the site emergency equipment and procedures are adequate
- store the chemicals properly
- check that a chemical is being used in the correct way for the job required
- decide whether any improvements or changes should be made to machinery or work practices
- decide whether any environmental monitoring should be done.

BORAL CEMENT
Safety Data Sheet

1. IDENTIFICATION OF THE MATERIAL AND SUPPLIER

1.1 Product identifier

Product name	GENERAL PURPOSE CEMENT
Synonym(s)	BERRIMA SL • BLUE CIRCLE GENERAL PURPOSE CEMENT • BLUE CIRCLE HIGH EARLY STRENGTH CEMENT • BLUE CIRCLE OFF WHITE CEMENT • BLUE CIRCLE ® SOUTHERN WHITE CEMENT • GP CEMENT • HE CEMENT • HIGH EARLY STRENGTH CEMENT • KOORAGANG GP • MALDON GP • OFF WHITE CEMENT • SHRINKAGE LIMITED CEMENT • SL CEMENT • SOUTHERN WHITE CEMENT • TYPE GP • TYPE HE • TYPE SL • TYPE SR • WHITE CEMENT • ISO-MENT • HARDIES CEMENT • HES CEMENT • CRÈME CEMENT • BRIGHTONLITE • SUNLITE

1.2 Uses and uses advised against

Use(s)	BINDING AGENT • CONCRETE • CONSTRUCTION • GROUT • INDUSTRIAL APPLICATIONS • MANUFACTURE OF CEMENTS • MASONRY • MORTAR • SOIL STABILISATION

1.3 Details of the supplier of the product

Supplier name	BORAL CONSTRUCTION MATERIALS LTD.
Address	Level 3, 40 Mount Street, Nth Sydney, NSW, 2060, AUSTRALIA
Telephone	(02) 9220 6300
Email	sds@rmt.com.au
Website	http://www.boral.com.au

1.4 Emergency telephone number(s)

Emergency	1800 555 477 (8am – 5pm WST)
Emergency (A/H)	13 11 26 (Poisons Information Centre)

2. HAZARDS IDENTIFICATION

2.1 Classification of the substance or mixture

CLASSIFIED AS HAZARDOUS ACCORDING TO SAFE WORK AUSTRALIA CRITERIA

GHS classification(s)
Specific Target Organ Systemic Toxicity (Repeated Exposure): Category 2
Serious Eye Damage / Eye Irritation: Category 2A
Skin Corrosion/Irritation: Category 2
Specific Target Organ Systemic Toxicity (Single Exposure): Category 3

2.2 Label elements

Signal word WARNING

Pictogram(s)

Hazard statement(s)

H315	Causes skin irritation.
H319	Causes serious eye irritation.
H335	May cause respiratory irritation.
H373	May cause damage to organs through prolonged or repeated exposure.

Prevention statement(s)

P260	Do not breathe dust/fume/gas/mist/vapours/spray.
P264	Wash thoroughly after handling.
P271	Use only outdoors or in a well-ventilated area.
P280	Wear protective gloves/protective clothing/eye protection/face protection.

www.boral.com.au — Page 1 of 7 — SDS Date: 21 Jan 2015

FIGURE 1.53 Page 1 of a sample SDS

Source: Courtesy of Boral Construction Materials Ltd.

Always ensure you obtain a copy of an SDS to fully understand the hazards and health effects of hazardous chemicals.

LEARNING TASK 1.4

PLAN AND PREPARE FOR SAFE WORK PRACTICES

Download an SDS for a 20 kg bag of concrete and answer the following questions:

- What are the health hazards associated with its use?
- What is PPE is required?
- What are the safe storage requirements?

COMPLETE WORKSHEET 4

Use safe work practices to carry out work

Managing hazards is everyone's responsibility and it is essential to employ safe work practices on site in accordance with company procedures. Working unsafely and carrying out tasks with poor technique will significantly raise the risk of illness and injury. When carrying out tasks, always employ the correct technique and be alert to potential hazards as the construction industry is high risk and the effects to the body can potentially last a lifetime.

Manual handling

Manual handling is an activity requiring a person to use force to lift, lower, push, pull, carry, move or hold any type of object. As manual handling is the most common hazard in the building industry, it is important for all workers to understand, and to be fully trained in, correct manual handling techniques.

Safe Work Australia provides guidance for manual handling with the *Model Code of Practice: Hazardous manual tasks* (see http://www.safeworkaustralia.gov.au). The Code of Practice aims to prevent injury and reduce serious injuries resulting from manual handling tasks at work. It requires employers, in consultation with their workers, to identify, investigate and control the risks coming from manual handling activities in the workplace.

Effects on the human body

Injuries to the human body, and diseases caused by accidents or exposure to an unhealthy environment through work-related hazards, will vary depending on the hazard and exposure.

The most common effects on the body are:

- *broken limbs, cuts, lacerations, crushing and bruising* caused by slips, trips and falls, misuse of tools, contact with sharp objects, being hit by falling or flying materials, hand tools or equipment, and accidents through poor concentration
- *electrocution and electrical burns* from coming into contact with live electrical equipment
- *burns* from hot objects, flames or contact with corrosive chemicals
- *crushing and suffocation* through excavation cave-ins, being crushed by materials or equipment, or lack of oxygen (asphyxiation)
- *hearing loss (temporary or permanent)* through exposure to excessive noise
- *work-related stress* from working under stressful work conditions
- *heat stroke, cramps, heat exhaustion and rashes* from exposure to excessive heat
- *heart conditions and high blood pressure* from exposure to harmful chemicals and body stress and strain
- *bone damage, slipped discs, hernias* from body stress and strain
- *stomach and digestive problems* from exposure to harmful chemicals
- *nervous system problems* from exposure to harmful chemicals and body vibration
- *white finger* from excessive hand vibration
- *poisoning (acute and chronic), allergies and irritations* from exposure to harmful chemicals
- *cancers, dermatitis and respiratory irritations* from exposure to harmful chemicals and radiation
- *diseases* from exposure to harmful chemicals, microorganisms, viruses, asbestos-containing material (ACM) and asbestos-contaminated dust or debris (ACD)
- *death* – in the worst cases, all workplace hazards could be fatal.

Causes and effects of bodily injury

Injuries to the body can occur when you least expect it. Having a greater awareness of what injuries may be sustained while performing manual handling tasks on site can help to reduce the likelihood of an injury occurring.

Back injuries

Most manual handling injuries are to a person's back. The spine consists of a series of vertebrae, separated by spongy discs or cartilage. These discs are called intervertebral discs (see Figure 1.54). They act as shock absorbers between the vertebrae. If the back is bent or twisted, the discs will be deformed by the vertebrae they support. Severe injuries occur when a load is so great that the disc ruptures (a slipped disc). However, painful injuries can occur without a rupture taking place.

FIGURE 1.54 The human spine

Some experts believe that serious back injuries may result from damage caused by years of poor practices that finally causes the injury to become apparent, rather than from a single lift, twist or other movement.

Fatigue

Fatigue caused by constant or heavy manual handling tasks can increase the chances of having an accident through loss of concentration.

Muscle injuries

Muscle or musculoskeletal injuries can be caused by strain to the legs, back, arms and tendons from overuse or by exceeding the capacity of the muscle to carry the load. These injuries may cause inflammation of the joints and surrounding nerves, spinal disc damage and hernias (rupturing of body tissue).

Always perform a light stretch of muscles before work commences or before lifting objects, as this may reduce the risk of injuries.

Heart and respiratory disease

Existing medical conditions can be aggravated by poor practices or excessive manual handling. Most of these injuries are caused by:

- using incorrect techniques for lifting objects
- being physically unfit
- not using mechanical aids such as forklifts and conveyors to eliminate the need for manual handling
- carrying out tasks without assistance
- undertaking the same task for long periods of time
- having work benches at an unsuitable height.

To help prevent injuries resulting from the lifting and carrying of objects, one should:

- use suitable mechanical equipment whenever possible
- redesign the task to minimise the risk of injury and eliminate the hazard
- use appropriate PPE
- learn the correct methods of lifting and carrying.

Methods of manual handling

Correct lifting and lowering techniques for manual handling are essential to avoid the risk of injury.

Lifting

Correct lifting methods require you to bend your knees, not your back. Never twist your body when lifting, carrying or moving a load. Protect your hands and feet with suitable PPE.

- *Size up the load.* Consider the shape and size of the load, as well as the weight. If the load appears too heavy, get assistance.
- *Position the feet.* Face the intended direction of travel. Place your feet comfortably apart, one foot forward of the other and as close as possible to the object to be lifted.
- *Obtain a proper hold.* Get a safe, secure grip, diagonally opposite the object, with the whole length of the fingers and palms of your hands.
- *Maintain bent knees and a straight back.* The knees should be bent before the hands are lowered to lift or set down a load. Keep the upper part of your body erect and as straight as possible.
- *Keep your head erect with your chin in.* Keeping the back straight, take a deep breath and begin to raise the load by straightening your legs. Complete the lift with your back held straight.
- *Keep your arms in.* Keep your arms close to the body. Keep your elbows and knees slightly bent. Hold the load in close to your body. Maintain flexible control over the load with your arm and leg muscles.

Ensure correct lifting and lowering techniques are employed when moving heavy or awkward objects. If an object is too heavy or awkward to move on your own, always employ a two-person lift method or use an appropriate mechanical aid.

Lowering

Setting down the load is the reverse of lifting. It is just as essential to keep the back straight and bend the knees while lowering the load as when lifting it.

FIGURE 1.55 Correct steps for lifting and putting down heavy objects

Dual lifting

When more than one person is required to lift and carry a load, the correct lifting methods (as shown in Figure 1.55) must be practised, and coordinated team-lifting techniques should be applied.

FROM EXPERIENCE

Effective communication skills and working as a team to achieve a common outcome will reduce the risk of injury when performing manual handling tasks.

Coordinating team lifting uses the following steps:

- One person should be nominated and give the orders and signals; this person must be able to see what is happening.
- The movements of the team members should be performed simultaneously (i.e. all lifting together).
- All people involved in the lift should be able to see or hear the one giving the orders.
- To enable load sharing, lifting partners should be of similar height and build, or lifters should be graded by height along the load.
- Persons should be adequately trained in team lifting and preferably have been trained together.

Pushing and pulling

Tasks requiring the pushing or pulling of loads are more effectively carried out if the force is applied at or around waist level (see Figure 1.56). When setting the load in motion, jerky actions should be avoided. Apply the force gradually to avoid overexertion and damage to the body.

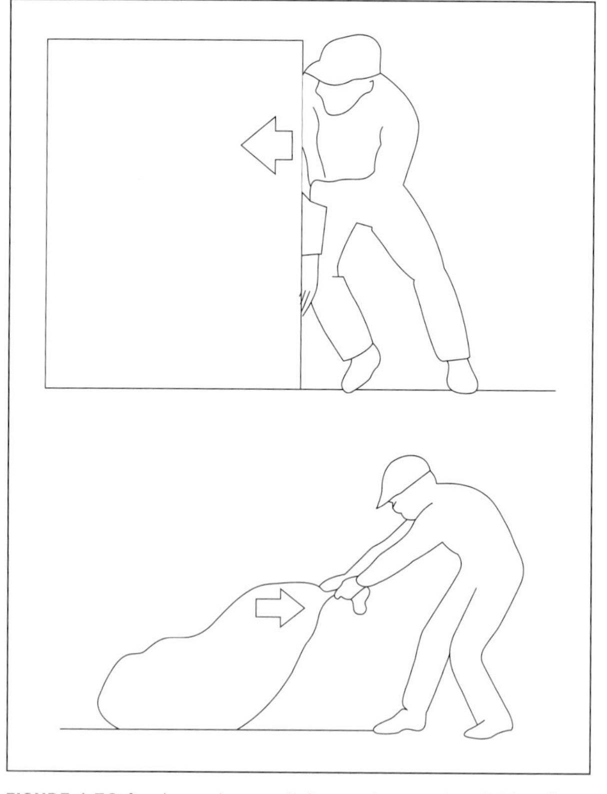

FIGURE 1.56 Apply push or pull force at around waist level

Shovelling

The selection of the correct type of shovel for the job is important. In all cases it is essential that the length of the handle is suitable to reduce the strain and exertion on the body (see Figure 1.57 and Figure 1.58).

FIGURE 1.57 Correct posture when using a long-handled shovel

FIGURE 1.58 Correct posture when using a short-handled shovel

Long-handle shovelling

Use the following techniques when using a long-handled shovel:

- Grasp the shovel with hands well apart.
- Place feet apart, one behind the other.
- Bend the forward knee.
- Use your body weight and pressure from the rear leg to drive the shovel forward and under the material.
- Lift the load by pressing down with the rear hand while using the front leg as an aid for lifting.
- Deliver the load by pivoting on the feet, using the front hand as a fulcrum (pivot point).

Short-handle shovelling

Use the following techniques when using a short-handled shovel:

- Grasp the shovel with hands well apart.
- Place the feet apart, one behind the other and bend both knees.
- Keep the back straight and inclined forward.
- Use your body weight in a forward and downward motion, with pressure from the rear leg and drive the shovel forward and under the material.
- Lift the load by straightening the front leg and back to a vertical position.

Safe and responsible manual handling

Always adhere to these general principles:

- Walk in an upright position and avoid bending the back either forward or backward.
- Do not carry a heavy load in one hand or under one arm, as this tends to bend the spine sideways. Distribute the load evenly so that the bone structure of the body can support the load without distortion. If this cannot be achieved on your own, get a helper or use carrying aids.
- Before attempting to move the load, check the route to be travelled. Make sure that there is nothing in the way on which you could slip or trip, that it is clear of other obstructions, and that there is no danger overhead or from vehicular traffic.
- Check the area where the load is to be placed for space, and that pallets or gluts are in place ready for stacking, before commencing to lift and carry the load.
- If supports are to be used to help carry the load, check that they are strong enough and are correctly placed to take the load.
- Dangerous substances or loads that are labelled corrosive or fragile should be handled with proper care.
- When it is necessary to change direction, move your feet and turn the whole body to avoid twisting your spine.
- Avoid manual handling in tight, constrictive positions.
- When carrying loads of separate units, divide the load evenly between both arms.

Clothing

Work clothes should be comfortable and allow freedom of movement and be suitable for the task to be performed. Clothing that is too tight will restrict movement and make safe manual handling more difficult.

Incident reporting

All workers have a responsibility to report any injury, illness, incident or near miss they are involved in or see. They should report these incidents to their immediate supervisor, site supervisor, workplace WHS committee member, union delegate or first aid officer. If there is any plant or equipment involved, they must make the operator aware of the problem immediately.

Once the supervisor becomes aware of an injury, illness, incident or near miss, they must determine whether it is a *notifiable incident*. If so, the necessary forms must be completed and forwarded to the appropriate authorities (see Figure 1.59). This reporting procedure allows for steps to be taken on a state and national level to help reduce workplace incidents and illnesses in the future. It also allows inspections of the workplace at which the incident or illness occurred to be made by the state or territory WHS authority inspectors, to see that steps are taken to prevent the problem from happening again.

Notifiable incident

Standards Australia's *AS 1885.1 National Standard for workplace injury and disease recording*, provides a National Standard for describing and reporting occupational injuries and disease. From this National Standard each state or territory develops an accident/incident reporting register and reporting procedures to suit its requirements under its own WHS legislation.

- AS 1885.1 National Standard for workplace injury and disease recording

The regulatory authorities of the various states and territories may require a notifiable incident to be reported on an **incident report form**. A notifiable incident relates to the death of a person, serious injury or illness of a person or a dangerous incident, and must be reported by law.

What is an incident?

An incident is when something happens unexpectedly and may often be referred to as an accident, perhaps occurring by chance or through poor management. An incident may cause a person pain and injury; however, they can be prevented. These events may occur as a result of unsafe acts, which may include the following:

- practical jokes
- using tools and equipment in a manner for which they were not designed

HAZARD/INCIDENT/REPORT FORM

Who uses this form?
Two people–the worker and his or her supervisor (from the PCBU).
Purpose?
When a hazard, incident or occurs, record what happened. what investigations occurred, and what was done to prevent furture injury or illness in relation to this incident or
What should happen?
The host PCBU/employer keeps the original and a copy is to be given to the lobour-hire agency, to be kept in a file with the host employer's name on it.

PART A – To be completed by worker

Name of employee: Name ______ Surname ______ **Date:** ______

Time of incident/accident: ______

Name of PCBU/employer: PCBU/Manager Name ______ **Work Area:** ______

1. Describe the hazard/detail what happened – include area and task, equipment, tools and people involved.

2. Possible solutions/how to prevent recurrence – Do you have any suggestions for fixing the problem or preventing a repeat?

FIGURE 1.59 Sample incident report form

- rushing, and taking short cuts
- not using PPE or other safety devices
- throwing materials or rubbish from roofs or upper floors.

An incident also may be a consequence of unsafe conditions, such as:

- little or no training in safety and proper use of tools and equipment
- poor housekeeping
- poor management of site safety issues
- poorly maintained tools and equipment
- damaged tools and equipment
- inadequate or no PPE
- poor site conditions and congestion due to lack of preparation.

Illness and injury

A reportable illness is one where a worker has a medical certificate stating that they are suffering from a work-related illness that stops the worker from carrying out their usual duties for a continuous period that is specified by the relevant WHS state or territory regulator.

A reportable injury is one occurring from a workplace incident where a fatality occurs or a person cannot carry out their usual duties for a continuous period that is specified by the relevant WHS state or territory regulator.

Employers or people in control of a workplace are normally required to send an incident report form to the WHS regulatory authority of the state or territory even if the person injured or killed is not one of their workers (e.g. is a subcontractor/worker or visitor to the site).

Near misses

These occur when the conditions are right for an incident but people don't get hurt and equipment is not damaged.

Near misses usually indicate that a procedure or practice is not being carried out correctly or that site conditions are unsafe. By reporting these near misses the problems may be looked at and rectified before someone is hurt or killed, or equipment is seriously damaged.

Incident report forms

The incident report may require the following information to be given, where appropriate.

Information about the employer or workplace:

- name of company
- office address
- address of site where accident happened
- main type of activity carried out at the workplace (e.g. building construction)
- major trades, services or products associated with this activity
- number of people employed at the workplace and whether there is a WHS committee at the workplace.

Information about the injured or ill person:

- name and home address
- date and country of birth
- whether the person is employed by the company
- job title and main duties of the injured person.

Information about the injury or illness:

- date of medical certificate
- type of injury or illness
- whether the injury resulted in death
- particulars of any chemicals, products, processes or equipment involved in the accident.

Information about the injury or dangerous occurrence:

- time and date that it happened
- exact location of the event
- details of the injury
- type of hazard involved
- exactly how the injury or dangerous occurrence was caused
- how the injury affected the person's work duties
- details of any witnesses
- details of the action taken to prevent the accident from happening again.

Details of the person signing, and the date when the incident report was signed are also required. When the report is completed, copies may be required to be sent to the state or territory regulatory authority and a copy is kept by the employer.

Injury management

The loss or disruption that a company can experience as a result of a hazardous incident can be severe when that incident leads to a worker being injured.

A comprehensive risk management system should include a well-thought-out plan to maximise the opportunity for injured workers to remain at work. This allows the worker to be productive in some capacity and assists with the recovery and rehabilitation process.

Therefore, the risk management system should cover the following points:

- early notification of the injury
- early contact with the worker, their doctor and the employer's insurance company
- provision of suitable light duties as soon as possible to assist with an early return to work
- a written plan to upgrade these duties in line with medical advice.

Workers' compensation

Workers' compensation insurance is a system that provides payment benefits and other assistance for workers injured through work-related incidents or illnesses. It may also provide their families with benefits where the injury is very serious or the worker dies.

Employers must take out workers' compensation insurance to cover all workers considered by law to be their workers.

Eligibility for compensation

To be able to claim workers' compensation a worker must have suffered an injury or illness. The injury or illness must be work-related and must have happened while working, during an allowed meal break or on a work-related journey.

The injury or illness must be categorised by one of the following:

- the death of the worker
- the worker being totally or partially unable to perform work
- the need for medical, hospital or rehabilitation treatment
- the worker permanently losing the use of some part of the body.

A claim for compensation is made by:

- informing the employer and lodging a claim as soon as possible
- seeing a doctor and obtaining a medical certificate. Ensure that you report to the doctor that the incident is 'work-related' so that the necessary paperwork can be generated for compensation purposes.

If any problems arise with the compensation claim the worker should contact the relevant state or territory compensation authority, or the worker's own union.

A record of *all* injuries that occur at a workplace must be entered in an injury register book or an applicable registry system that can be accessed.

Firefighting equipment and procedures

Learning how to prevent and fight fires is part of every worker's responsibility. It is important for the safety of every worker on a job to understand the procedures to follow in the event of a fire.

Large construction sites and buildings should have firefighting teams responsible for each floor or the whole building. The firefighting team must be specially trained staff members who are referred to as fire wardens, who can direct the evacuation, and firefighting operations, until the fire emergency services arrive.

Fire combustion triangle

The three elements necessary before there can be a fire are: fuel + heat + oxygen.

- *Fuel* can be any combustible material that is a solid, liquid or gas that can burn. Flammable materials are any substances that can be easily ignited and will burn rapidly.
- *Heat* that may ignite a fire can come from many sources, which can include flames, welding operations, grinding sparks, heat-causing friction, electrical equipment and hot exhausts.
- *Oxygen* comes mainly from the air. It may also be generated by chemical reactions.

If any one of the three elements are taken away, the fire will be extinguished (see Figure 1.60).

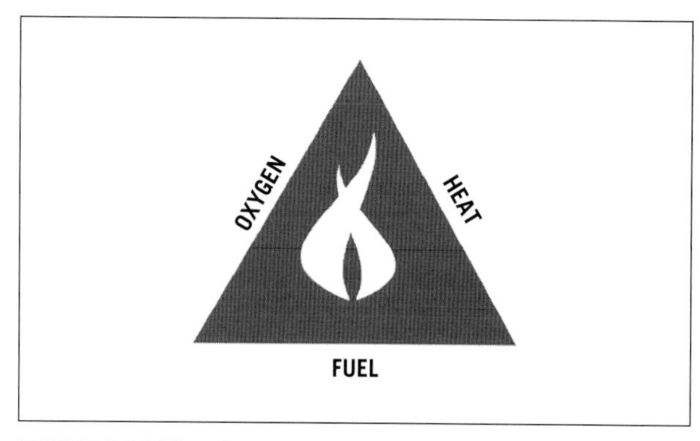

FIGURE 1.60 The elements necessary for a fire

On-site fires

Common causes of fires on site include the following:

- *Burning off rubbish* – site rubbish should be cleared away regularly and no burning off should take place, in accordance with EPA requirements.
- *Electrical fires* – these can occur due to overloading equipment, faulty equipment and faulty wiring. All equipment should be carefully checked, maintained and used correctly.
- *Use of naked flames* – workers such as plumbers or structural steel workers must ensure that they do not carry out naked flame operations within the vicinity of stored rubbish, paints, sawdust or any other highly flammable material.
- *Smokers* – carelessly disposing of cigarettes or matches can cause fires. Butane lighters may also be a source of ignition and should not be exposed to naked flames or other situations where ignition could occur.

How to prevent fires

Don't give fires the chance to start:

- Practise good housekeeping procedures in relation to your work area.
- Remove unwanted rubbish and waste materials from the workplace.
- Store fuels and combustible materials carefully, use approved pouring cans.
- Use only approved electrical fittings, keeping them in good order.
- Don't overload electrical circuits.
- Don't smoke at the workplace.
- Take special care if working with flammable liquids or gases.

- Be careful of oily rags; for example, turps- or linseed oil-soaked rags, which can ignite spontaneously.
- Avoid dust hazards. Many types of dust can be so highly flammable that they can explode when mixed with air or when they are exposed to a flame or sparks.

In the event of a fire

If there is a fire at your workplace:

- Do not put yourself or your work colleagues in danger.
- Don't panic – keep calm and think.
- Warn other people in the building.
- Those not needed should leave the building at once and assemble at the designated fire-assembly area.
- Arrange for someone to phone the fire emergency services.
- Have the power and gas supplies turned off if it is appropriate (some lighting may still be required).
- Close doors where possible to contain the fire.
- Stay between a doorway and the fire.
- Be aware of containers of explosive or flammable substances. Remove them from the area only if it is safe to do so.
- If it is safe to fight the fire, select the correct class of extinguisher, having others back you up with additional equipment.
- Know how to use the appropriate class fire extinguisher.
- If the fire is too large for you to extinguish, get out of the building and close all doors. Assemble at the designated area.

Note: It is recommended that construction workers undergo basic training in the use of fire-extinguishing equipment.

Classes of fires and extinguishers

Extinguishers have been grouped according to the class of fire on which they should be used. The class is determined by the type of material or equipment involved in the fire. Although there are several classes of fire, the *four* main classes are as follows.

Class A fires

Class A fires involve ordinary combustible materials; for example, wood, paper, plastics, cloth and packing materials (see **Figure 1.61**).

The correct extinguishers to use are:

- water type – most suitable
- any other type – except carbon dioxide (CO_2).

Class B fires

Class B fires involve flammable and combustible liquids; for example, petrol, diesel, spirits, paints, lacquers, thinners, varnishes, waxes, oils, greases, and many other chemicals in liquid form (see **Figure 1.62**).

FIGURE 1.61 Class A – ordinary combustible materials: wood, paper, plastics, cloth and packing materials

FIGURE 1.62 Class B – flammable and combustible materials: steel jerry can

The correct extinguishers to use are:

- foam type
- carbon dioxide (CO_2)
- dry chemical.

Class C fires

Class C fires involve flammable gases; for example, LPG and acetylene (see **Figure 1.63**).

The correct extinguisher to use is:

- dry chemical powder.

Class E fires

Class E fires involve 'live' electrical equipment; for example, electric motors, power switchboards or computer equipment (see **Figure 1.64**).

The correct extinguishers to use are:

- only extinguishers displaying an (E), as these will not conduct electricity; never use water or foam extinguishers
- dry chemical powder
- carbon dioxide (CO_2).

FIGURE 1.63 Class C – flammable gases: LPG

FIGURE 1.64 Class E – electrical switchboards

Identification and operation of extinguishers

Correct identification and operation of extinguishers is essential to ensure a fire is attacked appropriately and safely.

Water extinguisher (Figure 1.65)

Operation:

- The two main types of water extinguisher available are gas pressure and stored pressure.
- The range of operation is up to 10 m.
- Methods of activating the extinguishers are different for each type, so it is important to read the instructions on the container before attempting to use it.
- Once activated, the jet of water should be directed at the seat of the fire.

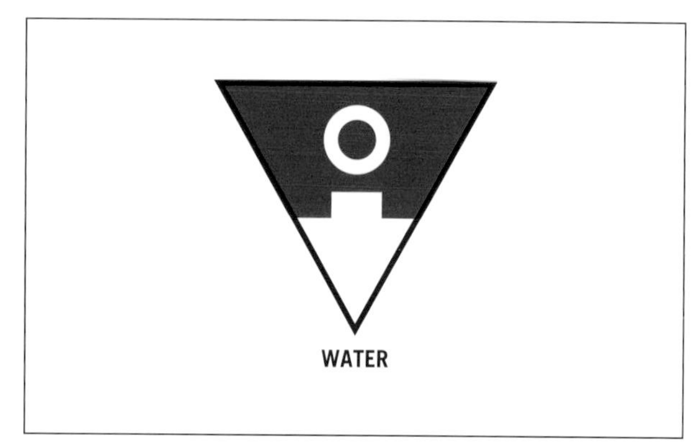

FIGURE 1.65 Water extinguisher indicator. Identifying colour: All Red

Foam extinguisher (Figure 1.65)

Operation:

- The three types of foam extinguisher available are gas pressure, stored pressure and chemical (being phased out).
- The range of operation is up to 4 m.
- Methods of activating the extinguishers are different for each type, so it is important to read the instructions on the container before attempting to use it.
- Once activated, the jet of foam is directed to form a blanket of foam over the fire. This stops oxygen from getting to the fire long enough to allow the flammable substance to cool below its re-ignition point.

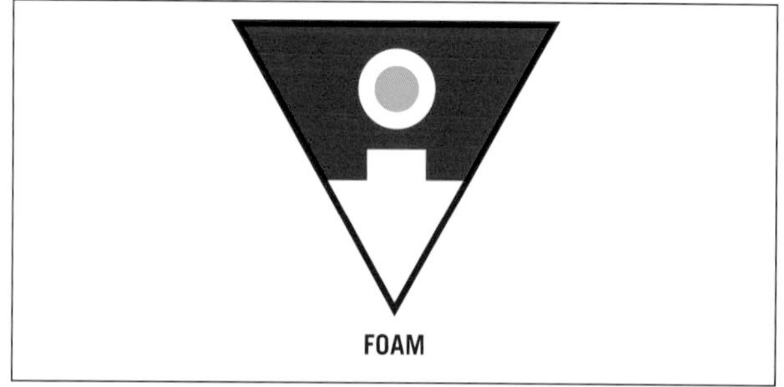

FIGURE 1.66 Foam extinguisher indicator. Identifying colour: pre-1999 – All BLUE; post-1999 – RED container with BLUE band

Dry chemical powder – ABE (Figure 1.67)

Operation:

- Small sizes have a range of 3 m, with larger types to 6 m.

- This is most effective for the extinguishing of large areas of burning liquid or free-flowing spills.
- Powder is discharged through a fan-shaped nozzle. It should be directed at the base of the fire, which should be covered with a side-to-side sweeping action.

FIGURE 1.67 Dry chemical powder indicator. Identifying colour: RED container with WHITE band

FIGURE 1.68 Carbon dioxide indicator. Identifying colour: RED container with BLACK band

Carbon dioxide (CO_2) (Figure 1.68)

Operation:

- Small sizes have a range of 1 m, with larger types to 2.5 m.
- These are useful for penetrating fires that are difficult to access.
- Use as close to the fire as possible. Aim the discharge first to the rear edge of the fire, moving the discharge horn from side to side, progressing forward until flames are out.

Exposure to carbon dioxide in a confined area could cause asphyxiation. Move clear of the area immediately after use and ventilate the area once the fire is out.

Note: Halon-type extinguishers (all yellow) have been found to have ozone-depleting potential, and their use has been phased out by government regulations.

Operating instructions for any extinguisher can be found on the label fixed to that extinguisher.

Other firefighting equipment

Fire blankets

Fire blankets can be used on most types of fires (see Figures 1.69 and 1.70). The action of the blankets is to smother the fire. These are very useful on burning oils and electrical fires. Keep the blanket in place until the fire is out and long enough to allow the flammable substance to cool below its re-ignition point.

FIGURE 1.69 A typical fire blanket packet

FIGURE 1.70 A fire blanket in use

Source: Newspix/Carmelo Bazzano.

Fire hose reels

Hose reels (where provided) are a very effective means of fighting Class A fires (see **Figure 1.71**). Extreme care is required to ensure that no live electrical equipment is in the area. They *should not* be used on liquid fires, as water may spread the burning liquid.

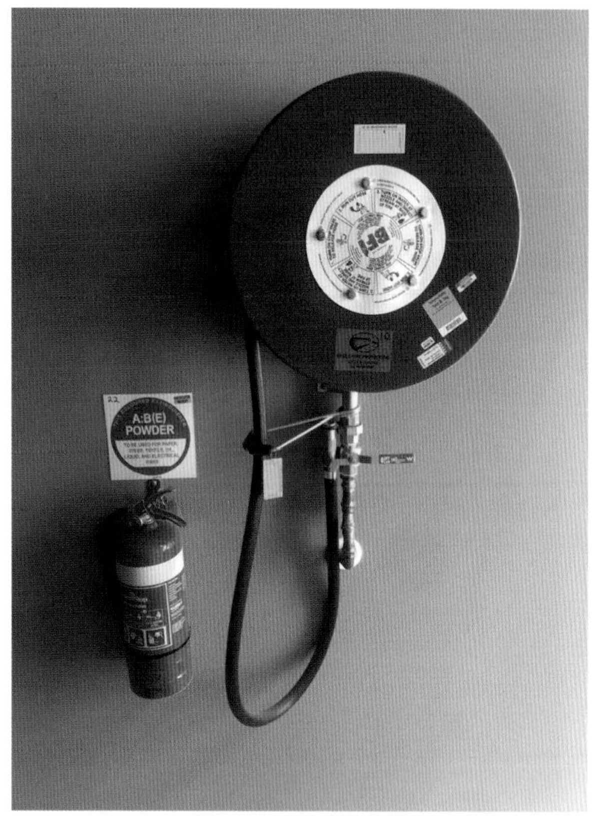

FIGURE 1.71 Hose reel

First aid

An employer must ensure that workers have access to first aid facilities that are adequate for the immediate treatment of common medical emergencies. Competent first aid personnel may include a person with a current approved first aid certificate, a registered nurse or a medical practitioner. For more on first aid, see Chapter 2 of this text.

An employer must also ensure that there are first aid facilities, a suitable first aid kit and sufficiently trained first aid personnel at their place of work.

The St John Ambulance Australia website (http://www.stjohn.org.au) contains a number of first aid fact sheets downloadable for the workplace. It is recommended that you contact your specific state or territory St John Ambulance office (freecall 1300 360 455) to enquire about what first aid kits are available to accommodate your specific needs.

Contents of a workplace first aid kit

The contents of a workplace first aid kit are listed in **Table 1.3** and a wall-mounted workplace first aid kit is shown in **Figure 1.72**.

TABLE 1.3 Contents required for a wall-mounted workplace first aid kit

Item	Quantity
Adhesive Strips hypo-allergenic, 50s	1
Bags: Bag resealable small 10 cm × 18 cm Bag resealable medium 15 cm × 23 cm Bag resealable large 23 cm × 30 cm	1 of each
Bandage Compression Extra Firm 10 cm	1
Bandage Conforming Light 5 cm	3
Bandage Conforming Light 7.5 cm	3
Combine Dressing 10 cm × 20 cm	1
Emergency Accident Blanket	1
Emergency First Aid Book	1
Eye Pad	4
Forceps Pointed 12.5 cm SS Sharp	1
Gauze Swabs 7.5 cm × 7.5 cm × 5	5
Gloves Nitrile Large pair (disposable)	5
Hydro Gel sachets 3.5 g	5
Instant Cold Pack	1
Non-adherent Dressing 10 cm × 10 cm	1
Non-adherent Dressing 7.5 cm × 10 cm	3
Non-adherent Dressing 5 cm × 5 cm	6
Notepad & Pencil in bag	1
Pen Ballpoint (black ink)	1
Resuscitation Mask	1
Safety Pins (assorted × 12)	1
Saline Steritube 15 mL	8
Scissors Medical 12.5 cm SS Sharp/Blunt	1
Splinter Probe (disposable)	2
Swabs Antiseptic	10
Swabs Iodine	10
Tape hypo-allergenic 2.5 cm × 9 m	1
Triangular Bandage 110 cm × 110 cm	2
Wound Dressing No. 14	1
Wound Dressing No. 15	1

Note: The above list is a recommendation only and it is strongly suggested that readers refer to the obligatory requirements of their own state or territory.

Source: WorkSafe Victoria, https://www.worksafe.vic.gov.au.

LEARNING TASK 1.5

USE SAFE WORK PRACTICES TO CARRY OUT WORK

Under the supervision of your teacher or a competent person, perform the following tasks while wearing appropriate PPE. Ensure the area is clear of hazards and risk assessed to prevent injury.

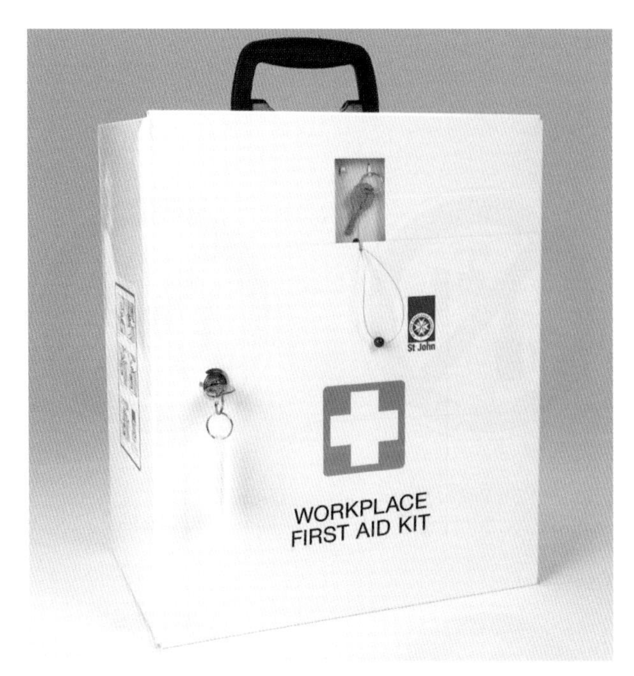

FIGURE 1.72 Wall-mounted workplace first aid kit

Source: St John Ambulance.

1 Using the correct lifting methods, lift and lower a box or object of a reasonable weight.
2 Practise dual lifting techniques with another person and alternate the person giving the orders.
3 Use a short- and long-handled shovel to excavate and backfill an area of a sand yard approximately 1 m² and 300 mm deep. Ensure the correct technique is used.

Maintain safety of self and others

Maintaining safety begins with proper identification of hazards and chemicals. One of the most effective ways to do this is through appropriate signage, and the identification, safe handling and storage of hazardous chemicals. Maintaining a safe site and reporting hazards in accordance with legislation and workplace procedures can prevent incidents and protect yourself and others.

Safety signs

Safety signs are placed in the workplace to warn of hazards or risks that may be present. They can also give information on how to avoid or manage the hazard or risk. If a safety sign is required to be placed on a piece of equipment it may be in the form of a safety tag.

Safety signs:

- prevent accidents
- signal health hazards
- indicate the location of safety and fire protection equipment
- give guidance and instruction in emergency procedures.

Standards Australia has three Standards covering the use of safety signs in industry:

- *AS 1216 Class labels for dangerous goods*
- *AS 1318 Use of colour for the marking of physical hazards and the identification of certain equipment in industry* (known as the SAA Industrial Safety Colour Code)
- *AS 1319 Safety signs for the occupational environment.*

- AS 1216 Class labels for dangerous goods
- AS 1318 Use of colour for the marking of physical hazards and the identification of certain equipment in industry (known as the SAA Industrial Safety Colour Code)
- AS 1319 Safety signs for the occupational environment

Placement of safety signs

Signs should be located where they are clearly visible to all concerned, where they can easily be read, and where they will attract attention. If lighting is not adequate, illuminated signs may be used.

Signs should *not* be located where materials and equipment are likely to be stacked in front of them, or where other obstructions could cover them (e.g. doors opening over them). They should *not* be placed *on* movable objects such as doors, windows or racks so that when the object is moved, they are out of sight or the intention of the sign is changed.

The best height for signs is approximately 1500 mm above floor level. This is at the normal line of sight for a standing adult. The positioning of the sign should not cause the sign itself to become a hazard to pedestrians or machine operators.

Regulation and hazard-type signs should be positioned in relation to the hazard to allow a person enough time to view the sign and take notice of the warning. This distance will vary for different signs; for example, signs warning against touching electrical equipment should be placed close to the equipment, whereas signs on construction work may need to be placed far enough away to permit the warning to be understood before the hazard is reached.

Care should be taken where several signs are intended to be displayed close together. The result could be that so much information is given in one place that little or no notice is taken of it, or that it creates confusion.

For more information on safety signs, refer to Chapter 3 of this text, your state or territory WHS regulator or *AS 1319 Safety signs for the occupational environment*.

Safety sign types

The four main types of safety signs covered in AS 1319 are regulatory, hazard, emergency information and fire.

Regulatory signs

Regulatory signs contain instructions that *control or limit certain actions*. Regulatory signs are subdivided into the following three categories:

- *Prohibition signs* indicate that the action identified on the sign is not permitted. Prohibition signs have a red border and crossbar overlaying a black symbol on a white background (see Figures 1.73–1.76).
- *Mandatory signs* indicate the action identified on the sign must be carried out. Signs have a white symbol overlaying a blue background and are symbolic, although sometimes text may be added (see Figures 1.77–1.80).
- *Limitation or restriction signs* indicate a defined or numerical limit on an activity (see Figure 1.81).

FIGURE 1.73 Smoking prohibited

FIGURE 1.74 Digging prohibited

Hazard signs

Hazard signs advise of hazards and are subdivided into the following two categories.

FIGURE 1.75 Water not suitable for drinking

FIGURE 1.76 Fire, naked flame and smoking prohibited

FIGURE 1.77 Eye protection must be worn

FIGURE 1.78 Head protection must be worn

FIGURE 1.79 Hearing protection must be worn

FIGURE 1.80 Face protection must be worn

FIGURE 1.81 Speed restriction sign

- *Danger signs* warn of a hazardous condition or hazard that is *potentially life threatening*. Signs have the word 'DANGER' in white letters on a red oval overlaying a black panel. The message or text of the sign will be in black letters on a white background (see Figure 1.82).
- *Warning signs* warn of a hazardous condition or hazard that is *not likely life threatening*. Signs have a yellow background overlaid with a black symbol and a black border that may sometimes have text added (see Figures 1.83–1.86).

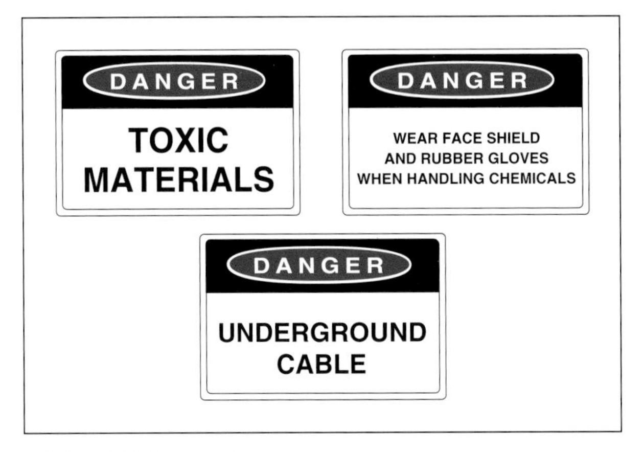

FIGURE 1.82 Danger signs

FIGURE 1.83 Fire risk

FIGURE 1.84 Toxic hazard

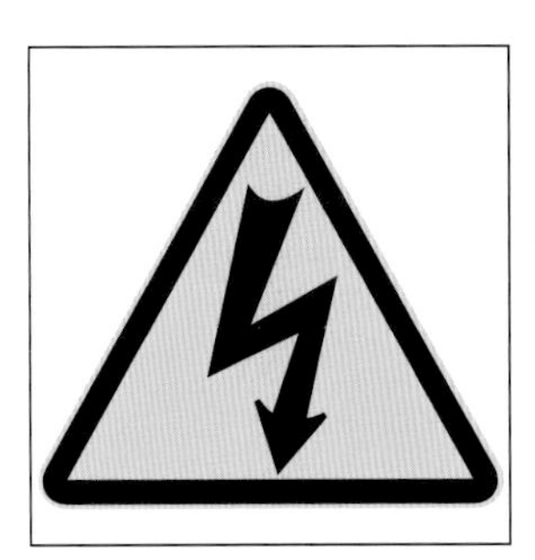

FIGURE 1.85 Electric shock risk

FIGURE 1.86 Forklift hazard

Emergency information signs

Emergency information signs carry emergency information that indicates the location or directions for emergency-related information. Signs may include emergency exits, first aid facilities, emergency eye washes, emergency showers and safety equipment. Emergency information signs have a white symbol overlaying a green background and may have text added (see Figures 1.87 and 1.88).

FIGURE 1.87 First aid

FIGURE 1.88 Emergency (safety) eye wash

Fire signs

Fire signs advise the location of fire-fighting facilities and fire alarms. Signs have white letters on a red background (see Figures 1.89–1.90).

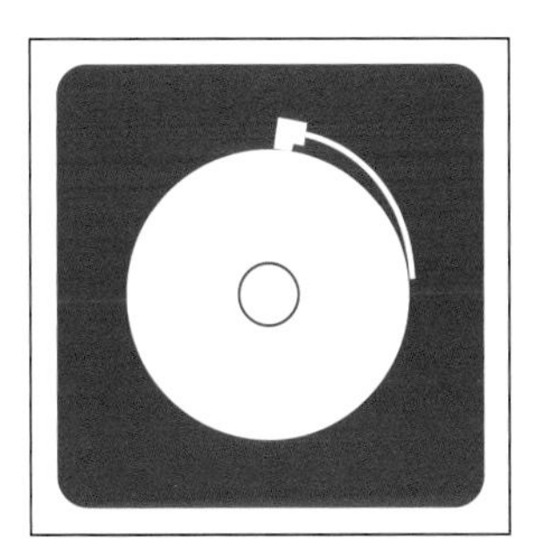

FIGURE 1.89 Fire alarm

FIGURE 1.90 Fire hose reel station

Fire text signs for fire extinguishers carry different colour codes that indicate the type of fire for which the extinguisher is intended.

To make sure the message reaches everyone at the workplace – including workers from non-English-speaking backgrounds and workers with low reading skills – picture signs should be used wherever possible. When this is not possible, it may be necessary to repeat the message in other languages. Picture signs can be made clearer with a short written message (see Figure 1.91).

Safety signs and tags for electrical equipment

Electrical wires and equipment that are being worked on or are out of service, or that are live or may become live, must have *Warning* or *DANGER* safety tags fixed to them to help prevent accidents. Once the hazard

FIGURE 1.91 Combined picture and word sign

has been removed, only the person who put the tag in place should remove it or authorise its removal.

Any standard safety sign may be made smaller and used as an incident protection tag (see Figure 1.92).

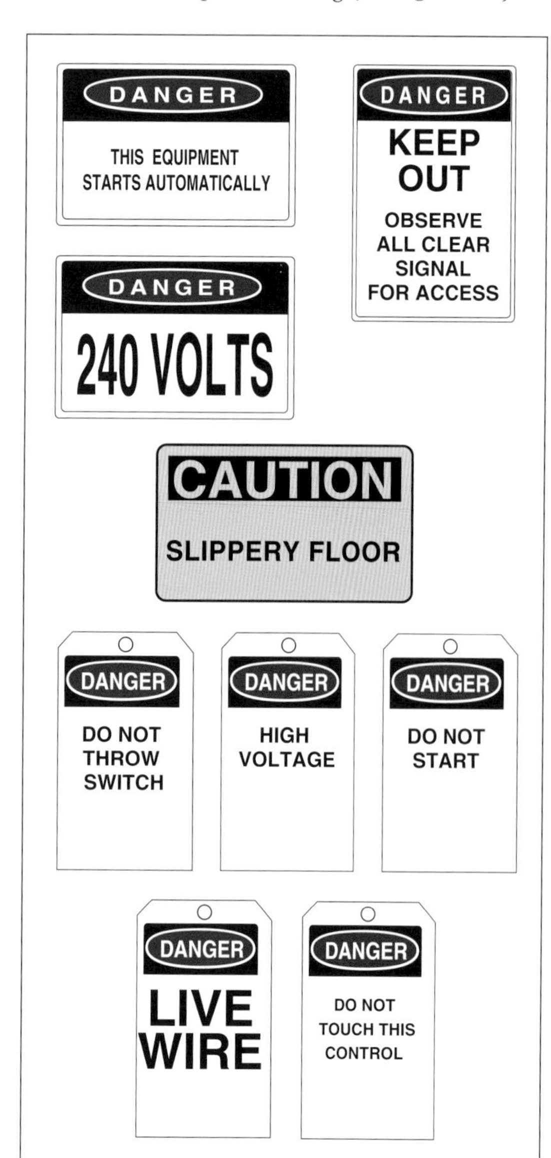

FIGURE 1.92 Electrical safety signs and tags

If words are to be used, this will generally be in the form of a danger sign. A tag should be at least 80 mm × 50 mm, plus any area required for tying or fixing the tag in place.

The background colour of the tag should be *yellow* for *warning* signs and *white* for *danger* signs.

Dangerous goods

Many of the chemicals used on building sites are classified as dangerous goods. Australia has adopted a system of classifying and labelling dangerous goods, which is the Globally Harmonised System of Classification and Labelling of Chemicals (GHS). The GHS is a single internationally agreed system of chemical classification and hazard communication through labelling and Safety Data Sheets (SDS). The system helps to globally classify chemicals, labels and SDS, and assists people to recognise dangerous goods and their properties and dangers quickly. The system was introduced in 2012 and made mandatory from 31 December 2016. The system for classification is detailed in the Australian Dangerous Goods Code.

Dangerous goods can be identified by a *diamond* sign or label (see Figure 1.93).

The nine classes of dangerous goods under this system are:

- explosive substances or articles
- gases
- flammable liquids
- flammable solids
- oxidising substances
- toxic and infectious substances
- radioactive materials
- corrosive substances
- miscellaneous dangerous materials.

The diamond-shaped sign or label shows which of the nine classes the dangerous substance belongs to. These signs have distinctive symbols and colouring. Not all hazardous substances have dangerous goods labels because the dangerous goods diamond indicates only an immediate hazard, and not necessarily a hazard that has only long-term health risks.

Not all dangerous goods have safe handling and storage instructions printed on them, and they may have only warning diamonds. The safe handling and storage instructions can be obtained from the SDS for the product, or in the relevant Australian Standard or Code of Practice for the substance.

If you find a product or substance with a diamond on the container, obtain and read the SDS for the material before storing, opening or using it.

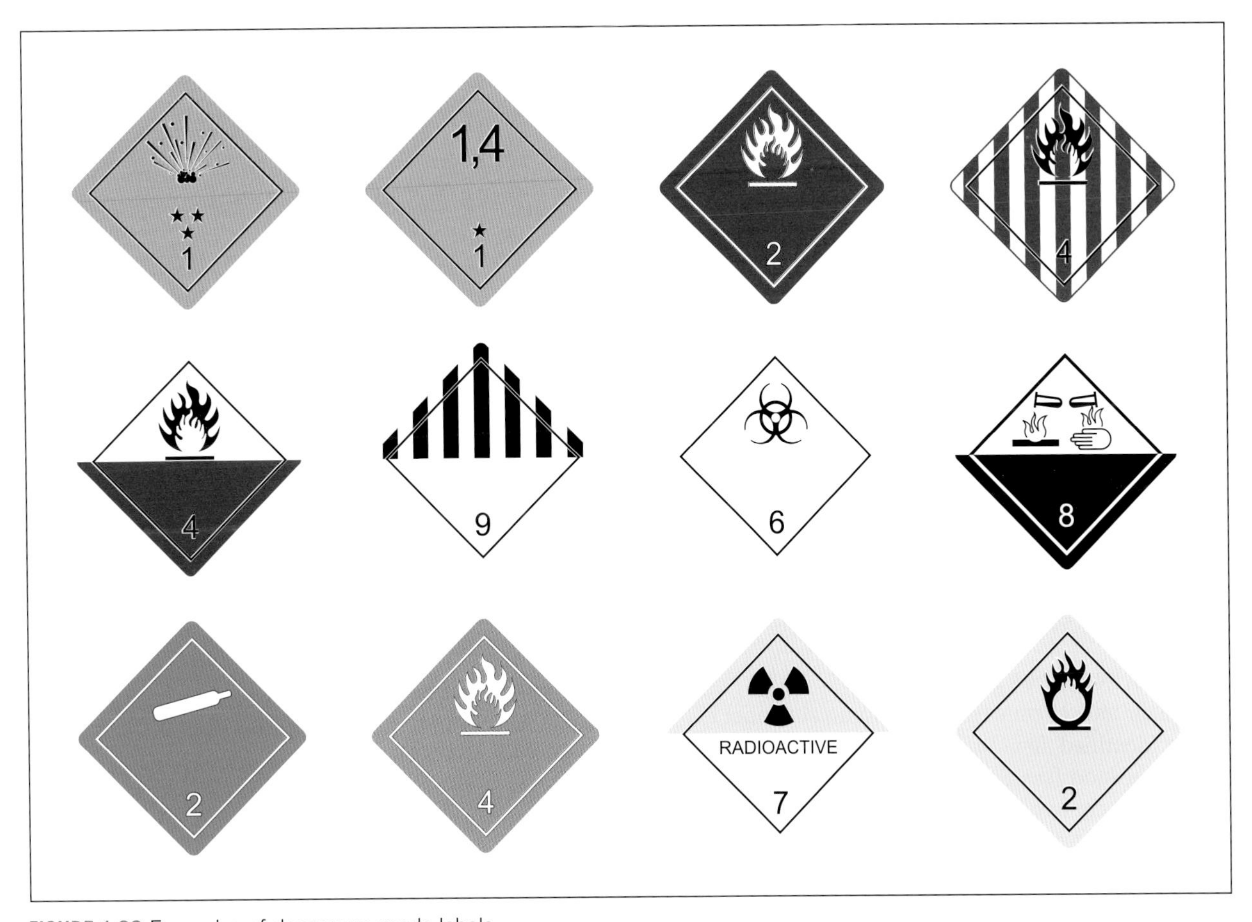

FIGURE 1.93 Examples of dangerous goods labels

Source: Shutterstock.com/popcic.

Signs and labelling

The details of the design and selection of signs and labels for the nine classes of dangerous goods are contained in the Australian Standard *AS 1216 Class labels for dangerous goods*.

- AS 1216 Class labels for dangerous goods

Correct labelling means dangerous goods don't have to be any more dangerous than they already are. You should not only learn to recognise the various symbols but also learn about the actual properties of the substances you may be exposed to.

It is important to know which goods can produce toxic gases, are highly flammable, are dangerous when wet, are dangerous on contact with air, and are harmful when they come into contact with your skin. What you don't know could hurt you.

Each label has a distinctive colour, has a symbol to make it easily recognisable, states the hazard in words and gives the classification number.

Chemical hazards in the construction industry

There are very large numbers of chemicals being used on construction sites and the possible effects associated with these chemicals are a major threat to construction workers' health and safety.

Some of the effects of exposure to chemicals are well known, such as dermatitis from cement and epoxy resins. The hazards of other materials are less well known, and the attitude in the industry is that most materials in use – such as adhesives, grouts, mastics and powders – 'are not really chemicals'. This is incorrect as the majority of these products *are* chemicals, and this incorrect assumption is causing most workers to use no protection, or very little protection, when using hazardous materials.

Health effects from hazardous chemical exposure

Health hazards can be categorised into two groups:

- Acute health effects relate to short-term exposure to hazardous chemicals. For example, if you swallow a poison, it could be fatal or you may fall very ill within 48 hours. Similarly, if you have acid splashed onto your skin, you might suffer burns immediately or within the next 48 hours.

- **Chronic health effects** relate to the long-term effects of exposure to hazardous chemicals. In general use, these effects may take years to become apparent. The chronic health effects may be just as serious as the acute health effects in the long term (e.g. exposure to cancer-causing agents or asbestos may take 20 years to become apparent, but may still be fatal).

Disposal of hazardous chemicals

It is extremely important to dispose of chemicals safely, as prescribed in the SDS for the product. Do not pour chemicals down a drain or dispose of them by throwing into a rubbish bin. You might inadvertently affect other people by doing this. Always wash yourself carefully, following the advice of the SDS, after handling any chemical. If you spill any chemical on your clothing, remove the clothing and wash the body part affected. If you experience skin problems or have difficulty in breathing, seek medical advice immediately.

Storage of hazardous chemicals

Any hazardous chemicals brought onto site must be clearly identified and stored safely in an appropriate storage location (see Figure 1.94) or properly constructed containers. It is critical that you adhere to the advice contained within an SDS. Correct signposting of the area should be carried out to warn of the hazards of the stored chemicals.

Stored chemicals can be dangerous to outsiders, such as rescue workers and firefighters. In incidents where chemicals are spilt or involved in fires, toxic fumes and/or gases could be emitted. These situations can present a hazard to members of the public, as well as to the emergency personnel in attendance. For this reason, a register and/or a manifest must be completed. An employer must ensure a register of hazardous chemicals at the workplace is kept up to date and readily accessible to workers or people who may be affected with the handling and storage of these chemicals. A manifest may be required where the quantities of the hazardous chemicals exceed the prescribed amounts allowed. The primary purpose of the manifest is to provide emergency services with the location, classification and quantities of hazardous chemicals. The manifest must also contain information such as emergency contact details and site plans. For any additional information refer to the *Model Code of Practice: Managing risks of hazardous chemicals in the workplace* through Safe Work Australia (see http://www.safeworkaustralia.gov.au).

FIGURE 1.94 Storage cabinet for hazardous chemicals

Good work practices

When working in or near hazardous chemical areas, you should:

- change your clothes daily
- shower after work
- check the workplace to ensure that:
  - adequate PPE is being used
  - there is medical monitoring of persons exposed to hazardous chemicals
  - dust-producing machines are isolated or enclosed
  - ventilation and other health and safety systems are effective.

Hazard report forms

Controlling the risks arising from hazards offers the greatest area of opportunity for reducing injury and illness in the workplace. All workplaces should have a hazard reporting system in place for workers to report potential hazards. This will bring problem areas to the attention of management as soon as the hazard has been identified.

A means of providing this system is to have standardised *hazard report forms* readily available to the workforce (see Figure 1.95). Workers should complete a form and give it to their immediate supervisor as soon as a potential hazard is identified. This will allow control measures to be put in place to remove the hazard at the earliest possible time.

Hazard/incident report form

Use this form in your workplace to report health and safety hazards and incidents. To notify SafeWork NSW of an incident, call 13 10 50.

Hazard/Incident

Brief description of hazard/incident: (Describe the task, equipment, tools and people involved. Use sketches, if necessary. Include any action taken to ensure the safety of those who may be affected.)

____

____

____

____

Where is the hazard located in the workplace?

____

____

When was the hazard identified? Date: ______/______/______ Time: ____________am/pm

Recommended action to fix hazard/incident: (List any suggestions you may have for reducing or eliminating the problem – for example re-design mechanical devices, update procedures, improve training, maintenance work)

____

____

____

____

Date submitted to manager: Date: ______/______/______ Time: ____________am/pm

Action taken

Has the hazard/incident been acknowledged by management? Yes/ No

Describe what has been done to resolve the hazard/incident:

____

____

____

____

Do you consider the hazard/incident fixed? Yes/ No

Name: ________________________ Position: ________________________

Signature: ________________________

Date: ______/______/______

SW09097 0918

FIGURE 1.95 A sample of the type of information that might be recorded on a hazard/incident report form

Source: https://www.safework.nsw.gov.au/__data/assets/pdf_file/0003/414192/Hazard-incident-repor-SW09097.pdf.

LEARNING TASK 1.6

MAINTAIN SAFETY OF SELF AND OTHERS

Under the supervision of your class teacher or a competent person, perform the following tasks while wearing appropriate PPE. Ensure the area is clear of hazards and risk-assessed to prevent injury.

1. Make a list the of the safety signs contained within the practical workshop.
2. Record which signs are no longer relevant and any signs that may be covered up by doors or in a poor line of sight.
3. If any signs are deemed to be a hazard in the work area, obtain a hazard report form.
4. Complete the hazard report form and hand to your class teacher.

COMPLETE WORKSHEET 6

Use electricity safely

Electricity is a major power source for equipment used in construction. This results in networks of temporary and semi-permanent cabling that present a potential hazard to all people on site. The risks involved can be as severe as electrocution, so the *Model Code of Practice: Managing electrical risks in the workplace* (see http://www.safeworkaustralia.gov.au) has been developed to provide guidance on safe work practices. Australian Standards and Licensing detail who can carry out electrical work, how that work will be configured, the equipment required and its performance standards. It also identifies the responsibilities of people connecting equipment to the construction power supply.

You should be familiar with your responsibilities for your local area.

Main safety considerations

The principal safety issues are as follows:

- Central to electrical safety is the residual current device (RCD; see Figure 1.96). The RCD monitors the flow and return currents in a circuit and will shut off the supply if a fault is detected. This function will stop the flow of electricity if there is a difference of more than 30 milliamps (mA), which is a level that should protect a person from serious harm. Always test the RCD by connecting to a power source and pressing the test button to ensure the safety switch trips (see Figure 1.97); if this does not occur, a competent person or electrician is required to test the RCD for faults.

FIGURE 1.96 A 4-way power box with an RCD fitted

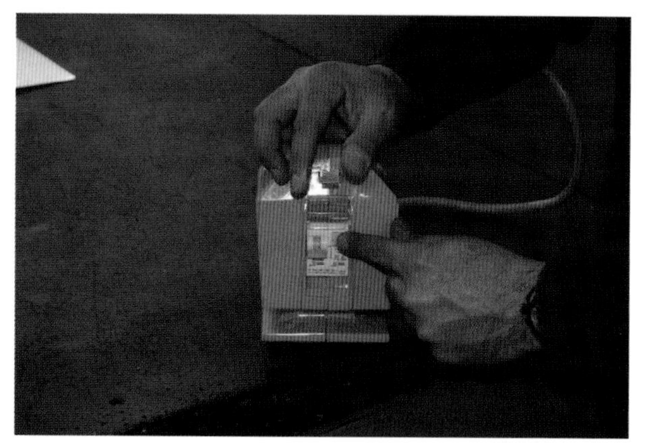

FIGURE 1.97 Testing the trip mechanism for an RCD

- Before connecting to an unknown power source, connect a power point tester (see Figure 1.98) to the power point to test if there is incorrect wiring, earth connection and/or polarity. Plumbers connect to power sources every day without knowing if it is safe to do so.
- Extension leads must also be of suitable quality, thickness and length for their application. They should be suspended away from work on insulated lead stands (see Figure 1.99) or hooks, and diverted away from traffic and water, and should be restricted to use on the same floor level as the power supply.
- In addition to these requirements, all electrical equipment and leads need to be maintained on a regular basis. This involves inspection and testing by a competent person who then tags the equipment and records the inspection details.

FIGURE 1.98 Testing a power point using a power point tester

FIGURE 1.99 Electrical lead on an insulated lead stand

Electrocution

An 'invisible' hazard that all plumbers need to be aware of is the fact that all metallic pipes, fixtures and appliances can potentially be 'live' electrically. This is because they may be connected to the electrical earthing system, which is also connected to the neutral wire at the switchboard under the multiple earth neutral (MEN) system used for supply in Australia. What this means is that faults within the premises or even in neighbouring properties can result in potentially lethal current flowing through the pipes. The person who breaks this circuit runs the risk of becoming the new path for this current. Take special care with electrical leads, particularly when working in wet or damp locations or on roofs.

Control measures

Various authorities outline the procedures to reduce the risk of electric shock when working on metallic water and gas piping systems. It is advised to always check pipework for stray electrical current with an approved voltage detector (see Figure 1.100) or multimeter (see Figure 1.101) before any work commences. Once the pipework has been checked and no current found, it is essential to connect a bridging conductor (see Figure 1.102) to effectively bridge (breach) any section of a metallic piping system (i.e. water or gas pipe) *before* the cut or disconnection is made.

FIGURE 1.100 Using an approved voltage detector to detect stray electrical current

FIGURE 1.101 Using a multimeter to detect stray electrical current

FIGURE 1.102 Bridging conductor

Before anyone attempts to cut through a metallic water or gas pipe or disconnect any part of the piping system (including the meter), the approved bridging conductor should be securely attached to the pipe on either side of the portion of pipe to be cut or disconnected. It is extremely important that the bridging cable be properly *earthed* on each end.

The reason for doing this is that if there is an electrical fault present the current will 'choose' to flow across the bridge, not the worker. The bridge should be maintained until after reconnection.

Any alteration that replaces a section of metallic piping material with plastic pipe needs to be checked by a qualified electrician. A multimeter or voltage detector can be used to detect stray current, and if found you must notify the electrical utility, qualified electrician and the homeowner/client of the situation.

When disconnecting water meters or cutting into pipework, always use an approved bridging conductor to prevent the risk of electric shock.

Note: Turning the power supply off at the main switches can reduce the risk of electrocution but it is no guarantee. If there is any doubt at all, have the system checked by a qualified electrician *prior* to commencing work on the service.

Apply emergency response

An emergency may develop due to a number of reasons, such as a fire, gas or toxic fumes leak, improper use of flammable materials, the partial collapse of a building (see Figure 1.103), a bomb threat, a crane overturning, unstable ground, materials improperly stored or a trench collapse.

Every organisation is required to have an emergency procedure in place and personnel appointed to control the safe exit of persons at the workplace.

FIGURE 1.103 Emergency caused by a partial building collapse

Source: Fairfax Syndication/Ken Irwin.

Responsible personnel

It is recommended that all persons have some basic training in 'life sustaining first aid' in an effort to deal with any emergencies that might arise.

On a large building site, the responsible personnel may include the emergency or fire warden, employer, worker, safety officer, head foreperson or site supervisor.

Small-building-site personnel may include the builder, foreperson, leading hand or a nominated tradesperson. Whichever the case, these people are responsible for following set procedures to get everyone else on site out of the danger area to a predetermined collection point so that in the event of an incident all persons may easily be accounted for. The nominated responsible personnel are protected against liability resulting from practice or emergency evacuations from a building where they act in good faith and in the course of their duties.

On large sites, these people may be identified by a coloured helmet they would wear, which would be determined as part of the organisation's emergency plan. Listed in Table 1.4 are the common roles used in the event of an emergency or incident.

TABLE 1.4 Common roles used in the event of an emergency or incident

Role	Responsibilities
Emergency coordinator	• Determine the nature of the emergency and the course of action to be followed. • Set off any alarm or siren to warn of an emergency. • Contact appropriate services such as police, fire and/or ambulance. • Initiate the emergency procedure and brief the emergency services when they arrive.
Warden or controller	• Assume control of the occupants or workers until the emergency services arrive. • Notify all persons regarding the nature of the emergency. • Give clear instructions and make a record of what was carried out. • Report all details to the coordinator as soon as possible.
Casualty control	• Attend to casualties and coordinate first aid. • Coordinate the casualty services when they arrive. • Arrange for further medical or hospital treatment.

Procedures for emergencies

An employer has the responsibility to ensure that in the event of an emergency, arrangements have been made for the following:

- the safe and rapid evacuation of persons from the place of work or to a designated collection area on the site
- emergency communications, such as a landline or mobile phone with emergency phone numbers clearly visible and accessible
- provision of appropriate medical treatment of injured persons by ambulance, medical officer or access to a suitable first aid kit.

Emergency phone numbers

The phone number to contact in an emergency is triple zero (000). There are two alternative numbers:

- The first is 112, which is the GSM (Global System for Mobile communication) international standard emergency number. It can only be dialled on digital mobile phones, provided there is GSM coverage.
- The second is 106, which is for people with speech or hearing impairment. This is directed through a TTY, which is a teletypewriter or textphone. It is an important service as SMS through mobile phones is not possible for contacting emergency services.

It is also recommended that each person with a mobile phone has an entry named ICE ('in case of emergency') on their phone. This should list the phone details of the person who should be contacted in case something happens to the owner of the mobile phone.

Note: In most cases people use passwords or codes to protect someone from accessing their mobile phone. Depending on the make of the phone, an emergency contact can be added to your phone and accessed through the lock screen in a similar manner that you can access emergency numbers through the lock screen. Type in 'emergency contacts' on a web search for the type of phone that you own to create your own emergency contact.

Emergency procedures

In the event of an emergency, any person discovering the emergency should:

- rescue any person in immediate danger if it is safe to do so
- alert other people in the immediate area
- extinguish or confine any fires before it takes hold if equipment is available
- dial 000 for emergency services to attend
- contact the emergency coordinator, or warden, as soon as possible.

Clean up work site area

Cleaning of work areas and removal of health hazards must be an ongoing operation on construction sites and other workplaces. It will help maintain a high standard of safety and a healthy environment in which to work. Before commencing the cleaning of a given work area, a planned approach must be formulated. An inspection of the area should be carried out to determine the extent of the work.

The planned approach should consider:

- the removal of all hazards
- the method of dust suppression to be used
- designated material storage areas
- cleaning and rubbish disposal methods to be used
- safe access to and from the area
- a systematic approach to the whole cleaning operation.

Housekeeping

Housekeeping of a building site involves maintaining the site in a *safe* and *clean* manner. This will improve:

- *safety* – by maintaining standards that will provide safe work areas
- *productivity* – by allowing work to proceed faster, improving production times
- *access* – by allowing safe access free of hazards to and from the work areas.

The tasks included under the term 'housekeeping' involve:

- *Sort and stack* reusable and unused materials neatly and in a safe manner. A list should be placed on the stack giving any necessary details, such as the number of items and lengths of materials.
- *Remove hazards* that may cause people to trip, slip or be cut. Some of these hazards include pipe offcuts, reo bars, timber offcuts, oil spills, water leakages and sharp pieces of materials.
- *Transfer waste materials* to designated waste bins or rubbish stockpile areas. Special areas should be provided for hazardous materials found on the building site; for example, asbestos, flammable liquids, oxyacetylene bottles, cleaning materials and solvents. Hazardous materials must be removed or isolated to avoid dangers to workers and delays to work schedules.
- *Safety rails and barricades* should be erected around the edges of floor areas, openings in floors, stairways and trenches. Place and fix sufficient safety covers over holes where people could trip.
- *Maintain safety equipment* in good condition so that it is ready to use. This should include cleaning and stocking first aid kits, making sure appropriate class fire extinguishers are in place and charged, and ensuring safety signs are in place.
- *Electrical leads* should be kept clear of work and access areas by the provision of insulated stands or hooks to keep them above the ground or floor.
- *Water hazards* should be drained, or barricaded off, to eliminate slippery conditions caused by spreading mud over walk areas.
- *PPE* should be worn at all times when carrying out housekeeping functions.
- *Correct lifting techniques* must be used.

Dangerous situations can occur from lack of good housekeeping if:

- combustible materials are left in areas where welding and grinding are being carried out
- spilt liquids are left on walk areas causing slippery conditions
- materials are stacked in an unsafe manner
- timbers and formwork materials are not de-nailed and stacked as they are dismantled
- unused materials are not stacked in a safe manner.

Non-toxic wastes are all wastes created on a building site that do not produce either a toxic or poisonous health hazard or a toxic threat to the environment. They may, however, cause hazards to workers and the environment in other ways.

Dust suppression

Dusts in the workplace can cause:

- chemical hazards
- respiratory problems
- explosive hazards.

Therefore, it is important to reduce to a minimum the amount of dust in the air. This is most important when clean-up operations are taking place, as large volumes of dust can be generated if care is not taken. Dust masks and eye protection may not keep all dust from entering the body tissue. Silica, asbestos-contaminated dust or debris (ACD), synthetic mineral fibre, cement and wood dusts are of particular concern on building sites.

The three most common methods of dust suppression are:

- *Wetting down* – the area to be swept is sprayed with a fine mist of water to dampen the dust particles before sweeping commences. This dampening of the dust stops particles from floating. Care must be taken not to cause a hazardous area through excessive amounts of water.
- *Damp sawdust* – dampened sawdust is spread over the area and when swept up the fine dust particles cling to the sawdust, preventing them from floating in the air.
- *Vacuum cleaners* – vacuuming is a very effective way of collecting hazardous dusts. It is particularly useful for those places that are difficult to reach with a broom. Ordinary household vacuum cleaners will not effectively trap the very fine dust particles and are prone to clogging after a short time. You may need an industrial-quality vacuum cleaner with a HEPA (high efficiency particulate air) filter. Where there is an explosion hazard, flame-proofed vacuum cleaners must be used. Some industrial-quality vacuum cleaners may be a 'wet and dry' type and will pick up water, allowing you to wet down the area before commencing vacuuming.

When working with or removing asbestos-contaminated dust or debris (ACD) or asbestos-containing material (ACM), a competent person who holds relevant removal industry practice and certification may be required.

Personal cleaning procedures

Cleaning operations bring workers into contact with many harmful substances, microorganisms and viruses that may harm a person's health. It is important to maintain personal hygiene at work.

The following minimum standards should be followed by all workers at all times:

- Wash hands and other exposed parts of the body before handling food or drink, before smoking and at the completion of the day's work.
- Wear proper clothing and footwear, which can be removed before leaving the work area, so that hazardous materials will not be spread away from the site. Clothing should be cleaned regularly.

- When working with hazardous dusts it is important to shower before leaving the site and have clothes appropriately cleaned or disposed of.
- Wash hands before leaving the toilet block. Use the soap and towels provided.
- Use rubbish bins provided for the disposal of food scraps.
- Don't spit.
- Apply a barrier cream to exposed areas of skin before handling harmful substances. This will prevent the absorption of the material into the skin and make it easier for you to wash it off.

Safety precautions

When working on site, think *safety first* at all times:

- Always look up and check what is overhead.
- Always respect any safety rules or regulations.
- Take the safest, most direct route from one place to another.
- Keep access routes clear of obstructions.

All waste products must be suitably identified and disposed of in accordance with regulatory authority requirements.

Tools and equipment

Tools and equipment will be selected to suit the clean-up job to be carried out. In most cases the items required include the following:

- *wheelbarrows* to provide a safe method of moving materials and rubbish
- *shovels and brooms* to sweep up and transfer rubbish to containers
- *cleaning equipment* to remove spills and stains
- *rubbish bins* for storage of rubbish until it can be removed from the site
- *rubbish chutes* to allow for a safe and easy method of transferring rubbish to ground level
- *vacuum cleaners* for safe collection of hazardous dusts
- *pallets and pallet trucks* for stacking of reusable and unused materials
- *timber gluts* to allow safe and orderly stacking of materials
- *ropes* to lift or lower materials or equipment from one level to another and stabilise equipment
- *PPE* safety equipment for cleaning operations, which may include:
  - safety boots or footwear
  - hard hats
  - safety goggles
  - earmuffs/plugs
  - protective gloves
  - protective clothing
  - respirators
  - dust protective masks.

Wearing these items of PPE will minimise hazards to health and safety.

At the completion of the clean-up, all tools and equipment should be cleaned and returned to their correct storage places. All documentation must be completed in accordance with workplace requirements and forwarded to relevant authorities.

COMPLETE WORKSHEET 7

REFERENCES AND FURTHER READING

Acknowledgements

Reproduction of the following resource list references from DET, TAFE NSW C&T Division (Karl Dunkel, Program Manager, Housing and Furniture) and the Product Advisory Committee is acknowledged and appreciated.

Texts

Graff, D.M. & Molloy, C.J.S. (1986), *Tapping group power: A practical guide to working with groups in commerce and industry*, Synergy Systems, Dromana, Victoria.

Web-based resources

Regulations/Codes/Laws

AustLII legislation database: **http://www.austlii.edu.au/databases.html**

Australian Codes of Practice: **http://www.safeworkaustralia.gov.au**

Globally Harmonised System (GHS) of classification and labelling of chemicals: **http://www.safeworkaustralia.gov.au/sites/swa/whs-information/hazardous-chemicals/ghs/pages/ghs**

Managing risks of hazardous chemicals in the workplace: **http://www.safeworkaustralia.gov.au/sites/swa/about/publications/pages/managing-risks-of-hazardous-chemicals-in-the-workplace**

NSW Codes of Practice: **http://www.workcover.nsw.gov.au**

Safe Work Australia – Model Codes of Practice: **http://www.safeworkaustralia.gov.au**

When and where was asbestos used?: **https://www1.health.gov.au/internet/publications/publishing.nsf/Content/asbestos-toc~asbestos-when-and-where**

WorkSafe Victoria: **https://www.worksafe.vic.gov.au**

Resource tools and VET links

NSW Education Standards Authority – Construction: **https://educationstandards.nsw.edu.au/wps/portal/nesa/11-12/stage-6-learning-areas/vet/construction-syllabus**

Training.gov.au: **http://training.gov.au**

Industry organisations' sites

Building Trades Group Drug & Alcohol Program: **http://www.btgda.org.au**

CITB (SA Construction Industry Training Board): **http://www.citb.org.au**

SafeWork NSW (1999), Hazard profile for demolition: **http://www.workcover.nsw.gov.au/Publications/Industry/Construction/demolition.htm**

WA Government OHS site – manual handling: **http://www.wa.gov.au**

Audiovisual resources

Short videos covering topics such as safe manual handling, workplace housekeeping, hazardous substances and accident investigation are available from the following organisations:

Safetycare: http://www.safetycare.com.au

TAFE NSW: **https://www.tafensw.edu.au**.

SafeWork NSW publications

'Applying the new safety regulations', Cat. no. 229 (also 100, 110, 1008, 2001)

'Back watch industry profile – construction trades', Cat. no. 718

'Manual handling', Cat. no. 9020

'Occupational protective gloves', Cat. no. 3017

'Personal protective equipment', Cat. nos. 032, 208, 310, 3003, 3010, 3012, 3017, 3019, 3029, 4005, 4007, 4500

'Protection from UV radiation in sunlight', Cat. no. 9017

'Protective helmets standard', Cat. no. 3012

'Reading labels on material safety data sheets', Cat. no. 400

'Safety helmets', Cat. no. 4500

'Skin cancer and outdoor workers', Cat. no. 116, 117

'Work method statements', Cat. no. 231

TAFE NSW resources

Resource list and order forms: Training and Education Support (TES), Industry Skills Unit Orange/Granville, 68 South St Granville NSW 2142 Ph: (02) 9846 8126 Fax: (02) 9846 8148

GET IT RIGHT

The photo below shows an incorrect practice that can be performed when installing pipe work near asbestos sheeting.

Identify the incorrect method and provide reasoning for your answer

WORKSHEET 1

WORKSHEETS 1

To be completed by teachers

Satisfactory ☐

Not satisfactory ☐

Student name: ____________________

Enrolment year: ____________________

Class code: ____________________

Competency name/Number: ____________________

Task: Review the section 'Participate in workplace induction' in this chapter and answer the following questions.

1. Describe what the term PCBU means.

____

____

____

2. List three categories of workers that an employer is responsible for on site.

____

____

____

3. What measures must a worker take regarding WHS Regulations?

____

____

4. List the two types of induction training that a plumber will undertake.

____

____

5. What is issued to a person after successful completion of the general construction induction training?

____

6. What could happen if a construction worker is working on a construction site without first undertaking the general construction induction training?

____

____

7. What is the purpose of a workplace induction?

____

____

8. Name both the Act and Regulation that govern WHS in the workplace of your relevant state or territory.

____

____

9. What are Codes of Practice?

____

____

____

____

10. Briefly describe three requirements that an employer must provide in relation to health, safety and welfare for their workers at work.

____

____

____

To be completed by teachers

Satisfactory ☐

Not satisfactory ☐

Student name: ______________________

Enrolment year: ______________________

Class code: ______________________

Competency name/Number: ______________________

Task: Review the section 'Assess risks' in this chapter and answer the following questions.

1. Briefly describe the difference between an 'acute hazard' and a 'chronic hazard'.

2. List the five major groups of common workplace hazard areas.

3. List three physical hazards and briefly describe how this would affect you in your workplace.

4. List four factors that may lead to stress disorders.

5. List the four steps that must be followed for managing health and safety risks.

6. What is the order of preference called for controlling hazards on site and what is the most effective control measure?

7. Why should control measures be reviewed?

8. Who would normally carry out safety inspections?

9. SWMS are a compulsory part of any site safety management plan and should form part of the planning for any site task. Using the blank SWMS below, complete the three columns to identify the tasks, risks/hazards and safety controls required to carry out a simple work-based project of your choice.

Safe Work Method Statement Student name: Workshop/college grounds: Practical project:		Signed off: (Teacher) Date: Accepted: YES / NO
TASK/PROCEDURE	**POTENTIAL HAZARDS**	**SAFETY CONTROLS**

WORKSHEET 3

To be completed by teachers	
Satisfactory	☐
Not satisfactory	☐

Student name: ____________________

Enrolment year: ____________________

Class code: ____________________

Competency name/Number: ____________________

Task: Review the section 'Identify hazards and hazardous materials on the worksite' in this chapter and answer the following questions.

1. List the two forms of ACM that may be found in the workplace.

2. List five types of ACM products that plumbers may find in the workplace.

3. Define what an asbestos register is used for.

4. What are the health effects if exposed to crystalline silica dust?

5. How would existing services be located before any excavation work commences?

6. List three factors that could contribute to trench collapse.

7. List three hazardous conditions associated with working in confined spaces.

8. List two Model Codes of Practice that could be referred to for preventing falls in the workplace.

9. List three pieces of equipment or PPE that would be used to prevent falls in the workplace.

10. Describe what suspension intolerance is, and how it would occur.

11. List two ways of working at a height of over 2 m.

12. Why shouldn't you walk on a skylight when working on the roof?

WORKSHEET 4

Student name: ______________________

Enrolment year: ______________________

Class code: ______________________

Competency name/Number: ______________________

To be completed by teachers	
Satisfactory	☐
Not satisfactory	☐

Task: Review the section 'Plan and prepare for safe work practices' in this chapter and answer the following questions.

1. Briefly describe the main purpose and function of PPE.

____

____

____

____

2. Who is responsible for providing PPE that is required for work?

____

3. List two items of PPE suitable to protect the following body areas.

   i. Head

   ____

   ____

   ii. Eyes/face

   ____

   ____

   iii. Hearing

   ____

   ____

   iv. Respiratory

   ____

   ____

   v. Hands

   ____

   ____

vi. Feet

vii. Body

4. What is the name given to the stretching impact barrier placed inside a safety helmet between the skull and shell of the helmet?

5. How can the back of the neck be protected from sunburn when wearing a safety helmet?

6. List the three hazard categories that eye protection is designed for.

7. State the two main PPE methods used to protect hearing.

8. State the main requirements of all safety footwear to provide maximum protection.

9. When gloves are too restrictive, what can be applied to the hands to protect from potential hazards?

10. What are dust masks generally used for?

____

11. State a suitable mechanical aid to use for the following situations.
    i. To carry concrete, bricks, tools, rubbish etc. around a building site

    ____

    ii. To lift loads too heavy for manual lifting techniques

    ____

    iii. To carry a heavy upright carton such as a hot water heater.

    ____

12. Who is an SDS prepared by?

____

WORKSHEET 5

To be completed by teachers	
Satisfactory	☐
Not satisfactory	☐

Student name: ______________________

Enrolment year: ______________________

Class code: ______________________

Competency name/Number: ______________________

Task: Review the section 'Use safe work practices to carry out work' in this chapter and answer the following questions.

1. List four common effects to the body when exposed to excessive heat.

____

____

____

____

2. State the four main bodily injuries that can occur due to poor or incorrect manual handling techniques.

____

____

____

____

3. List the six main steps to follow for the correct lifting and lowering of loads.

____

____

____

____

____

____

4. State the importance of the length of the handle of the shovel being used when shovelling materials.

____

____

____

____

WORKSHEETS 1

5. When an incident occurs in the workplace, who should it be reported to?

____

6. When would a notifiable incident be required to be reported?

____

____

7. What is workers' compensation?

____

____

____

8. State the three elements required to start and sustain a fire.

____

____

____

9. List three common causes of fires on site.

____

____

____

10. List three ordinary combustible materials that would be extinguished by a Class A fire extinguisher.

____

____

____

11. State the source or fuel and most suitable types of fire extinguishers for use on the following classes of fires.

Class B

Source/fuel ________________________________

Suitable extinguishers ________________________________

Class C

Source/fuel ________________________________

Suitable extinguishers ________________________________

WORKSHEETS 1

12. What type of extinguishers should never be used for Class E fires?

13. What class of fire should a fire hose reel only be used for?

14. With regards to first aid, what are three areas of concern that an employer must ensure are sufficiently covered at their place of work?

To be completed by teachers	
Satisfactory	☐
Not satisfactory	☐

Student name: ______________________________

Enrolment year: ______________________________

Class code: ______________________________

Competency name/Number: ______________________________

Task: Read through the section 'Maintain safety of self and others' in this chapter and answer the following questions.

1. State the purpose of safety signs.

   ____

2. Signs should be placed in a position that allows them to be clearly seen. State the ideal position for safety signs.

   ____

   ____

3. Briefly describe the four main categories of common safety signage used in the building industry.

   Regulatory ______________________________

   Hazard ______________________________

   Emergency information ______________________________

   Fire ______________________________

4. Briefly describe the background, symbol and colour found on an emergency eye-wash sign.

   Background ______________________________

   Symbol ______________________________

   Colour ______________________________

5. Briefly describe the background, symbol and colour found on a mandatory eye protection sign.

   Background ______________________________

   Symbol ______________________________

   Colour ______________________________

6. Identify the following signs, stating what they represent.

7. State the system for classifying and labelling dangerous goods.

8. How should hazardous chemicals be disposed of?

WORKSHEET 7

To be completed by teachers

Satisfactory ☐

Not satisfactory ☐

Student name: ______________________

Enrolment year: ______________________

Class code: ______________________

Competency name/Number: ______________________

Task: Review the sections 'Use electricity safely' and 'Clean up worksite area' in this chapter and answer the following questions.

1. Why should all electric power tools be connected through an RCD?

____

____

____

2. Why would a power point tester be used?

____

____

3. How would a plumber detect stray current on a metallic pipe?

____

____

4. What equipment would be required to safely disconnect a water meter?

____

5. State why it is necessary for an organisation to have an emergency procedure in place.

____

____

____

____

6. During an emergency, list four people who could be deemed responsible personnel.

____

____

____

____

7. In an emergency situation what does the term 'ICE' mean in regard to emergency details?

8. Housekeeping involves maintaining a worksite in a safe and clean manner. What are three main areas that this can improve as a result?

9. What are three areas of concern that dust in the workplace may cause?

10. State one method of suppressing dust on site.

11. Why is it important to maintain personal hygiene on the job site?

PROVIDE FIRST AID

2

Shaun Tinnion

This chapter addresses the following key elements for the competency 'Provide first aid':

- Respond to an emergency situation.
- Apply appropriate first aid procedures.
- Communicate details of the incident.
- Evaluate the incident and own performance.

It introduces the skills and knowledge required to provide a first aid response to a casualty and to provide resources to allow the reader to prepare for a formal first aid qualification. This applies to all workers who may be required to provide a first aid response in a range of situations, including community and workplace settings.

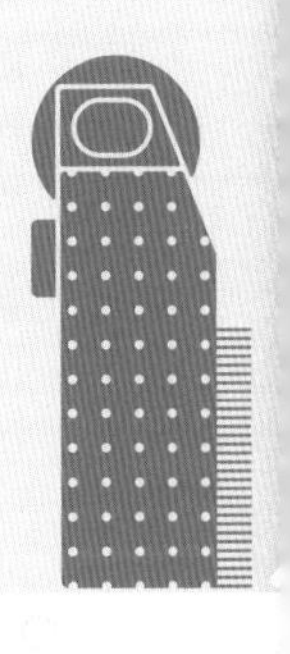

Respond to an emergency situation

First aid allows you to assist a person who is ill or injured. We encounter hazards every day and from our own experiences we risk-assess situations and act in a way that minimises the chance of injury, such as crossing the road in a safe manner. However, despite our best efforts, sometimes hazards result in injury or illness. In the workplace there can be additional hazards. The equipment used to carry out work duties may include chemicals or rotating machinery. The working environment may involve working at heights, in trenches or in confined spaces. First aid training provides the ability to respond to an emergency situation. Simple techniques are employed to ensure that the number of casualties does not increase and to promote the casualty's recovery.

First aid is the assistance given to someone who is ill or injured until medical assistance arrives or the person recovers. The objectives of first aid are the 4Ps:

- Preserve life.
- Protect the unconscious.
- Prevent a casualty's condition from becoming worse.
- Promote the recovery of the casualty.

First aiders should always keep these objectives in mind when carrying out treatment, as it is often simple and timely actions that can prevent the situation getting worse and give the casualty the best chance of recovery. The aim of the processes and techniques that follow is to support these four objectives.

Recognise an emergency situation

It is important to identify an emergency situation in order to apply the above mentioned 4Ps to any casualty. You may begin to suspect an emergency situation due to observations made of the worksite. For example, you may observe an odour, a liquid spillage or a machine that is operating but is unattended. The signs of an emergency situation may be more obvious, such as being alerted to a casualty who may be visibly hurt or bleeding, or seeing a person exhibiting unusual behaviour such as panic or lack of motion. You will not know if first aid is required until you approach the individual you have observed.

Identify, assess and manage immediate hazards to health and safety of self and others

Before you approach an individual, it is necessary to assess the situation for hazards. There may be more hazards than just the one that has caused the injury. For example, if you see someone fall over you do not know if they have slipped or if they have been overcome by fumes. It may not be immediately obvious that a confined space has a toxic environment or that the water pipes are connected to the building's main earthing conductors. Therefore, when you recognise an emergency situation it is important to act quickly to assess the situation for hazards using all of your senses. These hazards need to be controlled before you can approach the individual. If you do not control these hazards in the immediate area, you or another bystander may become injured, and reduce the chances of a positive outcome for the casualty.

Assessing the emergency situation can be compared to any other workplace risk assessment. Therefore, the hierarchy of control (see Chapter 1) becomes a useful tool. You may be able to eliminate the hazard; for example, turn off the electricity or gas, or turn the car engine off. However, an assessment of the scene might identify hazards that cannot be controlled safely, and specially trained assistance may be required to control the hazard before the casualty can be approached; for example, following a chemical spillage or if the casualty is in deep water and you cannot swim. You may need to isolate the hazard; for example, moving the casualty off the railway track, out of the river or away from falling rocks. The emergency situation may be worsening and the casualty may need to be moved to a safer area; for example, if a building is collapsing, a trench is filling up with water or a fire cannot be extinguished. (Movement of a casualty is covered later in this chapter.) You also may need to don personal protective equipment (for you, the casualty and bystanders); for example, it may be necessary to wear a life jacket to approach a casualty in deep water or head and eye protection in a factory.

FROM EXPERIENCE

In a workplace it is important to familiarise yourself with the lock-out procedures for machinery, and to know where the personal protective equipment (PPE) and first aid kit is stored

Infection control

Coming into contact with blood or other bodily fluids is a potential hazard due to the possible existence of a communicable disease. This is a hazard that the first aider needs to protect both themself and the casualty from. It is important to thoroughly wash your hands before and after administering first aid. PPE such as disposable gloves, eye protection and a mask (if required) also should be worn. A full list can be found in the appendix of the *Model Code of Practice: First aid in the workplace* (see http://www.safeworkaustralia.gov.au).

GREEN TIP

Dispose of contaminated first aid consumable equipment, including sharps, appropriately. Clean up the incident scene and put away any reusable first aid equipment.

Assess the casualty and recognise the need for first aid response

Once you have controlled any hazards you may approach the individual to assess if you will need to carry out first aid. The DRSABCD action plan is a useful aid in assessing if the casualty has any life-threatening conditions that require immediate first aid (see Figure 2.1). This action plan is sometimes referred to as the 'primary survey'.

Each of the steps in the action plan is discussed below. The Australian Resuscitation Council (https://resus.org.au) has guidelines on each of these steps for further reading.

Dangers?

Danger has already been explained above. By checking for, assessing and controlling any hazards before approaching the casualty, the danger of the situation is limited.

Responsive?

The first step in assessing the casualty is to determine their level of consciousness by observing if they are responsive. This can be done by assessing their response to external stimuli. Do this by gently touching the patient and talking loudly to them. A useful assessment method is to remember the acronym 'COWS'. This represents the following questions to ask the casualty:

- **C**an you hear me?
- **O**pen your eyes?
- **W**hat is your name?
- **S**queeze my hands?

If they are responsive and require further assistance, the first aider needs to consider the legal issues of consent, respect, duty of care and confidentiality (explained later in this chapter). The first aider may not need to continue with the primary survey other than sending for help.

Basic Life Support

D	Dangers?
R	Responsive?
S	Send for help
A	Open Airway
B	Normal Breathing?
C	Start CPR 30 compressions: 2 breaths
D	Attach Defibrillator (AED) as soon as available, follow prompts

Continue CPR until responsiveness or normal breathing return

FIGURE 2.1 Basic life support flow chart

Send for help

If there is no response, send for help by calling 000 (or 112 from a mobile phone) and request an ambulance. If there is another person present, ask them to make the call. If there are enough bystanders, ask one to go to the nearest road junction or entry to the premises to direct the ambulance to your location. The emergency operator will be available to answer questions and may ask a series of questions to help determine the likely severity of the injury or illness. This step of the assessment method is important as it is the first link in the 'chain of survival' (discussed later in the chapter).

Open airway

If the casualty does not respond to your questions, you will need to check for a clear airway.

First check their mouth for any loose items such as food, dentures or vomit. The finger sweep method is an effective method of clearing the casualty's mouth. To open the airway the head tilt–chin lift is a suggested technique (see Figure 2.2).

The casualty may require moving into the recovery position (see Figure 2.5 later in the chapter) to aid clearing the mouth.

Do not tilt the head back until the mouth is cleared of any foreign material.

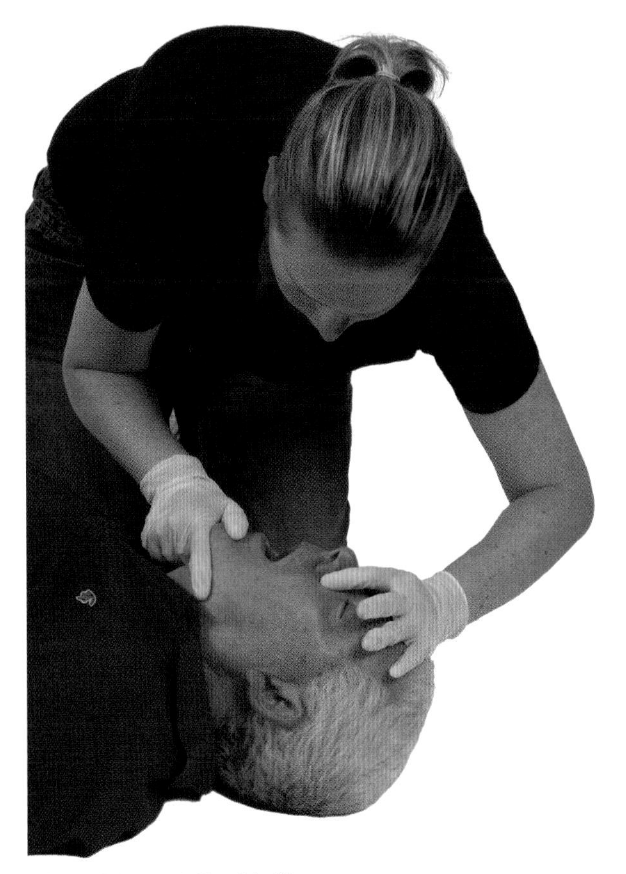

FIGURE 2.2 Head tilt–chin lift

Source: © St John Ambulance Australia Ltd

FIGURE 2.3 Look, listen and feel for breathing

Source: © St John Ambulance Australia Ltd.

Normal breathing?

Now that the casualty's airway is clear, check for breathing. In order to do this, you need to LOOK, LISTEN and FEEL (see Figure 2.3).

- Place your ear close to the casualty's mouth and nose and look down at their chest.
- Look for their chest rising and falling.
- Listen for the sound of air escaping from the nose and mouth.
- Feel for air against your cheek.
- If the casualty is breathing normally, place them in the recovery position.
- If the casualty is not responsive and not breathing normally, you will need to begin CPR immediately.

Important: Establishing and maintaining a clear airway always takes precedence over any injury, including a suspected spinal injury.

Start CPR

The Australian Resuscitation Council states that 'cardiopulmonary resuscitation (CPR) is the technique of chest compressions combined with rescue breathing. The purpose of CPR is to temporarily maintain a circulation sufficient to preserve brain function until specialised treatment is available.' It continues: 'Any attempt at resuscitation is better than no attempt.'

CPR should be started as soon as possible and continued until:

- the casualty begins to breathe normally
- medical assistance arrives and takes over
- it is impossible to continue.

Attach defibrillator (AED)

In the case of cardiac arrest, the time taken before defibrillation is a key factor that influences a person's chance of survival. An Automated External Defibrillator (AED) should be applied to the person who is unresponsive and not breathing normally as soon as it becomes available so that a shock can be delivered if necessary.

An AED attempts to restore the heart's normal rhythm. It does not replace CPR but rather works in conjunction with CPR.

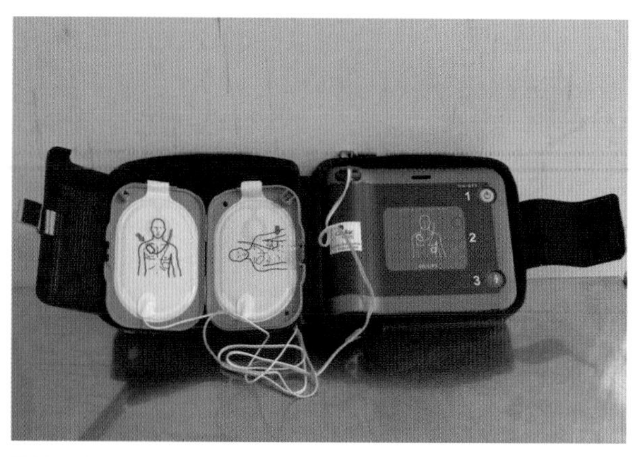

FIGURE 2.4 AED (Automated External Defibrillator)

Source: Alamy Stock Photo/Nsnap

The steps for using an AED (see Figure 2.4) are:

- Turn on machine.
- Follow the voice commands from the machine.
- Apply the pads as directed by the AED voice commands and images on the pads.

AEDs are electrical devices and the safety of the first aider and bystanders should be considered. Care should be taken not to touch the casualty while the AED is delivering a shock.

Recovery position

If the casualty is observed to be unconscious and breathing normally they should be moved into the recovery position (also known as lateral position). This position will help to keep the casualty's airway clear and open.

- Kneel beside the person.
- Put their arm that's furthest from you out at right angles to their body.
- Place their nearer arm across their chest.
- Bend their nearer leg up at the knee; the other leg should be straight.
- While supporting their head and neck, roll the person away from you.
- When they are on their side, keep their top leg bent at the knee, with the knee touching the ground.

Movement of the casualty

While carrying out the DRSABCD action plan, the first aider may decide to move the casualty from a hazardous situation, move them into the recovery position, or to perform CPR. The casualty may also need to be moved to treat other life-threatening injuries such as severe bleeding. However, the condition of a casualty may be affected by them being moved, especially if a spinal injury is suspected, so if you feel it is necessary to move the casualty, it is important to avoid bending or twisting their neck or back.

HOW TO

MOVE SOMEONE INTO THE RECOVERY POSITION

In order to move the casualty into the recovery position:

1. Kneel beside the person
2. Place the arm that is furthest from you away from the body (see Figure 2.5a).
3. Place the arm that is nearest to you over the casualty's chest (see Figure 2.5b).
4. Bend the leg that is closest to you up at the knee (see Figure 2.5b).
5. Support the head and neck with one hand; roll the casualty over by pushing on the knee with the other hand, keeping the casualty's head, neck and spine straight (see Figure 2.5c).
6. Ensure the casualty's bent knee and hand support them and tilt the casualty's head to open the airway (see Figure 2.5d).

FIGURE 2.5 The recovery position

Source: © St John Ambulance Australia Ltd

Chain of survival

The chain of survival (see Figure 2.6) highlights the process to achieve the best possible outcome for a casualty in cardiac arrest. Any delay in the timing of the DRSABCD action plan may affect the chain of survival. The four links in the chain are as follows:

- Early recognition and call for help – contact the ambulance service by calling 000 and provide the operator with as much information as you can. This will provide early access to advanced skills and equipment.
- Early CPR – to help preserve brain function. This maintains blood flow and therefore the supply of oxygen to vital organs.
- Early defibrillation – to attempt to stop dangerous heart rhythms and assist the heart to regain a normal rhythm.
- Post-resuscitation care – to provide advanced life support. This medical and rehabilitation care will be provided by trained medical professionals.

The earlier each of the four links in the chain occurs, the greater the chances of survival are for the casualty.

Assess the situation and seek assistance from emergency response services

Once immediately life-threatening conditions and injuries have been dealt with, the first aider may need to assess the casualty further to understand the casualty's condition. This is sometimes referred to as the 'secondary survey'.

Medical history

Knowledge of the casualty's medical history may provide the information required to manage the situation and provide first aid. If the casualty is conscious, ask open-ended questions such as, 'What work activity were you carrying out prior to the incident?'

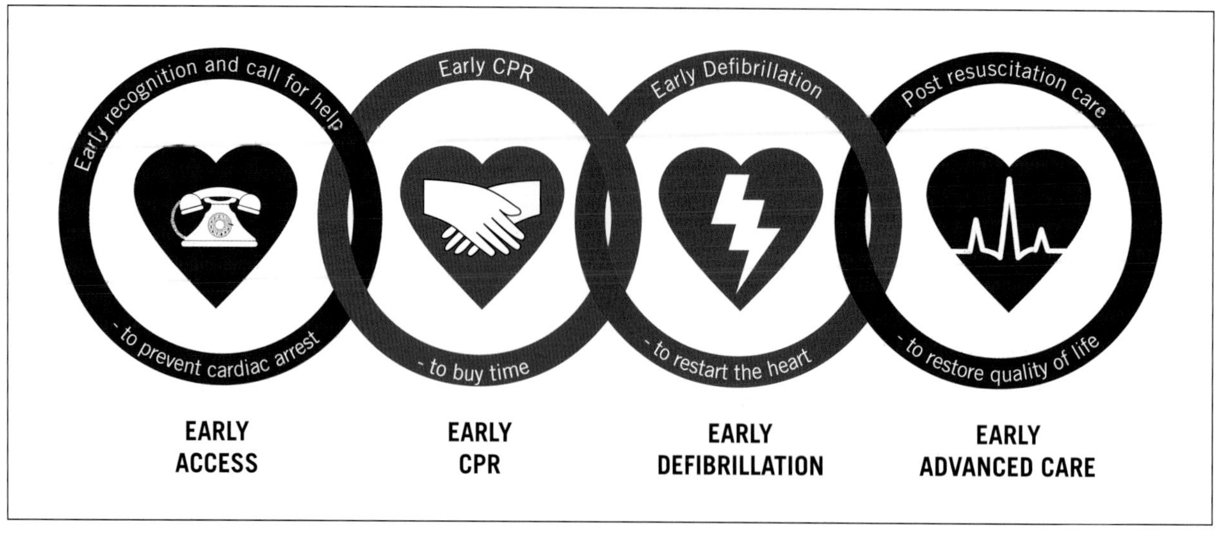

FIGURE 2.6 Chain of survival

Source: https://www.firstaidforfree.com/what-is-the-chain-of-survival.

A useful tool for assessing medical history is SAMPLE. Ask questions about the following:

- **S**igns/symptoms
- **A**llergies
- **M**edication
- **P**ast illnesses
- **L**ast oral intake
- **E**vents leading up to the present illness/injury.

Bystanders may be friends, family or colleagues, and so may be able to provide information about the casualty's medical history.

If you suspect poisoning, ring Poisons Information Hotline (13 11 26).

Head-to-toe examination

A head-to-toe examination of the casualty is used to find any injuries that have not yet been identified. Explain to the casualty the purpose of the examination and pay attention to their reactions during the examination. Verbal reactions or facial reactions from the casualty may help with this process. Refer to the 'Respectful behaviour' section later in the chapter.

Duty of care

Once you have commenced the administration of first aid treatment, you have a duty of care to continue to administer treatment until:

- the casualty refuses treatment
- the casualty recovers
- medical care arrives
- the first aider is physically unable to continue or the situation becomes unsafe.

People trained in first aid also may have a duty of care under certain circumstances; for example, if they are involved in a road traffic accident or are at their place of work. First aiders should know the specific legislation that applies to their state, workplace and industry. Safe Work Australia laws and Codes of Practice are models only and may not have been wholly adopted by your state government. First aiders have a duty of care to ensure that they work within the guidelines of their current first aid training.

LEARNING TASK 2.1

RESPOND TO AN EMERGENCY SITUATION

Choose one of your work areas and familiarise yourself with this area, then identify the following:

- potential hazards in this area
- injuries these hazards may cause
- lock-out procedures for equipment in this area
- location of the first aid kit
- PPE that is available in this work area.

COMPLETE WORKSHEET 1

Apply appropriate first aid procedures

Applying first aid procedures means more than physically caring for the casualty, and includes issues of respect and consent, along with proper use of equipment and the need to monitor the casualty until professional help arrives.

Perform CPR in accordance with Australian Resuscitation Council guidelines

The purpose of CPR has already been explained earlier in this chapter. The practical assessment for this qualification will include a simulation of carrying out the DRSABCD action plan and performing CPR on a resuscitation mannikin for at least two minutes. In order to promote standardisation in the techniques of performing CPR, the first aid trainer will demonstrate the method established by the Australian Resuscitation Council.

HOW TO

PERFORM COMPRESSIONS

1. Ensure the casualty is laid on their back on a firm surface with the airway open.
2. Position yourself at the side of the casualty in line with their chest.
3. Place the heel of one hand in the centre of the lower half of the chest and then place the other hand on top (see Figure 2.7).
4. Push directly downwards approximately one-third of the chest depth 30 times at a rate of 100–120 compressions per minute. Some first aid trainers suggest humming the song 'Staying Alive' (Bee Gees) to maintain the correct rate.
5. Give two rescue breaths.
6. Continue the cycle of 30 compressions followed by two rescue breaths.

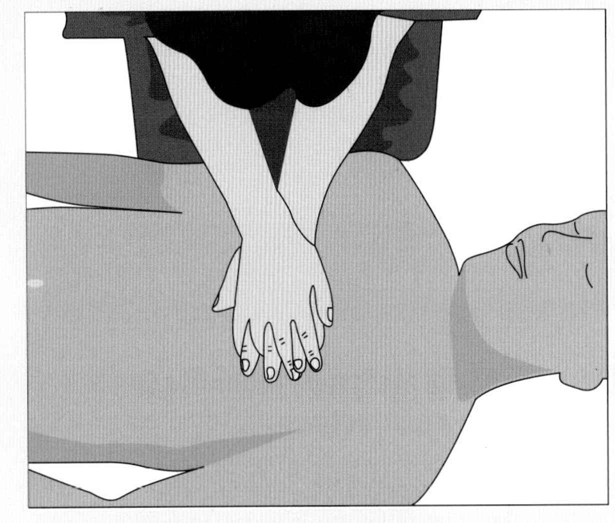

FIGURE 2.7 Location of chest compressions

Provide first aid in accordance with established first aid principles

The first aid principles and techniques required to complete your examination will be explained in detail during an intensive first aid course. These will include basic anatomy and physiology instruction to provide deeper understanding of first aid treatments. The course will also incorporate a practical assessment where students will be required to respond to first aid scenarios and make appropriate use of bandages and dressings. Students will need to demonstrate the skills and knowledge required to manage the treatment of a variety of injuries and illnesses, such as respiratory and cardiac emergencies, fractures and bleeding wounds, to name just a few.

As well as published first aid books, there are a number of useful resources available to first aiders either online or as mobile phone apps. Some apps provide the phone's location to facilitate a timely response from the emergency services, and some provide step-by-step guides to the treatment of certain illnesses and injuries. The appropriateness of these resources may be dictated by your workplace or personal circumstances. Links to resources that may be useful are listed at the end of this chapter.

Display respectful behaviour towards the casualty

So far we have covered instances where the first aider may need to interact with a potential casualty, including verbal and physical assessment of the casualty, moving them (if required) or treating their condition. However, just as important as the physical care is the need to treat the casualty respectfully at all times. They may require assistance, but it is important that the first aider's approach to the situation does not cause undue stress to the casualty. The first aider should endeavour to keep the casualty calm, instil confidence regarding the first aider's capabilities, and reassure them that their loved ones will be notified of the situation. Consideration should also be given to the fact that the casualty may have cultural or religious beliefs that may affect the way the first aider manages the situation.

Obtain consent from the casualty where possible

If the casualty is conscious, the first aider should identify themselves and obtain consent to carry out first aid treatment. This is important because mentally competent adults have the right to refuse treatment, and any first aid treatment without consent could be considered assault. If the casualty is unconscious and therefore unable to give consent, then consent is inferred by the first aider, who may provide the necessary treatment.

If a first aider acts in good faith and within the limits of their training, in most states and territories their actions and any consequences of these have some form of protection from liability under legislation such as the Victorian *Wrongs Act 1958* (Good Samaritan legislation).

Use available resources and equipment to make the casualty as comfortable as possible

Emergency situations may not occur in close proximity to a first aid kit. Therefore, it may be necessary to use the resources that are immediately available to assist in making the casualty comfortable. A sheet of plasterboard may need to be used to shade the casualty from the sun, although on a windy day that same piece of plasterboard may cause further harm. Items of clothing may need to be used to form a cushion to support the casualty's head, or a lunchbox ice block wrapped in a towel may need to be utilised as an ice pack.

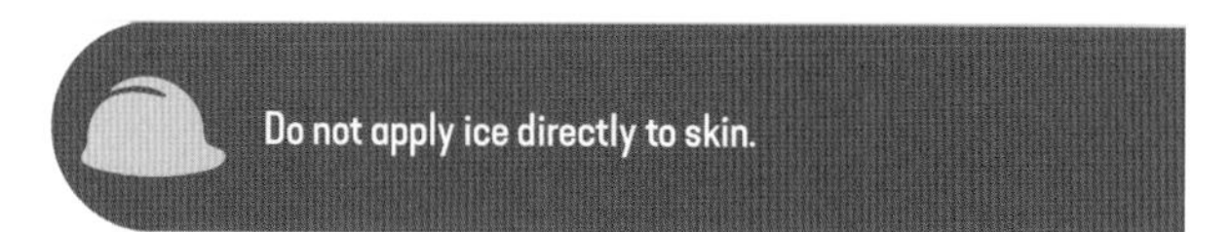

Operate first aid equipment according to manufacturer's instructions

Some emergency situations involve the use of first aid equipment, such as an AED. The first aider should operate the equipment in accordance with the manufacturer's instructions. This is to ensure it is operated in a safe manner and to maximise the effectiveness of the equipment. In addition to the AED and resuscitation manikins already discussed, the practical element of the first aid qualification will include a demonstration and practice in the use of:

- placebo bronchodilator (see Figure 2.8).
- adrenalin auto-injector training device (see Figure 2.9).

It is important that first aiders familiarise themselves with the first aid supplies and resources available to them in order to use them effectively. For example, the pressure bandage is an elasticated bandage that helps provide pressure to an area of the body. It is utilised in various first aid treatments including soft tissue injury, snake bite injury and dressing wounds. The bandage is stretched in order to achieve compression. Pressure bandages are currently available with an indicator pattern on the back that helps the first aider generate the correct amount of compression. The indicator comprises of a row of rectangles that change shape into squares when the bandage is stretched to the correct tension for a snake bite injury.

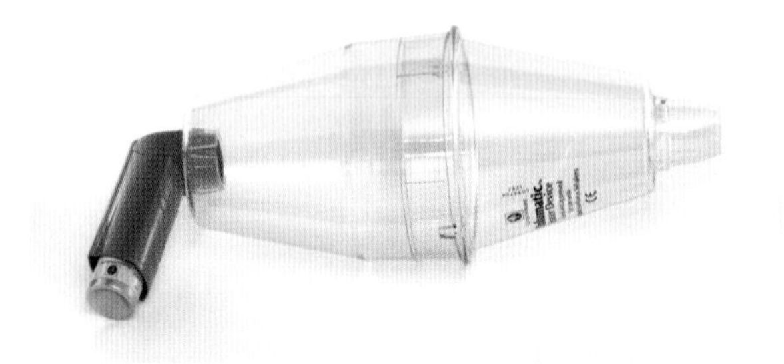

FIGURE 2.8 Bronchodilator and spacer

Source: Alamy Stock Photo/Science Photo Library.

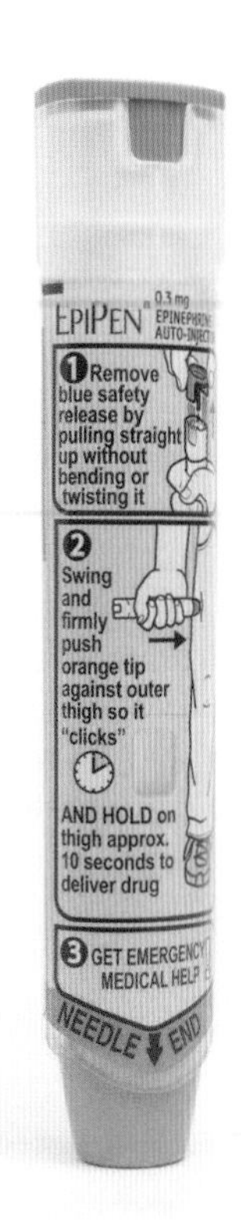

FIGURE 2.9 Adrenalin auto-injector training device

Source: Shutterstock.com/Amy Kerkemeyer.

Monitor the casualty's condition and respond in accordance with first aid principles

As already stated, a duty of care exists once first aid treatment commences. The first aider will need to monitor the casualty until medical assistance arrives and takes over. By monitoring changes to the following vital signs, the first aider will be able to respond with the appropriate first aid treatment:

- level of consciousness
- breathing
- pulse
- skin colour.

Where possible, record the vital signs. If you have no writing material, tell the 000 operator what the vital signs are or use the voice recorder or memo app on your mobile phone if available.

Communicate details of the incident

If you need to ring 000 it is important to speak clearly when answering the operator's questions. The operator will ask if you need police, fire or ambulance services – remember that you may need more than one service. They will ask for your location in order to dispatch assistance as quickly as possible. Your

LEARNING TASK 2.2

APPLY APPROPRIATE FIRST AID PROCEDURES

1 You are going to be working in a rural area that is known to be inhabited by snakes. Prior to going to the worksite carry out the following:
 - Find and download a mobile phone app that would guide you through the treatment of snake bite injuries.
 - Find a website with a printable fact sheet for snake bite injuries that you can display in your first aid area.
2 Using online research, identify the acronyms used in the treatment of:
 - sprains
 - stroke.

location can be communicated as a street address or as longitude and latitude coordinates, as displayed on a mobile phone or GPS device. After conveying all the information you have, do not hang up – the operator will be able to give you advice and assist you until medical help arrives. If you have a speech or hearing impairment, or have limited English language skills, visit the government Triple Zero website (https://www.triplezero.gov.au/Pages/default.aspx) to identify the best way to communicate with the emergency services. This website also contains information regarding the Commonwealth smartphone app that helps provide the caller with information to assist emergency services get to your location quickly.

Accurately convey incident details to emergency response services

When emergency services arrive, you will need to pass on as much relevant information as possible. Explain clearly and calmly how the emergency situation occurred, what injuries you identified and what you did to assist the casualty. You will also need to communicate any information regarding the vital signs that were monitored.

Report details of incident to workplace supervisor as appropriate

As well as providing the emergency services with information, you will need to complete an incident report or provide the information to your workplace safety representative or supervisor. This will be in line with your current workplace and state or territory regulations.

Maintain confidentiality of records and information in line with statutory and/or organisational policies

Once you have handed over care of the casualty to emergency services, and completed any required paperwork, you have no need to document any information regarding the incident. Any information that could identify the casualty is regarded as confidential, and this information should not be disclosed without the person's written consent. This may involve deleting any photographs and voice recordings from your phone and destroying any paperwork no longer required.

LEARNING TASK 2.3

COMMUNICATE DETAILS OF THE INCIDENT

1 Physically identify or use a smartphone to trace your current location, and supply:
 - the street address
 - the latitude and longitude.
2 Identify where you would ask a bystander to wait to meet emergency services.

COMPLETE WORKSHEET 3

Evaluate the incident and own performance

First aid skills are obtained through a combination of knowledge, training and experience. Often the experience of providing first aid is physically and emotionally draining, especially if the situation involves a close work colleague, friend or family member. Being involved in a highly stressful situation can have unpleasant or unexpected effects on people.

Recognise the possible psychological impacts on self and other rescuers involved in critical incidents

Some of the common responses to emergency situations are:

- crying for no apparent reason
- difficulty making decisions
- difficulty sleeping
- disbelief, shock, irritability, anger, disorientation, apathy, emotional numbing, anger, sadness and depression
- excessive drinking or drug use
- extreme hunger or lack of appetite
- fear and anxiety about the future
- feeling powerless
- flashbacks
- headache and stomach problems.

These effects may not become evident until some time after the event. It is important to recognise these physical or emotional responses in order to deal with them. The process of dealing with these issues is often called psychological first aid. This may take the form of self-help by exercising or using relaxation techniques. It may require speaking with close friends or support groups or seeking professional medical assistance.

Participate in debriefing to address individual needs

Your company's first aid procedure should identify how to access debriefing or counselling services to support first aiders and workers after a serious workplace incident. This is intended to minimise any lasting or detrimental impact on you.

LEARNING TASK 2.4

EVALUATE THE INCIDENT AND OWN PERFORMANCE

Make a copy of your company's first aid procedure. Highlight the section that identifies the debriefing or counselling services

Maintain currency of qualifications

First aid techniques are reviewed by the peak bodies at regular intervals. It is important to understand that resources available become out of date. Therefore, it is important to ensure your training is current. This chapter is intended as an overview to support students undertaking their first aid qualification. More detailed knowledge will be gained from the intensive first aid course and information from the additional resources in the reference list. Every effort has been made to reference current first aid techniques at the time of writing.

COMPLETE WORKSHEET 4

PART 1

2

REFERENCES AND FURTHER READING

Texts

Queensland Ambulance Service (2013), *Pre-Course Reading: Excerpts from the QAS First Aid Manual*, Queensland Department of Community Safety.

Web-based resources

Australian Government – Triple Zero: **https://www.triplezero.gov.au**

Australian Resuscitation Council: **https://resus.org.au**

Legal issues in first aid: **http://www.activepublications.com.au/surflifesaving/unit/legal-issues-in-first-aid/?id=1260**

NSW Poisons Information Centre – Factsheets: **https://www.poisonsinfo.nsw.gov.au/Factsheets.aspx**

Queensland Ambulance Services – First Aid Workbook: Pre-course study: **https://bookings.qld.gov.au/services/firstaid/files/HLTAID001-007%20Pre-Course%20Workbook%20 6MAY14.pdf**

Royal Life Saving Australia – YouTube channel: **https://www.youtube.com/user/stjohnambulance/videos?disable_polymer=1**

St John Ambulance – First Aid Facts: **https://stjohn.org.au/first-aid-facts**

St John Ambulance – YouTube channel: **https://www.youtube.com/user/stjohnambulance/videos?disable_polymer=1**

Training.gov.au – Unit of competency details: HLTAID003 - Provide first aid: https://training.gov.au/Training/Details/HLTAID003

Work Safe Australia: **https://www.safeworkaustralia.gov.au**

Apps

St John Ambulance Australia Corporation mobile phone app

Australian Red Cross mobile phone app

GET IT RIGHT

The following photographs show incorrectly tensioned and correctly tensioned compression bandages.

How do you know that the bandage in the first photograph is tensioned incorrectly?

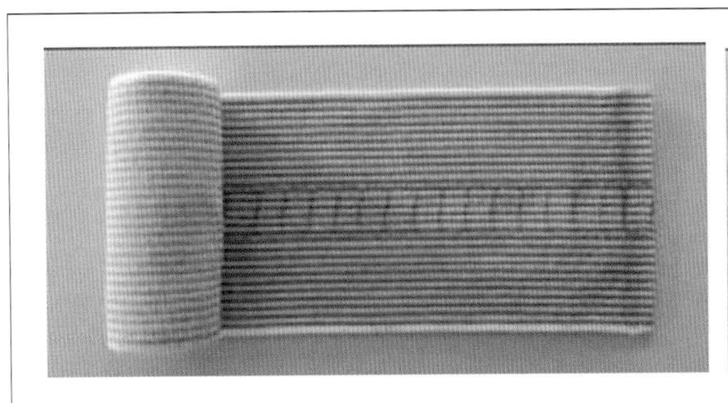

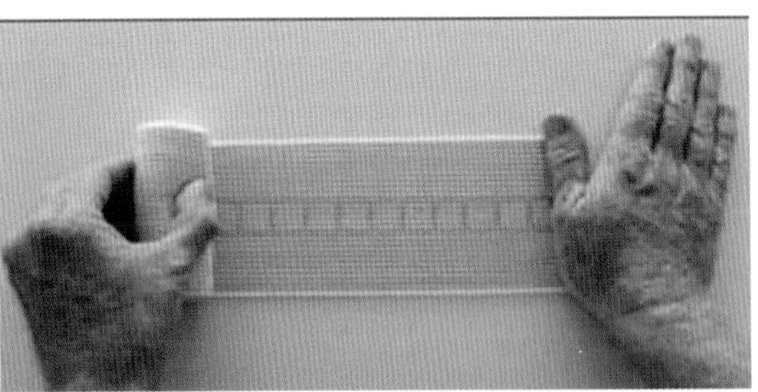

WORKSHEET 1

To be completed by teachers	
Satisfactory	☐
Not satisfactory	☐

Student name: ______________________

Enrolment year: ______________________

Class code: ______________________

Competency name/Number: ______________________

Task: Review the section 'Respond to an emergency situation' in this chapter and answer the following questions.

1. List the four objectives of first aid.

____

____

____

____

2. What questions would you ask a casualty to check for a response?

____

____

____

____

3. List the links in the chain of survival.

____

____

____

____

4. What is the purpose of CPR?

____

____

____

5. List instances where a first aider may have a duty of care to a casualty.

____

____

____

WORKSHEET 2

To be completed by teachers	
Satisfactory	☐
Not satisfactory	☐

Student name: ______________________

Enrolment year: ______________________

Class code: ______________________

Competency name/Number: ______________________

Task: Review the section 'Apply appropriate first aid procedures' in this chapter and answer the following questions.

1. What is the ARC-recommended ratio of compressions to rescue breaths?

____

____

2. What is the ARC-recommended rate of compressions per minute?

____

____

3. List three items of first aid equipment.

____

____

____

4. Why should a first aider obtain consent to carry out first aid treatment?

____

____

5. List four of a casualty's vital signs that you would monitor.

____

____

____

____

WORKSHEET 3

To be completed by teachers	
Satisfactory	☐
Not satisfactory	☐

Student name: ______________________

Enrolment year: ______________________

Class code: ______________________

Competency name/Number: ______________________

Task: Review the section 'Communicate details of the incident' in this chapter and answer the following questions.

1. List three pieces of information that would be useful to pass on to the emergency services.

____

____

____

2. List two ways of identifying your location to the emergency services.

____

____

3. Once all of the necessary paperwork regarding an emergency situation is complete, why should you dispose of any remaining written information about the casualty?

____

____

____

WORKSHEET 4

To be completed by teachers	
Satisfactory	☐
Not satisfactory	☐

Student name: ____________________

Enrolment year: ____________________

Class code: ____________________

Competency name/Number: ____________________

Task: Review the section 'Evaluate the incident and own performance' in this chapter and answer the following questions.

1. List three common responses that a first aider may have to emergency situations.

____

____

____

2. List three ways to address psychological first aid issues.

____

____

____

WORKSHEETS 2

3 WORK EFFECTIVELY IN THE PLUMBING AND SERVICES SECTOR

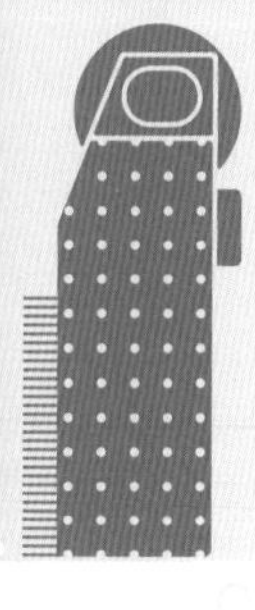

This chapter addresses the following key elements for the competency 'Work effectively in the plumbing and services sector':

- Identify the industry work context and setting.
- Organise and accept responsibility for own workload.
- Work in a team.
- Participate in identifying and pursuing own development needs and processes.
- Participate in workplace meetings.
- Observe sustainability principles when preparing for and undertaking work processes.

It introduces the requirements in preparing for and carrying out effective work in accordance with work health and safety principles, and public health and safety requirements within the plumbing and services sector. It also addresses organising and accepting the responsibility of work to be undertaken in accordance with organisational requirements; participation and identification of learning needs for career path progression; and effective participation in a plumbing and services workplace to meet organisational requirements and promote a harmonious and efficient team-work environment. The chapter finishes with a discussion of sustainability principles as they relate to the planning and undertaking of work processes.

PART 1

3

Identify the industry work context and setting

In Australia, the number of registered tradespeople and licensed practitioners in the plumbing and services sector continues to grow. This growth is due to a continual change in the plumbing market, with a demand for the installation of sustainable products and new materials, and a continual demand for essential skills for existing services.

What do plumbers do?

The scope of what plumbers do is extremely broad and diverse, ranging from the delivery of potable drinking water and providing quality sanitation and disposal facilities, to the installation of hot water and heating for comfort, and roofing and cladding for dry and comfortable shelter. All areas of plumbing require new installation, with most products requiring periodic or routine maintenance.

Public health and safety

Plumbers contribute significantly to the quality of potable water and sanitation, which relates directly to public health and safety, economic growth and employment opportunities. Potable water is described as 'fit for human consumption' and is a basic but essential requirement for survival.

Kofi Annan, the former United Nations Secretary-General, said the following in relation to the importance of water and sanitation as a basic need:

> We shall not finally defeat AIDS, tuberculosis, malaria, or any of the other infectious diseases that plague the developing world until we have also won the battle for safe drinking water, sanitation and basic health care.

Dr Lee Jong-wook, the former Director-General of the World Health Organization, stated the following in 2004:

> Water and Sanitation is one of the primary drivers of public health. I often refer to it as 'Health 101', which means that once we can secure access to clean water and to adequate sanitation facilities for all people, irrespective of the differences in their living conditions, a huge battle against all kinds of diseases can be won.

In 2020, at the time of writing, the world was in the grip of a global pandemic and shutdown due to the infectious COVID-19 respiratory illness, which is a new form of coronavirus. The COVID-19 disease is spread from person to person through close contact with an infected person by coughing or sneezing, or by touching surfaces that have droplets of saliva from an infected person and then touching your face or mouth. Transmission of the disease is slowed by social distancing, self-isolation and good hygiene.

One form of good hygiene is the regular washing of hands, and plumbers play a fundamental role in this as they are responsible for ensuring both the supply of potable water and the installation of sanitary plumbing and drainage.

In addition to the potential spread of infectious disease, Australia is one of the driest continents on Earth and has at times struggled to ensure a reliable water supply during times of drought and erratic seasonal rainfalls. In 2019, Australia suffered drought, followed by devastating bushfires to a significant portion of the country. Water is vital and an invaluable resource to the survival and prosperity of this country.

These needs have driven the development of several complex and innovative water supply schemes, beginning with the use of natural streams to present-day desalination plants.

GREEN TIP

As plumbers, it is essential we commit to engaging in sustainable and innovative practices to ensure we continue to secure our water supply for many years to come.

Severe water shortages affect many parts of Australia and are brought about by an increase in the size of all Australian cities. There are many efficiency schemes in operation, such as recycled water reticulation systems, deep pumps in dams, reuse of 'grey water' or 'black water' and rainwater harvesting. Delivering and maintaining water for consumption to the population is now an extremely skilled trade.

Technological advances are allowing society to keep pace in solving the problems brought about by a growing world population, shrinking natural resources and a changing climate. The need to conserve is reflected in social, economic and infrastructure policies.

FROM EXPERIENCE

Keeping up to date with new, emerging and innovative technologies will allow you to learn and grow to ensure you are on a prosperous career path in the plumbing trade.

Industry structure

The building industry is essentially divided into two groups: the commercial, industrial or large business sector; and the residential or smaller domestic sector.

Commercial buildings can be classified as:

- car parks
- factories
- hospitals
- hotels
- multistorey buildings
- offices
- schools
- shopping centres
- warehouses.

Residential buildings can be classified as:

- boarding houses
- flats
- guesthouses
- hostels
- single dwellings
- townhouses
- villas.

Plumbing services are required in all classes of these buildings and must be scheduled in the construction program along with many other trade categories, such as:

- bricklayer
- builder
- carpenter and joiner
- concreter
- electrician
- glazier
- labourer
- machine operator
- metalworker
- painter and decorator
- plasterer (render and plasterboard)
- rigger
- roof tiler
- scaffolder
- stonemason
- trades assistant
- wall and floor tiler.

The plumbing industry embraces many skills areas and requires plumbers to ensure that any work carried out is in accordance with relevant Codes of Practice, Australian Standards, Regulations and plumbing legislation. All documents should be accessed through a variety of mediums and be well understood and implemented. The services that plumbers provide are outlined in the following sections.

Water services

The plumber is responsible for the installation of water service pipework in buildings and dwellings in compliance with the various Codes of Practice and Australian Standards, such as *AS/NZS 3500.1 Plumbing and drainage Part 1: Water services*, *AS/NZS 3500.4 Plumbing and drainage, Part 4: Heated water services* and *AS 2845 Water supply – Backflow prevention*. This can involve the installation and maintenance of:

- household taps and supply systems to bathrooms, kitchens, laundries, toilets, reticulated water mains, industrial and commercial complexes, and large-scale residential and industrial subdivisions
- backflow prevention devices, thermostatic mixing valves, hot- and cold-water systems, recycled water systems, rainwater systems, treated water for special purposes and many more applications.

In addition to the installation of these water services there is also the associated flushing, testing and both regular and emergency maintenance of these systems.

- AS/NZS 3500.1 Plumbing and drainage Part 1: Water services
- AS/NZS 3500.4 Plumbing and drainage, Part 4: Heated water services
- AS 2845 Water supply – Backflow prevention

Gas services

Gasfitters install and maintain consumers' gas supply pipework and appliances to ensure compliance with current relevant Codes of Practice and Australian Standards (which include *AS/NZS 5601.1 Gas installations – Part 1: General installations* and *AS/NZS 5601.2 Gas installations – Part 2: LP Gas installations in caravans and boats for non-propulsive purposes*) and local gas utility network requirements. Gas appliances may be used for room heating, water heating and cooking purposes in domestic dwellings, while also being used on a larger scale in commercial and industrial buildings.

3.2 STANDARDS

- AS/NZS 5601.1 Gas installations – Part 1: General installations
- AS/NZS 5601.2 Gas installations – Part 2: LP Gas installations in caravans and boats for non-propulsive purposes

Fire services

Plumbers, along with the sprinkler fitting trade, help to protect residential buildings, offices, factories, educational facilities and shopping centres. Many diverse systems are installed and maintained where occupants or contents of buildings need protection from fire. Systems may include fire hose reels, fire hydrants, ring mains and suppression systems.

Sanitary plumbing and drainage

The removal of human and industrial waste to centrally located treatment plants and approved disposal facilities, and the installation of trade waste on-site treatment systems, are essential to the health and safety of the population. The licensed plumber and drainer – following very strict regulations, Australian Standards and using many trades and building skills – ensures that our precious water supplies are not polluted by our sewage.

Irrigation

Irrigation systems provide water to plants and vegetation and are commonly used on farms for food and plant production. Irrigation is also used in parks, golf courses, sporting fields, large gardens and residential homes. Due to sustainability principles and regulatory requirements, plumbers are installing systems for the collection and reuse of storm water, and the connection to recycled water mains and artesian bores, in addition to providing potable water supply to meet these demands.

GREEN TIP

The use of reclaimed or recycled water in irrigation systems can reduce the reliance on potable water systems. Plumbers must be aware of state or territory and local regulatory requirements for the installation of irrigation systems with water other than a potable supply.

A plumber with the right skills can design, carry out, test and commission an irrigation system. Service of the irrigation system is needed to keep it operating properly and to make adjustments as the seasons or plants require.

Stormwater drainage

Buildings and properties require stormwater to be collected and removed to an approved point of discharge or diverted to an on-site detention (OSD) system where required for appropriate disposal. Rainwater is increasingly being collected and reused for potable water supply, gardens and sanitary flushing, both in rural areas and major metropolitan cities. Plumbers can design and install stormwater systems that may include piping systems, pits, pumps, OSD systems and arrestors.

Mechanical services

Mechanical services plumbers require specialised skills to install and maintain the pipework and equipment that may provide treated, heated, cooled, high pressure and potable water, as well as equipment and piping systems to carry fuel gas, medical gas, compressed air, steam, ventilation and many other fluids. Mechanical service plumbers often use heavy gauge material, which may include copper tube, stainless steel and black mild steel, along with a variety of jointing processes. Mechanical systems may often be referred to as HVAC (heating, ventilation and air conditioning).

Roof plumbing and drainage

All buildings require a form of weatherproofing and may also require rainwater to be removed and conveyed to stormwater systems.

The roof plumber installs and maintains metal roofing products, including roof sheets, guttering, rainwater downpipes (spouting) and flashings, to provide a sound structure that is comfortable to occupy.

Due to the increase in popularity of water tanks, the plumber's role in water conservation, rainwater harvesting and reuse also includes providing an 'inert' catchment area for water to be collected and stored.

A career in plumbing

Plumbing is a continually developing trade, with demands placed on it by population increase, the effects of a changing climate, health requirements, demographic shifts of people to cities, social expectations and rapid changes in technology. It is important to identify and follow new trends in technology, as the plumbing services sector continues to evolve, and to ensure you stay competitive in a changing market. It is essential to keep track of these changes and identify any learning or training requirements and make a commitment to lifelong learning for any future work requirements in this trade.

Undertaking a career in plumbing requires specific skills sets, which may include:

- practical skills
- communication and teamwork skills
- technical skills
- learning skills
- computing and technology skills
- literacy and numeracy skills
- initiative and enterprise skills.

Depending on the specific tasks that a plumber is involved in, they are generally required to have a good knowledge in the following areas:

- bricklaying – support for piping, and repairs
- carpentry/building – understanding structural supports, and noggins in frames
- concreting/rendering – repair of paths or masonry walls
- formwork – repair of penetrations/concreting
- sheet metal work – roofing
- tiling – maintenance
- welding/brazing – water, waste and gas pipes.

Plumbers require good hand skills, the ability to work hard physically and mentally, determination to complete difficult tasks, and the ability to solve mathematical and arithmetical for tendering, ordering of materials and solving design problems. There is a lot of paperwork involved complying with the relevant state or territory workers' compensation schemes, Australian Taxation Office (ATO) requirements, contract and tender documents, progress reports, accountancy records, bills and invoices.

An apprenticeship in plumbing

Apprenticeships and traineeships are jobs that combine work and structured training, and, although they vary from one industry to another, all apprenticeships and traineeships include the following:

- paid employment under an appropriate industrial arrangement (e.g. an award or enterprise agreement)

- a training agreement/training plan or registration that is signed by both the employer and apprentice, or trainee, in conjunction with a registered training organisation (RTO), and then registered with the relevant state or territory authority (e.g. an Australian Apprenticeships Network provider)
- a training program delivered by an RTO that meets the requirements of a declared apprenticeship or traineeship and leads to a nationally recognised qualification.

The Australian Apprenticeships System offers apprenticeships in over 500 occupational/work areas. Within the building industry sector, they include plumbing, bricklaying, carpentry, joinery, construction carpentry, stonemasonry and wall and floor tiling. All are aligned to a current award or industrial agreement.

The duration of an apprenticeship is normally four years and is available for young people as well as mature-age and special target groups, such as people with disability. Subsidy levels may also be available in certain target areas for employers. Further details may be obtained from an Australian Apprenticeships Network provider (see http://www.australianapprenticeships.gov.au).

As plumbing is a trade and part of the building industry, all plumbers must learn this trade through an apprenticeship in the employment of a licensed plumber. The minimum age to undertake an apprenticeship may vary between state and territory, but is generally 15 years.

Many employers prefer apprentices to have completed secondary schooling as a minimum, with good results in mathematics and English.

Generally, a plumbing apprentice is required to attend an RTO, such as a TAFE college, to gain the broad-based underpinning knowledge and practical skills necessary for their trade certificate and subsequent licensing qualifications.

Attendance at TAFE can be via either day release or block release for the trade component, and the employer is legally obliged to pay the apprentice for this attendance.

The apprentice, in turn, works on the job under the supervision of the licensee for the remaining time, during which they can observe and practise the skills necessary to work effectively in the trade.

You will need to register with an Australian Apprenticeships Network provider and enter a contract with your employer.

In this contract you will agree to:

- remain in the service of your employer and abide by their instructions
- carry out your duties, learn about the plumbing trade and work safely in the industry
- attend an RTO to complete your course.

Your employer will:

- teach you the plumbing trade to the best of their ability
- abide by the relevant industrial award
- be responsible for your safety and welfare while under their care.

A great deal of information relating to apprenticeships in Australia can be found at http://www.australianapprenticeships.gov.au.

Your 'trade certificate' is awarded after successfully completing your first four years of on-the-job training, which includes completion of a nationally recognised Certificate III in Plumbing qualification. Once a Certificate III in Plumbing is obtained, and you have reached the prescribed amount of years in work, you can register as a tradesperson, although this does not qualify you to obtain a contractor's licence. The current licensing/qualification requirements are such that an apprentice may need to be involved in a variety of aspects of the trade with their employer and/or to move around the plumbing industry. This is so that they can gain enough experience and exposure to the many facets of the trade in preparation for applying for a licence.

Licensing

When you successfully complete your Certificate IV in Plumbing and have gained the minimum amount of acceptable industry supervised trade experience, you are able to apply for your plumbing contractor's licence. Issuing a licence ensures that all contractors/workers are registered with their relevant state or territory's licensing regulator. This gives the client the ability to check on the contractor's credibility and history. It also provides the client with some protection against faulty production and workmanship.

It is also the licence holder's responsibility to undertake work only for the purpose for which the licence is held, and they must also show their respective licence numbers on all advertising, stationery and signage. Should either of these conditions not be adhered to, it could be deemed a direct infringement of the legislation, and the licence holders could be fined as a result.

Note: It is recommended that potential plumbing licence holders become familiar with their relevant state or territory licensing requirements as the length of experience required may vary. Contact can be made through visiting the appropriate regulatory authority's website, as detailed in Table 3.1.

Plumbing sector employment and conditions

Plumbing apprentices may spend their plumbing career working in the one area or find themselves gaining experience in several areas. It is ideal to gain as much experience as possible due to the diversity of the plumbing industry. Plumbing apprentices generally work in the following two areas of the plumbing industry.

TABLE 3.1 Regulators for plumbers, drainers and gasfitters

State/territory	Regulator	Website
Australian Capital Territory	Access Canberra	https://www.accesscanberra.act.gov.au
New South Wales	NSW Fair Trading	http://www.fairtrading.nsw.gov.au
Northern Territory	Plumbers and Drainers Licensing Board (NT)	https://plumberslicensing.nt.gov.au
Queensland	Queensland Building and Construction Commission	http://www.qbcc.qld.gov.au
South Australia	Consumer and Business Services	https://www.cbs.sa.gov.au
Tasmania	Consumer, Building and Occupational Services	https://www.cbos.tas.gov.au
Victoria	Victorian Building Authority	http://www.vba.vic.gov.au/
Western Australia	Plumbers Licensing Board, Department of Commerce (WA)	http://www.commerce.wa.gov.au

Source: National Plumbing Regulators Forum (NPRF), http://www.plumbingregulators.org

Small businesses

Most of the plumbing industry is made up of small companies or businesses run by an owner/manager, and employ fewer than 20 staff members. These companies may specialise in one or more of the many areas that make up the industry, which may include:

- fire hydrant and sprinkler systems
- gasfitting
- general plumbing maintenance
- hot- and cold-water systems
- mechanical services and heating – HVAC (heating, ventilation and air-conditioning)
- roofing installation and water management systems
- sanitary plumbing and drainage
- sustainable plumbing systems
- urban irrigation systems.

Large companies

It is usual practice for large construction companies to contract work out on the building projects that they are managing. The contracted firms employ a large portion of tradespeople in the construction industry. Most of these companies employ fewer than 100 people and operate on construction sites, while a smaller portion of these large companies concentrate on services and maintenance.

Responsibilities and obligations

Apprentices are an integral part of the building industry and it is important for them to be aware of their rights and responsibilities and those of their employer, and to become aware of the dynamics of different building sites and the role that the apprentice may play in certain situations. At times apprentices may become involved in site meetings, site inductions and workplace meetings, or may simply be trying to put forward a point of discussion about a circumstance relevant to their own welfare.

There are a several issues that may affect individuals, either directly or indirectly, when working in the industry. These issues can affect management, a supervisor, an employee or a subcontractor. Therefore, it is essential that everyone has a basic understanding of how these issues may affect them, which will allow for workplace harmony and individual improvement.

Industrial relations

Industrial relations (IR) is about people and organisations working together within the social and political systems of our society. Employment makes up a large part of our lives and determines our living standards, while the industrial relations process determines the employment conditions of the environment in which each employee works.

Industrial relations issues may include work health and safety (WHS), welfare, childcare, new technology, social welfare, unemployment, illness, redundancy, wages, award restructuring, multi-skilling, career paths and early retirement.

There are two industrial relations systems in Australia: the federal system and the state system. The federal system is independent of all the state systems and is intended to cover industrial problems that are larger and further-reaching than those in any state. The system was first established by the *Conciliation and Arbitration Act 1904*.

Each state has its own industrial relations system and there are no constitutional limitations. As a result, state systems have a broader scope of operation than the federal system. The most recent Act is the *Fair Work Act 2009*, which led to the creation of the Fair Work Commission, which is the national workplace relations tribunal. Matters of importance and public interest are referred to the Fair Work Commission.

Awards

An award is the law that establishes the minimum wages and conditions of employment in defined industries or occupations. Conditions for workers can include overtime, sick leave, annual leave loading and WHS requirements.

Resolution of disputes

A dispute exists when conflict arises out of a disagreement over the rights and interests of two parties. A dispute can occur when a job done by one member of a union should in fact belong to a member of another union. This is commonly referred to as demarcation.

Grievance procedures

It is usually in the interests of the worker, the employer and the government to settle disputes as quickly as possible. If unions and employers cannot settle their differences by discussion among themselves, they can make use of conciliation and arbitration procedures.

One of the principal ways of resolving disputes or disagreements in the workplace is through grievance or dispute-settling procedures. These include:

- negotiation
- collective bargaining
- enterprise bargaining
- mediation
- conciliation
- arbitration
- other tribunals.

The resolution of industrial disputes often involves changes to the wage and/or non-wage aspects of the employment relationship.

Trade unions

A trade union is an association formed by employees to act for them inside and outside the workplace. Historically, trade unions arose in Australia as organisations for the defence and improvement of the conditions of various sections of the workforce. Unions also exert influence in arguments over environmental issues and the need for public facilities. Unionism is a useful tool in the area of negotiations between worker and employer.

Unions usually employ organisers whose job covers promoting membership, contact with delegates/union representatives, assisting delegates/union representatives in difficult negotiations at the workplace, and representing workers in negotiations and court appearances on award matters. Unions will also liaise with employer associations, other unions, state or territory labour councils, the ACTU, the media and politicians.

Obligations and benefits for trade union members

Union members are obliged to:

- pay regular fees
- abide by union rules
- pay levies for specific purposes
- encourage other employees to join the union
- support fellow members discriminated against or victimised by employers
- improve working conditions where necessary
- ensure that wages are paid
- support any stop-work meetings or strikes that may be called in members' interests
- encourage fellow unionists to attend union meetings off the job.

Union members have:

- access to all services provided by the union
- legal aid and funeral benefits
- protection from unfair dismissal and unlawful termination.

Unions, employers and governments aim to find common ground to improve industry and workplace efficiency. Members will also benefit from taking an interest in issues not directly connected with the workplace, such as conservation, politics, and the rise and development of technology.

Employer associations

Employer associations are organisations formed by management to act for them outside the workplace and to provide information and advice. The Australian Chamber of Commerce and Industry (ACCI) is the largest single organisation representing industry and commerce in Australia.

Relevant industrial awards

Industrial awards will generally continue to have relevance, as enterprise bargaining will operate in tandem with the award system for the foreseeable future. A registered enterprise agreement forms part of a common law contract of employment for employees who are bound by it.

TABLE 3.2 Rates of pay for apprentices, including overtime allowance as of 1 November 2020

Classification	Weekly pay rate	Hourly pay rate	Saturday – first two hours	Saturday – after two hours	Sunday	Public Holiday
1st year	$379.88	$10.00	$15.00	$20.00	$20.00	$25.00
2nd year	$542.01	$14.26	$21.39	$28.52	$28.52	$35.65
3rd year	$680.97	$17.92	$26.88	$35.84	$35.84	$44.80
4th year	$866.26	$22.80	$34.20	$45.60	$45.60	$57.00

Source: Fair Work Ombudsman, https://www.fairwork.gov.au/pay/minimum-wages/pay-guides, CC BY 3.0.https://creativecommons.org/licenses/by/3.0/au/legalcode

Enterprise bargaining

In its broadest sense, enterprise bargaining involves an employer negotiating directly with their employees about wages, conditions and work practices for the workplace. The result of enterprise bargaining is an enterprise agreement.

Enterprise agreements

An enterprise agreement is a contract between an employer and employees on wages and conditions of work in the employer's business. Enterprise agreements, or workplace agreements, fall into one of the following categories:

- a collective agreement involving a group of employees and an employer
- an individual agreement or contract between an employer and an employee covering employment matters on an individual basis
- independent contractor/worker agreements or contracts, which are a special form of individual contract covering independent contractors/workers who are sometimes engaged by a business to do work that otherwise would be done by employees.

Mechanism for obtaining agreements

If negotiations result in agreement on the content of an enterprise agreement, the following steps must then occur:

- If there is a negotiator or negotiating team representing employees, then agreement from employees is required.
- An agreement containing the agreed conditions should be drawn up on the appropriate form.
- The agreement should be made official by lodging it for registration with the appropriate organisation.

Changes in vocational education and training

The vocational education and training (VET) system is led by governments in consultation with industry, and there have been several training reforms in recent years with the aim of building a better system. Such reforms are often recommended and implemented by governments to support change in the VET sector and improve the quality of learning outcomes. In 2014, the Australian Government implemented the VET Reform Agenda with a view to enhance the quality of training providers and their courses, improve the outcomes of job prospects for students and the competitiveness of Australia's economy, and increase the status of VET in industry and the wider community. The vision of the VET Reform Agenda was to achieve six objectives:

1. better governance
2. qualifications that meet the needs of industry
3. trade apprenticeships that are valued and utilised as career pathways
4. more responsive and fair regulation
5. better access to consumer information
6. better targeted funding.

The objectives are aimed at improving VET from an educational perspective, and at making it more responsive to the needs of industry. These improvements include:

- training and skills development at all levels of the workforce
- a diverse and efficient training market
- an emphasis on competence rather than time served
- more flexible approaches to training
- nationally consistent arrangements for standards and qualifications
- improved access for target groups
- improved delivery arrangements within and across sectors.

As part of these reforms, additional apprenticeships and traineeships are being offered. From an employer's perspective, apprenticeships and traineeships are a good way to expand a business, increase its skills base, and keep up with technological changes so that existing customers' needs are met, and the demands of new and emerging markets are addressed.

Training packages

Nationally endorsed training packages are developed by Skills Service Organisations (SSOs) to meet the growing training needs of the building and construction sector. The training package forms the foundation of VET in the industry and provides a range of flexible training options that can be used by employers and RTOs to train apprentices and trainees.

Competency-based training

Skills are very important in the construction industry, and there are many skills that you may already have and skills you may need to acquire. Competency-based training refers to the concept of ensuring people have the right skills to do a job competently.

Recognition of prior learning (RPL) may also be used as an important tool for assessment as you may have your existing skills or knowledge gained through previous work recognised formally. Skills can be potentially assessed on site by an accredited workplace assessor, or off-site in a formal vocational education setting through registered RTOs.

General construction induction training

Employers and principal contractors/workers must ensure that persons carrying out work have relevant training, including WHS training. This is an existing obligation outlined within the relevant WHS legislation.

It applies to all persons carrying out work in the residential, commercial and high-rise sectors. It is also a requirement under the relevant Acts and Regulations that principal contractors/workers and employers must not direct or allow a person to carry out construction work unless that person has completed WHS induction training as follows:

- WHS general construction induction (meeting the criteria for the card system relevant to your state or territory), consisting of a broad range of safety awareness instruction
- workplace-specific induction.

The Regulation also requires that self-employed individuals must not carry out construction work until they have personally completed WHS general construction induction training.

The person responsible for the general construction induction training must also provide a written declaration for each person inducted, stating that the person concerned has satisfactorily completed the training, listing the activities covered in the training, specifying the dates on which the training occurred, and specifying the name and qualifications of the person who conducted the training. It must be signed by the person who conducted the training. Each person who has successfully completed the general construction induction should keep the written statement, until such time as they receive their plastic induction card (see Figures 1.1–1.4 in Chapter 1), which must be produced if a safety officer asks for it.

Workplace-specific induction

A workplace-specific induction, often referred to as a site-specific induction, must be carried out for every worker or visitor before work starts or before they enter an operational construction zone. The induction is to inform people of the hazards and risks that need to be identified at the workplace. This induction training is to be provided for all workers at no charge. For further details and information, readers can contact their respective WHS regulatory authority, or access its website (see Table 1.1 in Chapter 1.)

Workplace committees

Committees in the workplace take many forms; listed below are some committees that may be present when working on site.

Safety (WHS) committees and Health and Safety Representatives (HSRs)

Safety committees exist to improve the health and safety of workers in the workplace. They are established when the workplace consists of five or more employees, at the request of a Health and Safety Representative (HSR) or as per WHS Regulations according to individual states/territories. Permanent safety committees normally consist of representatives from management and the corresponding workforce, and are responsible for carrying out workplace safety inspections and meeting at least every three months to discuss relevant worksite WHS issues. HSRs are elected to represent specific work groups, which they themselves must be a member of, at their respective workplaces.

All Australian states and territories now legislate to provide for workplace consultation via HSRs and safety committees. The legislation provides a systematic approach by which management and employee representatives can regularly discuss, prevent and resolve WHS issues.

Note: It is recommended that readers become familiar with their relevant state or territory guidelines for the establishment of WHS committees, by either getting in contact or visiting their respective WHS regulatory authority's website (see Table 1.1 in Chapter 1).

Consultative committees

These committees consist of employees and members of middle management who consider problems and make suggestions for policies in areas such as safety, health, social activities and amenities.

Works committees

During negotiations with management over issues such as enterprise agreements and work conditions, it may be appropriate to have a team of employee negotiators representing different areas of the workplace. The team might consist of a full-time union official as team leader, supported by employees from different work areas. If a union is not involved, the team leader and other team members should be elected from among the employees.

Workplace quality assurance requirements

Any product that is manufactured or processed must maintain a quality or tolerance that is consistent over time. Inspections and testing are carried out to ensure that this happens.

Quality assurance has been included in building contracts to ensure that the clients get what they pay for. The contractor/worker takes more responsibility for the final product, and this results in quality control for the building industry.

Quality assurance

Quality assurance can be expressed as providing confidence that quality is represented in a product or service. It is the skill and commitment that you and everyone you work with bring to their job, each time, all the time. When products or services are to be supplied or used as they were intended, this is described as 'fitness for purpose' and provides the consumer with some protection.

Contracts should include quality assurance, and this puts the responsibility back on the supplier in much the same way as a guarantee or warranty would apply if you bought a new car or computer. It puts more meaning into ensuring the work undertaken is performed correctly the first time, and that the correct tool or product is always used to complete the job.

Quality assurance is of benefit to everyone, from the consumers, who expect and demand quality in the products and services they receive, to management, who know it will increase profits, to the employee, who benefits from job security and better pay, and of course to the whole country, which will be more competitive in a globalised market.

Quality management principles

The seven key principles of quality management are:

1. Customer focus – enhancing customer satisfaction and ensuring it is maintained
2. Leadership – demonstrating leadership and commitment
3. Engagement of people – respecting and involving people at all levels

4 Process approach – understanding and managing processes for consistent results
5 Improvement – determining and selecting opportunities for improvement
6 Evidence-based decision making – decisions based on evaluation and analysis of data
7 Relationship management – maintaining relationships with relevant parties.

Quality assurance and quality management are addressed in the Australian Standards series ISO 9000:2016 *Quality management systems – Fundamentals and vocabulary* and ISO 9001:2016 *Quality management systems – Requirements*.

3.3 STANDARDS

- ISO 9000:2016 Quality management systems – Fundamentals and vocabulary
- ISO 9001:2016 Quality management systems – Requirements

Cost benefit

Cost benefit will be achieved if staff are trained to be more skilful and more productive. Each member of staff will have job security and will know that they can do their job better, therefore being a valuable human resource for the company.

Quality assurance, when applied to all the materials supplied to a building, results in less waste and a better-quality product. When standards are specified, suppliers are responsible for orders placed. This guarantees the quality of the materials. Having quality assurance in place improves the standard of work, which is required of the final product. The benefit of quality assurance will be a saving in cost, as the material supply and the quality of the work will be of an acceptable standard the first time.

Standards and Codes of Practice

Standards and Codes of Practice are used extensively on the construction site to ensure work is carried out to a safe and minimum standard.

Australian Standards

Australian Standards ensure that goods, services and products are manufactured and installed to minimum standard, and are accurate, safe and fit for purpose. They are compiled by a panel of experts in the specific area of the scope of the standard. They may be representatives of safety committees, manufacturing industries, suppliers, customers and government departments.

Its work is conducted solely in the national interest and its principal functions are to prepare and publish Australian Standards and to promote their adoption.

Examples are:

- *AS/NZS 3500 Plumbing and drainage*
- *AS/NZS 5601.1 Gas installations – Part 1: General installations*
- *AS 2845 Water supply – Backflow prevention devices*.

3.4 STANDARDS

- AS/NZS 3500 Plumbing and drainage
- AS/NZS 5601.1 Gas installations – Part 1: General installations
- AS 2845 Water supply – Backflow prevention devices

Model Codes of Practice

Model Codes of Practice provide detailed information and give practical guidance on how to perform specific tasks in industry. They do not replace standards or laws but give workers a comprehensive guideline to ensure standards required under relevant WHS Acts and Regulations are complied with. Model Codes of Practice cover, for example, confined spaces, excavation work, and managing the risks of falls in the workplace.

Copies of Model Codes of Practice can be sourced through relevant state and territory WHS regulatory authority websites (see Table 1.1 in Chapter 1).

LEARNING TASK 3.1

IDENTIFY THE INDUSTRY WORK CONTEXT AND SETTING

Access the Australian Apprenticeships Network provider website that applies to your specific area of work and produce the terms of your apprenticeship contract.

Note: You may be able to obtain this information from the organisation that signed you as a registered apprentice.

COMPLETE WORKSHEET 1

Organise and accept responsibility for own workload

It is important that each worker is organised and accepts responsibility for their own workload. Work activities should be planned, and appropriate time management skills used to ensure that timelines and the people working around you will not be affected by delays. If an employer is confident that each worker has a positive attitude and pays attention to detail in every area of job performance, then personal responsibility can be an advantage for the employer.

Problems can be prevented by maintaining standards and building quality assurance into procedures, products and design at the start of a job. Rather than only fixing mistakes, one should take the time to analyse failures in order to avoid them next time. Always review work procedures on a regular basis and ensure appropriate action is taken to report issues to management in accordance with organisational requirements.

Take the initiative when it comes to accepting new challenges and stick with them. Set goals to do your job better and aim to do it efficiently. Share with your employer any ideas you might have to improve procedures, save money or increase productivity. All these personal responsibilities will lead to a quality improvement for yourself and those around you.

FROM EXPERIENCE

For a project you're currently working on, develop a personal vision by setting some short-term goals that lead up to completing one larger goal. When the project is finished, analyse how the project went. Which goals worked, which didn't, and how would your analysis change your approach in the future?

Purpose of planning

If you are not very good at planning and organising, it probably does not matter too much if you take the wrong clothes on a holiday, or if you forget to buy some items at the supermarket. If you are a poor planner and organiser in your private life you mainly cause problems for yourself.

FIGURE 3.1 Poor preparation and planning can lead to conflict, which can prevent your team from doing their job effectively.

But if you are a poor planner and organiser in your working life, you could jeopardise the success of your projects, which could put strain on your working relationships (see Figure 3.1). If you cause enough disruption for your team at work, they may not consider you for work in the future.

FROM EXPERIENCE

If you demonstrate to your team that you're willing to communicate clearly and professionally, and plan effectively for a job, you'll quickly become an indispensable team member.

Aim to work efficiently and safely

In the construction industry it is important that everybody is a good work planner and organiser, because poor work organisation can not only waste time and materials, but can also cause incidents and injuries. You must be able to work efficiently and safely.

To work efficiently means not wasting your own or other people's energy by creating unnecessary work. Work in the building industry can be very physically demanding. Wasted energy means less work completed. Less work completed means less money earned. Construction workers do not like having to do unnecessary work because someone else has not done their job properly.

It also means not wasting materials, tools or equipment – costs that must come out of the price of the job. Money spent on replacing materials and replacing or repairing lost or damaged tools and equipment means either less money for wages or higher prices quoted to do jobs. The higher the prices quoted to do jobs, the fewer jobs are won and the less work is available.

This does not mean that working efficiently means working unsafely. In contrast, work injuries lead to lost work time, higher costs and higher prices quoted to do jobs. Working unsafely for an imaginary rise in your pay packet will never make up for your (or another worker's) damaged back, lost sight or hearing, or damaged lungs or hands. More money in your pay packet cannot make up for you having to leave the building industry to work in a lower-paid job because your injuries make you unfit for building work.

Working efficiently has specific benefits that will allow you to work skilfully and productively so that you produce a high-quality product that is good value for money and that will ensure the success of the company and continuity of work. Working safely means to work in such a way that you create a healthier and safer workplace for everybody. Planning and organising your work have general benefits for everybody on a construction site by increasing efficiency and safety.

When you plan and organise a task you have been given on a construction site, you can:

- reduce problems later by making important decisions beforehand
- get the cooperation of other workers when you inform them about the task

- have materials, tools and equipment ready when and where you need them
- save your time, materials and tools
- eliminate or reduce hazards for your safety
- easily adapt to overcome problems you did not foresee.

On a busy construction site these are important benefits that can improve the quality of your day. Planning your tasks can make your work more satisfying and less stressful.

Make sure you know what you are required to do

Before you begin any task, it is essential that you make sure that you know exactly what you are required to do and complete any risk assessments and Safe Work Method Statements (SWMS). Your supervisor may give you spoken instructions, written instructions, drawings, or a combination of all three.

While your supervisor is doing this, it is important for you to be patient and to check that you have clearly understood exactly what is required. Utilise your communication skills to check that you know – not just think you know – what the task is that you must carry out. Always seek and offer feedback and confirm what is required to be completed.

FROM EXPERIENCE

When carrying out a task it is vital that you communicate effectively. Many apprentices assume they understand how the task is to be undertaken and do not evaluate what is required, which can cause injuries and costly mistakes.

FIGURE 3.2 Prepare a checklist

It is very easy – through eagerness, enthusiasm or impatience – to just get on with the task, to not listen properly, to ignore what is being said, or to dismiss information that does not seem important to you. Many a task has been undertaken incorrectly right from the start because attention has not been paid to crucial information. Because it is difficult to remember a lot of spoken instructions, several different sizes or dimensions, or a series of different steps in a procedure, always be prepared to write down what is required.

For simple tasks, it is a good idea to write a small reminder on a piece of board or sheet material to keep you focused on what is required and stop you from making unnecessary mistakes. It is advisable to carry a pocket notepad and pen/pencil to record important information.

Before you begin any tasks, be clear in your mind about what it is you are trying to achieve. When you have a clear idea of exactly what the task is that you are required to complete, the next step is to decide the sequence of actions that you will follow.

FROM EXPERIENCE

It is important to listen to and understand instructions, and not just hear them. Ask questions and make notes to ensure that you're listening actively, and if you don't have a clear idea of the task to be performed, seek feedback until you are sure.

Time management

Time management is very important in building and construction projects, simply because it relates directly to money. It is therefore the project manager's responsibility to manage this very valuable resource to the best of their ability by prioritising tasks that require completion and ensuring deadlines are established in conjunction with people working on the project.

Planning methods

Workplace planning has a direct effect on the whole job and enables tasks and systems to be completed methodically. Construction planning applied to these principles would involve many sound decisions made by the building manager, from the decision to tender to the final completion of a project. Allowing adequate time for planning is extremely important and must begin before site operations commence to ensure methods and equipment are decided on, and the correct materials are ordered in the correct quantities.

The labour requirements of subcontractors and tradespeople will determine the programming of a job. Stages of construction and sequencing of trades should commence as soon as possible, without necessarily waiting for the completion of the preceding work and should continue without interruption. Labour and time should be allowed for in proportion to the amount of work being performed. Brickwork is a labour-intensive trade, as is the erection of formwork for reinforced concrete. Buildings that have these trades will require detailed planning for the job to be done.

Forecasting

To be able to stay in business requires long-term planning as a business cannot look after itself. Planning for the future, as well as for the ongoing commitments of a business, is the responsibility of management. When deciding on the type of contracts that will be undertaken it is necessary to consider the financial outlay and the risks involved for financial gain. The success of a competitive business is dependent on its personnel and its equipment. This must be kept in mind during the planning process. Although it is difficult to predict financial and government policies, they must be considered when planning for the future.

Time charts

Planning will involve charting programs to look at when workers will be needed, when orders are to be placed, when supplies are to be delivered and to show the sequence of operations and provide a guide for progress and costing. These charts can be used to show time, progress and financial positions. They can be weekly or monthly charts, or for the duration of the total contract. The charts should be adjusted to consider any changes or variations to the contract, or any other disruption or difficulties to the on-site work. They are used to compare the current work with planning prior to the commencement of the contract, and for the delivery of materials and the commencement of different stages of the job.

Table 3.3 shows a typical sequence of events, with the time given in days. It also shows the relationship of activities to each other. Figure 3.3 shows the same information, but graphically highlights the overlaps in activities.

Sequencing of tasks

Work activities require significant planning and preparation, which will include sequencing of work activities to a set timeline. The sequence will depend on the type of building to be constructed, which could be an upper-storey conversion, an addition to an existing building, or a new building on a vacant site. With multistorey buildings it might begin with hoardings or excavations for basements, or the removal or addition of foundation material. On other sites it might be the stockpiling of materials or the erection of compounds for security. Services may have to be diverted, extended or capped off before work commences on the actual building. Therefore, the site and the type of building will present their own sequence of building activity.

The sequence of events will be started by the client, who will provide the finance.

Method of sequencing

There are several possible combinations of sequence relating to the need and reason for the building to be constructed. The development application and building application must be approved by the local council before the work commences on site.

TABLE 3.3 Schedule for a brick-veneer cottage on a concrete slab

Construction of brick veneer cottage on a concrete slab							
Ref.	Activity	Time (days)	Preceding activity	Ref.	Activity	Time (days)	Preceding activity
1	Site establishment	2	—	16	Insulation	1	8,10,13,14
2	Setting out of house	1	1	17	Plasterer	3	16
3	Excavate for slab	1	2	18	Waterproofing	2	17
4	Concrete slab	3	3	19	Carpenter internal joinery	5	17
5	Drainer sewer/stormwater	2	4	20	Tiler wall and floor	4	18
6	Carpenter frame/roof	2	5	21	Painter	4	17,19,20
7	Metal worker fascia/gutter	1	6	22	Concrete paths and driveway	3	5
8	Roofer	2	6,7	23	Landscaping/fencing	5	22
9	Carpenter windows	1	8	24	Electrical final fit-off	2	21
10	Bricklayer perimeter course	1	8	25	Plumbing final fit-off	2	21
11	Termite protection	1	10	26	Floor coverings	1	24,25
12	Bricklayer main	4	11	27	Window dressings	1	26
13	Electrician rough-in	3	6	28	Internal clean	1	26,27
14	Plumber rough-in	3	6	29	External clean	1	23
15	Carpenter eaves	2	12	30	Hand over	1	28,29

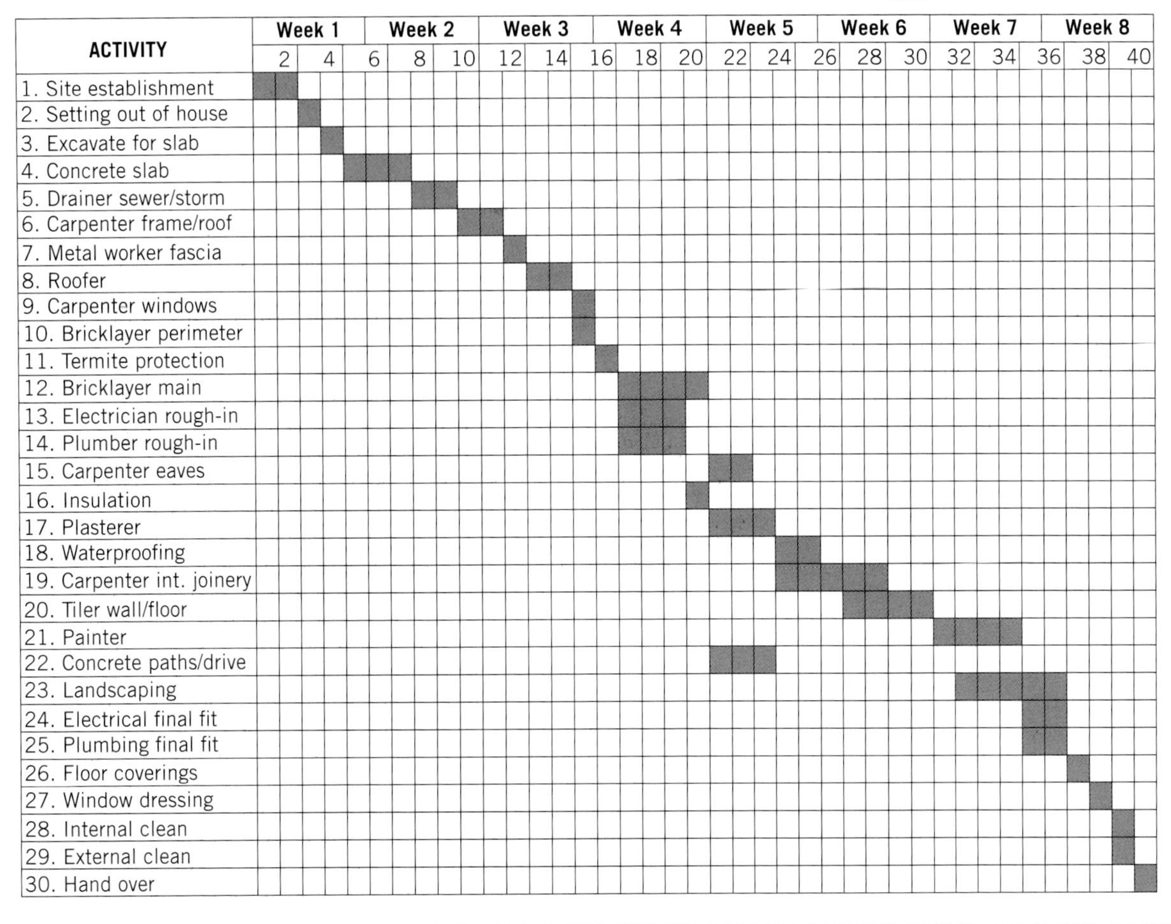

FIGURE 3.3 Typical Gantt or bar chart for a brick-veneer cottage on a concrete slab

The architect, together with the draftsperson, engineers and consultant, prepares the drawings and specification for the building to present to the local council for building approval. The quantity surveyor prepares a bill of quantities so that the building can be costed. The land surveyor sets up levels, reference points and set-out points to enable the builder to construct the building and assists in checking the vertical alignment for tall buildings.

The amount of pre-planning a builder does before commencing on site will depend on the type of building, as mentioned above. Once a contract is won it can be carried out. The builder will then sequence the order of trades, tradespeople, plant and equipment for the construction of the building. Depending on the size of the job, the builder will employ either a project manager or a site manager to look after the on-site construction management.

Staff at head office will include a construction manager, who controls the entire building operations for the company. A chief estimator assists the construction manager and is responsible for costing the building works.

Smaller construction sites and residential construction may not need the same force of professionals to run the job. It may simply be the builder who controls the site and coordinates the activities. A typical sequence of events in construction for a simple brick-veneer cottage on a concrete slab is depicted in Figure 3.4.

In the building industry, construction follows a logical order. The construction of an entire building can be broken down into stages. Each of these stages can be broken down into a series of steps, then followed by tasks and finally broken down into a series of actions.

Some stages must be completed in preparation for the next stage, and some steps must be completed in preparation for the next step.

Some tasks require completion at a specific point because that is the only possible time. Tasks that are not completed at the correct time may never be able to be done, which will have serious consequences,

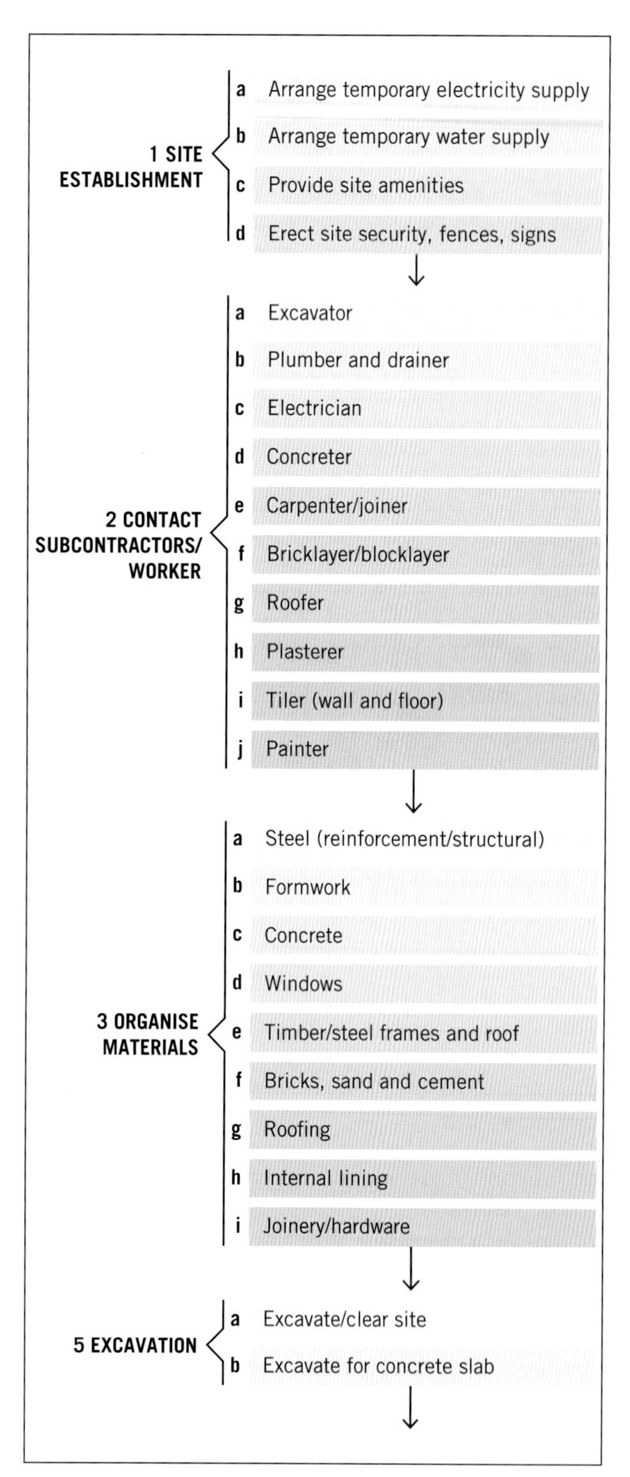

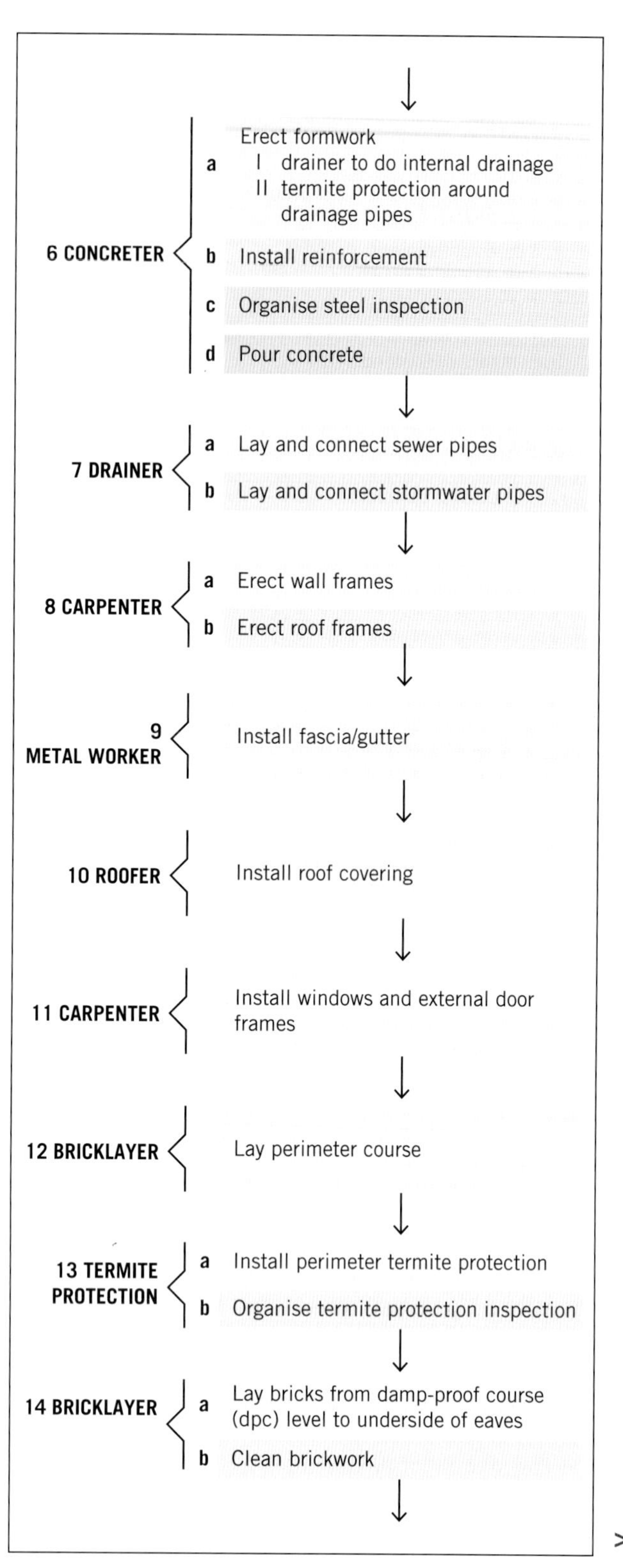

FIGURE 3.4 A typical construction sequence

>>

>>

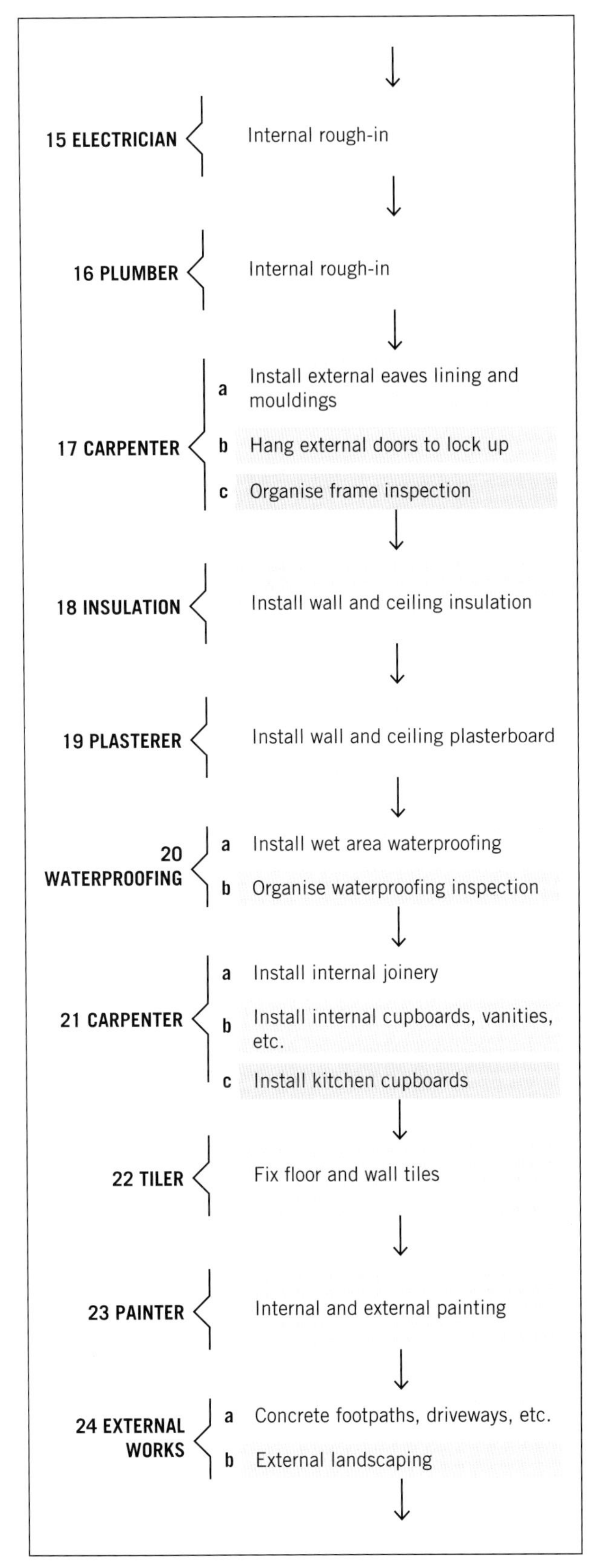

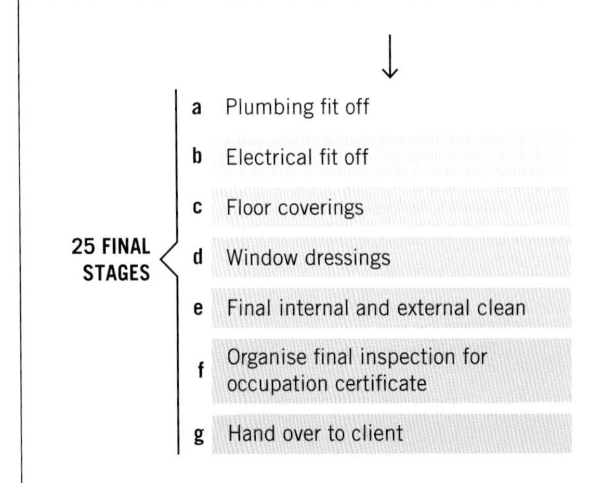

FIGURE 3.4 (*Continued*)

whereas tasks that are completed too soon may have to be done again.

Mistakes can not only be very expensive but also delay a job and make it unprofitable, which could send the builder or construction company bankrupt. If bankruptcy occurs, many people will be out of a job.

When you are carrying out a task, make sure that you are clear in your mind about the sequence of actions you will follow (see Example 3.1).

EXAMPLE 3.1

To complete a building to ground-floor level, plumbers must follow the following sequential steps: obtain plans and set-outs of ground-floor rooms containing plumbing fixtures, set out drainage turn-up points, excavate and install drainage pipework, and arrange an inspection before termite chemical pre-treatment is applied. The pipework is backfilled and capped off, then concreters can install underlay, tie the steel reinforcement and pour the concrete.

If the plumbing/drainage is held up, the sequence can be broken, and the program delayed.

Complete sequence

- Step 1: Set out accurately.
- Step 2: Excavate.
- Step 3: Lay drainage pipes.
- Step 4: Arrange inspection.
- Step 5: Backfill correctly.
- Step 6: Cap off and clean up.

Incomplete sequence

- Step 1: Set out accurately.
- Step 2: Excavate.
- Step 3: Lay drainage pipes.
- Step 4: Backfill.
- Step 5: Cap off and clean up.

FIGURE 3.5 Preparing the drainage before pouring of the concrete slab

EXAMPLE 3.2

Similarly, bricklayers must install damp-proof courses, flashings and brick ties as they build a wall, because these materials are impossible to install after the wall is finished. If this is not completed, the wall will have to be demolished and built again.

FIGURE 3.6 Preparing the wall frame ready for the brick veneer

EXAMPLE 3.3

When a house is being built, the plumber and electrician must rough in their pipes and cables through the walls before the plasterboard is put on. If the plasterboard goes on first it will have to be removed, the pipes and cables installed, and then the plasterboard replaced.

FIGURE 3.7 Rough in the services before fitting linings

Costing

As time is money, the construction plan considers the duration of the job, and the costing is linked to the bill of quantities, working drawings, specifications and the contract documents. Once the contract is signed and the project is managed financially, a close watch is kept on the budget for the job. Progress is matched to the budget and costing on a monthly basis to keep track of the overall contract.

One of the project manager's greatest responsibilities is to plan for a building to be constructed at or under budget. The financial control of a contract is most important, and all materials, labour and time are related to money. A manager with financial expertise will mean financial gain for the company.

Labour is the major cost for a building, and intensive planning will be required to cover this, as statutory and award costs can change over the life of a contract. It will depend on the type of contract, whether payment will be through a lump sum or staged, and whether an allowance has been included for rise and fall in material and labour costs. The costing of the building during its construction will also keep a check on the financial state of the company and must be conducted monthly.

The choice of plant and equipment and its operators is also of great concern for management. Breakdowns and the wrong choice of equipment can result in delays and lower-than-expected outputs. Considerable planning is needed for this, as these choices are linked to the costing of the job, and if mistakes are made at this stage the company will have difficulty with meeting its budget.

During the progress of the job it is necessary to secure materials stored on site, especially when the job is nearing completion. If valuable materials or equipment are damaged or stolen, this can disrupt the progress of the job and add to costs, which will affect profit. Therefore, security may have to be considered, even if it is only for the night and weekend hours. Security fencing, lighting or a surveillance system are some options that may be considered.

Insurance can be taken out to provide compensation for unexpected happenings, such as fire, burglary or accidents. This is short-term, and premiums may vary from year to year. Businesses in building and construction quite often deal with insurance brokers to get the best deal and coverage. During the planning stage the insurance cover is another budgetary consideration.

Use of bonuses and incentives

The greatest asset of an organisation is its labour resource. Calculating the expected labour is an important consideration when costing a job, but this can change due to many factors. The most efficient way to complete a job is to maximise production and increase job satisfaction. To guarantee this, an incentive scheme could be put in place, which can provide extra motivation for the worker.

Some factors that contribute to job satisfaction or motivation for work are achievement, recognition, the work itself, responsibility, advancement and money. This can be achieved by providing bonuses, profit-sharing schemes and non-financial incentives.

Causes of delays

With the planning process, it is not possible to predict delays that may happen on a job, but it is common for delays to happen. Delays are expensive on building sites, and this adds to the problem of planning and can affect the progress of the job. Lost time on site can be difficult to catch up on and mistakes potentially could be made.

The project manager must overcome any problems, whether it be industrial disputes, material shortage, variations, the weather or government and council requirements. The construction plan should be adjusted when these delays are experienced, and the appropriate people notified.

Hot weather can be a problem when placing concrete, as it dries out the concrete too quickly. Wet weather can cause water stains on finished surfaces, make timber swell and flood excavations. All of this requires extra work that may not have been allowed for in the planning stage and may require replacement or rectification.

Industrial disputes can cause a site to close, resulting in no production on site at all, and there may be a financial outlay for labour. This can have a disastrous effect if the increase in labour has not been allowed for in the contract.

The supply of material can be interrupted at any time for various reasons, and unless other supply lines can be obtained the output of labour at different stages of the job is seriously affected. The supply of materials could be held up on the wharves in containers, due to a breakdown in the manufacturing stage or because it is a specialty product imported from overseas. Any such cause is difficult to foresee. Planning could therefore involve stockpiling materials or obtaining alternative sources of supply.

Variations to design can potentially have financial incentive; however, this may cause delays to other areas of the project or to future work that may have been scheduled.

Government and council requirements can cause delays at the beginning of a job by holding up the approval to commence work. Complying with demands such as WHS requirements from relevant state or territory WHS regulators, pollution control from the Environmental Protection Authority (EPA) and traffic control from relevant state or territory roads and traffic authorities can all take time and can cause delays when waiting for approvals or when complaints arise.

FROM EXPERIENCE

It is essential to have good planning and organisational skills, while showing initiative and being creative in new situations and in tasks that you may face. Start planning by simply carrying a diary and writing down the tasks that you have performed throughout the day. Keeping track of where you have been working and the tasks that you have undertaken will make you more efficient.

Prepare a safe site

A safe worksite starts with adequate preparation. Listed below are common ways of ensuring that the public and people working on site are safe from potential dangers.

Barricades

These include any physical barriers placed to prevent entry and signify that a danger exists. They may be as strong as a timber or steel hoarding or as lightweight as a coloured plastic strip or tape.

Hoardings

There are two main types of hoardings used on worksites:

1. *Type 'A'* – fence-type hoardings that are used for residential, commercial or industrial sites where there is a minimal risk of objects falling on the heads of passers-by
2. *Type 'B'* – overhead-type hoardings that are primarily used on commercial or multistorey construction sites where there is a high risk of objects falling on passers-by.

Fence-type hoardings may be constructed of timber and sheeted with plywood or galvanised steel posts and chain wire. They range in height from 1.8 m for most residential sites, and up to 2.1 m for commercial or industrial sites and may be described as a high temporary fence or structure enclosing a demolition site, or a building site during construction or alterations, to restrict access, prevent theft, and provide side but not overhead protection to passers-by (see Figure 3.8). The plumber may need to install additional barricades within the main fencing or in public areas.

FIGURE 3.8 Typical temporary fence

Other barricades

These include any form of physical barrier used to draw attention to the existence of a hazard or danger zone. They may be in the following forms:

- *barricade tapes* – similar to those used by emergency services to close off an area
- *warning tapes* – placed on top of underground wiring or pipes to indicate their presence
- *concrete barriers* – used to block off and define lanes on roads; fencing can also be placed on top of the barrier for additional security and protection
- *plastic water-filled units* – hollow plastic interlocking units used to create a lineal barrier to prevent the entry of vehicles
- *board and trestles* (see Figure 3.9) – similar to those found on the side of the road where roadworks are underway; that is, a horizontal timber board painted with black and yellow angled stripes held up at either end by a metal 'A'-frame trestle
- *bollards* – solid vertical barriers of approximately 1 m high to visually or physically deter the entry of vehicles into an area of free pedestrian movement (similar to ram-raid posts); these are also used as protection for vertical plumbing stack work and downpipes in car parks, and are often larger-diameter steel pipes filled with concrete.

FIGURE 3.9 Board and trestle barricade

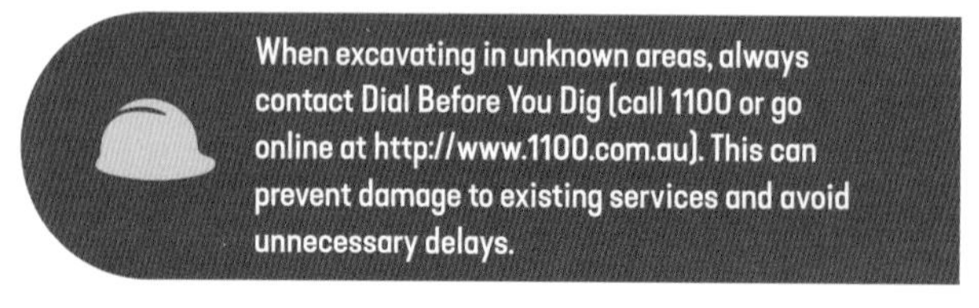

Determine the tools and equipment required

You should list the tools, equipment and PPE necessary to complete the task and list them in the order in which you will use them (see Table 3.4). If specialised equipment is required, keep in mind that it can be hired. Therefore, you must remember to order specialised equipment in time for it to be delivered to the site so that it is there when you need it.

After you have prepared your list of tools and equipment (see Figures 3.10 and 3.11), you must plan how you are going to get them to your destination where the task is to be performed.

TABLE 3.4 Checklist for hot- and cold-water rough-in

Tool checklist	
Project: Rough-in hot- and cold-water pipes	
Check	Install support for pipes and fittings
✓	• Timber saw, hammer, electric drill, chisel, tape, square
	Install pipework and fittings
✓	• Hacksaw, tube cutter, file, pipe benders, multigrips, tube expanders, poly pipe cutters, crimping tool, heating equipment
	Clip, saddle, insulate and level up
✓	• Hammer, screwdriver, knife, sealant gun, spirit level
	Test and clean up
✓	• Test bucket, broom, shovel, rubbish bags, wheelbarrow

FIGURE 3.10 Determine tools required

FIGURE 3.11 Determine equipment required

Most tools and equipment will be carried from the work vehicle, from the store or from another part of the site. In some situations, you may even have to bring them from another site or from home. Most tradespeople store their equipment at home in a shed or garage and take what is needed to the job only when it is required. It would be impractical for most tradespeople to carry every tool they own to every job they attend. They would need a very large vehicle to cart around tools and equipment, which they did not need at the time. This would not only be a waste of time and effort but also a waste of fuel and money.

Therefore, it is important to plan what equipment you will need each day. If you do not take a tool to the job when you need it, you will either have to collect it, buy a replacement or set the job back by having to do another task that you do have the tools for.

Once you have the correct tools and equipment on the site, you must make sure that you have the tools you need with you for completing the task. This means you must have an appropriate means of transporting the tools from where they are stored to where you will use them.

Check tools and equipment

You cannot afford to waste time and energy going back and forward all day to the work vehicle or store just because you are not mentally organised enough to plan and take what you need to where you will be working. Check that the tools are working properly. Check that all equipment will do what it is designed to do. Ensure that all PPE is correct for the task at hand.

If some time has elapsed since you have been to a job where you have been working previously, make sure that the path is still clear for you to carry the tools, equipment and PPE. Building sites are busy places and workers are continually shifting and moving materials around and out of their way. In doing this they may unwittingly put things in your way. Always check that your task has not been made more difficult in such a situation.

Site security

Tools and equipment make up a large part of a worker's overheads, so loss or theft will add a considerable cost to future quoting to cover replacement. Also, costs are incurred due to rises in insurance premiums and payment of excess on premiums when a claim is made. Therefore, a system or checklist needs to be followed so that tools and equipment are accounted for and locked up or secured at the end of each day.

It is also important to cover materials or restack them at the end of the day to avoid spoiling or loss. Providing a security compound or a chain wire fence around the job with lockable sheds and containers is advisable to secure equipment that cannot be taken home each day.

Procedures for leaving the worksite clear and safe at the completion of each day's activities could include:

- clearing site of debris using appropriate bins and recycling materials where possible
- keeping all work areas clean and tidy

- cleaning tools and equipment
- storing tools in sheds
- rolling up electrical leads
- locking up sheds and the site
- turning off power and water
- putting barricades in place
- covering materials in case of rain.

Worker may be responsible for the care and securing of the following items of plant and equipment at a building site:

- crane/forklift for hoisting and loading
- air compressor for a variety of pneumatic tools
- bobcat, with a variety of other attachments for minor excavation tasks, piers, trenching and site cleaning
- assorted power tools, chainsaws and hand tools for a variety of construction tasks
- generator for power supply
- assorted shovels, rakes and brooms for site work and cleaning
- concrete mixer for mixing mortar
- wheelbarrow for transporting materials
- elevator/conveyor belt for transporting materials.

Determine materials

After you have worked out the sequence in which you will complete a task, you should have an idea of the materials you will require. You must list the materials necessary to do the task and the order in which you will use them. All materials need to be on the site when required to be incorporated into the building. Therefore, it is important to not order materials too early or in larger-than-required quantities.

While it is important to have materials delivered just before they are needed, it is just as important to not have materials on the job that are not needed at the time. This is because the site can become congested with materials that are not required at that point in the construction stage and therefore get in the way. The materials may be damaged, lost, stolen, vandalised, or simply slow the entire job down because they are required to be repeatedly moved so that other people can continue with their work.

Therefore, you must have your materials on the site, ready to be used, in the sequence in which you will need to use them.

The following material list may be used for a hot and cold water rough-in:

- timber for noggins to support pipe, fittings etc.
- nails, screws and fixings
- various sizes and gauges of copper tube
- various sizes and classes of polyethylene pipe
- various tap bodies, mixers and fittings
- clips, saddles and fixing brackets
- insulation material
- duct tape
- plugs and caps.

An example of a rough-in is shown in Figure 3.12.

FIGURE 3.12 Rough-in timber-framed walls

Material storage location

The next step is to make sure that the materials you will need are where you will need them. Do not make the mistake of picking up what you need and start walking off and carrying it to where you will be doing the task. If the material you need is large, heavy or bulky, check the route you will be following through the job site. Carrying large and awkward items of building materials can be difficult. If you are going upstairs, along corridors, through doorways, up ramps, along planks, along scaffolding, around corners, through tight spaces, over trenches or through mud, you can place an unnecessary strain on your body by walking through hazardous areas while you are under load. It is important to inspect your route beforehand to make sure that all is clear and that you are not going to injure yourself or others.

When you reach the destination where you will be carrying out your task, make sure you locate your materials in a safe and secure position out of harm's way (see Figure 3.13).

There are many types of building materials, and each type must be transported, stacked or stored in a specific way. Stack or store your materials in the approved way for each material. To find out the best way to stack each type of material, pay attention to the way the material is delivered from the manufacturer and repeat this process. The manufacturer's stacking or storage method will be intended to maintain the material in the best possible condition in the safest possible way, and to ensure that the warranty is not voided or the materials damaged.

For more information refer to Chapter 6.

Carry out the task correctly

Now that you have your materials, tools, equipment and PPE in the location where you will be working, it is time to take care and follow your instructions.

To do the task properly you must follow a safe and efficient sequence of work. Make sure that you have your SWMS if one is required for the task.

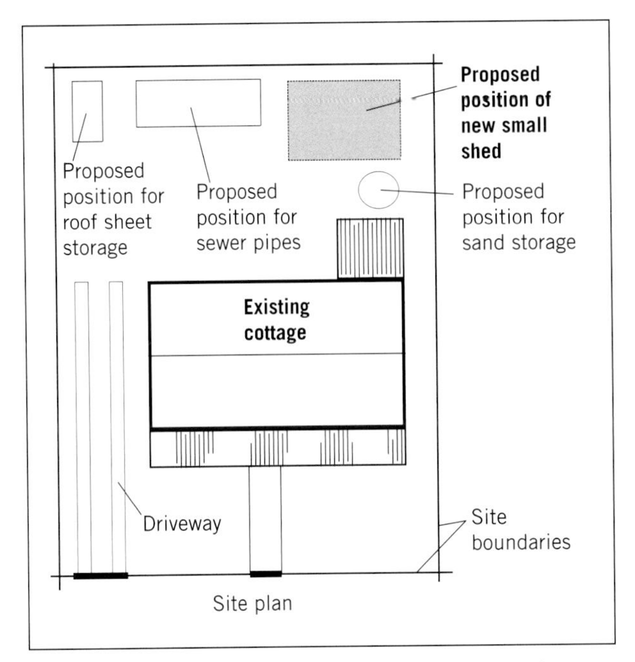

FIGURE 3.13 Identify material placement on the site plan

If your instructions were spoken, think back to when you received them and be clear about what it is you were instructed to do.

If your instructions were written, or in the form of a drawing, make sure you read them again (see Figure 3.14). Do not make the mistake of relying on your memory. Expensive errors are made by people who think they can remember the written instructions or what was on the drawing. Never be too lazy to double check as this may help you to keep your job or save a lot of money.

FIGURE 3.14 Double check details before you start

Finally, it is time to perform the task. Make sure you follow the logical sequence of work. Check your SWMS for the steps in the procedure. It is important that you carry out your work in the correct sequence and not rush the task. If you rush the task you are likely to carry out the job incorrectly. Speed is only good in building if the task is done correctly the first time. Having to go back and redo or demolish work costs much more than some imagined small saving in time gained by rushing. A good rule to remember when setting out materials for cutting is: 'Measure twice, cut once'.

Regularly take the time to make sure that you keep your work area clean and clear of debris. A few moments taken to ensure good housekeeping will save you having to stumble over waste, save you lost time, and may even save you or another worker from having an accident.

While you are working, you will not be using all your tools and equipment all the time. Therefore:

- make sure that the tools and equipment not in immediate use are safely located
- do not block walkways
- never stand up, stack or lean tools or equipment in such a way that they can slip, slide or fall over and injure another worker, or be damaged.

Solve problems as they arise

Working on a construction site is not like working in a factory. Construction workers are not assembling parts that have been manufactured elsewhere. A building is like a huge jigsaw puzzle, in which the people who are trying to put the puzzle together are also making the parts as they go along (see Figure 3.15).

FIGURE 3.15 A building is like a huge jigsaw puzzle

Source: Shutterstock.com/mikeledray

To an observer it may look as though all the workers are part of a mass production process, but this is not the case. The building is probably of an original design that has never been constructed before and is being erected on that site for the first time.

Although the workers have a great deal of experience and may be repeating tasks that they have performed many times before, they are performing that particular task on that site for the first time. In doing so they are dealing with different people and different materials in different situations. This means that problems and difficulties can arise. How the workers deal with these problems and difficulties will depend on their communication skills, training, knowledge and experience.

FROM EXPERIENCE

Issues can arise on building sites due to previous trades not installing products correctly. The most successful construction workers have highly developed problem-solving skills that allow them to rectify issues safely and efficiently.

In the construction process, it is important to solve each construction problem as it arises and to not allow problems to accumulate at the end of the job when it is too late to do anything about them (see Figure 3.16). The accumulation of errors on a job can lead to structural failure, faulty finishes on the building, and accidents and injuries to other workers.

FIGURE 3.16 Discuss and solve problems as they occur

Expect problems to arise and be prepared to overcome them. Problems can arise with the procedure you had intended to follow. Because of work going on nearby, you may not be able to follow the procedure you had initially planned. You will have to change your sequence of work to accommodate what is going on around you. When problems arise, they should be reported to appropriate personnel for rectification and followed by a regular review process in accordance with organisational requirements to ensure problems in the future are minimised.

Problems can also arise with the materials you had intended to use. The materials may not be available in the order you need or at the time you need them. Alternatively, some of the materials may be incorrect or faulty and you will have to arrange for replacement material but keep the work progressing, which could potentially require you to change your sequence of work. Weather may affect your task sequence and your materials. This would require a major rethink and reorganisation of the sequence of the task that you are undertaking.

Construction processes rarely go according to plan, but this does not mean that a plan is unnecessary. A plan is important because it tells us when we must start and when we must finish a task. It also identifies key points along the way. It tells us when various important sections must be finished. But along the way we must be flexible. The important thing to remember is that the task must be completed safely, efficiently, to the required quality standard and in an appropriate timeframe.

Clean up when work is complete

Working effectively in the plumbing sector always involves maintaining a safe and clean working environment.

Cleaning up and good housekeeping is an essential and vital part of any work project and is just as important as the preparation phase. It is best practice to clean up as you go.

Good housekeeping is essential to reduce on-site injuries.

A good tradesperson will automatically clean up as they carry out their work, irrespective of whether they are working on something relatively simple, like changing a tap washer in someone's home, or working on a large building site on a longer-term basis. The following four situations consider what may be involved in a clean-up phase:

- small jobs in private homes
- major work in the private sector
- maintenance work other than in private homes
- learning skills while attending educational institutions.

Small jobs in private homes

This could include anything from replacing a tap washer, repairing or replacing a cistern or toilet suite, or unblocking a sewer to re-laying a drain or replacing a water service. The location of your work, whether you are working inside or outside the house, will govern the overall process, but either way it is important to leave the premises in a similar or better condition than when you entered. Some items to consider are:

- obtaining permission to enter the premises
- explaining the procedure to the client in case any modification to normal practice by the client is required (e.g. allowing them to fill the kettle or ensuring the dishwasher has finished its cycle before the water supply is isolated)
- respecting property and taking all necessary steps to prevent any damage
- observing safe operating procedures with all types of work
- being clean and neat with your attire, footwear etc.
- working cleanly
- removing/recycling waste material
- using a dustpan and brush and/or vacuum cleaner to clean up any material deposited as a result of your work
- explaining how to safely operate any newly installed devices (e.g. gas stove, gas room heater or different tapware)

- repositioning any moved items such as furniture or pot plants that may have been moved as a result of your work.

FROM EXPERIENCE

As part of your company's quality assurance approach, it is vital that you instruct the client on how to operate recently installed appliances to reduce any potential damage from misuse.

You will find that the more thorough you are with the cleaning-up process (and the higher the quality of your work), the greater potential you will have for obtaining more work in the future. Good tradespeople tend to get more work by 'word of mouth' rather than advertising if they are meticulous with their work ethic.

FIGURE 3.17 Clean the work area daily

Major work in the private sector

This involves longer-term projects such as project homes for which you may install services such as drainage, water services, gas services and guttering. This situation generally involves you working with other tradespeople such as carpenters, bricklayers and electricians, and therefore means that you must coordinate your activities with the building foreperson and the other tradespeople. Items to consider throughout the project and in the clean-up phase include the following:

- Risk control measures should be monitored to ensure that they are effective and appropriate to the task and work environment.
- Risk control measures should be reassessed, as required, in accordance with changed work practices and/or site conditions, and alterations undertaken within scope of authority. These may be addressed or altered during regular 'toolbox talks' (discussed later in this chapter).
- Daily clean-up should take place by all trades (see Figure 3.17). During this process you must allow time to:
  - ensure the site is safe for other tradespeople during the progress of work
  - clear the worksite of excess materials and trip hazards
  - remove material that is hazardous
  - clean up food and drink waste, wrappers and containers
  - securely and safely store materials and fittings that may be used later
  - ensure that tools and equipment are cleaned, checked, maintained and stored in accordance with the manufacturer's recommendations and workplace procedures (see Figure 3.18)
  - remember that you have a duty of care to all other tradespeople and any other visitors to your workplace.

FIGURE 3.18 Clean, maintain and store tools and equipment

- The work area should be cleared, and materials disposed of or recycled in accordance with state and territory legislation and workplace procedures.
- Documentation should be completed in accordance with workplace requirements. This may include the use of a SWMS.
- The work area should be restored in accordance with workplace procedures.
- Any trenches should be backfilled, consolidated and the ground surface returned to its state prior to excavation. The local government (council) authority may have specific backfilling and restoration requirements, which should be identified in the planning stage.
- The safety system, if used, is dismantled in the correct sequence and removed from the worksite.

Maintenance work other than in private homes

This type of work involves you having to enter another person's place of work, which could include an office or factory, and you therefore may be subject to workplace procedures. The procedures may require you to follow company protocol to enter and work in a specific area and may involve the use of specialised equipment and PPE.

Learning skills while attending educational institutions

Where individuals are in a learning environment and are still developing skills, cleaning up is as important as any other situation and is normally factored into each of the tasks addressed. The fact that several different learning activities may take place in any one week will mean that any new activity should commence in a safe and clean environment and that documentation must be completed in accordance with workplace requirements. This will include the use of JSA or SWMS.

Remember that, as above, you have a duty of care to all staff, other students and visitors to your workplace. You and your colleagues will be involved in ensuring that:

- risk control measures are monitored to ensure that they are effective and appropriate to the task and work environment
- risk control measures are reassessed, as required, in accordance with changed work practices and/or site conditions, and alterations are undertaken within the scope of authority. These may be addressed or altered during regular toolbox talks.

In summary, cleaning up is a best practice approach and should be a natural part of any activity addressed in the building trade. It leads to a safer working environment and is normally factored into the pricing of any activity.

Report on completed work

After you have carried out all your responsibilities, it is important to report to your supervisor on the completion of the task (see Table 3.5). Make sure that you tell your supervisor about any problems or difficulties you experienced. It is also important to report any damaged or faulty equipment so that it can be repaired before it is required again. It is very frustrating when you are undertaking a job to find that the tools or equipment you need are not working. Advise your supervisor of any unsafe or dangerous behaviour or procedure you have experienced or witnessed, as this can risk everybody's health, safety and welfare. Your supervisor needs to know about WHS breaches because the unsafe worker, supervisor, manager and company can all be heavily fined for failing to follow WHS Regulations.

TABLE 3.5 Create a checklist for reporting

✓	Make a report for the supervisor
✓	List any problems or difficulties you experienced
✓	Report any damaged or faulty equipment
✓	Report any unsafe or dangerous behaviour
✓	Report any uncooperative workers who may have hindered your progress
✓	List any other matters you feel are relevant

Finally, tell your supervisor about any lack of cooperation or assistance from other workers. On building sites, many tasks are carried out by subcontractors who are paid to come onto the site, do a task and then leave. Some subcontractors want to get the task done as quickly as possible so that they can get on with another task somewhere else. This can lead to a lack of consideration, cooperation and assistance for other workers on site, which may generate friction and lead to arguments and disputes. These actions can disrupt the smooth running of a job, slow the job down, lead to faulty or unsafe work, and waste time, materials and money. Your supervisor needs to know anything that can affect the efficient running of the job.

LEARNING TASK 3.2

ORGANISE AND ACCEPT RESPONSIBILITY FOR OWN WORKLOAD

Prepare a simple site hierarchy chart for your own workplace and briefly describe the roles and responsibilities of each person in the chart that would be undertaken daily.

COMPLETE WORKSHEET 2

Work in a team

Creating an environment that supports teamwork has numerous advantages and creates a positive culture in the workplace. When working in a team you must identify what contributions and workplace goals are required, and work towards avoiding barriers and situations of disharmony.

Value of teamwork

Teams of workers can prove very useful when building site tasks require rapid completion, such as constructing and covering a new roof on an existing cottage by carpenters and plumbers, or when work needs to be completed within a given timeframe, such as a plumbing rough-in so walls can be sheeted. Teamwork is also useful to complete a number of related parts or elements, such as installation of below-ground drainage before sanitary plumbing stack work can proceed. Once the team members understand the role they play in the overall construction of the building, they will be able to contribute more effectively to meet the 'site goals' and to work more efficiently with other teams or individuals.

Individuals within the team

To allow smooth operation and cohesiveness within a team, each person should be allocated a job or function that will complement other team members and not duplicate their effort. A simple team meeting or toolbox talk (discussed later in this chapter) would help to identify and establish the role of each member within the team and allow for a review of contributions periodically where required. Ideally, one person in the team should be nominated and appointed as the 'team leader' to ensure consistency of effort and communication.

Regular rotation of roles within the team is also important to avoid one or two members doing all the work or dominating the learning process. This rotation of roles and responsibilities also reduces the risk of individual egos getting in the way of a productive team effort.

Individual strengths and weaknesses

Rotation of roles and responsibilities is important for effective functioning, but there is also room for specialisation within the team. If individuals have special skills or talents, they should be able to maintain a role or position in the team and allow the other team members to assist them. This may occur where only one member of the team holds a qualification or has a licence to do specific work.

Where a member of a team lacks skills or knowledge to carry out a task on their own, this person may 'team up' with another member who can supervise and assist them in a mentoring capacity. This could occur, for example, where some members of the team are labourers and the others are tradespeople.

Enhancement of roles

Team roles may be enhanced by giving individuals more responsibility than they would normally receive, or by encouraging and assisting individuals to undertake further structured training to gain a qualification or enhance their role within the team. This may mean enrolling in a trade or post-trade course or perhaps attending training that targets specific skill sets for the benefit of the individual.

Roles

When people work together in an organisation, it is essential that work is distributed appropriately, and any requirements are communicated effectively. This requires planning and delegation and can be implemented through organisational policies, guidelines and requirements. It is important that organisational policies are relevant to the work role to ensure harmony and efficiency in the organisational structure.

An organisation chart is drawn up to give a pictorial representation of the roles in the business structure. Senior positions are shown at the top, descending to the least experienced positions. They can extend horizontally with multiple positions of equal status (see Figure 3.19).

Individual levels of authority will depend on the size of the company or site, with each person being responsible for the person or persons on the next level below them.

This arrangement of site personnel can be referred to by several descriptions, such as:

- site hierarchy
- site lines of authority
- site organisation diagram.

A large building firm is structured in the same way as firms involved with smaller individual sites, with the person having the most authority at the top of the firm and the people with the least authority at the bottom. It is also possible to have several people with the same authority or responsibility at the one level.

In addition to roles within an organisation, there are a variety of roles that workers can engage with on the worksite and it is important to understand the role they play and how they can benefit the team. These are outlined in Table 3.6.

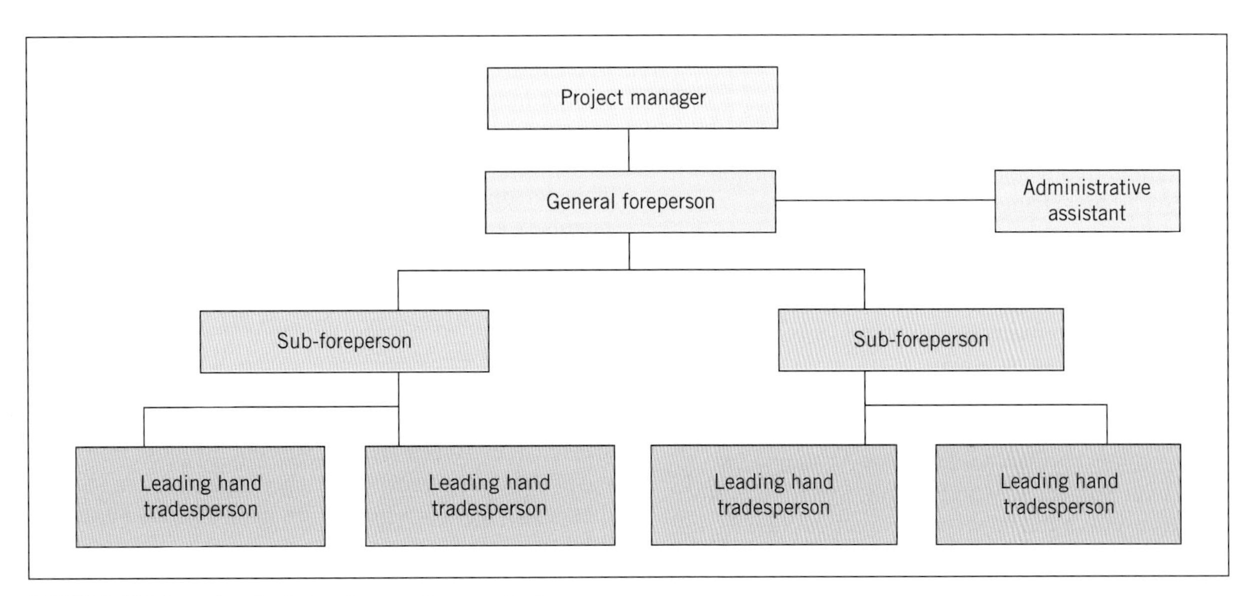

FIGURE 3.19 Hierarchy chart showing on-site organisation

TABLE 3.6 Roles in the building and construction industry

Role	Description
Client	Finances the project. This may be a developer or persons involved in a joint venture who require a building for a specific purpose.
Architect	Engaged by the client to design and control the building on the client's behalf. Other responsibilities include lodging the development application with designs or models, supervising the drafting of plans and the specification, preparing tender documents to engage a builder, and engaging consultants for the hydraulic, structural, mechanical and other specialist areas.
Quantity surveyor	Responsible for preparation of a bill of quantities for the architect to enable the building to be priced by an estimator. They are also required to prepare quantities for variations to the contract and assess progress claims.
Land surveyor	Responsible for setting out boundaries and levels of the site and control of the vertical alignment of a building.
Draftsperson	Engaged by the architect, or other professional designers/consultants, to prepare plans and details before and during construction.
Builder	The main contractor/worker selected by a tender process to carry out construction as per plans and specifications. The builder will also engage subcontract labour to deal with specific parts of the construction.
Construction manager	Responsible for all building contracts the company has won or is tendering for. Usually working from the main office, the construction manager forms tender documents and controls the building activities for the company.
Project manager	Usually employed on very large sites to control the running of the site and to make sure that projects run to schedule and to budget.
Site manager	Runs the day-to-day activities and liaises directly with the foreperson and the project manager.
General foreperson	Responsible for the delegation of duties to workers on site and for site safety. Other responsibilities are the coordination of subcontractors/workers and the progress of work.
Quality assurance officer	Engaged to control the quality of construction on site and ensure that the work meets the standards laid down in the specification and tender documents.
Building inspector	For domestic construction, the inspector visits the site on behalf of the council to check structural work and pass completed stages of a job.
Plumbing inspector	For water, waste and gas services, the inspector visits the site to inspect the services to ensure that they meet the requirements of Australian Standards and Codes of Practice.
Principal certifying authority (PCA)	For domestic construction, the PCA carries out similar duties to the local government building inspector but is engaged directly by the client to conduct the inspections.

Team rules

Each team should discuss and set ground rules by which all members must abide. This is important to maintain harmony and allow the team to function effectively. Any changes to these rules, and/or improvements made, must be discussed and agreed on by all the members. All members should have equal rights and responsibilities and be able to suggest changes and improvements without fear of being ridiculed or vilified by others.

Disharmony

Where conflict within a team arises, it should be dealt with immediately and not allowed to intensify and grow into a full-blown confrontation. Ideally, all conflicts should be resolved within the team so that other unwanted persons are not involved, as this may lead to individuals feeling resentful. If a resolution cannot be found, another person from outside the team should be brought in to mediate the situation. This person should be someone that all the team members know and respect; otherwise an agreeable resolution may not be found, or bitter resentment may result.

The key to successful teamwork lies in the selection of members with suitable skills who are willing to do their bit and abide by the team rules, have an equal say in the running and organising of the team, and hold skills that are valued and complement the other team members. It also involves putting a team structure in place that encourages participation and appreciation of each member.

PART 1

3

LEARNING TASK 3.3

WORK IN A TEAM

Carry out a simple team activity, as directed by your teacher/instructor, and follow the processes outlined with a toolbox talk. The activity could be a group practical task that requires participation and coordination by everyone to achieve completion.

Each person should be allocated a job/role in the group, with one person nominated as the team leader. Set ground rules for your team to follow and rotate roles as often as you can.

Participate in identifying and pursuing own development needs and processes

Identifying and pursuing the needs of personal development is an important process to develop and learn the skills required for a continually changing workplace environment.

Workplace competencies

Before a person can progress beyond any given academic level in the building industry, a list of required competencies must be identified. In most trade qualifications, the individual is required to attempt and successfully complete a range of 'core' or compulsory competencies plus a minimum number of elective competencies. Several of these core competencies may be 'common units' to a wide range of trade and non-trade areas. These core and elective competencies, which would be selected from a training package, are grouped together to form the structure of the 'qualification'.

As discussed earlier in this chapter, the competencies undertaken as a part of the plumbing apprenticeship go towards completion of the qualification for a Certificate III in Plumbing. Once this qualification is successfully completed, a Certificate IV in Plumbing can be undertaken and will give the plumber the opportunity to apply for a plumbing licence.

Changing nature of work

The nature of work has changed in Australia during recent times, reflecting the changing needs of business and the growth of a globalised economy. Many industries are meeting these changes by varying employment patterns, such as changing from the traditional full-time employment to casualisation of the workforce through offering part-time, casual, contract, outsourcing and labour hire as alternatives. Apprentices, however, are employed fulltime under a training contract with an Australian Apprenticeships Network provider and have far greater protection of conditions within their workplace.

Career pathways for workers are also changing, and in some industries are breaking down, leaving little or no option of advancement. Some employers expect their employees to contribute new knowledge and innovation to their organisation to assist the growth, effectiveness and competitiveness of the business. Opportunities in the workplace may present themselves in different ways, so it is important to look at your own learning needs to be able to develop and grow within the organisational structure.

Career paths and development opportunities

There are numerous career paths and opportunities that a plumber may undertake in the building and construction industry. Conducting a plumbing business is always a key and primary goal to achieve for many apprentices, as it has the potential to bring job satisfaction, freedom of choice and financial reward for those who work efficiently. There are, however, other options for career progression, which will require you to identify your own learning needs and engage in professional development to ensure that any future work requirements are met. A career path choice can be made using the chart in Figure 3.20 and can be mapped according to the learner's preferences.

Identifying own learning needs

Individuals may need to seek advice from a range of personnel to identify their own learning needs and find out what they need to do to move in another direction. Good advice and opportunities are usually gained from the most experienced people with a good understanding of the trade or industry, who may include:

- the foreperson on the job
- architects and/or engineers
- progressive builders
- teachers/lecturers from vocational education organisations
- environmental experts
- designers
- building material manufacturers.

LEARNING TASK 3.4

PARTICIPATE IN IDENTIFYING AND PURSUING OWN DEVELOPMENT NEEDS AND PROCESSES

1 Investigate and note the title, code and list of competencies for the course that you are currently undertaking.
2 Select a career path from Figure 3.20 and identify the learning needs required to achieve this career path change.

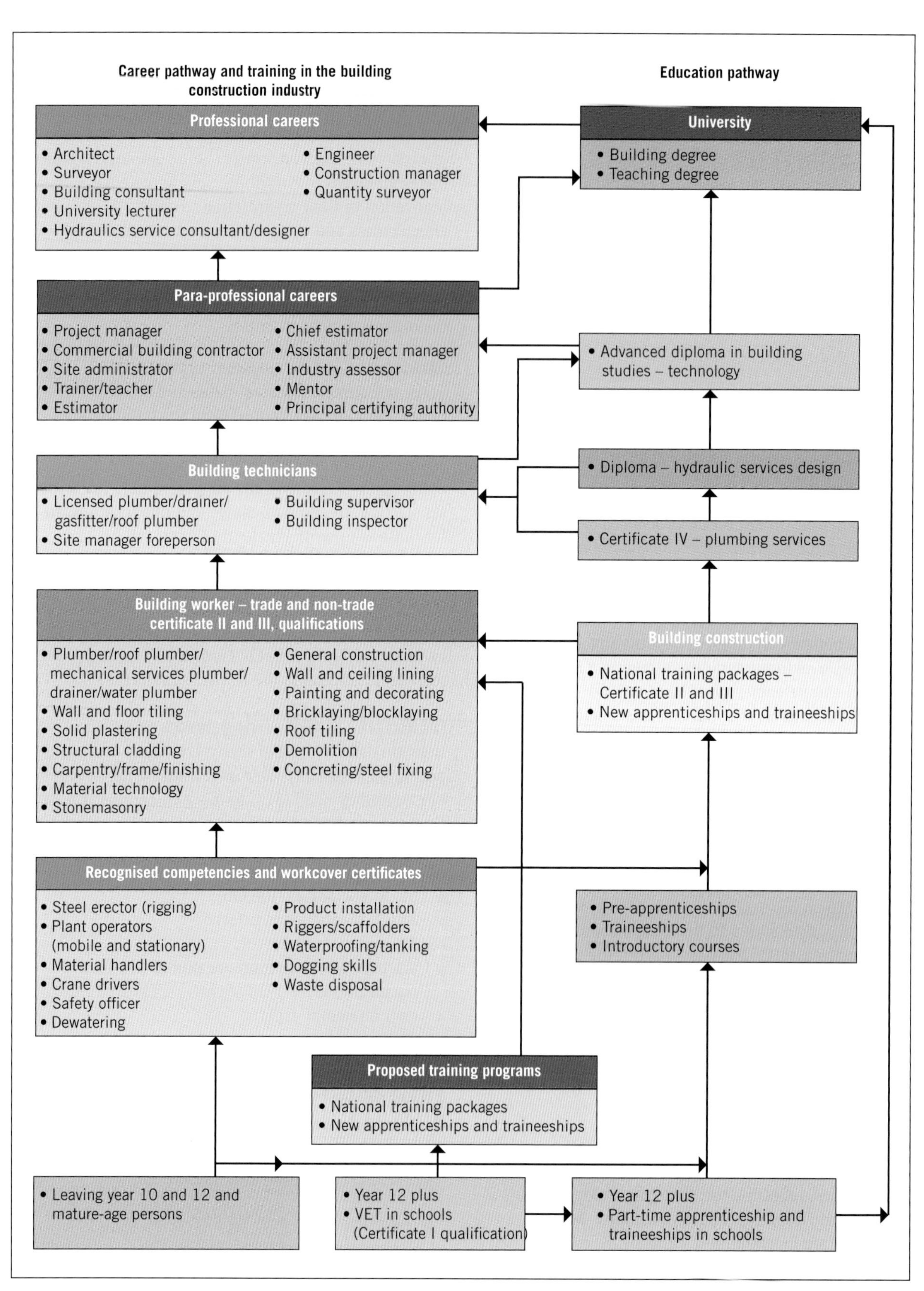

FIGURE 3.20 Chart showing training and career path progression

COMPLETE WORKSHEET 3

Participate in workplace meetings

Meetings in the workplace are an important process to identify objectives or procedures that can impact everyone involved in the construction process. They provide a platform to convey information and allow people to exchange ideas and provide points of view, facts and processes. There are a number of methods used to raise issues or discuss problems on a building site, with most meetings being either formal or informal site meetings. Formal meetings consist potentially of all members of an association or their representatives, while informal meetings are generally conducted with simplified procedures. Special meetings may be held on site, involving the builder, client, architect and subcontractors/workers, to consider important matters relating to the project. Where deals involving variations are made, all affected parties may sign a statement on the spot, confirming changes that have been agreed to. When a meeting is called it is essential to have an agenda about what is to be discussed to ensure the meeting finishes in the allotted time period.

Purpose of meetings

It is important to be quite clear about the purpose of the meeting or committee discussion so that each member is fully aware of the reason for their participation. Common reasons for calling a meeting would be to:

- pass on information
- seek ideas to solve a problem
- coordinate information gathered over a period of time
- negotiate changes to contracts
- resolve disputes
- form new policies or revise old ones to improve the running of the site
- plan for future work
- decide on proposals or outstanding issues
- act as quickly as possible on decisions made to ensure that enthusiasm and credibility are ongoing.

Types of meetings

There are several different types of meetings that may be held on a plumbing and construction site. These are:

- general staff meetings
- union meetings
- committee meetings
- team meetings
- social club meetings
- toolbox talks.

General staff meetings

General staff meetings are used to inform staff about issues such as company performance, direction, future and policies. All employees are expected to attend.

Union meetings

These meetings may be called by either union members, union delegates or union organisers. At these meetings members may discuss pay and conditions, unresolved safety problems, disputes with employers, problems on the site, support for other employees, or support for victims of tragedies. Trade unions have meetings, such as annual, delegate, branch and committee meetings. The branch meeting may take place once a month, but the executive committee may meet once a week.

Committee meetings

A committee has a limited membership that has been elected or appointed to perform specified functions. The most common committee meetings on the construction site are as follows:

WHS committee meetings

Work Health and Safety committees are formed on sites as required by the *Work Health and Safety Act 2011*. The members of the committee are generally elected Health and Safety Representatives (HSRs), and they regularly inspect the site and hold meetings where they make policies and recommendations about health and safety on the job. Committee members and HSRs represent both management and general staff.

Special-purpose committee meetings

Special-purpose committees are formed to undertake tasks or responsibilities within the workplace. Meetings will usually be conducted according to a program or timetable that has been established to achieve an objective. When this has been achieved, the committee will generally be disbanded. The committee may be composed of specialists, experts, or volunteers who want to get something done.

Team meetings

Team meetings are held by work teams that have been formed to increase efficiency on the job (see Figure 3.21). The teams meet regularly with the objective of planning, organising and carrying out the work as successfully as possible.

FROM EXPERIENCE

Working together as a team and ensuring tasks are well planned and organised will foster a safe and efficient worksite.

Social club meetings

These meetings are usually held by volunteers to arrange social functions after work for interested staff. Activities

FIGURE 3.21 Typical team meeting

can include barbecues, concerts or visits to restaurants. The success of the social club depends on the enthusiasm and shared interests of the staff.

Toolbox talks

These meetings are used to help manage safety on plumbing and construction sites by providing an open informal forum where site workers can voice their opinions and help identify potential safety hazards or other issues that can then be eliminated or minimised before injuries occur. Toolbox meetings ensure that safety is a key priority throughout the duration of a plumbing project.

FROM EXPERIENCE

Toolbox talks create a platform to work from to problem-solve in teams and create strategies and solutions.

Conducting meetings

Meetings may be conducted either formally or informally.

Formal meetings

Formal meetings are run by elected office bearers called the chairperson, secretary and treasurer.

Each meeting will follow an agenda (see Figure 3.22). The agenda is prepared by the secretary and sent to the people attending the meeting before the day of the meeting. It will include the time and location of the meeting, as well as what will be discussed and the order in which it will be discussed.

The meeting will be controlled by the chairperson. When each item is raised by the chairperson it will be discussed, and the meeting will follow a set procedure for making decisions using movers, seconders, speakers for and against the motion, then finally a vote by the meeting for or against the motion. During the meeting the secretary will write the minutes of the meeting, which are summaries of the discussions and the decisions agreed on (see Figure 3.23). The minutes then become the formal record of the meeting. Formal

Flyingfox constructions

Project: renovation of sports bar

Project meeting

Venue: site office
Date: Monday 8 October, 2021
Time: 9 am

Agenda

1. Welcome, introduction and apologies

2. Project update:
– plumbing work
– floor tiling and carpet laying
– painting and decorating

3. Proposed variations to plans

4. General business

Next meeting:

FIGURE 3.22 Typical example of a meeting agenda

minutes should be kept of all site meetings arranged and conducted by the project manager or general foreperson. Copies of these minutes, which represent a report of proceedings, can be distributed to all people present at the meeting and any other people who were absent or need to be informed.

These procedures are usually followed only when the items to be discussed affect a lot of people, have legal implications, are required by law, or need to be accurately recorded for future reference.

Company meetings, union meetings and WHS meetings are usually conducted as formal meetings.

Informal meetings

Informal meetings are usually conducted with simple procedures and without the election of office bearers. A group of people will get together to discuss an area of interest (see Figure 3.24). There may not be a formal agenda, but everyone will have a general idea of the focus of the meeting.

Someone will facilitate the meeting, and, if there are to be several meetings, group members may take turns at facilitating the meeting.

There will be no set procedures for conducting discussions and making decisions. The success of the meeting will depend on the rationality, fairness, self-discipline and depth of desire to achieve a successful outcome of each of the members of the group.

All members will generally take their own notes about issues that affect them directly. During the meeting, individuals may volunteer or be persuaded to undertake tasks that have been decided on by the group.

Work Health and Safety
Committee meeting

Date: Friday 12 October 2021
Venue: Hut 18
Time: 10 am

Agenda

1. Welcome and apologies
2. Minutes of previous meeting
3. Business arising from minutes
4. General business
   4.1 Non-tagging of electrical leads
   4.2 No handrails on stairwells
   4.3 Bricks falling from fourth floor
5. Other business
6. Next meeting
7. Close

(a)

Work Health and Safety
Committee meeting

Date: Friday 12 October 2021
Venue: Hut 18
Time: 10 am

Minutes

Present: A Adams, W Calper (chair), N Couri, G Boon, F Jones, J Melendez, P Nguyen (secretary)

Apologies: B Grune, N Prkic

1. Minutes of the previous meeting were read and accepted.
2. Business arising from minutes:
   2.1 Missing safety signs have been replaced on entrance gate.
3. General business:
   3.1 Untagged leads. N Couri reported that all untagged leads have been removed from the site.
   3.2 No handrails on stairwells. J Melendez is to organise a team to install handrails on all stairwells.
   3.3 Bricks falling from fourth floor. A Adams reported that toe boards had been removed from the bricklayer's scaffold. These have been replaced, which should prevent any further incidents.
4. Other business.
5. Next meeting was set down for Friday 16 December at 10 am.
6. The meeting closed at 11 am.

(b)

FIGURE 3.23 Typical examples of meeting documents: (a) Formal agenda; (b) Minutes of the meeting

FIGURE 3.24 An example of an informal meeting in progress

Informal meetings are much more dependent than formal meetings on the ability of the members to work as a team.

Team meetings, social club meetings and toolbox talks are most often conducted as informal meetings. In the workplace, depending on the situation, meetings may be a combination of both informal and formal meetings.

Participation in on-site meetings

Regardless of whether a meeting is formal or informal, it is important to participate. When you attend a meeting, make sure that you:

- understand the purpose of the meeting
- contribute to relevant discussion
- are prepared to listen
- offer only corrective feedback, and do not criticise
- deal with issues or problems, not with people or personalities
- are prepared to resolve problems
- are prepared to accept and carry out the decisions reached by the group, as the group's decision is binding on everyone.

Use of committees

There are three main ways of dealing with a situation or settling a problem:

- by imposing a solution from either above or below
- by accepting a compromise reached through bargaining
- by agreeing to the best outcome arrived at after joint discussion.

The first method implies the use of force, which emphasises a permanent division, as it potentially represents either a victory to be maintained or a defeat to be avenged at the next opportunity. Equally, the second method is an uneasy resolution decided by relative power, rather than on true assessment of the facts.

By contrast, the employment of discussion ensures a genuine solution based on conviction, with the maximum chance of acceptable and harmonious action. Therefore, committees or informal group meetings can play an important part in the organisational structure. Committees can be used for different purposes, with appropriately varied membership as follows:

- as an advisory body, such as a site WHS committee with elected Health and Safety Representatives (HSRs) from each activity, to deliberate on problems and assist an executive by combining the total knowledge and experience of its members
- as a means of consultation, such as a planning meeting where members are appointed by reason of their individual functions and contribute their different viewpoints, ensuring that all aspects are properly considered
- as a channel for information and a method of communication, usually between a supervisor and their managers, where reports are tabled, and resources are pooled to find a suitable solution to a problem
- as a process of coordination, usually between a specialist and several production units, such as a transport manager allocating their daily vehicle tasks to meet the requirements of the respective general foreperson.

All these elements may go into forming an effective working committee that is able to resolve problems and make judgements.

Participation and representation

These two roles are not the same. Voting for a representative to do what they think would be best for you is different from being there yourself to make your own alliances and express your own thoughts. Sending a delegate instructed to vote in a specific way is different again.

Rights

In a democratic society we have the right to be represented in our national government, to access health care, to gain an education, to vote and so on. These are rights that people have argued and fought for, which we continue to assert, which are accepted by our society, and which are continually subject to readjustment as our society changes. Therefore, each person, representative of a committee or participant in a meeting has the right to voice an opinion on the subjects raised. You should not sit and say nothing if you disagree strongly, as you may have to live with the decision. Your say may sway the argument in your favour by convincing others who are undecided.

Communication and consultation

Representation or participation is seen as a means of finding out what people want, and of communicating information and decisions to them. This aids in understanding, freeing processes of decision-making, and action. It may be as much a matter of persuasion as sharing in making decisions.

Efficacy

People are more likely to implement a decision effectively if they have shared in making it, possibly because it is felt that their contribution makes it a better decision, they feel responsible for it, and they trust and understand it, and therefore will work to produce the desired result.

Group dynamics

To gain the greatest contribution from people at a meeting, it is important to have a room layout where everybody can see and be easily seen. This will promote a more inclusive environment than having members pushed to a corner and making them feel insignificant. Two of the most important requirements for effective meetings are that communication takes place in a well-ordered manner, and that content is relevant. Too often it is assumed that these requirements are the responsibility of the chairperson, who achieves them by controlling the meeting. There are obviously instances where such control is necessary, but responsibility for orderly communication and relevance lies with each individual member. The chairperson does not need to intervene if every member adopts the self-discipline needed to achieve an effective meeting. It is easy to get off track when many people have their own contribution, so it is necessary to outline a plan at the beginning of the meeting to which everyone agrees. Everyone should have the opportunity to contribute to the discussion, which may mean that some people will have to be restrained while others may need to be encouraged.

During the meeting, it is important to make sure that:

- everyone sits where they can see each other
- everyone has a place to put forward their agenda and take notes
- everyone has an opportunity to contribute
- no one can dominate the meeting and control the outcomes.

Effective participation

If the meeting or committee is to function effectively, the following areas will need to be addressed:

- The aims and objectives should be clearly formulated and understood. They must also be accepted and applied by all members.
- Discussion should be relevant, with virtually everybody contributing.

- Everyone should be prepared to listen to each other and consider the points made. Members should not be afraid to put forward views and ideas.
- The atmosphere should be informed and relaxed, with all members of the group being involved and interested.
- Only corrective feedback should be offered and accepted.
- Disagreements should be dealt with openly in an orderly manner and examined to find a compromise.
- Members should feel free to express their own feelings and attitudes towards a problem.
- Decisions should be arrived at by consensus and individuals should not be afraid to disagree and should be given fair consideration.
- Decisions and follow-up action should be determined, with everyone fully aware of what has been agreed upon. Jobs should be allocated clearly and appropriately.
- The chairperson should not dominate, while the members should not defer unquestioningly to that person.
- The group should be aware of and be prepared to discuss its own deficiencies.

Limiting factors

Things that may limit a person's performance at a meeting or within a committee are:

- experience
- knowledge
- fear
- behaviour
- dominance.

People's lack of experience and knowledge of how meetings and committees operate may inhibit their performance. Also, fear of conflict, hidden hostilities and underlying personal factors are difficult to overcome.

Another inhibiting factor in the success and effectiveness of a meeting is the misconception that the effectiveness of a group rests solely with the leader. The greatest danger of all is that the person with the loudest voice or strongest personality will become the chairperson and completely dominate the meeting, so that those who have relevant knowledge and opinions are not heard.

Outcomes of the meeting

Because meetings are often called to resolve problems or deal with difficult situations, you are likely to be required to help resolve a conflict. In such a situation, the most common approaches are to:

- withdraw from the situation, which allows others to win, and, because the conflict is not resolved, may allow it to grow out of control
- suppress your feelings and refuse to acknowledge the problem, which does not allow others to recognise your feelings and have the opportunity to behave differently
- compromise, which can lead to dishonesty and either haggling or exaggerated claims
- confront the other, which can lead to win-lose ego-fired battles of will, which have nothing to do with the pros or cons of the issue at hand.

These approaches of response can lead to a win-lose situation. In time, successive win-lose situations can produce a culture of people retaliating, where it is more important to win and get even than to solve a problem. This leads to a breakdown of harmonious and cooperative work relationships and the creation of an unhappy, unsatisfactory and unproductive workplace.

The ideal response to conflict is the win-win solution, which is one that meets everyone's needs. Not only do we get what we want but all involved also get what they want. We give up trying to persuade or convince others that we are right, and we give up trying to destroy the argument that they are using to try to convince us that they are right. We set out to cooperate and to find a solution that will benefit everyone.

LEARNING TASK 3.5

PARTICIPATE IN WORKPLACE MEETINGS – TOOLBOX TALKS

Create a toolbox talk meeting scenario with the class and discuss the WHS and work requirements for a practical activity to be undertaken in the practical workshop.

One person is to be nominated to be chairperson and a second to take minutes. The meeting should take place in an environment where members are seen and heard. Each member is to contribute to the meeting and discussion should be relevant to the suggested activity.

COMPLETE WORKSHEET 4

Observe sustainability principles when preparing for and undertaking work processes

The plumber's role in sustainability and energy awareness is increasing dramatically. Those who are well informed in these areas will be able to provide current and relevant information to their clients and will therefore present a much more professional approach.

Building sustainability and energy efficiency ratings

Buildings in Australia are being designed and built to increasingly stringent guidelines. This is having a significant impact on reducing our environmental footprint through innovative design and reduction in use of critical resources such as water, electricity and raw materials. The Building Code of Australia (BCA) stipulates building sustainability and energy efficiency ratings for housing and should be used to ensure compliance when undertaking projects.

GREEN TIP

Having a broad-based knowledge on sustainable plumbing techniques and products may give you a competitive edge over other plumbers in the workplace.

More information relating to these topics can be addressed on the following websites:

- Australian Building Codes Board (ABCB): http://www.abcb.gov.au
- Nationwide House Energy Rating Scheme (NatHERS): http://www.nathers.gov.au.

The latter website will refer you to the various state- and territory-specific websites (e.g. NSW: http://www.basix.nsw.gov.au).

Building Sustainability Index (BASIX)

BASIX, the Building Sustainability Index, is a web-based tool developed by the NSW Government to assess the potential performance of new homes against a range of sustainability indices including stormwater, landscape, water, energy and thermal comfort. Since 1 July 2004, all new homes in NSW are required to have a BASIX certificate.

BASIX requires new residential buildings located in NSW to have a potable water savings target ranging between a 0% and 40% reduction, and an energy savings target ranging from 10% to 50% of the average existing NSW benchmark homes of the same type prior to 2004. Regional areas of NSW have varied targets based on the location and local conditions. Water conservation measures include installation of rainwater tanks and 'grey' and 'black' water treatment devices in homes. Energy efficiency can also be improved in new buildings through the installation of new technologies to meet percentage targets.

How to minimise waste

When renovating or demolishing a house, or constructing it from new, a large quantity of waste is often generated, but approximately 80% of this waste material has the potential to be reused or recycled by adopting thoughtful processes and procedures. In doing so, vast quantities of energy, water, resources and, of course, money can be saved.

GREEN TIP

Always look at reusing materials where possible and recycle any materials that cannot be reused.

Averaged nationally, each Australian produces in excess of one and a half tonnes of what would be classified as 'waste' annually, with approximately 40% of these waste products resulting from renovation, demolition and/or new construction activities. It would be safe to suggest that a major percentage of this specifically discarded construction waste, if managed carefully, could be segregated for future recycling and reuse.

By recycling we are saving the Earth's resources, which are the raw materials of all the products we buy, such as minerals, oil, petroleum, plants, soil and water. We also reduce our consumption of energy, limit pollution and lessen global warming by cutting down on the harvesting, construction, transportation and distribution of new products.

Recycling one tonne of aluminium saves four tonnes of bauxite and 1000 L of petroleum. It also prevents the associated emissions (which would include 35 kg of the toxic air pollutant aluminium fluoride) from entering our air. In addition, it would reduce the load on our already stretched landfill sites.

In order to minimise waste, follow the waste minimisation hierarchy – avoid, reduce, reuse and recycle – in that order.

1. *Avoid*. Building and construction waste often enters our waterways through stormwater drains, which is a major cause of water pollution. A typical instance of this could be a stockpile of sand. If not properly stored or sensibly located, a sudden downpour of rain could see the sand washed away into the stormwater system. Ensure that you do not contribute to the pollution of our waterways by adopting a careless attitude when undertaking the simple task of storing building products on a construction site. The two obvious advantages gained are:
   - minimal damage to our water system as a result of unnecessary contamination
   - minimal replacement of building materials already purchased.
2. *Reduce*. The general trend these days is to build from boundary to boundary. Over time the typical house in Australia has evolved from having three bedrooms, one bathroom and separate living areas into a more open plan design, including extra bedrooms and en suites. All this has had obvious consequences, notably the overall increase in size of the modern home. In fact, according to the Australian Bureau of Statistics, nationally since the mid-1980s, residential houses have grown by

approximately 40% (from 162.2 m^2 to 227.6 m^2). This simply equates to the fact that 40% more energy, resources and materials, across the board, are required to construct a house today than were required in the mid-1980s.

You can considerably reduce waste by planning carefully and sensibly. Determine exactly what it is that you require within your own residential home. Don't overindulge, and design to accommodate standard sizes of materials and use prefabricated frames and trusses, as these are proven ways of reducing waste. Reducing the size of your construction will see a reduction in the use of energy, resources, materials and generation of waste.

3 *Reuse*. Make a commitment to use, wherever possible, recycled materials. The following items are the most common recyclable materials and are therefore generally easy to locate:
   - steel
   - recycled or plantation timber
   - recycled concrete
   - second-hand bricks
   - soil and fill.

   Remember that there are many fittings and fixtures (such as doors and windows) that are also available second hand. A list of recycled building products and/or suppliers can be sourced by visiting online services.

   Buying recycled products increases the market for them, making it more viable for businesses to supply them.

4 *Recycle*. Some materials can be recycled directly into the same product for reuse. Others can be reconstituted into other usable products. Unfortunately, recycling that requires reprocessing is not usually economically viable unless a facility using recycled resources is located near the material source. Many construction waste materials that are still usable can be donated to non-profit organisations. This keeps the material out of landfill and supports a good cause.

   The most important step for recycling construction waste is on-site separation. Initially, this will take some extra effort and training of construction personnel. Once separation habits are established, on-site separation can be done at little or no additional cost.

GREEN TIP

Ensure that all waste is segregated and recycled to ensure sustainable reuse.

The initial step in a construction waste reduction strategy is good planning. Design should be based on standard sizes, and materials should be ordered accurately. Additionally, using high-quality materials such as engineered products reduces rejects. This approach can reduce the amount of material needing to be recycled and strengthen profitability and economy for the builder and customer.

The following list details the most common building products that can be recycled and suggests pathways for which they could be reused.

- Concrete:
  - un-set: used on future projects
  - set: crushed and used for future concrete works, or as road base or fill.
- Bricks and tiles:
  - cleaned and used on future projects
  - cleaned and sold on
  - crushed and used as backfill or gravel.
- Steel:
  - retained for use on future projects
  - sold on as second-hand product
  - recycled into new products.
- Aluminium products:
  - retained for use on future projects
  - sold on as second-hand product
  - recycled into new products.
- Gypsum plasterboard:
  - large sheets retained for use on future projects
  - recycled into new plasterboard product
  - used as a soil conditioner or for composting.
- timber/green waste:
  - large beams/sheets/etc. retained for use on future projects
  - reprocessed into other timber building and landscaping products
  - used as firewood (untreated materials only)
  - chipped and used as mulch either on site or at other projects.
- Plastics – recycled into new products.
- Clean fill/soil:
  - utilised in on-site landscaping
  - stockpiled for use on future projects
  - sold on as a landscaping material.
- Glass:
  - large sheets retained for use on future projects
  - crushed and used as aggregate in concrete
  - recycled into new products.
- Carpet:
  - large pieces retained for use on future projects
  - sold on as second-hand product
  - used on site to prevent erosion, dust mobilisation and weed invasion
  - shredded and used as fill in garden beds or composted (natural fibre carpets only).

Also consider the food and drink containers used by on-site workers, as one recycled aluminium can saves half a can of petroleum and 20 L of water.

You can also take the materials yourself to your local recycling centres or transfer stations. Your local council will be able to provide details of your nearest station. Your local waste facility or landfill operator might also handle some recycled products, so it might be worthwhile giving them a call.

Why is this action important?

It is possible to recycle and reuse up to 80% of demolition and construction materials. This would greatly alleviate the huge and growing pressure on the Earth's resources: on our forests, our land and our human effort. In addition, instead of carting hundreds of tonnes of material from mine to house to landfill, we would simply use what is at hand, saving transport and carbon emissions and their contribution to climate change.

Towards resource efficiency and waste minimisation

The following steps will assist in the initiation of a resource recovery program on a typical construction site.

1 Commit to responsible waste management:
   a Develop and implement a business waste minimisation policy.
   b Involve all staff members in this process.
   c Incorporate waste minimisation into position descriptions.
   d Request that subcontractors/workers sign project waste minimisation plans.
   e Incorporate waste minimisation into site induction programs.
   f Provide positive feedback to staff successfully minimising waste.
2 Identify resource pathways:
   a Review materials utilised during construction and demolition activities.
   b Assess volumes of resources currently going to waste.
   c Assess material avoidance, reduction, reuse and recycling options.
   d Determine the avoidance, reduction, reuse and recycling methods of each material.
3 Develop a project waste minimisation plan, using the following key steps:
   a Committed key field staff:
      i Ensure that all staff understand what is to be achieved.
      ii Keep staff informed of progress.
      iii Ensure the provision of up-to-date training.
   b Project-specific planning:
      i Conduct a site assessment.
      ii Prepare a project waste minimisation plan.
      iii Determine resources for the project, pathways for excess and/or waste materials.
      iv Determine the location of both waste and materials recovery stations.
      v Set targets and objectives for each project.
      vi Require subcontractors/workers to recycle and dispose of their own waste.
   c Understand options and limitations:
      i Identify collection, sorting and resource utilisation options.
      ii Determine the suitability of each option for each job.
   d Establish a monitoring and reporting program:
      i Quantify results and identify shortfalls.
      ii Provide a record for comparison across worksites and methods.
      iii Set targets and objectives.
      iv Determine financial outcomes.
   e Focus on high potential materials and practices:
      i Use materials that are high volume and can be readily separated and collected.
      ii Select those that have a viable economic value for recovery.
4 Educate staff about the waste minimisation plan:
   a Communicate with staff.
   b Inform staff of the methods and objectives of maximising resource recovery.
   c Provide copies of the project waste minimisation plan.
   d Involve staff during development and review of the project waste minimisation plan.
5 Implement waste avoidance strategies:
   a Limit the types of resources being consumed on the worksite.
   b Design works to avoid waste generation.
   c Use modular/prefabricated frames and fit outs when possible.
   d Request minimal packaging from material suppliers.
   e Ensure that materials that will generate minimal waste are used.

Note: It is recommended that readers become familiar with their relevant state or territory Environmental Protection Authority (EPA), by either contacting them directly or by downloading any relevant information from their website, as detailed in Table 3.7.

Sustainable housing

Sustainable housing is the design and construction of homes that are comfortable and practical to live in, are economical to maintain, and cause the least possible burden on the environment. This balanced approach considers the social, economic and environmental aspects of housing development, and ensures that all the key issues are considered together at the design stage.

Economically sustainable homes are cost-efficient over the lifespan of the dwelling. The design balances upfront construction and fit-out costs against ongoing running and maintenance costs. The building may be

TABLE 3.7 Current state and territory regulating environmental protection authorities

State/territory	Regulatory authority	Current EPA Act	Website and contact numbers
ACT	Environment Protection Authority	*Environment Protection Act 1997*	https://www.accesscanberra.act.gov.au 13 22 81
NSW	NSW Environment Protection Authority	*Protection of the Environment Administration Act 1991*	http://www.epa.nsw.gov.au 131 555
NT	Environmental Protection Authority	*Northern Territory Environment Protection Authority Act*	http://www.ntepa.nt.gov.au 08 8924 4218
Qld	Department of Environment and Science	*Environmental Protection Act 1994*	http://www.des.qld.gov.au 137 468
SA	Environment Protection Authority	*Environmental Protection Act 1993*	http://www.epa.sa.gov.au 08 8204 2004
Tas	Environment Protection Authority	*Environmental Management and Pollution Control Act 1994*	http://www.epa.tas.gov.au 03 6165 4599
Vic	Environment Protection Authority Victoria	*Environment Protection Act 2017*	http://www.epa.vic.gov.au 1300 372 842
WA	Environmental Protection Authority	*Environmental Protection Act 1986*	http://www.epa.wa.gov.au 08 6364 7000

constructed of low-maintenance materials and feature efficient fittings and appliances.

Environmentally sustainable homes are resource-efficient in terms of materials, waste, water and energy. They are designed for water efficiency in the house and garden, waste reduction during construction and occupancy, and energy efficiency in terms of orientation and energy consumption. An environmentally sustainable home can reduce household running costs by up to 60%, saving over three tonnes of greenhouse gases and more than 100 000 L of water a year. Environmentally sustainable elements include water-efficient and energy-efficient appliances, solar hot water, insulation and efficient lighting.

Does a sustainable home cost much extra?

It may cost more initially for a new home builder to install the following items:

- a solar or gas hot water system instead of electric
- low-flow taps and fittings
- a water tank for the garden or toilets
- a dual-flush toilet
- bulk ceiling and wall insulation.

However, in time the householder will recoup their initial outlays while enjoying greater living comfort and the knowledge they are helping the environment. At the same time, they are adding value to their home.

Sustainable alternatives

In Victoria, the Sustainability Victoria website (http://www.sustainability.vic.gov.au) contains water- and energy-saving information and rebate details for installations. For example, rebates are available for solar hot water systems, solar panels and solar batteries. Consult the website for more details.

The environmental protection agency in Queensland (http://www.des.qld.gov.au) recommends you consider the following design features, product decisions and site management practices when designing or renovating a home. These features would be advisable for use in all states and territories.

General

- An open plan and northerly orientation will maximise breezes and avoid the western sun.
- The bathroom, kitchen and laundry should be located close to the hot water system.
- Ensure living areas are positioned to capture winter sun and summer breezes.
- Plan window size, style and location to optimise protection against the summer sun and access to the winter sun.
- Minimise windows on the western side to avoid the afternoon sun.
- Use materials with low long-term maintenance costs.
- Install awnings and eaves to reduce heat.
- Install insulation in the roof, ceiling and walls.
- Consider an insulated skylight to let in natural light and not heat.
- Install compact fluorescent lighting including down-lights with efficient 12V task lighting.
- Incandescent lighting can be used for shorter-duration lighting in selected areas.
- Paint the exterior of your house and roof in a light colour to help cooling.

Kitchen

- Install double sinks so you can rinse in a second sink and not under a running tap.
- Install high star-rated water-efficient taps.
- Provide task lighting over sink, stove and work surfaces.
- Choose water-efficient and energy-efficient white goods: oven, dishwasher, refrigerator and freezer. Look for the highest star rating on water products and energy-efficient appliances.
- Place your fridge in a cool spot away from the stove and direct sunlight.
- Stove range-hoods should be vented to the outside.

Bathroom and laundry

- Use rainwater for toilet flushing.
- Grey water diversion from laundry and bathrooms may be used for the garden irrigation system (depending on state and territory legislative requirements).
- Use high-star-rated taps and shower roses for water efficiency.
- Install mixer taps in showers to reduce hot water loss while you adjust the temperature.
- Install 4.5 L/3 L dual flush toilets to reduce water use.
- Choose a high-star water-conservation-rating and high-star energy-rating front-loading washing machine.
- If you must install a clothes dryer, choose an energy-efficient one, but it's best to use the outdoor clothesline.

Finishes

- Use non-toxic paints, renders and floor finishes with either no or low VOC (volatile organic compound) emissions to give superior air quality compared to a standard house.
- Consider using floor tiles in rooms reached by the winter sun.
- Bamboo flooring is an efficient renewable resource with low VOC emissions.
- Ensure that your carpet underlay is fully recyclable, and the carpet has some natural fibre.

Hot water systems and energy supply

- Install a gas, solar or heat-pump hot water system for the greatest energy efficiency. Governments may offer rebates on selected hot water systems.
- A solar photovoltaic (PV) electricity system converts sunlight into electricity. This will eliminate electricity bills for the life of the system, and you may be able to sell any excess electricity back to the utility.

Garden and outdoor areas

- Position trees to maximise shade on your property.
- Plant local native plants in well-mulched gardens to minimise the need for external watering.
- Install an automatic underground irrigation system to minimise water use.
- Where practical, create porous surfaces outside the house to allow stormwater to soak into the soil.
- Use recycled timber for outside decking.
- Install an external clothesline.
- Use compost bins and worm farms to encourage the recycling of all food wastes.
- Use pervious materials such as rocks and pebbles for driveways and paths to slow water run-off into gutters and stormwater drains.

Rainwater tanks

- Install a rainwater tank to supply water for purposes such as toilet flushing, hot water, washing and garden irrigation. (See Standards Australia's *Rainwater tank design and installation handbook*: HB 230–2008.)

During construction

- Use renewable resources and materials with low VOCs.
- Work around established trees rather than cutting them down.
- A site management plan will help control stormwater and waste, minimise soil loss and ensure that materials are handled efficiently and that the site is clean and safe.
- Recycle construction waste where possible.
- Direct stormwater to stormwater drains, not to the sewerage system.

COMPLETE WORKSHEET 5

REFERENCES AND FURTHER READING

Acknowledgement

Reproduction of the following resource list references from DET, TAFE NSW C&T Division (Karl Dunkel, Program Manager, Housing and Furniture) and the Product Advisory Committee is acknowledged and appreciated.

Texts

Graff, D.M. & Molloy, C.J.S. (1986), *Tapping group power: A practical guide to working with groups in commerce and industry*, Synergy Systems, Dromana, Victoria.

Web-based resources

Regulations/Codes/Laws

Fair Work Ombudsman: **https://www.fairwork.gov.au**

NSW Industrial Relations: **http://www.industrialrelations.nsw.gov.au**

SafeWork NSW: **https://www.safework.nsw.gov.au**

Vocational Education and Training Reform, Department of Education, Skills and Employment: **http://www.education.gov.au/vocational-education-and-training-reform**

Resource tools and VET links

Education Network Australia: **http://www.edna.edu.au**

Training Services NSW: **http://www.training.nsw.gov.au**

Training.gov.au: **http://training.gov.au**

Industry organisations' sites

Australian Government – Department of Health: **https://www.health.gov.au**

CITB (SA Construction Industry Training Board): **https://citb.org.au**

Long Service Corporation: **https://www.longservice.nsw.gov.au/bci**

World Health Organization: **https://www.who.int**

Audiovisual resources

Videos on workplace security and safety awareness are available from Safetycare: **http://www.safetycare.com.au**.

TAFE NSW resources

Training and Education Support (TES), Industry Skills Unit, Orange/Granville 68 South St Granville NSW 2142 Ph: (02) 9846 8126 Fax: (02) 9846 8148 (Resource list and order forms).

GET IT RIGHT

The photo below shows an incorrect practice that can be performed when undertaking a meeting.

Identify the incorrect method and provide reasoning for your answer.

WORKSHEET 1

To be completed by teachers	
Satisfactory	☐
Not satisfactory	☐

Student name: ____________________

Enrolment year: ____________________

Class code: ____________________

Competency name/Number: ____________________

Task: Review the section 'Identify the industry work context and setting' in this chapter and answer the following questions.

1. What benefits do plumbers provide to public health and safety?

____

____

2. Define the term 'potable water'.

____

____

3. The building industry is divided into two main sectors: commercial and residential. List seven types of structures that sit within the residential sector.

____

____

____

____

____

____

____

4. List six possible skill areas in plumbing other than 'water services' and 'gas services' that you may become involved in.

____

____

____

____

____

____

5. List five skill set areas that a plumber requires.

6. List three areas other than welding/brazing in which a plumber should have good knowledge.

7. Who does an apprentice need to register with to enter a contract with their employer?

8. What qualification is required to receive a plumbing contractor's licence?

9. Industrial relations address many areas of employment. List six issues relating to industrial relations.

10. What is the current federal workplace relations Act?

11. Apart from setting minimum wages, state three conditions within an 'award'.

12. Give three examples of procedures used to resolve grievance disputes.

13. What group of people acts for a trade union inside and outside of the workplace?

14. List three improvements aimed at improving Vocational Education and Training to meet the needs of industry.

15. Briefly describe an 'enterprise agreement'.

16. What is the main function of a safety committee?

17. List the seven key principles of quality management.

18. What is the main function of Australian Standards?

WORKSHEET 2

To be completed by teachers	
Satisfactory	☐
Not satisfactory	☐

Student name: ______________________

Enrolment year: ______________________

Class code: ______________________

Competency name/Number: ______________________

Task: Review the section 'Organise and accept responsibility for own workload' in this chapter and answer the following questions.

1. Why should work activities be planned and time management skills be used on the worksite?

____

____

2. Define the term 'working efficiently'.

____

____

3. List the specific benefits from working efficiently.

____

____

4. Before you start a task, what two documents are you required to complete?

____

____

5. What is the purpose of a time chart?

____

____

6. List three common delays that can occur on the worksite.

____

____

____

7. Briefly describe in what situation would a type 'A' fence hoarding be used?

____

____

8. Briefly describe the purpose of a bollard.

9. Prior to excavating or digging in a new or unidentified area, what must be done to avoid possible damage to underground cables and pipes?

10. Why would it be impractical for most tradespeople to carry every tool they own to every job they attend?

11. Why is it important to check tools, equipment and PPE before you transport them?

12. Why is it necessary to list all the materials required for the job?

13. Why is it important to not order materials too early?

14. Once a material storage location is chosen, how would you determine the best way to store materials?

15. Why should problems be solved as they arise?

WORKSHEET 3

To be completed by teachers

Satisfactory ☐

Not satisfactory ☐

Student name: ______________________________

Enrolment year: ______________________________

Class code: ______________________________

Competency name/Number: ______________________________

Task: Review the section 'Work in a team' up to and including 'Participate in identifying and pursuing own development needs and processes' in this chapter and answer the following questions:

1. What is the benefit of working in a team?

____

____

2. Why should individuals within a team be allocated a job?

____

____

3. How can team roles be enhanced?

____

____

4. Why should conflicts be dealt with immediately?

____

____

5. List three potential career paths that may be obtained once a Certificate III in plumbing has been obtained.

____

____

____

6. List three individuals that you may seek to determine a career path and identify your own learning needs.

____

____

____

To be completed by teachers

Satisfactory ☐

Not satisfactory ☐

Student name: ______________________

Enrolment year: ______________________

Class code: ______________________

Competency name/Number: ______________________

Task: Review the section 'Participate in workplace meetings' in this chapter and answer the following questions.

1. List three purposes of holding a meeting.

____

____

____

2. List six types of meetings.

____

____

____

____

____

____

3. State the purpose of a toolbox talk.

____

____

4. List the two main methods used to conduct a meeting.

____

____

5. Describe the purpose of an agenda for a meeting and who prepares it.

____

____

____

____

6. Describe the purpose of minutes at a meeting and who may record them.

____

____

____

7. Circle or underline the type of society in which we have the right to free speech and an individual opinion (such as the way a site meeting should be conducted).

Autocratic Hippocratic

Democratic Hypocritical

8. State the importance of room layout during a meeting.

____

____

9. List four factors that may limit a person's performance at a meeting on a building site.

____

____

____

____

10. Name and describe the ideal outcome to a conflict.

____

____

WORKSHEET 5

To be completed by teachers	
Satisfactory	☐
Not satisfactory	☐

Student name: ______________________

Enrolment year: ______________________

Class code: ______________________

Competency name/Number: ______________________

Task: Review the section 'Observe sustainability principles when preparing for and undertaking work processes' in this chapter and answer the following questions.

1. When renovating or demolishing a house, how could 80% of construction waste material be reused?

____

2. By recycling we can save on the demands for the Earth's natural resources such as minerals, oil, petroleum, plants, soil and water. What are three other distinct advantages gained by recycling?

____

____

____

3. What are the four steps to follow in a waste minimisation hierarchy?

____

____

____

____

4. What are the 10 most common building products that can be recycled?

____

____

____

____

____

____

____

____

____

____

5. What are the five steps to follow in achieving resource efficiency and waste minimisation?

6. Define 'sustainable housing'.

7. What are three general design features that will assist in achieving a sustainable home?

4

CARRY OUT INTERACTIVE WORKPLACE COMMUNICATION

This chapter addresses the following key elements for the competency 'Carry out interactive workplace communication':

- Understand good communication.
- Apply oral communication.
- Apply visual communication.
- Apply written communication and signage.
- Understand alternative forms of communication.

It introduces the requirements for using oral, visual, written and workplace signage communication. It also addresses accepted industry practices when communicating in the workplace and the ways in which to send and receive communication successfully.

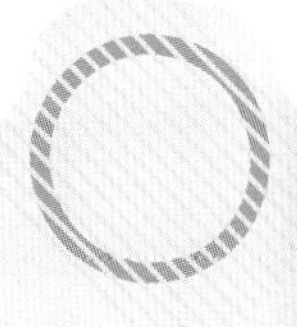

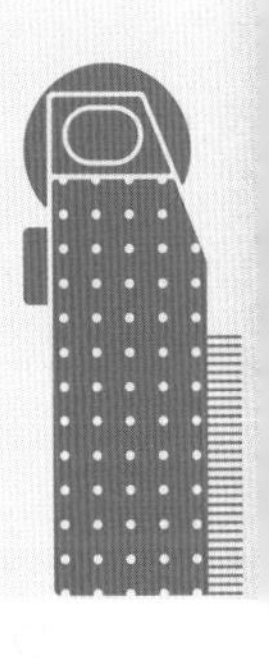

Understand good communication

Communication can be understood as the transmission of information or a message to another person or group of people. Communication takes a variety of forms, which can include writing, talking, gesturing, signalling, drawing, hugging and crying. These communication methods carry a message of some kind and are all personalised by the emotions, wants, needs or opinions of the communicator. The act of smiling and talking enthusiastically about new responsibilities in the workplace, in addition to the information you are conveying, imparts the message that you are pleased about your promotion. In contrast, a blank expression or frown accompanied by a dull unexcited tone of voice used to convey the same information would send a message of displeasure about the burden of the extra responsibility.

Communication is not only important on the worksite but is also an integral part of everyday life, and the ability to communicate effectively is equally important in both spheres.

FROM EXPERIENCE

Good communication skills are essential on plumbing and construction sites to ensure harmonious relations, which can increase productivity on site.

All forms of communication both convey and receive information. Whether the information is received and understood depends on how clearly and accurately it is given.

Oral communication in a workplace may be conducted by face-to-face discussion (see Figure 4.1), mobile telephone conversation or over a loudspeaker. Visual means of communicating on a construction site may include flashing lights, barricade tapes or hand signals. Written means of communication may include hydraulic plans and specifications, safety signage, incident reports or regulatory authorities' documentation. Selection of the appropriate means of communication is essential to ensure that the message is understood and confirmed in a clear and accurate manner.

Good communication skills

Good communication skills are important in every workplace as nobody wants to be injured or waste their time or materials. Good communication benefits everyone involved, with many acknowledging the need for good communication. However, few people ever think of themselves as being poor communicators.

So, if people think they are communicating properly, why are people being injured, and why is faulty work being undertaken? This is often because people think that if they can talk, read and write, they can communicate effectively. Unfortunately, this is not the case.

These skills are an important part of being a good communicator but are not sufficient on their own. Some people who are not expert at talking, reading and writing are excellent communicators, as they understand that in order to communicate they must create understanding in the mind of the person they are communicating with. No matter how good you are at talking, reading or writing, you will not get your message across if you are not creating a clear understanding.

Good communicators create a clear understanding by carefully considering the context in which the communication will take place, applying the most appropriate mode of communication and identifying any influences that may affect the message, such as personal emotions or physical barriers.

Barriers to communication

Unfortunately, successful communication is not as simple as Figure 4.2 implies, and there are many barriers to effective communication. The inability to read technical plans and specifications correctly can lead to expensive mistakes, while poor communication could lead to WHS issues. These barriers to effective

FIGURE 4.1 Communicating by talking

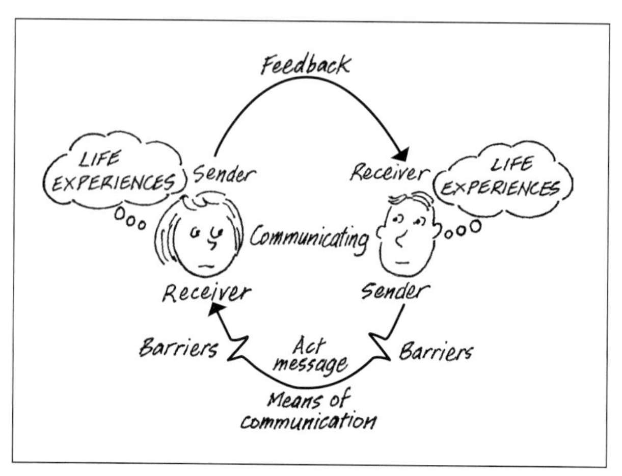

FIGURE 4.2 Communicating = creating understanding

communication can exist in the environment in which the people are communicating, in the system they are communicating within, or with the people they are trying to communicate with.

Barriers to communication can be conveniently categorised as physical, emotional, psychological, knowledge-based and diversity-related.

Physical barriers

Physical barriers in the environment or the workplace, such as loud machinery noise or obscured safety signage, can affect clear communication. In a telephone or two-way radio communication system, interference (static noise) can interrupt contact. In people, disabilities such as deafness, blindness or failing eyesight can create a barrier to communication.

Physical barriers such as excessive noise from heavy machinery can reduce effective communication on site and can have a negative impact on the safety of yourself and those around you.

Emotional barriers

Emotional barriers occur in people and can include anger, resentment, frustration, love and affection. The presence of these emotions in the sender can override the content of the message and affect good communication. The presence of these emotions in the receiver can distort the perception of the message.

Psychological barriers

Psychological barriers also occur in people and can influence the communication process. If people are continually aggressive, arrogant, judgemental, manipulative, confrontational or guilt-inducing in their communication behaviour, they are unlikely to encourage the feedback that is essential to successful communication. Similarly, people who avoid, refuse to acknowledge or divert communication messages are avoiding communicating honestly about an issue.

Knowledge-based barriers

Knowledge-based barriers may limit successful communication. A person's lack of knowledge can limit their effective use of complicated communication systems such as computers and new technologies. Not understanding the language of signs or instructions might jeopardise a person's safety. The lack of a shared language could cause serious misunderstanding between a sender and a receiver. In the workplace, the use of language that is too technical for beginners, or the use of jargon with the inexperienced, can lead to problems.

Consideration should also be given for people from a range of varying cognitive abilities.

Diversity-based barriers

Diversity-based barriers may be present in effective communication as there are many diverse backgrounds globally. Language and concepts should be appropriate to people from a range of cultural backgrounds and may include languages other than English.

People who understand the benefit of clear communication anticipate the potential barriers affecting their message and alter the communication to suit the receiver within the environment.

FROM EXPERIENCE

Learning and adapting to new technologies can assist in carrying out tasks effectively.

Apply oral communication

Oral communication is a two-way process between a sender and a receiver (see Figure 4.3). The message must be received as it was intended for the communication to be deemed successful. As senders we tend to focus exclusively on the message we wish to convey. The receiver of your information may understand only a portion of your message or miss your point entirely. This message may be, in your opinion, completely transparent.

Feedback of understanding

Sender (speaker)

Receiver (listener)

FIGURE 4.3 The sender/receiver method of communication

Everyone has different life experiences that shape their character, personality and view of the world, and this affects the way everyone decodes a message from someone else.

Both the sender and receiver have the responsibility for successful communication and must work together to create understanding. This can be achieved by checking that the message received is the same as the message sent, by seeking and giving feedback (see Figure 4.4).

The sender checks by:

- asking for the message to be repeated
- asking questions
- asking for the message to be clarified or confirmed.

FIGURE 4.4 The feedback process

The receiver checks by:

- repeating the message
- asking questions
- restating the message.

To make sure that you have overcome the barriers to communication, always seek and offer feedback.

Communication is a circular process that requires checking and rechecking information to ensure that the message received matches the message sent. Without feedback, failure to communicate is highly likely.

Forms of oral communication

Oral communication, which is also referred to as spoken language, may be:

- face-to-face
- via a telephone, either a landline or mobile telephone
- via a two-way radio
- via web conferencing.

Oral communication face to face

When you are being spoken to face to face, always listen actively by:

- looking at the person talking to you
- maintaining an attentive posture
- showing interest
- encouraging the speaker
- asking questions
- summarising to check your understanding.

Be aware of the barriers to communication that cause misunderstanding and make sure they do not interfere with the message. Always speak with a civil tone and treat people with politeness and respect.

When speaking to someone face to face, make sure your message is clear and concise:

- Make 'eye contact' to ensure that the receiver is listening.
- Speak clearly and don't mumble.
- Choose your words carefully, focusing upon the positive outcome you want to achieve.
- Use the right tone of voice and make your body language match your words. Often, it's not what you are saying but how you say it that transfers the message. If your tone does not match your meaning it can be confusing for the receiver.
- Always treat people with politeness and respect. Make sure you:
- state the overall goal you require
- describe the main steps in the task in a logical order
- explain the details of each step slowly
- emphasise the critical points
- seek and offer feedback to check the receiver's understanding
- summarise the main steps of the task in a logical order.

Oral communication via a telephone

When you answer a business telephone:

- Give a polite greeting.
- State your company name.
- State your name.
- Offer assistance.

If you are not able to help the caller:

- Write down the caller's name and where they are from.
- Write down the caller's number and message.
- Repeat the message to the caller.
- Write down the time and date.
- Give a polite farewell.
- Write down your name.
- Deliver the message responsibly (see Figure 4.5).

Message

To:
Mr/Mrs/Ms ....
Date .... Time ....

From:
Mr/Mrs/Ms ....
Of ....
Phone no. ....

☐ Please phone ☐ Urgent
☐ Wants to see you ☐ Will phone again
☐ No message ☐ Returned your call

Message:
....
....
....
....
....
....
....
....
....
....
....

FIGURE 4.5 Record accurate messages

Remember to speak at a moderate pace. The caller may have never spoken to you before, or may not speak English as a first language, and may need time to tune into your way of speaking.

Oral communication via a two-way radio

When you are using a two-way radio:

- Use an individual identity or call sign to identify yourself.
- Say 'Over' to indicate you have finished speaking so the other person can reply.
- Turn your microphone off after saying 'Over' or you will not be able to hear the other person.
- Spell out important words using the international alphabet: **A**lpha, **B**ravo, **C**harlie, **D**elta etc.

Speak clearly and at a moderate pace and remember that radio frequencies are public, so be careful about what you say.

Oral communication via a web conference

Web conferencing is a form of oral communication that is similar to face-to-face communication. A web conference connects two or more people through their phone or computer with the convenience of not having to meet in person, while still having the ability to see each other through their screen. Although the people within the conference are in a different location or on the other side of the world, the same guidelines and etiquette of face-to-face communication should be followed.

Be aware that in the construction and plumbing industry you may have to combine spoken language, technical language, written language and diagrams to make your instructions clear and understandable to the person you are trying to communicate with.

LEARNING TASK 4.1

APPLY ORAL COMMUNICATION

Collate a material list for a water rough-in located in a new residential home. Through oral communication via a telephone, contact a plumbing supplier to obtain a quote for the materials required. This will allow for practice in telephone technique in receiving and conveying information.

Note: A similar project or material list could be used that is less complex and requires less materials.

COMPLETE WORKSHEET 1

Apply visual communication

Visual communication is an important process on the construction site, as barriers such as noise and distance between the sender and receiver can limit or prevent effective oral communication. Successful visual communication can at times be challenging, so it is essential to obtain the receiver's attention, acknowledgement and confirmation of the communication that is sent. Where possible, a set of communication signals or gestures should be decided upon before communication takes place to prevent unnecessary mistakes or injuries from occurring. Unclear and ambiguous visual communication can be unhelpful and dangerous, and if unclear communication has occurred, a review should be followed to avoid problems in the future.

Forms of visual communication

Bodily visual communication can take several forms:

- body language
- hand signals
- gestures (see Figure 4.6)
- facial expressions
- movements
- posture, colour and breathing.

FIGURE 4.6 Communicating using hand gestures and facial expression

Body language

Be alert to the messages conveyed by people's body language and ensure your body language is clear and concise. It is very easy to confuse others with an inaccurate or ambiguous gesture. In your normal interpersonal communication, make sure that your body language and facial expressions are consistent with your spoken language to prevent sending conflicting messages.

Hand signals

Specific hand signals are used to guide:

- crane operators
- excavator operators
- surveyors (see Figure 4.7)
- truck drivers.

FIGURE 4.7 Communication using hand signals

Learn and practise the hand signals commonly used in the plumbing and construction industry.

Your hand raised with your palm outwards will usually mean 'Stop where you are' (see Figure 4.8).

FIGURE 4.8 Telling others to stop

Your hand and forearm rotated in a circular movement towards your chest will usually mean 'To me or towards me' (see Figure 4.9).

A forward pushing movement with your hands, palms outwards in front of your chest, will usually mean 'Move away from me' (see Figure 4.10).

Your right arm extended to the right with the hand and forearm swinging in a horizontal arc, gesturing to the right with your index finger pointing to the right, usually means 'Move to the right'. Your left arm extended to the left with the hand and forearm swinging in a horizontal arc, gesturing to the left with your left finger pointing to the left, usually means 'Move to the left' (see Figure 4.11).

FIGURE 4.9 Telling others to move forward

FIGURE 4.10 Telling others to move away

FIGURE 4.11 Telling others to move left or right

Crossing and re-crossing your right and left hands and forearms horizontally in swinging movements in front of your chest usually means 'Cease what you are doing' (see Figure 4.12).

Remember, you can use head and body movements to get help, warn others and facilitate teamwork.

FIGURE 4.12 Telling others to cease what they are doing

All the channels of communication are important. We use some more often than others, which makes us more familiar with them. But we cannot afford to ignore the channels we are not familiar with. If we need to get our message across, we must choose channels that will give us the greatest chance of success. If necessary, decide on a set of gestures or guidelines between yourself the sender, and the receiver to ensure successful communication is achieved.

Gestures

Other workers may use hand or body gestures to tell you to:

- 'come and help quickly'
- 'come over here'
- 'go over there'
- get out of the way quickly
- lift or lower something as a team
- stay where you are
- watch out above you
- watch out behind you
- watch out below you
- watch out for danger (see Figure 4.13).

Facial expressions

Facial expressions may tell you if another worker is:

- angry
- concentrating on the task at hand (see Figure 4.14)
- confused
- frightened
- happy
- having difficulty
- ill
- in pain.

Movements

Workers' body movements may indicate that they are in trouble or are struggling and need assistance. Their movement could also indicate that an emergency is unfolding.

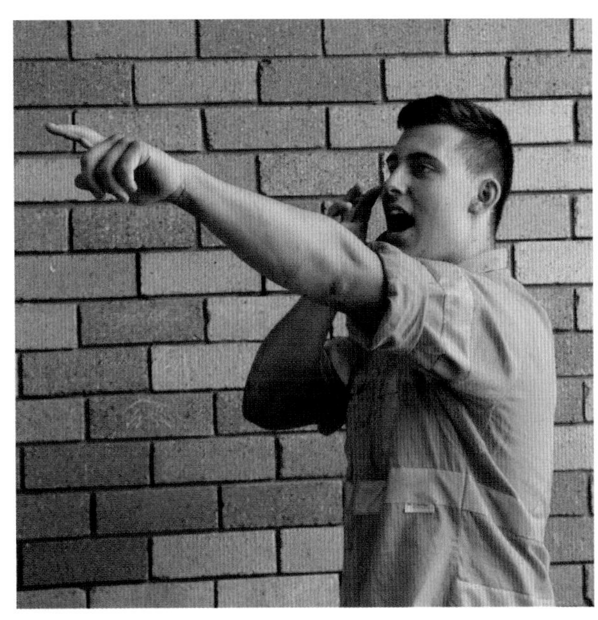

FIGURE 4.13 Warning others of danger

FIGURE 4.14 Communication using facial expressions

Posture, colour and breathing

Posture, colour and breathing may indicate:

- asphyxiation
- dehydration
- illness (see Figure 4.15)
- injury
- intoxication
- poisoning
- poisonous bite/sting
- sunburn.

Body language and posture can indicate when a worker is unwell or hurt. A worker who is staggering may need immediate assistance.

Barricades, warning tapes, barricade mesh and bollards

Barricades and warning tapes give a visual indication of areas that require limited access to authorised personnel only on a construction site. In addition,

FIGURE 4.15 Communication through body posture

coloured warning tapes where required are laid on top of existing services in the ground and give a visual and early indication of the presence of services for when excavation is undertaken.

Barricade tape is a roll of coloured polyethylene plastic tape (see Figure 4.16) printed with a warning message that can be quickly run out around large areas. The tape can be easily be tied or wrapped around posts, poles, stakes, railings or any convenient support, then rolled up again at the end of the job for later reuse. Barricade tape is manufactured in many different colours, styles, lengths and widths to suit specific requirements. Common choices in industry are yellow and black diagonal stripes or red and white diagonal stripes, which come with a variety of different warning messages in contrasting letters.

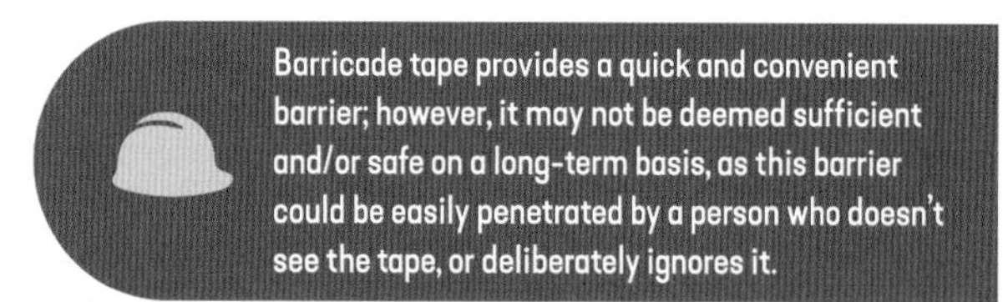

Barricade tape provides a quick and convenient barrier; however, it may not be deemed sufficient and/or safe on a long-term basis, as this barrier could be easily penetrated by a person who doesn't see the tape, or deliberately ignores it.

FIGURE 4.16 Barricade tape

Barricade mesh is a brightly coloured safety fence (see Figure 4.17) constructed from woven polyethylene that is generally installed around a series of star pickets to prevent pedestrian entry to building sites or excavations.

FIGURE 4.17 Barricade mesh

Bollards are generally manufactured from moulded fluorescent orange polyethylene (see Figure 4.18) incorporating a weighted baseplate to keep them in a vertical position. Safety bollards or cones are generally used to create a traffic diversion or for pedestrian control.

FIGURE 4.18 Bollards

Underground warning tapes

Underground pipes and cables are susceptible to damage during excavation, so it is important to be able to identify their exact position in the ground. Although above-ground signs can give some idea as to the approximate location of underground utility lines, they do not accurately indicate these along their entire length. An underground warning tape (see Figure 4.19) buried above a pipe or cable will give a visual indication of the installation at every point along its length.

FIGURE 4.19 Underground warning tape

Underground warning tapes are plastic tapes that are usually buried 150 mm above and along the length of the underground installation and must comply with *AS/NZS 2648.1 Underground marking tape – Non-detectable tape*. Generally printed with the name of the installation, the warning tape will provide an excavating machine operator or plumber with sufficient visual warning of the existence of the utility line below to avoid expensive damage and service interruption.

4.1 STANDARDS

- AS/NZS 2648.1 Underground marking tape – Non-detectable tape

There are two types of underground warning tapes available. The first type is a colour-coded plastic tape for visual identification of the installation. The second type is a detectable warning tape and is for use with non-metallic installations such as plastic or concrete pipes. The warning tape is traced with a detector from above ground due to being embedded with stainless steel trace wire. This tape is also brightly coloured for easy visual identification should a detector not be utilised.

Common colour-codes that may be encountered for different utilities are outlined in Table 4.1.

TABLE 4.1 Colour-codes for underground warning tape

Communications	White
Electricity	Orange
Fire	Red
Gas	Yellow
Potable water	Green
Recycled water	Purple/lilac
Sewer	Cream/beige
Stormwater	Blue

A close watch must always be kept for any signs of buried installations during excavation, with a good indicator being the presence of embedment material or sand surrounding existing services. Warning tape must always be reinstated during excavation to ensure that any workers in the future are able to detect and avoid damage to existing services. Damage to underground services is not only likely to be expensive to repair and cause great inconvenience to customers but also may jeopardise the health and safety of the workers involved if underground electricity, fibre optic cable, gas or sewer lines are cut.

Always remember to contact Dial Before You Dig (call 1100 or go online at http://www.1100.com.au) to locate underground pipes and cables.

LEARNING TASK 4.2

APPLY VISUAL COMMUNICATION

Working in pairs, select a basic plumbing task and create a step-by-step sequence of visual communication techniques to complete the task from start to finish. The sender is to communicate the process to the receiver to see whether the communication techniques are clear and easy to follow. The receiver may offer visual feedback to confirm or ask for more information.

For example, a task could be for the sender to communicate to the receiver to excavate a trench to a certain depth and length using only hand signals.

COMPLETE WORKSHEET 2

Apply written communication and signage

Written communication in the workplace is accessed daily, as we are required to read plans and specifications, record and report workplace documentation, and read technical instructions for job processes and procedures. Before work is to be undertaken, workplace documentation such as a risk assessment or a Safe Work Method Statement (SWMS) should be completed in accordance with regulatory authorities' requirements. Workplaces often require workers to record job details and times to ensure workers receive payments and clients are invoiced correctly. When written documents are used as a communication method it is important that the correct message is interpreted and received as it was intended.

Forms of written communication include:

- applications (apps) on smartphones and tablets
- bill of quantities
- circulars
- contracts
- delivery dockets
- emails
- plans and specifications (see Figure 4.20)
- information bulletins
- invoices
- letters
- plumbing standards
- quotations
- reports
- Safe Operating Procedures (SOP)
- Safe Work Method Statements (SWMS)
- Safety Data Sheets (SDS)
- text messages (SMS)
- social media
- technical instructions
- tender documents
- websites.

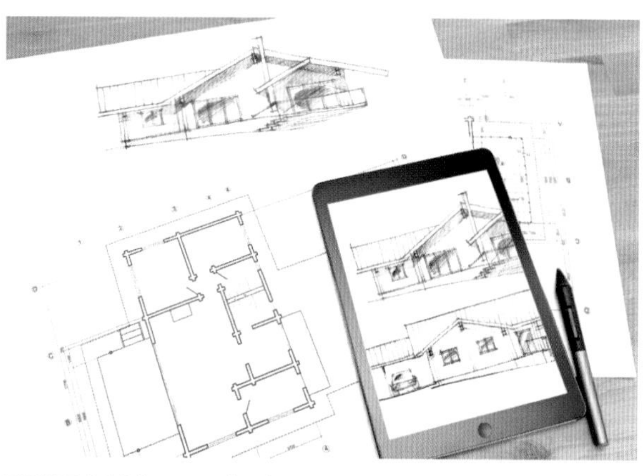

FIGURE 4.20 Communicating through plans and specifications

Source: Shutterstock.com/Mr Twister.

FROM EXPERIENCE

New technologies such as smartphones, tablets and social media platforms allow information to be accessed quickly and made available to a larger proportion of people. This information can range from WHS requirements to technical information from manufacturers, right at your fingertips.

Reading written communication

When you are seeking information from a document, you can:

- predict the content from the title or pictures
- skim quickly through the document to get a broad idea of the content
- scan the document to find a specific piece of information
- read for the main ideas
- read in detail for deep understanding.

How you read will depend on your purpose and it may not be necessary to read in detail for every document. Choose a way of reading to suit your purpose for the type of document you need to read.

Writing to be understood

When you write, keep in mind that you are writing to be understood. If your readers fail to get your message you have wasted your time and theirs. Your objective must be to get your message across as clearly and quickly as possible. You do this by writing in plain English.

Write your document:

- in a logical order, from the least important point to the most important point, or vice versa
- in an active voice ('I need more tap washers') not a passive voice ('More tap washers are needed')
- in short sentences, with one main idea to a paragraph.

Avoid being misunderstood

Ensure your message is clearly understood by avoiding the following:

- Ambiguities: 'I put the pipe bender in the truck that had been damaged'.
- Clichés: 'Someone spat the dummy' when you mean 'Someone became upset', or 'down the track', when you mean 'in the future'.
- Jargon: 'Optimise output by maximising labour input' when you mean 'Achieve more by working harder'.
- Slang: 'mud' when you mean 'mortar', or 'eel' when you mean 'drain cleaning machine'.
- Technical terms, when you are writing for non-technical readers: 'Place a 100 mm, 20 MPa slab with F72 mesh' when you just mean 'Pour a concrete plinth'.

Jargon, slang and technical terms may also be referred to as 'metalanguage', and if used with people new to the industry or not familiar with them, these terms may be confusing, and their meaning may be lost.

Note: The points listed above can also be used in context for oral communication methods; instead of writing to be understood, you would be talking to be understood.

Sketches and drawings

Sketches and drawings are an effective method of communicating a large amount of information quickly when working on site.

When a sketch is required, a piece of plasterboard, paper or cardboard torn from a carton will provide a handy drawing medium (see Figure 4.21). It is necessary to have both the ability to draw and read sketches, which is much easier than giving long and involved verbal instructions.

When working on site, supervisors, leading hands and other tradespeople will often do a quick sketch to show what needs to be done because it is quicker,

FIGURE 4.21 Communication by sketching

easier, more accurate and more easily understood than a spoken description.

Diagrams

Diagrams that contain relevant information will reduce conflict resulting from misunderstandings.

Diagrams may appear in:

- dangerous goods labels
- instructions for use
- safety tags
- signs (see Figure 4.22).

FIGURE 4.22 Communication using pictograms

Safety signage in the workplace

Safety signs in the workplace place have many purposes and communicate messages of health and safety. They are designed to prevent injury and warn others of significant risk and harm, and in many cases are mandatory in specific areas of the workplace.

In all categories, signs may be designed as either symbolic signs, text signs or a combination of the two. *Symbolic* signs can be circular, triangular, square or rectangular in shape, and carry a symbol or picture that is intended to carry a message without the use of words. *Text* signs also come in different shapes and contain only words to convey the message simply and accurately.

Sign manufacturers have a large range of signs in their catalogues, and new signs can be prepared to suit any situation or circumstance for which an existing sign is not suitable. New signs must always be designed and manufactured according to *AS 1319 Safety signs for the occupational environment*.

- AS 1319 Safety signs for the occupational environment

Safety signs in the plumbing industry fall into four categories – regulatory, hazard, emergency information, and fire – and are used according to their functions, as outlined in Table 4.2.

General information signs

In addition to the categories listed Table 4.2, many workplaces display *general information* signs that communicate information applicable to the workplace.

TABLE 4.2 Safety signage categories

Sign type	Explanation	Visual example
Regulatory signs contain instructions that control or limit certain actions.	Regulatory signs are subdivided into three categories: • *Prohibition signs* indicate the action identified on the sign is not permitted. Prohibition signs have a red border and crossbar overlaying a black symbol on a white background (see Figure 4.23). • *Mandatory signs* indicate the action identified on the sign must be carried out. Signs have a white symbol overlaying a blue background and are symbolic, although sometimes text may be added (see Figure 4.24). • *Limitation or restriction signs* indicate a defined or numerical limit on an activity, particularly speed.	FIGURE 4.23 Prohibition signs FIGURE 4.24 Mandatory symbolic signs

TABLE 4.2 (*Continued*)

Sign type	Explanation	Visual example
Hazard signs advise of hazards.	Hazard signs are subdivided into two categories: • *Danger signs* warn of a hazardous condition or hazard that is potentially life threatening. Signs have the word 'DANGER' in white letters on a red oval overlaying a black panel. The message text of the sign is in black letters on a white background (see **Figure 4.25**). • *Warning signs* warn of a hazardous condition or hazard that is not likely life threatening. Signs have a yellow background overlaid with a black symbol and a black border that may sometimes have text added (see **Figure 4.26**).	 **FIGURE 4.25** Danger signs  **FIGURE 4.26** Warning symbolic signs
Emergency information signs carry information that indicates the location or directions for emergency-related information.	Signs may include emergency exits, first aid facilities, emergency eye washes, emergency showers and safety equipment. Emergency information signs have a white symbol overlaying a green background and may have text added (see **Figures 4.27** and **4.28**).	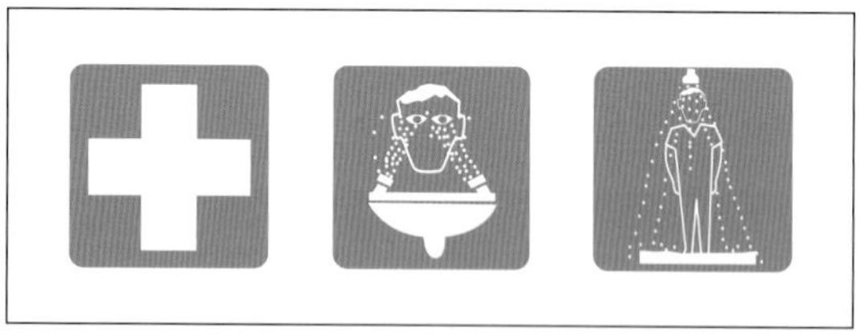 **FIGURE 4.27** Emergency information symbolic signs 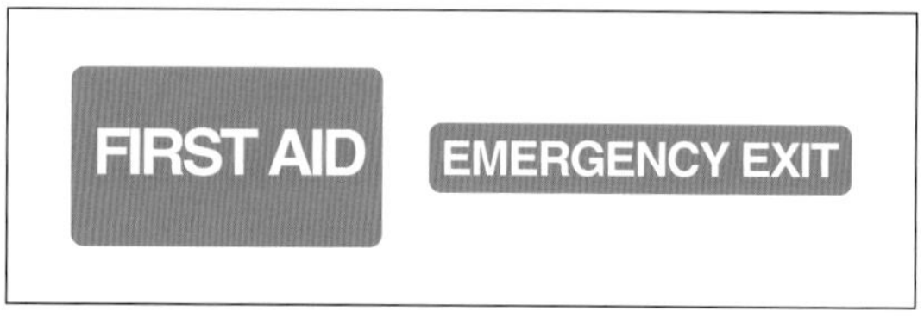 **FIGURE 4.28** Emergency information text signs
Fire signs advise the location of fire-fighting facilities and fire alarms.	Signs have white letters on a red background (see **Figure 4.29**). Fire text signs for fire extinguishers carry different colour codes that indicate the type of fire for which the extinguisher is intended. • *Fire instruction symbolic signs* have a white symbol overlaying a red background (see **Figure 4.30**).	 **FIGURE 4.29** Fire text signs  **FIGURE 4.30** Fire symbolic signs

A common example is the *caution text* sign, which displays the word 'caution' in yellow letters on a black background (see Figure 4.31). The message text of the sign is displayed in black letters on a yellow background.

FIGURE 4.31 Caution text signs

Common signs on site

Following is a list of many of the signs that may be encountered in the plumbing and construction industry.

Note: The wording may vary from manufacturer to manufacturer, although the intent of the wording must remain the same.

Regulatory – prohibition signs

- No naked flame (symbolic)
- No pedestrian thoroughfare (symbolic)
- No smoking (symbolic)
- Water not fit to drink (symbolic)

Regulatory – mandatory signs

- Use protective equipment
- Wear eye protection
- Wear face protection
- Wear foot protection
- Wear hand protection
- Wear head protection
- Wear hearing protection
- Wear protective clothing
- Wear respiratory protection

Hazard – danger signs

- Danger – asbestos dust or removal
- Danger – authorised personnel only
- Danger – buried cable
- Danger – construction site, unauthorised persons keep out
- Danger – crane working overhead, keep clear
- Danger – deep excavation
- Danger – demolition work in progress
- Danger – explosives
- Danger – flammable materials
- Danger – hard hat area
- Danger – high voltage
- Danger – keep hands clear of moving machinery
- Danger – keep out
- Danger – live wires
- Danger – power tools in use
- Danger – toxic material
- Danger – underground cable

Hazard – warning signs

- Biological material hazard (symbolic)
- Caution – buried cable
- Caution – do not watch welding arc, protect your eyes
- Caution – explosive power tools in use
- Caution – flammable liquid
- Caution – highly flammable
- Corrosive material risk (symbolic)
- Electric shock risk (symbolic)
- Explosion risk (symbolic)
- Fire risk (symbolic)
- Ionising radiation risk (symbolic)
- Laser beam in use (symbolic)
- Toxic material risk (symbolic)

Emergency information signs

- Emergency direction-indicating arrows (symbolic)
- Emergency exit
- Emergency eye wash (symbolic)
- Emergency shower (symbolic)
- Emergency use only equipment
- First aid (symbolic)
- First aid equipment
- First aid room
- First aid station
- Stretchers

Fire signs

- Fire alarm (symbolic)
- Fire bucket
- Fire door
- Fire equipment
- Fire escape
- Fire exit
- Fire extinguisher (various classes)
- Fire hose (symbolic)
- Fire hydrant
- Sand for fire only
- Sprinkler valve

General information signs

- Caution – beware of crane
- Caution – beware of hoist
- Caution – beware of traffic
- Caution – crane working overhead
- Caution – keep clear
- Caution – low clearance
- Caution – people working below
- Caution – respirator required
- Caution – slippery underfoot
- Caution – watch your step
- General risk of danger (symbolic)

Placement and location of signs

It is important to remember not to place or erect a sign in such a way as to create a hazard or cause an obstruction. An injury may be caused by an insecurely fixed sign being knocked over onto someone. A badly placed sign may obscure someone's vision at a critical moment, which may result in creating an accident rather than preventing it. A sign that protrudes into either pedestrian or vehicular traffic is likely to be run into by someone or something.

It is equally important to remove signs that are no longer appropriate as soon as the warning or information set out on a sign is no longer relevant. This is particularly important where a sign has been used to warn of a temporary danger, which may include demolition work in progress, explosive power tools in use or people working above. Failure to remove signs appropriately may encourage a tendency to disregard signs.

Generally, the best height for signs is at approximately 1500 mm, which is the line of sight and eye level for most adults (see Figure 4.32). They should be placed in such a way that they are clearly visible and are unlikely to be covered by stacks of material, machines or doors and windows. Sign placement at heights other than eye level will vary and may be more suitable, as there could be large stationary objects or moving machinery that could block clear view at eye level.

FIGURE 4.32 Place signs just above average eye-height

Signs that are for regulatory, mandatory and hazard purposes should be located as close as possible to the hazard that the sign is drawing attention to. These signs should be placed at an appropriate distance and location to the hazard to allow those seeing the sign enough time to react before they are placed in danger.

Dangerous goods labels

Many of the materials stored and used on building sites are classified as dangerous goods and must meet the requirements of the *Globally Harmonised System of Classification and labelling for workplace hazardous Chemicals* (GHS). There are nine classes of dangerous goods, and each class is identifiable by specific standard labels.

1. *Class 1 – Explosive substances or articles*. There are six subdivisions in this class. They range from mass explosion hazard to may mass explode in fire.
2. *Class 2 – Gases*. There are three subdivisions in this class. The subdivisions are flammable gases, non-flammable and non-toxic gases, and toxic gases.
3. *Class 3 – Flammable liquids*.
4. *Class 4 – Flammable solids*. The subdivisions are flammable solids, spontaneous combustible substances, and substances that in contact with water emit flammable gases.
5. *Class 5 – Oxidising substances*. This class contains two subdivisions, which are oxidising agents and organic peroxides.
6. *Class 6 – Toxic substances and infectious substances*. This class contains two subdivisions, which are toxic substances and infectious substances.
7. *Class 7 – Radioactive materials*. There are no subdivisions in this class.
8. *Class 8 – Corrosive substances*. There are no subdivisions in this class.
9. *Class 9 – Miscellaneous dangerous materials*. There are no subdivisions in this class. No label is required for this class.

All dangerous goods must be stored, handled and used safely. The information necessary to do this correctly is contained in Safety Data Sheets (SDS), which are prepared by product manufacturers. SDS give information about the product's constituents, possible health effects, first aid instructions, precautions for use, and safe handling and storage.

By law, employers are required to provide employees with information about the safe use, handling and storage of dangerous substances at work. Having an SDS available for use on the job is an effective way of achieving this.

For further information on dangerous goods labels, refer to your relevant Australian state and territory WHS regulatory authority (see Table 1.1 in Chapter 1).

Understand alternative forms of communication

In addition to the three mains forms of communication, workers may also receive communication in the following ways:

- sounds
- lights
- touch sensations
- odours.

Sounds

Sounds may come from:

- approaching machinery
- bells (see Figure 4.33)
- buzzers
- collapsing scaffolding
- collapsing trenches
- horns
- other workers
- overhead cranes
- reversing machinery
- sirens
- straining machinery
- whistles.

FIGURE 4.33 Communication through sounds

Lights

Lights may come from:

- approaching machinery
- operating machinery
- overhead cranes
- reversing machinery
- rotating amber beacons (see Figure 4.34)
- rotating red beacons
- traffic control lights.

Rotating amber beacons are used to warn you to be cautious of working or moving machines such as excavators, backhoes, bobcats, forklifts, trucks and cranes. Rotating red beacons may be used to warn you of danger in situations of fire, emergency evacuation or blasting about to begin.

FIGURE 4.34 Communication using flashing beacons

Touch sensations

Touch sensations may come from:

- faulty machinery
- hot or cold water
- other workers (see Figure 4.35)
- pipework
- sharp objects
- straining machinery
- unstable scaffolding
- unstable structures.

FIGURE 4.35 Communication through touch

Odours

Odours may come from:

- engine exhausts
- fuels
- gases
- glues and adhesives
- liquids (see Figure 4.36)
- paints
- sealants
- sewer gases
- solvents.

FIGURE 4.36 Communication through smell

In the plumbing and building sector you must be aware that odours can be a form of communication that is vital not only for your safety and the safety of others but also for the prevention of faulty work, waste of time and waste of materials.

An unusual odour could be communicating a potentially hazardous situation. Inform your team and supervisor of any odours that you may have encountered to prevent any injuries or potential asphyxiation.

COMPLETE WORKSHEET 3

REFERENCES AND FURTHER READING

Acknowledgement

Reproduction of the following resource list references from DET, TAFE NSW C&T Division (Karl Dunkel, Program Manager, Housing and Furniture) and the Product Advisory Committee is acknowledged and appreciated.

Texts

Access Series (2001), *Communication for business*, McGraw-Hill, Sydney

Access Series (2001), *Communication for IT*, McGraw-Hill, Sydney

Australian Code for the Transport of Dangerous Goods by Road & Rail Edition 7.6, 2018

Basic Work Skills Training Division, NSW TAFE Commission (1995), *Workplace communication, NCS001, A teaching/learning resource package,* Basic Work Skills Training Division, South Western Sydney Institute of TAFE

Eagleson, R.D. (1990), *Writing in plain English,* Australian Government Publishing Service, Canberra

Elder, B. (1994), *Communication skills,* Macmillan Education, Melbourne

Graff, D.M. & Molloy, C.J.S. (1986), *Tapping group power: A practical guide to working with groups in commerce and industry*, Synergy Systems, Dromana, Victoria

National Centre for Vocational Education Research (2001), *Skill trends in the building and construction trades*, National Centre for Vocational Education Research, Melbourne

NSW Department of Education and Training (1999), *Construction industry: Induction & training: Workplace trainers' resources for work activity & site OHS induction and training*, NSW Department of Education and Training, Sydney

NSW Department of Industrial Relations (1998), *Building and construction industry handbook*, NSW Department of Industrial Relations, Sydney

WorkCover Authority of NSW (1996), *Building industry guide:* WorkCover NSW, Sydney

Web-based resources

Resource tools and VET links

NSW Education Standards Authority – Construction: **https://educationstandards.nsw.edu.au/wps/portal/nesa/11-12/stage-6-learning-areas/vet/construction-syllabus**

Safe Work Australia: **http://www.safeworkaustralia.gov.au**

Training.gov.au: **http://training.gov.au**

GET IT RIGHT

The photo below shows an incorrect practice that can be performed when backfilling a trench with an existing service.

Identify the incorrect method, and provide reasoning for your answer.

WORKSHEET 1

To be completed by teachers	
Satisfactory	☐
Not satisfactory	☐

Student name: ______________________

Enrolment year: ______________________

Class code: ______________________

Competency name/Number: ______________________

Task: Review the section 'Understand good communication' up to and including 'Apply oral communication' in this chapter and answer the following questions.

1. How does good communication benefit everyone in the building industry?

____

____

2. How do good communicators create a clear understanding?

____

____

3. List the five barriers that people may experience during the communication process.

____

____

____

____

____

4. Who are the two main parties in the oral communication process?

____

____

5. Who has the responsibility for successful oral communication?

____

6. What do we call the procedure for checking whether communication has been successful?

____

7. List the procedures for checking whether communication has been successful.

The sender checks by:

The receiver checks by:

8. When being spoken to in face-to-face communication, what should you do to make sure you get the message?

9. What four steps would you follow to answer a business telephone?

10. If you cannot help the caller, what seven steps should you follow?

11. How do people communicate through web conferencing?

WORKSHEET 2

To be completed by teachers	
Satisfactory	☐
Not satisfactory	☐

Student name: ______________________

Enrolment year: ______________________

Class code: ______________________

Competency name/Number: ______________________

Task: Review the section 'Apply visual communication' in this chapter and answer the following questions.

1. List two forms of visual communication.

____

____

2. State the importance of understanding body language while on site.

____

____

____

3. Describe a plumbing job or situation where hand signals could be used.

____

____

4. Why is it important to use hand signals when working on site?

____

____

5. Describe the hand signal that means 'Come towards me'.

____

____

6. When using hand signals or gestures, how can you ensure the message will not confuse others?

____

____

7. Describe the purpose of underground warning tape.

____

____

8. What is the importance of reinstating underground warning tape if it has been removed?

____

____

9. When identifying underground warning tape, what is the colour coding for the following services?

Electricity ______________________________

Recycled water ______________________________

Gas ______________________________

10. Which agency should be contacted prior to underground excavation? How can it be contacted?

____

____

WORKSHEET 3

To be completed by teachers	
Satisfactory	☐
Not satisfactory	☐

Student name: ______________________

Enrolment year: ______________________

Class code: ______________________

Competency name/Number: ______________________

Task: Review the sections 'Apply written communication and signage' and 'Understand alternative forms of communication' in this chapter and answer the following questions.

1. What are the three forms of 'written communication'?

____

____

____

2. When writing to be understood, what is your objective and how do you do this?

____

____

____

____

____

3. List the types of language you must avoid using if you do not want your meaning to be misunderstood.

____

____

____

____

4. State the importance of having the ability to draw and interpret sketches.

____

____

____

5. Diagrams are used on safety signs to highlight the action required. Describe prohibition safety signs and state the colour of the sign.

____

____

____

____

6. The signs in Figure 4.23 tell you that you *must not:*

7. Describe mandatory safety signs and state the colour of the sign.

8. The signs in Figure 4.24 tell you that you *must* wear:

9. List three emergency safety signs.

10. Describe the main difference between 'fire instruction' symbolic signs and 'fire instruction' text signs.

State the colours used for the following parts of the signs.

Symbol background: ______________________________

Symbol: ______________________________

Text: ______________________________

Text background: ______________________________

11. At what level should signs generally be placed?

____

____

____

12. List the nine classes of dangerous goods.

____

____

____

____

____

____

____

____

____

13. Employers are required to provide employees with information about dangerous goods. State the most effective way of ensuring this information is made available.

____

____

14. In addition to oral, visual and written communication, list the three *other* forms of communication that you may receive in the workplace.

____

____

____

PART 2

SETTING OUT

Part 2 of this textbook deals with the setting out requirements for new workers that are undertaking training and education within the plumbing services sector. It will help you build and enhance the required knowledge of reading plans and calculating plumbing quantities on the worksite.

Part 2 is based on one unit of competency, as described in Chapter 5. The key elements for the following competency are as follows:

5 Read plans and calculate plumbing quantities
- Prepare for work.
- Identify types of plans and drawings and their functions.
- Identify commonly used scales, symbols and abbreviations.
- Locate and identify key features on a services plan.
- Read and interpret job specifications.
- Obtain measurements and perform calculations.
- Calculate material quantities.
- Clean up.

Work through Part 2 and engage with your teacher and peers to prepare for the setting-out skills that will enable you to read plans and calculate plumbing quantities effectively on the worksite.

The learning outcomes for each chapter are a good indicator of what you will be required to know and perform, what you will need to understand, and how you will apply the knowledge gained on completion of each chapter. Teachers and students should discuss the knowledge and evidence requirements for the practical components of each unit of competency before undertaking any activities.

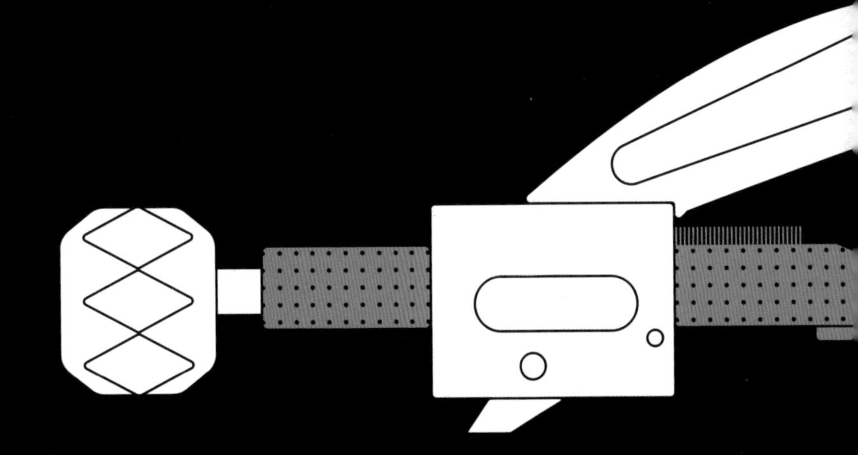

5 READ PLANS AND CALCULATE PLUMBING QUANTITIES

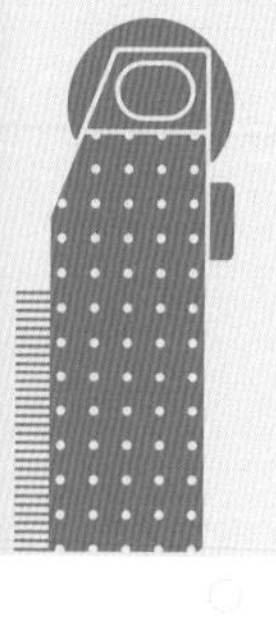

This chapter addresses the following key elements for the competency 'Read plans and calculate plumbing quantities':

- Prepare for work.
- Identify types of plans and drawings and their functions.
- Identify commonly used scales, symbols and abbreviations.
- Locate and identify key features on a services plan.
- Read and interpret job specifications.
- Obtain measurements and perform calculations.
- Calculate material quantities.
- Clean up.

It introduces the requirements for reading and interpreting plans, the specifications related to them and the knowledge required to extract critical information regarding plumbing installations and the calculation of quantities.

Prepare for work

The majority of communication and instructions in the building industry come from plans, drawings and the specifications associated with them; this information is passed on to people so that they can perform and quote for work. To be successful when preparing for work, it is critical to obtain and be able to correctly read and interpret plans, drawing details, specifications and standards. In addition to these requirements, work health and safety and environmental procedures must be adhered to, and quality assurance processes must be identified in accordance with workplace requirements.

Plans and drawings are presented in a scaled format showing views from above, front, sides and sectional views inside the building or structure.

The work shown on plans and drawings is also described in words within a document called the specification. The plans, drawings and specifications must be submitted to the local council, the local water/sewerage authority and the state or territory plumbing regulator where required, for approval before contracts are signed and work commences.

The details and measurements shown or described in these documents allow for quoting procedures and material orders to be quantified and placed. Reading and interpreting plans and specifications form the basis of successful completion of building contracts. The skill of reading plans is best acquired by practice and exposure to a variety of situations.

Drawings are used to convey a great amount of technical information from the designer of a building to the builder and their subcontractors, including the plumber. This information must be conveyed without the risk of any misunderstanding, and can be achieved only if the technical language of drawings is universally understood by everyone who is required to use them (see Figure 5.1). The technical language of drawings is expressed by using standardised drawing layouts and abbreviations of terms and symbols that are used in the construction industry (see Figure 5.2).

FROM EXPERIENCE

Understanding the technical language used in the construction industry is important to ensure effective communication is achieved.

Identify types of plans and drawings and their functions

Plans and drawings, together with a specification, are the instructions for different trades to follow, minimising the need for further referral. This enables construction methods and concepts to be undertaken during the construction process. This process requires several people to be involved, so the users and the functions of each drawing must be identified.

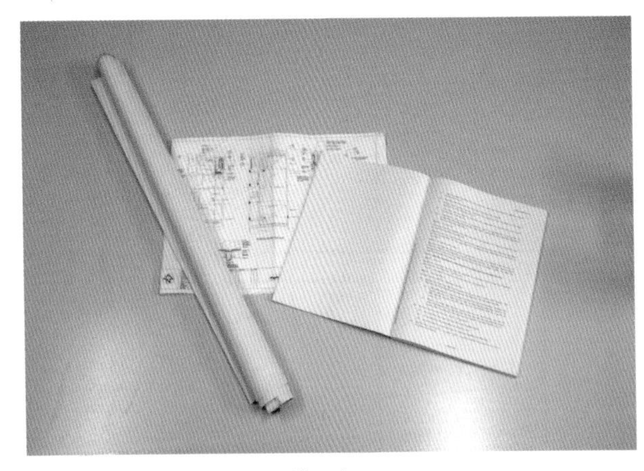

FIGURE 5.1 Plans and specifications

Plans and drawings in the plumbing industry

Completed plans and drawings must be submitted and approved by the local council, local water/sewerage authority and the state or territory plumbing regulator (where required) before contracts are signed and work begins. Plumbing inspectors or regulators may carry out inspections on buildings during and at the completion of projects for approval, and will use drawings as the basis of their approval. Drawings come in a variety of types and must be in a standard format for accurate interpretation by builders and tradespeople from any area or background.

Drawings in construction are generally divided into two groups:

- pictorial representations
- working drawings.

Pictorial representations

A pictorial representation is used by architects and designers to determine, with the client, the final design or appearance of the project.

A perspective view (see Figure 5.3) or a pictorial representation (see Figure 5.4) are often used to assist in visualising the object. The faces or sides of the object appear to taper away or recede to a vanishing point. Parallel lines appear to come together, similar to the effect when viewing railway tracks in the distance.

Perspective drawings are the closest to what the eye would see and are often used by builders to present a project for sale.

Isometric projections (see Figure 5.5) are also pictorial views, with lines drawn parallel to the axis at 30° because of the ease in using 60°/30° set squares.

Perspective view and isometric projection are used as schematic or freehand sketches for ease of interpretation for house design and building construction.

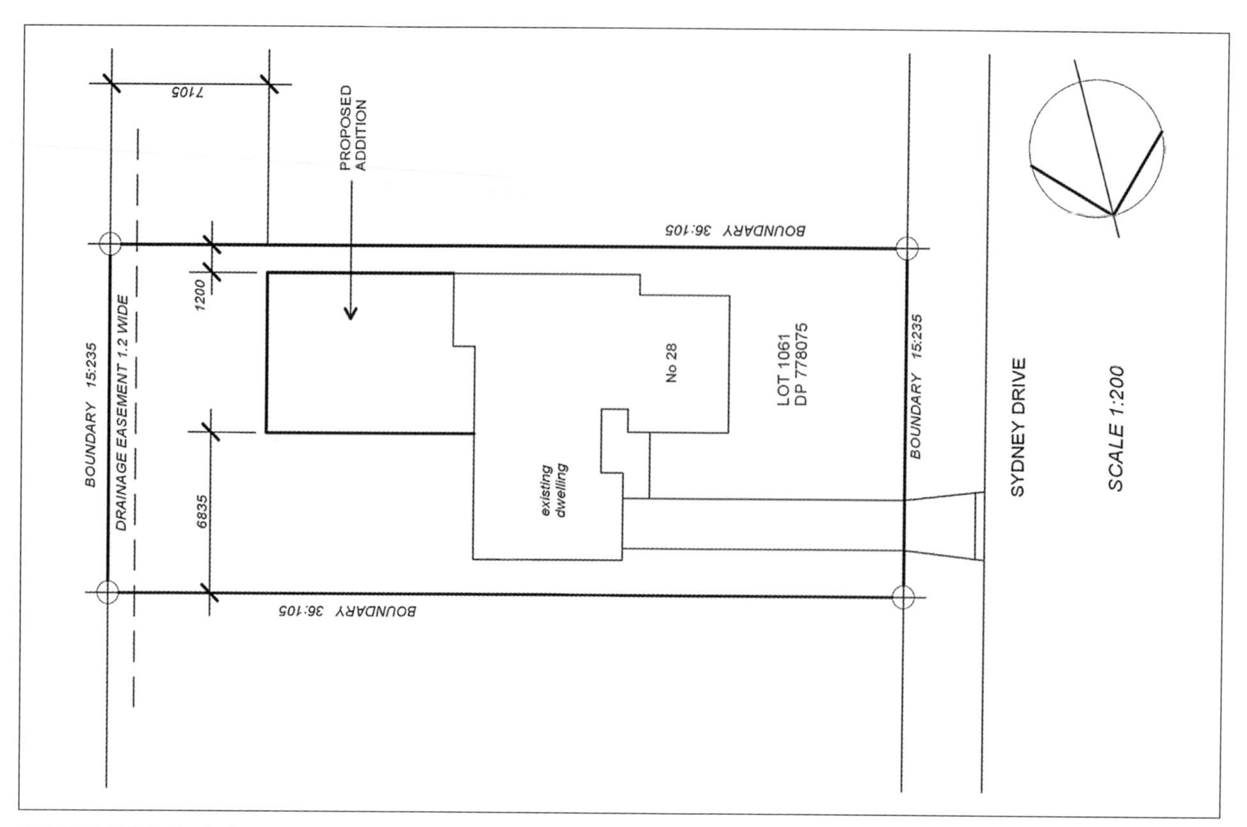

FIGURE 5.2 A typical plan

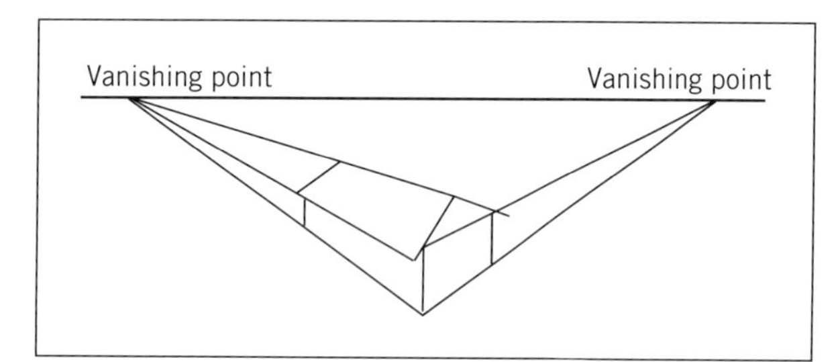

FIGURE 5.3 Perspective view

FIGURE 5.4 Pictorial representation

Many people in the building industry will use a basic freehand form of isometric projection to explain specific details or to give a three-dimensional view. This style is useful for estimating quantities of pipework.

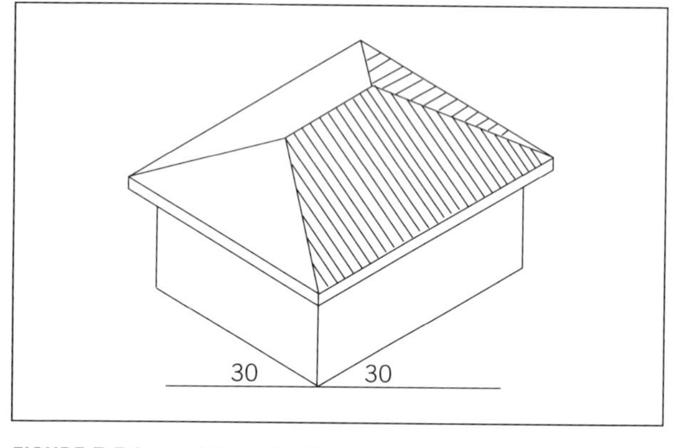

FIGURE 5.5 Isometric projection

Working drawings

Working drawings show technical details of a building and are commonly referred to as a building plan or plans, and are produced so that users may:

- gain an overall picture of the layout and shape of the building
- determine set-out dimensions for a building
- locate and identify areas of the building (e.g. rooms, doors, windows and drainage).

A basic projection form used for working drawings is orthographic projection, which consists of three related views: plan, elevation and section. These three views are used to give a complete understanding of the building (see Figure 5.6).

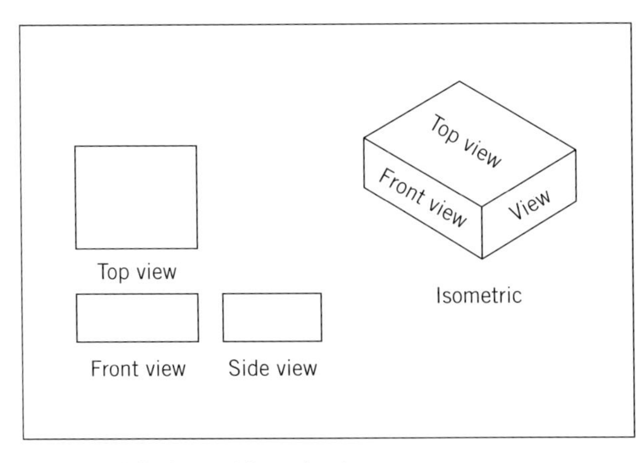

FIGURE 5.6 Orthographic projection

Floor plans

The floor plan is a horizontal section of the building as viewed from above with the roof taken off, and is the most important of the related views, as it contains most of the information for construction (see Figure 5.7). Floor plans are usually drawn to a scale of 1:100, but can be up to 1:500 if the building is large.

Items shown on a floor plan can include:

- overall dimensions to the outside of the walls
- door and window positions and opening sizes
- thickness of external and internal walls
- internal room dimensions
- position of cupboards, stoves, laundry tubs, baths, basins and showers
- function of each room, such as kitchen or bathroom
- floor surface and type of floor covering
- position and direction from which section lines are taken for sectional elevations (north point)
- section lines
- scale.

A plumber is required to accurately read the plan view so that items such as 'core hole' penetrations or points can be set out, and specific plumbing services can be located. It is also important to refer to the specification while reading the plans to ensure the correct materials and fixtures are used.

Elevations

Elevations provide information relating to vertical measurements and external finishes. Each view is identified according to the direction it faces in relation to the points of a compass, based on the north point shown on the site plan.

Elevations give a projection of the building at right angles, and show:

- height of finished floor level (FFL), which relates to a datum and depths of all ground works
- finished ceiling level (FCL)
- design of the building
- roof shape, width of roof overhang, position of gutters and downpipes
- position of doors and windows
- window sill height above FFL
- type or function of windows
- roof covering and slope
- FFL height above ground level
- the finish of external walls.

Standard working drawings in orthographic projection generally require a minimum of two elevations, which are the front and side view of the building. This assists in correct interpretation of design. Typical information may be indicated on one or more elevations (see Figure 5.8).

Section elevations

Section drawings are elevations cut through the building in the position and direction indicated on the floor plan. The section is a cross-section from the bottom of the footings, through walls, ceilings and roof structure (see Figure 5.9). Sections give information such as:

- footing sizes
- subsoil drainage position
- wall thickness and construction
- design of subfloor
- floor construction
- roof construction (e.g. trussed)
- roof pitch
- section sizes and spacing of structural members
- scale.

The reference points for the sectional views are indicated on the floor plan as A-A, B-B etc.

Details

Details are sectional views drawn to a larger scale than sectional elevations, and detail specific requirements that cannot be drawn accurately to scale on sectional elevations, such as a bathroom fixture set-out (see Figure 5.10) or noise reduction floor drain details (see Figure 5.11). Common scales that are used are 1:50, 1:20, 1:10 and 1:5.

Location plans

These plans (see Figure 5.12) are used as a quick reference by the builder or worker to direct employees to a site, to estimate travel costs and to locate sites for specific information from utilities and local authorities (e.g. Sydney Water). They may also show details of proposed developments in surrounding properties.

Site or block plans

Site or block plans are essential for determining the location of the building on the building block. Information contained on a site (see Figure 5.13) or block plan includes:

- boundary dimensions of the block
- distance from street to boundary
- set-back distance from front boundary to building line
- distance from side boundary to building
- contour lines and their height
- position of paths and driveways

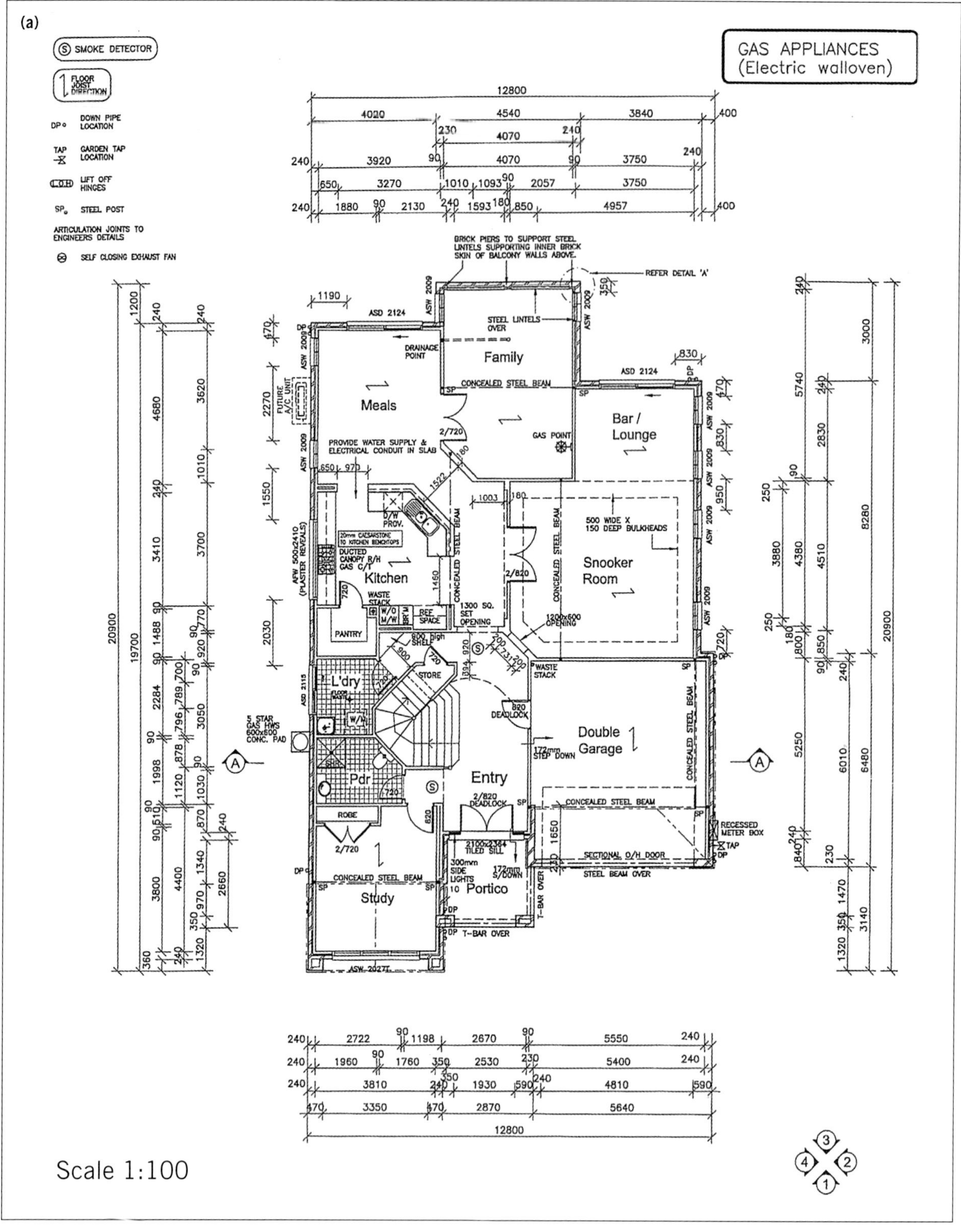

FIGURE 5.7 Typical floor plan: (a) Ground floor plan; (b) First floor plan (*continued*)

- location of trees
- direction of north
- lot number
- scale
- datum point
- reduced level
- invert level.

The plumber must be able to read the site plan to establish where services such as water, sewerage, gas and stormwater are best connected. This information may be required to install temporary site services for water and waste, and to locate any existing services.

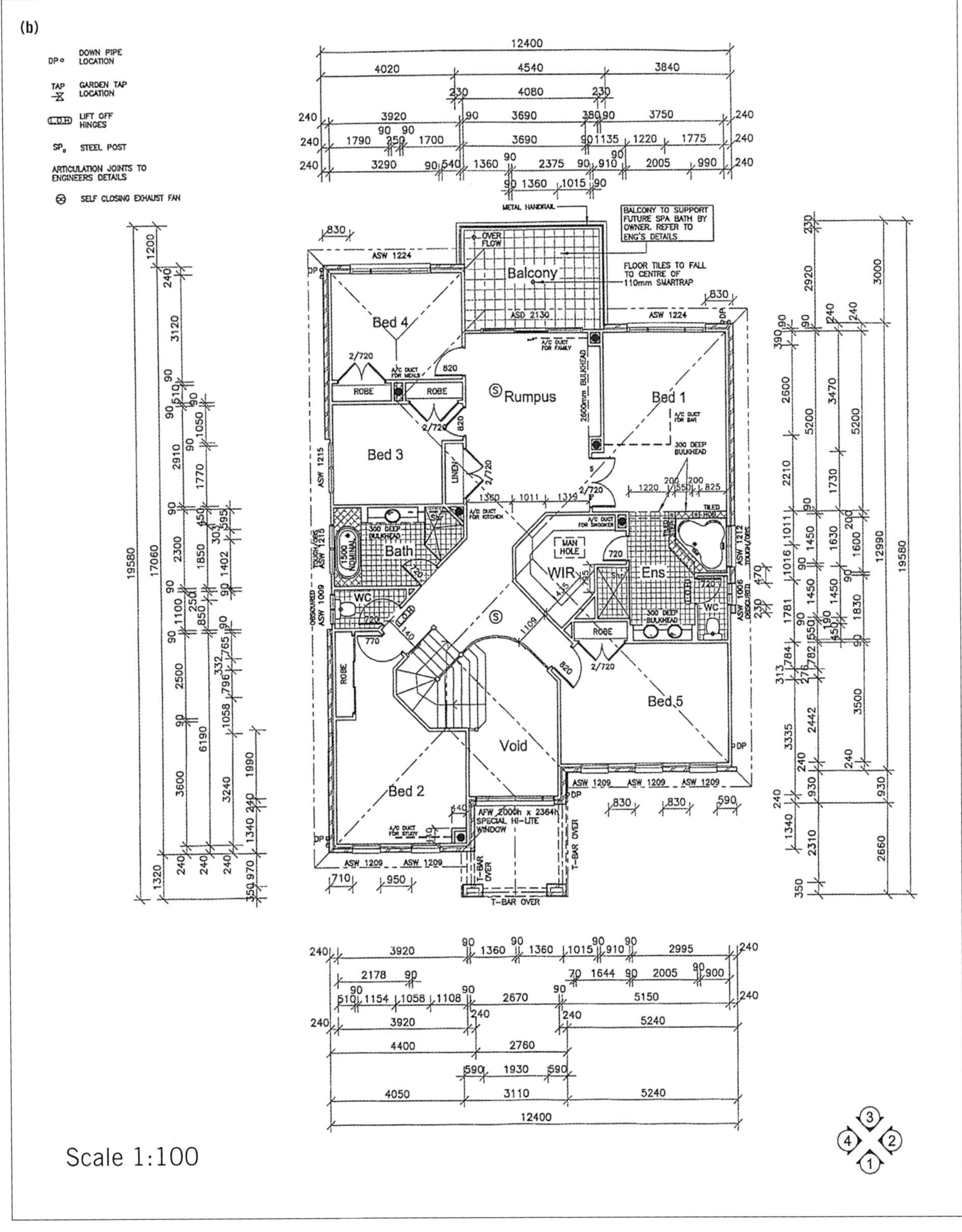

FIGURE 5.7 *(continued)*

Note: A reduced level (RL) is a vertical height or elevation taken in relation to a datum point. An invert level (IL) is a vertical height taken from the bottom internal surface of drainage pipework when performing levelling with a staff. Understanding the concepts of ILs is an important concept when connecting different drainage pipe materials. Not all materials have the same thickness due to their material properties, but it is essential that, when joined, both materials line up horizontally to avoid blockages (see Figure 5.14). Both heights are used to determine heights, connection points, falls and levels on building sites.

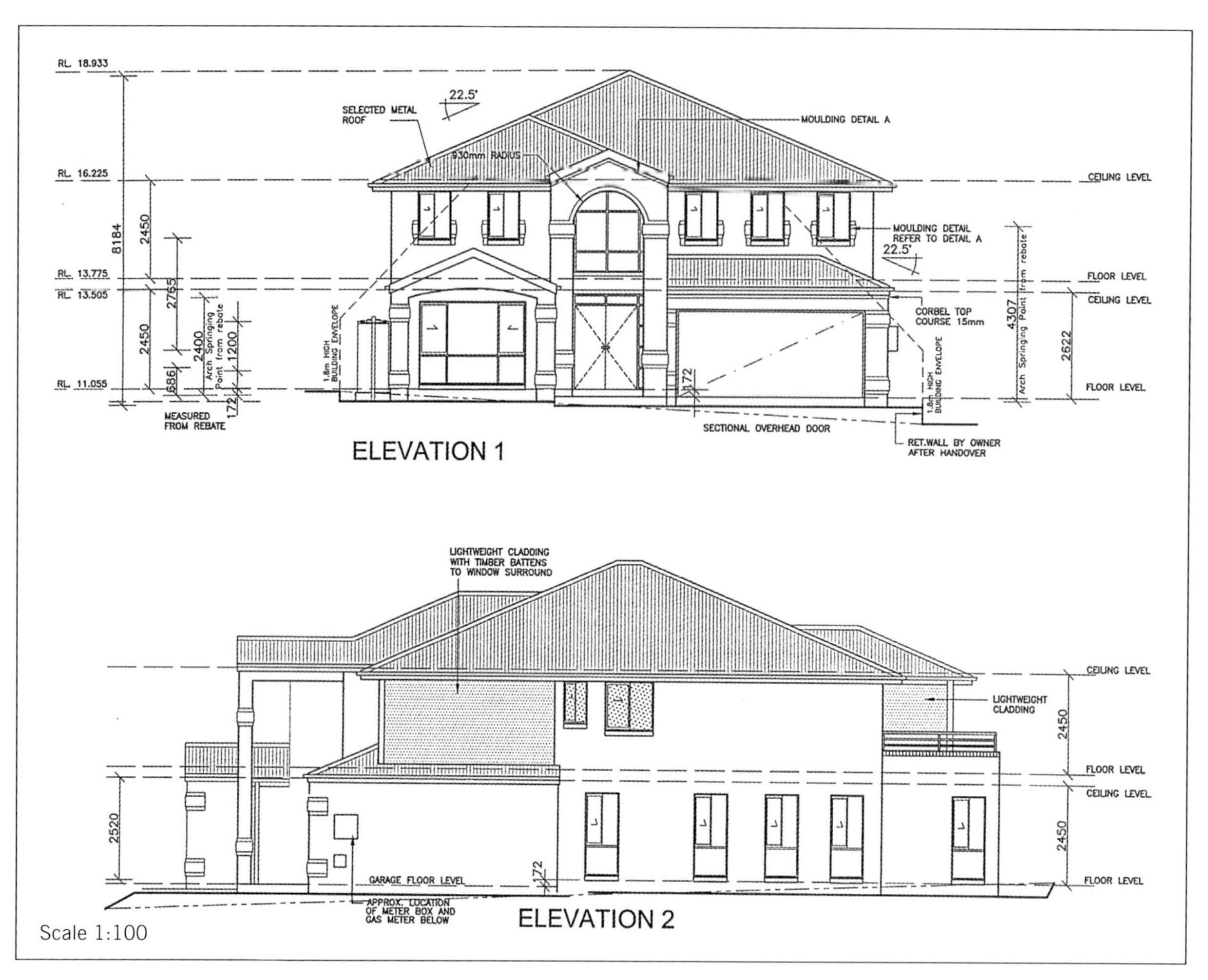

FIGURE 5.8 Typical details of elevations

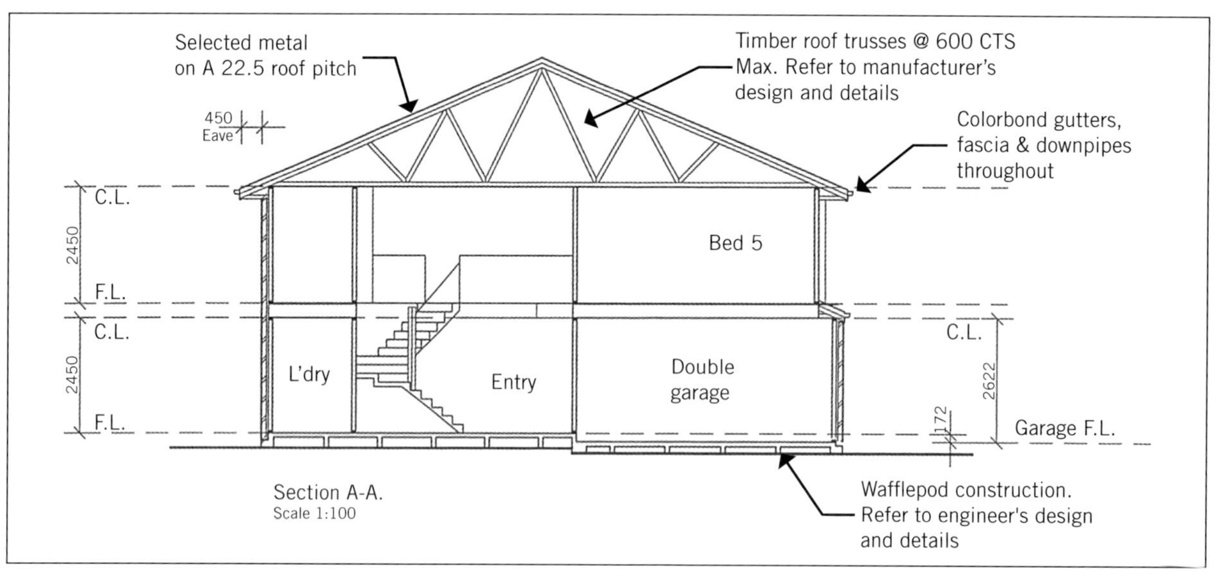

FIGURE 5.9 Sectional elevations

FIGURE 5.10 Bathroom fixture detail

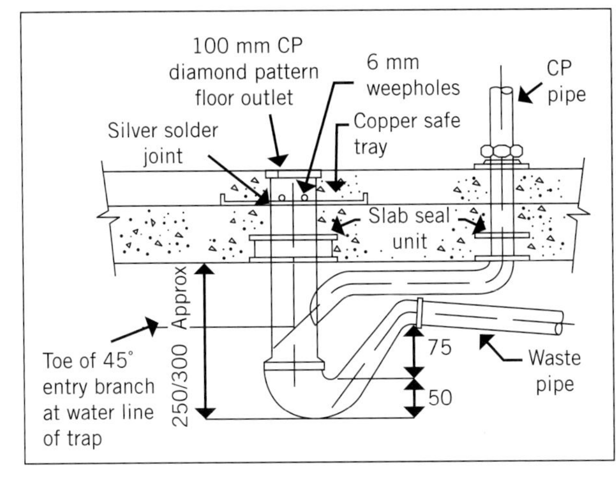

FIGURE 5.11 Noise reduction floor drain detail

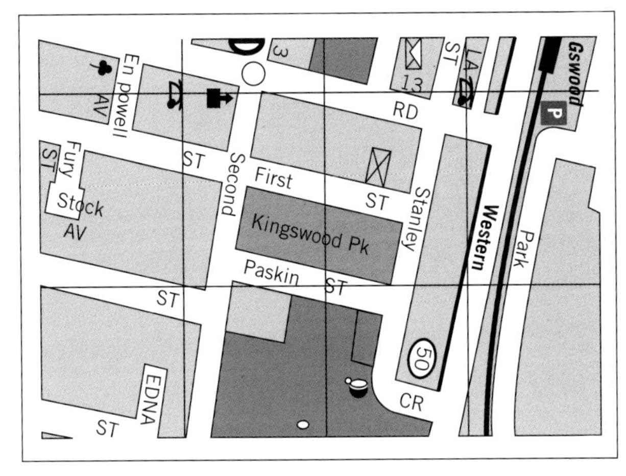

FIGURE 5.12 Location plan

Special details

These details may include:

- authorities' service easements (e.g. sewer lines, stormwater connection points, water mains and electrical services)
- a site plan to indicate landscaping details
- details of retaining walls and subsoil drainage
- plumbing services proposed for the site and environmental compliance
- large-scale details of special construction requirements.

Sewerage service diagram

A sewerage service diagram (see Figure 5.15), often referred to as a 'helio', is a drawing issued by the sewerage authority or an authorised agent, after a fee has been paid. It contains the details of the drainage to a property and can highlight the following:

- authority's point of connection
- outline and position of buildings/structures
- access chambers along the sewer main and invert levels
- house drainage layout
- fixtures within the property connected to the sewer
- easements located near or within the property
- size and type of material used for the sewer main
- size and type of material used for the house service line
- direction of flow within the sewer main
- approximate depth of the sewer main measured at access chambers
- north point to show orientation
- street name
- property lot and street numbers.

Sewerage service diagrams may be drawn with differing scales, so it is important to check the individual scale used for the identified property. Preferred scales used for diagrams include 1:100, 1:200, 1:250 and 1:500, depending on the size of the building located on the diagram.

Key users of drawings

When a detailed drawing of a proposed building or structure is prepared, several copies are issued to the people who will use them:

- The owners of the proposed building require a copy of the drawing to see that the design is to their satisfaction.
- Hydraulic plumbing consultants, structural, electrical and mechanical engineers require copies so they can design their part of the structure.
- Council health and building surveyors require copies of drawings and specifications to ensure that the building conforms to building codes and council regulations.
- Water and sewerage authorities require copies of council-stamped and -dated plans to ensure their easements will not be built over or concrete-encased (where applicable).
- Bank or building society officers require copies of the drawing before giving approval for finance for construction.
- A builder requires copies of the drawing to cost the building, to prepare a quotation and then to construct the building. The builder will pass copies of the drawing to subcontractors/workers, such as plumbers, to obtain their prices to carry out work during construction.

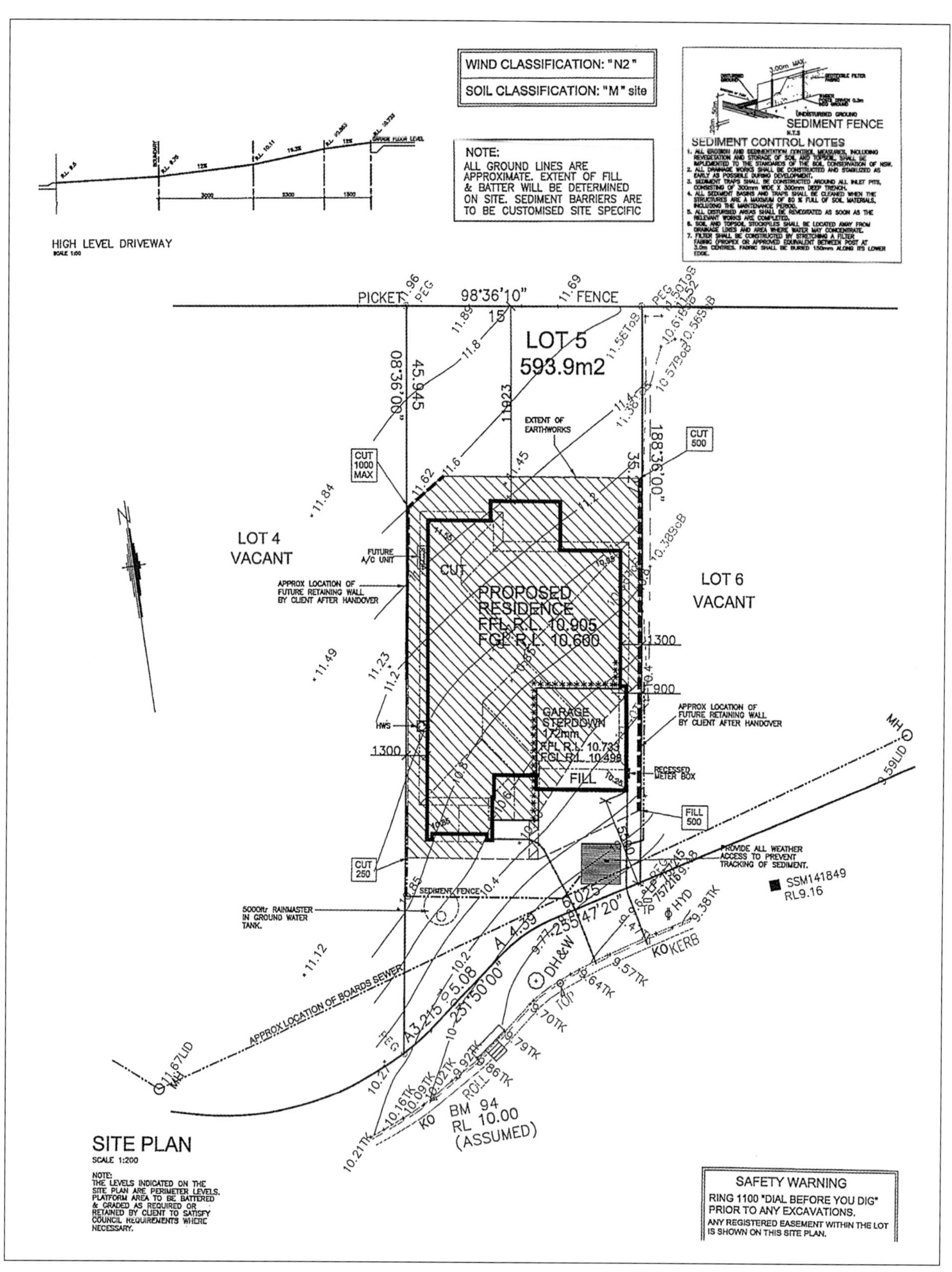

FIGURE 5.13 Typical site plan

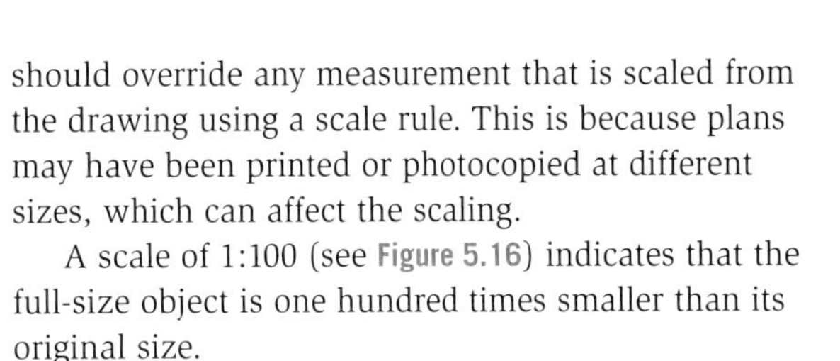

FIGURE 5.14 Invert level on two different materials

- Suppliers of prefabricated components such as hot water systems, pump sets, air-conditioning and heating systems, as well as material and fixture suppliers, require copies of the drawing in order to calculate a price for their part of the job.

LEARNING TASK 5.1

PLAN INTERPRETATION

Obtain a set of plans from the job you are currently working on now, or use a set from a previous job (ask your employer or teacher for a copy), and complete the following:

1. Identify the scale of the site plan.
2. Identify the direction the front of the building faces and the street name it faces.
3. Is the plan the original version or has it been amended?
4. What access is provided from the roadway?
5. Do the plans state a reduced level measurement, a benchmark measurement or an invert level measurement?
6. What do each of these measurements denote?

Identify commonly used scales, symbols and abbreviations

Plans and drawings commonly use scales, symbols and abbreviations to minimise congestion and to allow a large amount of information to be drawn on the plan. A legend is often used and placed on the plan to define a symbol or abbreviation.

Scale drawings

Plans are drawn to scale, which means that a full-sized object is proportionally reduced to a suitable scale allowing it to be reproduced on drawing sheets of a manageable size. Working drawings state the scale that have been used on the drawing; however, in all cases the written measurement appearing on the drawing should override any measurement that is scaled from the drawing using a scale rule. This is because plans may have been printed or photocopied at different sizes, which can affect the scaling.

A scale of 1:100 (see Figure 5.16) indicates that the full-size object is one hundred times smaller than its original size.

Table 5.1 shows the standard scales that may be used to produce working drawings.

TABLE 5.1 Standard scales used to produce working drawings

Site plans	1:500	1:200	
Plan views	1:100		
Elevations	1:100		
Sections	1:100		
Construction details	1:20	1:10	1:5

Calculating scales

When a scale rule with the appropriate scale cannot be found, or an unusual scale needs to be used, it may be necessary to make your own (see Examples 5.1 and 5.2). This may be done simply by using a calculator and a rule with standard size millimetres or a scale of 1:1 (which is full-size).

EXAMPLE 5.1

Scale required = 1:25

Measurement to be scaled = 6.200 m

1. Change the measurement from metres to millimetres: 6.200 m changes to 6200 mm.
2. Divide the millimetre measurement by the desired scale: 6200 ÷ 25 = 248.
   Therefore, the 1:25 scaled measurement of 6.200 m = 248 mm (full-size millimetres).

EXAMPLE 5.2

Scale required = 1:75

Measurement to be scaled = 1.500 m.

1. Change the measurement from metres to millimetres: 1.500 m changes to 1500 mm.
2. Divide the millimetre measurement by the desired scale: 1500 ÷ 75 = 20.
   Therefore, the 1:75 scaled measurement of 1.500 m = 20 mm (full-size millimetres).

Dimensions

Dimension lines on drawings enable scales to be used to determine lengths that are not shown. The forms of dimension lines vary (see Figure 5.17), but all are shown as a line parallel to the drawing. Lines at right angles to the main line indicate the position at which the dimension is taken. It is best practice to use the same type of dimension lines throughout the drawings.

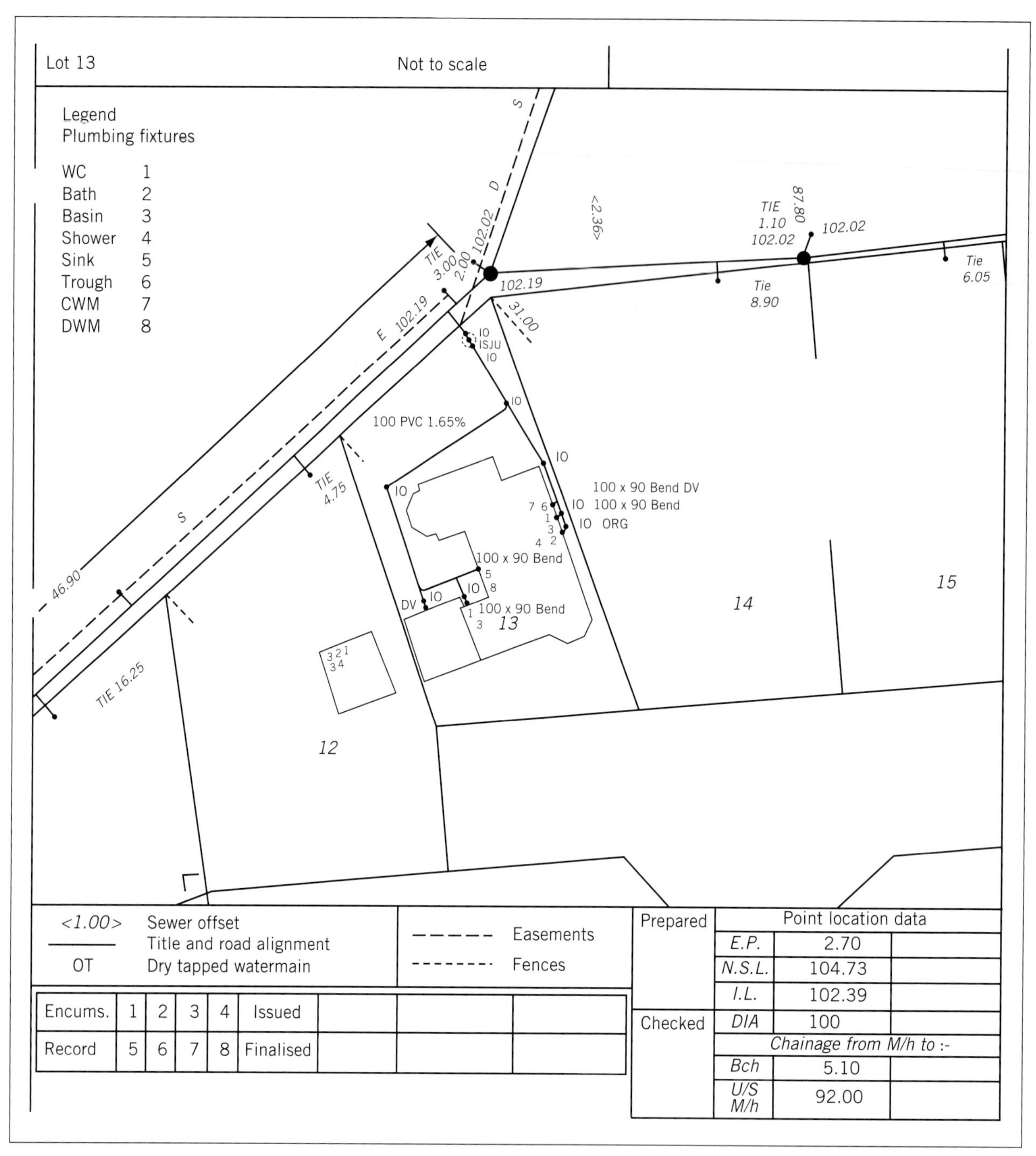

FIGURE 5.15 Sewerage service diagram

Drawing sheet sizes

Plans such as working drawings are produced to scale on standardised drawing sheets, which range in size from A0 (1189 × 841 mm) to A4 (210 × 297 mm), with margin or border lines and title blocks. The margins assist in the folding or filing of the plans, while the title block contains information essential to the project (see Figure 5.18).

Drawing sheet sizes decrease by half starting from A0. If you fold an A0 sheet in half, you will get an A1 sheet size. Half of A1 will give an A2, half an A2 will give an A3, and half of A3 will give you A4. (These are common sized sheets used for drawings.)

Title block

The title block on a set of plans for a house may contain a minimum amount of information, such as the name of the owners, the lot number and street number, street name, suburb, and scales used in the drawing (see Figure 5.19).

The title block for a commercial project may contain:

- the name of the client or company for whom the project is to be constructed
- the lot number and address of the project
- scale or scales used on the plans (e.g. 1:100)

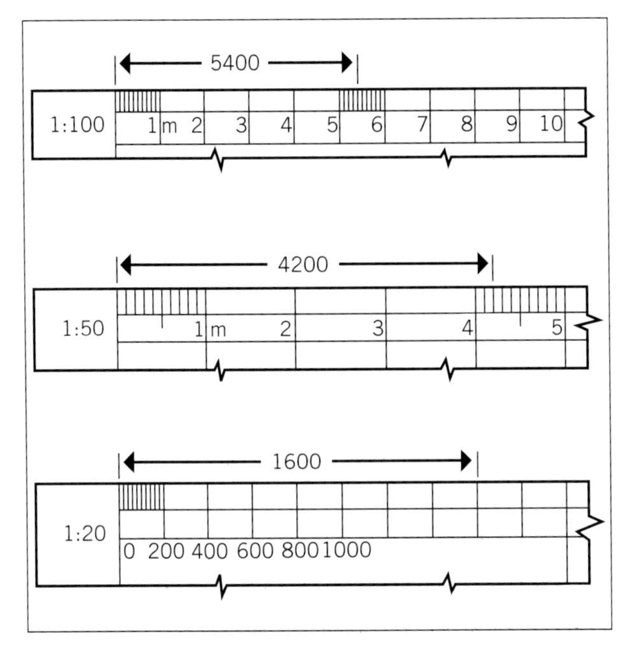

FIGURE 5.16 Common reduction scales

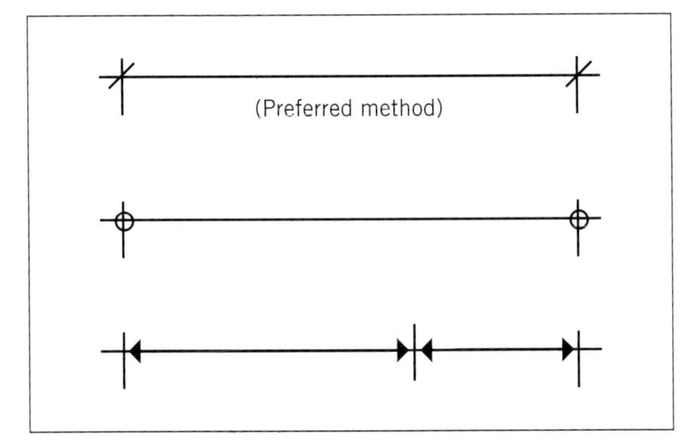

FIGURE 5.17 Examples of dimension lines

- numbers on the drawing sheets, if more than one is used (e.g. sheet 1 of 3)
- the name of the person or drafting service that has prepared the drawings
- a filing system in the form of numbers or letters for use by the person who prepared the drawings
- warnings (in some cases) against scaling from drawings to prevent incorrect measurement when using scales rather than using figured dimensions
- the date the plans were drawn.

Best practice is to have the title block on the right-hand side of the page. The main reason is that large sets of plans are generally fixed together on the left-hand side so that information such as drawing numbers or amendments is readily accessible.

Drawing symbols and abbreviations

Symbols and abbreviations appear on plans, elevations and section views. Common symbols in use are shown in Figures 5.20–5.22. They represent common construction practice for standardised details that are common knowledge in the building industry. Understanding symbols allows the transfer of technical information from the designer to the hydraulic consultant and plumber, as they reduce the risk of any misunderstanding of the designer's intentions.

Symbols may be used to show the following:

- direction of flow of fluids, grade of pipework, and the type of fluids being carried by that pipework
- pit and sump construction, such as brick, reinforced concrete or plastic
- foundation soil composition, earth, sand or other backfill material
- the types of fixture being used.

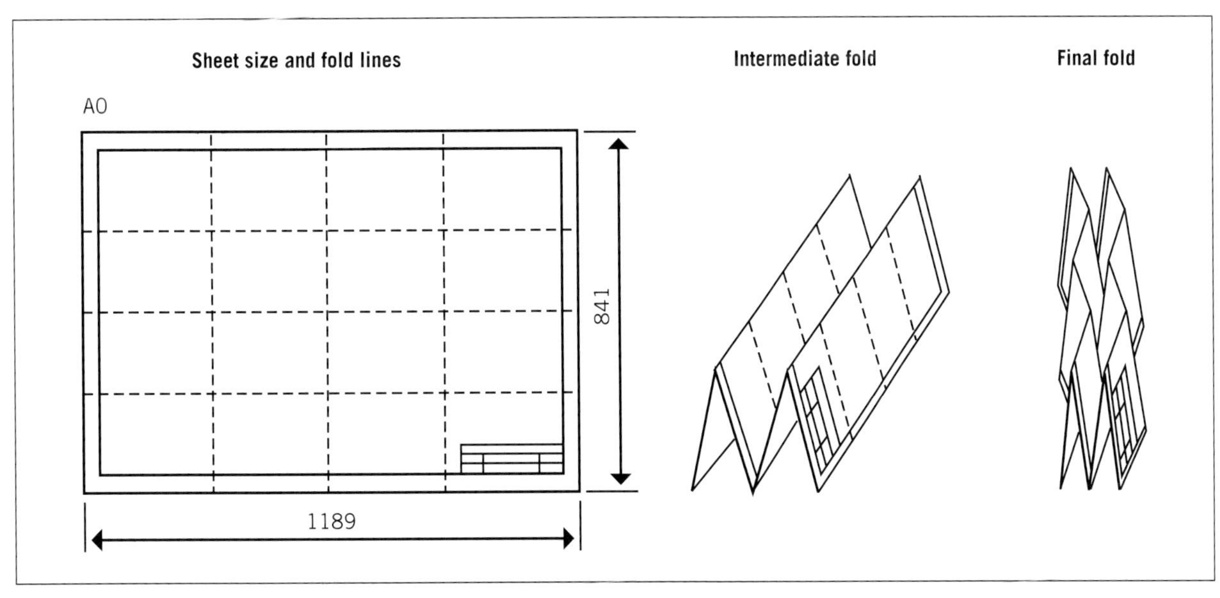

FIGURE 5.18 Typical drawing sheet size and format

Proposed dwelling for: Mr & Mrs B. Good, lot 12 Aldgate St Prospect NSW 2149		
Scale: 1:100	Fly-by-night Builders, Lic. No: 4054	Sheet 1 of 3
Drawn by: R. Joven Date: Jan 2021	• Do not scale off drawings • All dimensions shown are theoretical only and are subject to site measure during construction.	Job No: RJ 06/01/57

FIGURE 5.19 Typical title block showing required information

Fitting	Symbol	Fitting	Symbol
Kitchen fixtures and fittings all coloured french ultramarine		Water closet	WC
Stove	S	Bidet	B B
Wall oven hot plates	O	Laundry fittings all coloured french ultramarine	
Refrigerator	R	Sink/Tub	
Dishwasher	DW	Washing machine	WM
Sinks (with drainers)		Clothes dryer	CD
		Miscellaneous fittings all coloured french ultramarine	
Sanitary fittings all coloured french ultramarine		Cleaner's sink	CS
Bath		Slop sink	SS
Basin		Hot water unit alternatives	HW HW
Shower recess		Rainwater tank	RWT 2000 litre
Urinal		Fire hose rack/reel recessed	FHR
Urinal stall		Fire hose rack/reel free standing	FHR

FIGURE 5.20 Symbols for kitchen and hydraulic fixtures

Abbreviations are used on drawings to reduce the written content, which minimises the congestion of information necessary to convey the correct interpretation of the drawings. The abbreviated form should be used only where confusion or misinterpretation is not likely to occur.

FROM EXPERIENCE

Clear communication when producing plans is essential as mistakes can occur.

Element	Symbol
Sewer line	S Green
Industrial sewer	I S Green
Soil pipeline	SP Blue
Waste water pipeline	W Yellow
Vent pipeline	V&VP Red
Acid or chemical waste	AW Green
Vent (soil vent) pipe	V(SV)
Agricultural pipe drain	APD
WATER SUPPLY	
Water main (pipeline)	
Domestic water service cold water	CW
Domestic water service hot water	HWS
Fire water service	FWS Red

FIGURE 5.21 Graphic line representations

Tables 5.2 and 5.3 contain lists of common hydraulic abbreviations in use.

TABLE 5.3 Examples of hydraulic services (plumbing) abbreviations

Element	Abbreviation
Autopsy table	AT
Bain-marie	BM
Basin	B
Bath	Bth
Bidet/bidette	Bid
Clothes washing machine	CWM
Dental unit	DU
Dish washing machine	DWM
Drinking fountain	DF
Floor waste gully	FW
Glass washing machine	GWM
Induct pipe mica flap	IPMF
Junction (for future use)	JN
Mica flap	MF
Shower	Shr
Sink	S
Slop hopper	SH
Tea sink	TS
Trough (ablution)	Tr (A)
Trough (laundry)	Tr (L)
Urinal	Ur
Vent pipe	V
Water closet pan	WC

TABLE 5.2 Hydraulic fluid abbreviations

Element	Abbreviation	Element	Abbreviation
Air	A	Low-temperature hot water	LTHW
Condensate	C	Nitrogen	N
Cold water	CW	Oxygen	OX
Condenser cooling water	CCW	Nitrous oxide	NO
Chilled water	ChW	Refrigerant (show number if any)	R
Chilled drinking water	Ch DW	Steam	S
Distilled water	DW	Superheated steam	SHS
Drain or overflow	D	Vent	V
Feed water	FW	Fire water service	FWS
Liquefied petroleum gas	LPG	Fire foam	FF
Hot water supply	HWS	Fire sprinkler	FS
High-temperature hot water	HTHW	Fire CO_2	F CO_2
Medium-temperature hot water	MTHW		

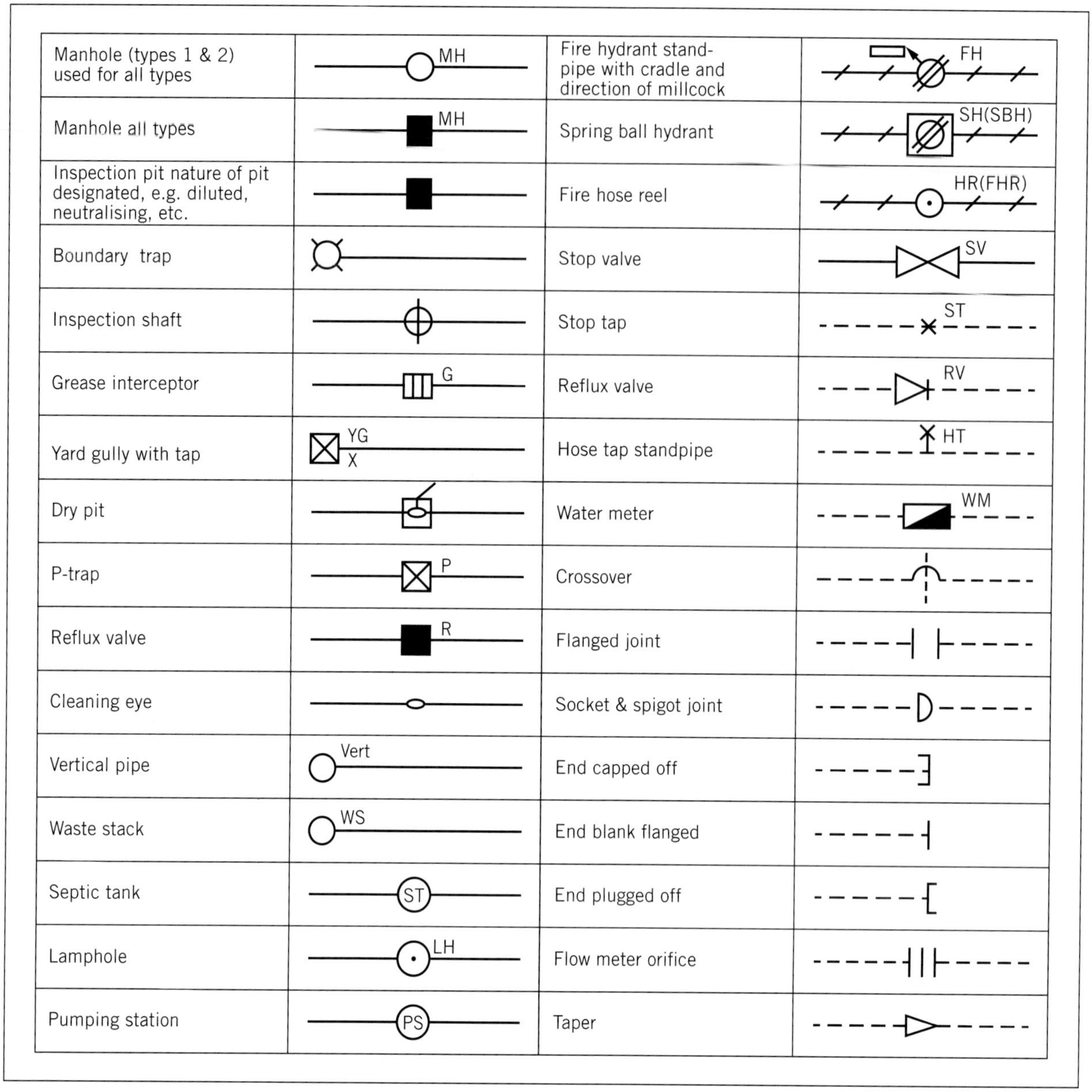

Manhole (types 1 & 2) used for all types	MH	Fire hydrant stand-pipe with cradle and direction of millcock	FH
Manhole all types	MH	Spring ball hydrant	SH(SBH)
Inspection pit nature of pit designated, e.g. diluted, neutralising, etc.		Fire hose reel	HR(FHR)
Boundary trap		Stop valve	SV
Inspection shaft		Stop tap	ST
Grease interceptor	G	Reflux valve	RV
Yard gully with tap	YG X	Hose tap standpipe	HT
Dry pit		Water meter	WM
P-trap	P	Crossover	
Reflux valve	R	Flanged joint	
Cleaning eye		Socket & spigot joint	
Vertical pipe	Vert	End capped off	
Waste stack	WS	End blank flanged	
Septic tank	ST	End plugged off	
Lamphole	LH	Flow meter orifice	
Pumping station	PS	Taper	

FIGURE 5.22 Hydraulic services fixtures and fittings symbols

A more detailed list can be obtained from the following Australian Standards:

- *AS 1100.101 Technical drawing, Part 101: General principles*
- *AS 1100.301 Technical drawing – Architectural drawing* (plus supplements)
- *AS/NZS 3500.2 Plumbing and drainage, Part 2 – Sanitary plumbing and drainage.*

5.1 STANDARDS

- AS 1100.101 Technical drawing, Part 101: General principles
- AS 1100.301 Technical drawing – Architectural drawing
- AS/NZS 3500.2 Plumbing and drainage, Part 2 – Sanitary plumbing and drainage

Legend

A legend is often included on plans to define the symbols and abbreviations that have been used. Symbols that a user may be familiar with can be depicted differently, as they may be drawn to a different standard. It is important to note that whenever you produce a drawing you should provide a legend or key to identify any symbols, abbreviations or line types that have been used.

LEARNING TASK 5.2

IDENTIFYING THE CORRECT SCALE

Obtain a set of plans from the job you are currently working on now or use a set from a previous job (ask your employer or teacher for a copy), and identify the correct scale:

>>

>>

1 Locate the scale shown on the site plan.
2 Determine if the site plan is to the scale that is listed on the plan. This can be achieved by using a scale rule to measure a dimension shown on the plan. If the measurement shown on the plan does not match the measurement on the scale rule, this indicates that the plan has been reproduced through photocopying at a scale smaller or larger than the original.

COMPLETE WORKSHEETS 1 AND 2

Locate and identify key features on a services plan

Plumbing services are found in all classes of buildings, which range from industrial and commercial buildings to farms and marine craft. Plumbing services may include:

- water supply
- gas supply
- sanitary plumbing and drainage
- stormwater collection, disposal and reuse
- roof coverings
- fire suppression systems
- mechanical services
- specialised fluid delivery and disposal.

Plans depicting these services are essential to allow plumbing services to be located and installed in the correct sequence within the general construction schedule.

Hydraulic services consultants design most plumbing systems to fit in with general construction design, along with civil, structural, electrical and mechanical services that are normally found on large building projects. However, hydraulic consultants may also be required for smaller building projects if requested by a client.

The hydraulic services consultant's role may incorporate, but is not limited to, the following:

- carrying out site investigations to establish the location of various existing services, which can include water, waste, gas, stormwater, electrical and communication
- preparing preliminary designs and cost estimates for the plumbing portion of the project
- liaising with authorities to confirm available water and gas pressures, location of easements etc.
- completing all documentation required for a project, including final contract designs and specifications
- carrying out tender and variation assessments for the project
- carrying out construction inspections to ensure that services are installed correctly, and the project runs to plan
- completion of 'as built' drawings when the project is finished.

Part of the plumber's role in these projects is to help coordinate the installation of the plumbing services in an approved manner and in conjunction with other trades to prevent activities occurring out of sequence, which may lead to costly delays in construction.

Plumbers may use plans to:

- find a location of a job – see the location plan in Figure 5.12
- find positions of main pipe runs in ground works – see the site plan in Figure 5.23
- determine the position of services to fixtures – see the schematic drawing in Figure 5.24
- determine position of authority's sewer mains – see the sewerage service diagram in Figure 5.15
- determine where existing services might be in public places – contact Dial Before You Dig (phone 1100 or go online at http://www.1100.com.au)
- determine falls, inverts, depth of pits, amount of spoil to be excavated and removed, and/or backfill to be ordered and delivered – see contour plans and site plans
- determine structural features of a building
- locate horizontal and vertical measurements
- compile estimates for tenders – see site plans, floor plans (see Figure 5.7) and elevation plans (see Figure 5.8)
- measure quantities for orders – see floor plans.

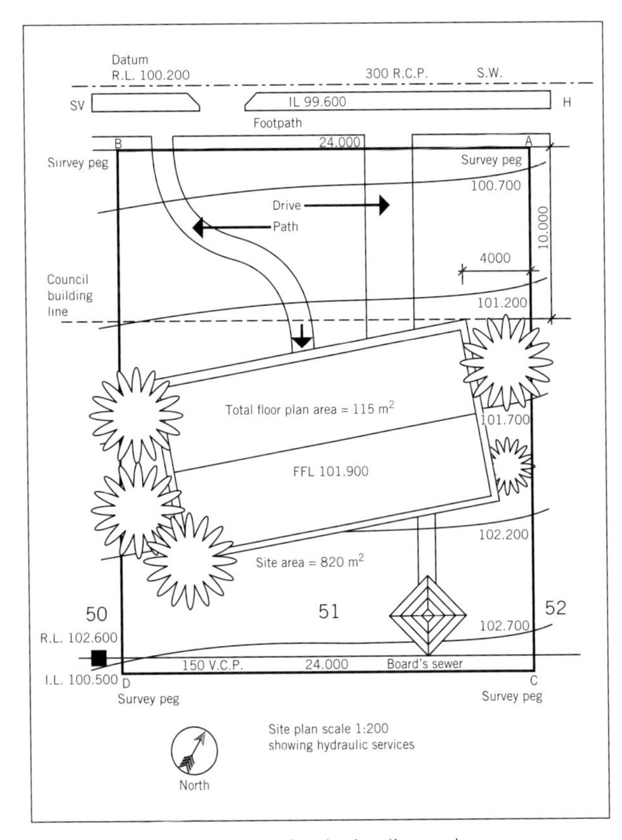

FIGURE 5.23 Site plan showing hydraulic services

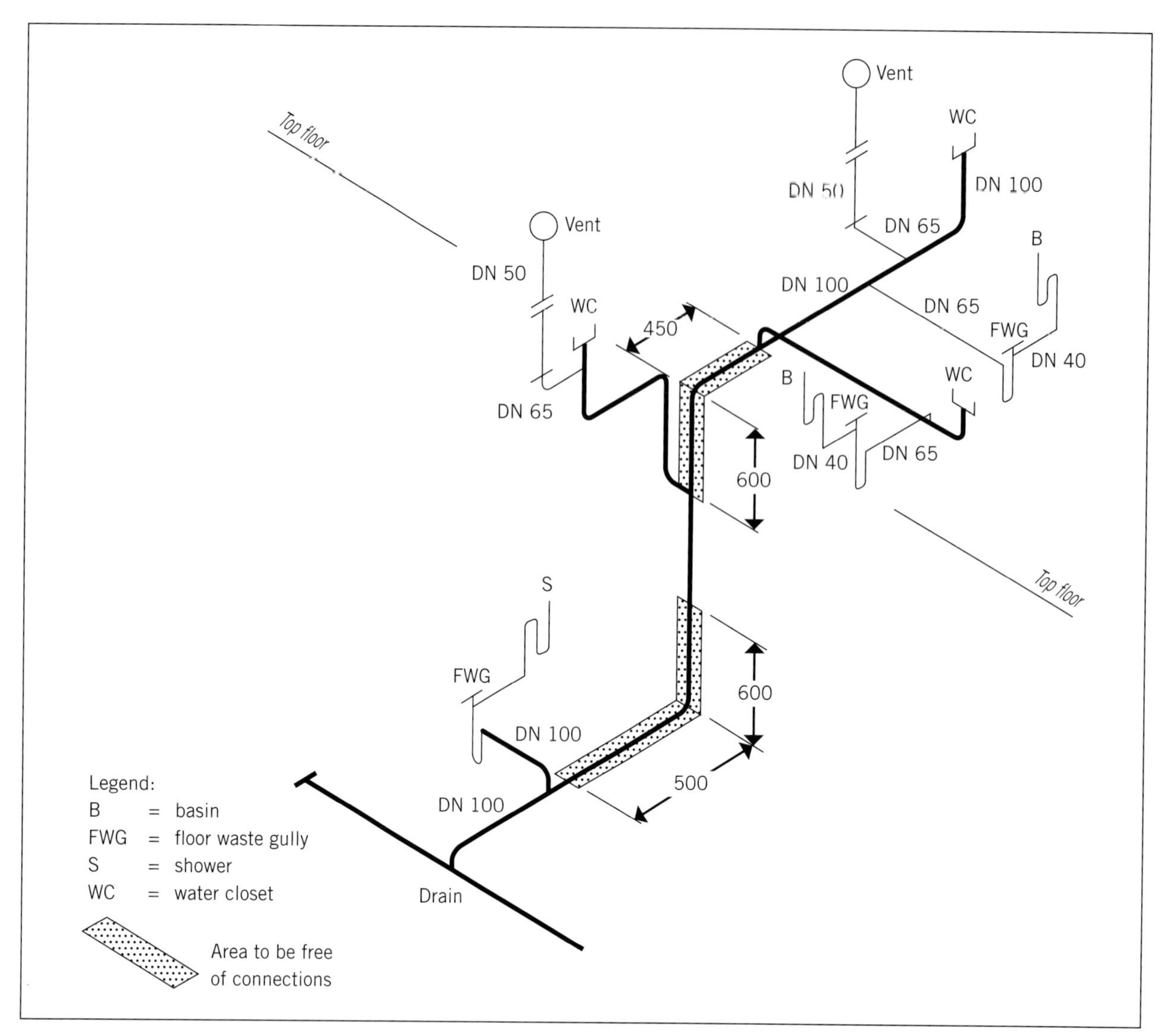

FIGURE 5.24 Single stack systems

Plumbers may also draw plans for the following purposes:

- as-executed drawings where original plans could not be adhered to
- workshop drawings for manufactured fixtures and fittings
- instructional sketches to employees
- verification of work completed (sectional plans that show invert levels and falls for installed sewer services required by the local sewerage authority)
- applications to commence work
- for future maintenance records.

LEARNING TASK 5.3

LOCATE SYMBOLS AND ABBREVIATIONS

Obtain a set of plans from the job you are currently working on now or use a set from a previous job (ask your employer or teacher for a copy).

List all the symbols and abbreviations on the plan, then identify the meaning of each.

Read and interpret job specifications

Reading and interpreting specifications and documents is a skill that requires practice. Accurate and quick interpretation of specifications requires a sound understanding of basic drawing symbols and abbreviations. Knowing where to look for the information required is a definite advantage. For example, the external size of the building is found on the floor plan; the position of the building on the building block is found on the site plan; the heights of walls and window heads are found on the elevation plan; and specific construction practice can be found on section or detail plans. The technical information, such as material size and type, that is not found on the plans will be described in the specifications.

GREEN TIP

Obtaining plans and documents through emails, tablets and smartphones can reduce your carbon footprint by reducing your printing requirements.

Specifications

A specification is a precise description of all finishes and workmanship that are not shown on the drawings. A hydraulics specification can include:

- types of materials
- types of fixtures
- colours of products
- brands of tapware
- brands of appliances
- size and type of hot water system
- number and position of garden hose taps.

The specifications are arranged according to the trade and cover all information relevant in the sequence of construction for that trade. Specifications must be kept with the drawings and are read in conjunction with them. If there is a conflict between the details on the working drawings and the details in the specification, then the details in the specification should take precedence (see Figure 5.25). However, if there is any conflict discovered, the architect or principal contractor should be consulted for verification.

LEARNING TASK 5.4

DETERMINE MATERIALS REQUIRED

Obtain a set of specifications from the job you are currently working on now or use a set from a previous job (ask your employer or teacher for a copy).

Make a list of the materials required based on the specification.

Obtain measurements and perform calculations

Reading and using measurements forms part of the everyday activities within each section of the building industry. A plumber is constantly measuring and cutting pipe to length; a bricklayer will continually measure the length and height of brickwork and determine the position of openings; a concreter will measure and cut timber formwork and quantify the volume of concrete required; and a dry wall plasterer will measure the length and height of walls and the position of windows and doors.

Measurements and dimensions obtained from plans and drawings are used to perform simple calculations and to determine material quantities for tasks or jobs within the general construction and plumbing industry. Accurate measurement will save time and wastage of materials, and reduce cost.

Correct units of measurement

The units of measurement adopted for the building industry are metric units taken from the *Système International d'Unités* (International System of Units, or SI). The SI units commonly used in construction, together with an example of typical use, are shown in Table 5.4. When using SI units, small letters are used for all symbols except where the value is over one million, such as the megapascal (MPa), which is used to measure how much pressure concrete is designed to withstand, or the pressure measured within a vessel or pipe.

Measuring tools are manufactured to match SI units. Accurate measuring using appropriate tools is critical when working on site (see Figure 5.26), as mistakes can affect job timelines and profit may be lost.

Note: Working drawings of houses and buildings generally have their dimensions shown in millimetres, although some long dimensions on site plans may be in metres. Centimetres are not recognised as true SI units and are therefore not used in the building industry. For example, a tradesperson may take a measurement of 5432 mm, and will call the measurement 5.432 m or five metres, four hundred and thirty-two millimetres.

Basic measuring tools

Measuring tools are essential items in construction and come in a variety of sizes, costs and uses. Most tradespeople carry the retractable metal tape on the job; however, for precise measurement a four-fold rule or steel rule may be used (see Figure 5.27). You must be able to use these measuring tools accurately.

Scale rule

The scale rule is a plastic rule, which measures 150 mm or 300 mm in length and is used to scale off dimensions on plans that are not given on the drawing (see Figure 5.28).

Four-fold rule (folding rule)

The four-fold rule is 1 m long and is made in four hinged sections so that it can be folded for convenience (see Figure 5.29). The rule is best used for measuring and marking out dimensions of less than 1 m.

Made of plastic, the rule blades are approximately 4 mm to 5 mm thick. The rule should be used on its edge so that the graduations marked on the blades of the rule are in contact with the surface of the material being measured or marked out.

Section 5 drainage

5.01 Regulations, inspections, tests

Comply with the acts, regulations, etc., of the local authority and all other relevant Authorities and pay all fees. Arrange for all necessary inspections and approvals. Supply all necessary apparatus for and carry out all tests required by the relevant authorities. Obtain and hand over to the Architect on completion certificates showing satisfactory installation of drainage from sanitary fittings, together with Council's official drainage diagram.

5.02 Drawings

Refer to drawing No. XYZ.

Drawings supplied are diagrammatic and show general layout only. Make detailed layout drawings and submit for approval if required before commencing work.

Make WORK AS EXECUTED drawings of all drainage work on completion and supply two copies to the architect. Show accurately thereon the depth and location of all pipes, fittings, cleaning eyes, pits, etc. The architect will supply, if requested at the commercial rate for printing, reproducible negative prints of any contract drawings needed to facilitate this work.

5.03 Generally to all drainage

The builder must consult with the architects as to the position of all pipes, vents, etc. before proceeding with the work. No drainage work to be covered up until it has been inspected and passed and when this has been done all trenches, etc. are to be filled in, thoroughly consolidated and all surfaces made good.

The work in this section shall generally comprise but not be limited to the following:

i Removal of all redundant works.

ii Provision of new 100 mm dia sewer line and connection to *existing vented stack* together with connection and installation of a new 100 mm dia vented sewer stack.

iii Provision of a boundary trap and connection to existing sewer line inside site boundary.

iv Provision of a gully trap and connection to sewer line.

v Provision of 100 mm stormwater line.

5.04 Excavate and fill

All relevant clauses in EXCAVATOR shall apply generally to this section.

Excavate to the extent, widths, grades and depths required for the correct laying of lines and fittings. Backfill with filling A.B.S. and mechanically compact.

5.05 Sewer drainage

5.05.01 uPVC PIPES – LAYING AND JOINTING

FIGURE 5.25 An example of a building specification showing the method in which drainage from the hydraulics part of the specification would be set out

Retractable metal tape

The retractable metal tape is available in lengths of approximately 1 m to 10 m, with the most common size being 8 m, and is the most suitable tool for measuring or marking dimensions and shorter distances (see Figure 5.30).

Metal retractable tapes have a hook on the end, which adjusts in or out depending on whether the measurement taken is internal or external. The hook slides a distance equal to its own thickness, so that an internal measurement, such as between two walls, begins from the outside of the hook. But an external measurement, such as from the end of a piece of timber, begins from the inside of the hook.

Care must be taken when returning the blade to the case to not allow the hook to slam against the case, which will stretch or distort the hook, making the tape inaccurate. It is also important to prevent the hook being torn from the end of the tape, making it unusable and resulting in a replacement being required.

Measurement and calculations

Construction workers regularly make calculations to obtain areas, volumes, quantities, and costing of materials.

Linear measurement

Linear measurement in the building industry is expressed in metres and millimetres, with 1 m equal

TABLE 5.4 SI units commonly used in construction

Unit	Function	Symbol	Use
Length			
• metre	unit of length	m	length of a building block, length of timber
• millimetre	1000 millimetres = 1 metre	mm	section, dimensions and lengths
Area			
• square metre	unit of area (m × m)	m^2	surface area of wall, area of floor
• hectare	unit of area (10000 m^2)	ha	land subdivision, building site area
Volume			
• cubic metre	unit of volume or capacity (m × m × m)	m^3	excavation of soil, concrete quantity
• litre	1000 litres = 1 cubic metre 1 litre of water = 1 kilogram	L	capacity of water heaters, rain water tanks Multiply m^3 by 1000 to obtain litres
Mass			
• gram	unit of mass	g	quantity of small nails
• kilogram	1000 grams = 1 kilogram	kg	mass of building material (e.g. a bag of cement)
• tonne	1000 kilograms = 1 tonne	T	safe working load of lifting equipment (e.g. cranes)
Force			
• newton	unit of force	N	calculation of design loads of buildings
Pressure			
• pascal	unit of pressure = 1 N/m^2	Pa	calculation of design loads to floors
• kilopascal	1000 pascals	kPa	gas-line operation and test pressures
• megapascal	1000000 pascals	MPa	calculation and specification of design strength of concrete

FIGURE 5.26 Accurate measuring and marking is critical

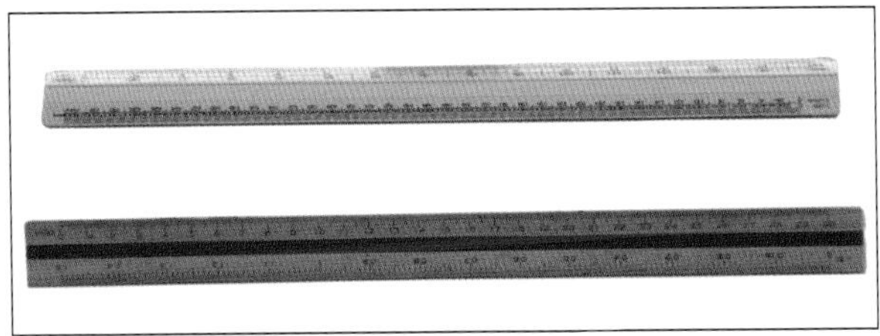

FIGURE 5.28 A typical scale rule

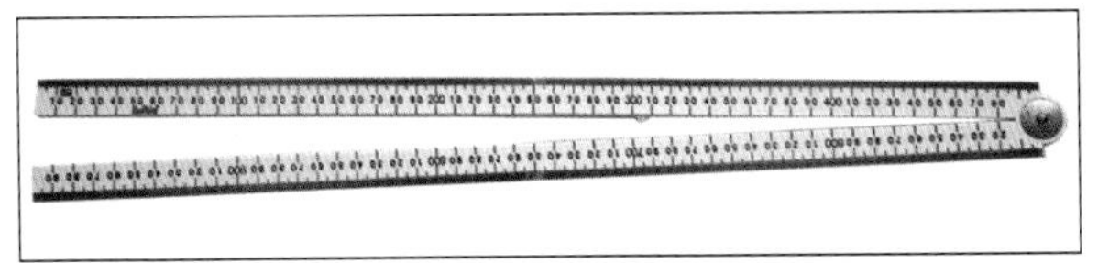

FIGURE 5.29 The four-fold rule

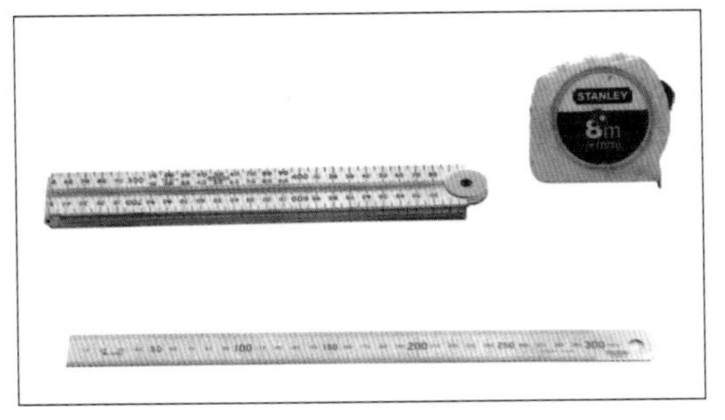

FIGURE 5.27 Select the correct measuring tools

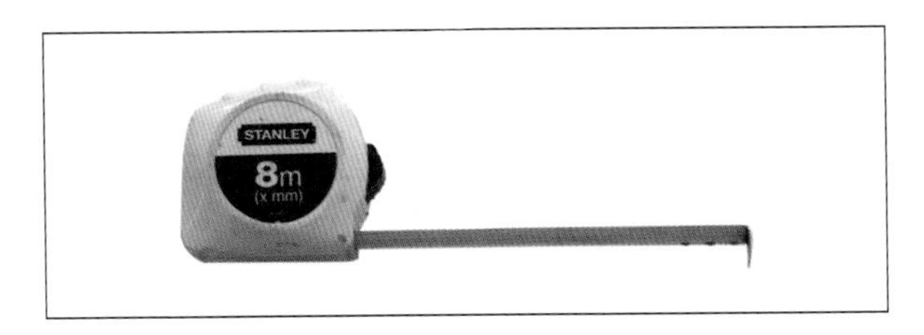

FIGURE 5.30 A typical retractable metal tape

to 1000 mm. Plans and drawings may use either measurement and may express linear measurement in metres by use of a decimal point to separate the metres from the millimetres, without the need for the symbol (m). For example, if a figure of 3600 is used, it is accepted as millimetres; however, 3.600 would indicate metres.

Plumbing pipe, guttering and sheet metal is sold in set multiple metre lengths (e.g. 20 mm type B copper tube is sold in 6 m lengths).

Linear measurement has many uses, as shown in Example 5.3 and Example 5.4.

EXAMPLE 5.3

Find the total number of linear metres for 115 mm quadrant gutter when a plumber requires 21 lengths @ 6.0 m, 12 lengths @ 2.4 m, 10 lengths @ 1.8 m.

21×6.0 m = 126.0 m,
12×2.4 m = 28.8 m,
10×1.8 m = 18 m
= 172.8 m

EXAMPLE 5.4

Find the total length (perimeter) of safety fencing required to enclose a site measuring 35.750 m + 23.500 m + 35.750 m + 23.500 m
= 118.5 m of fencing

Area measurement

Surface area measurements must be accompanied by the symbol m^2 and volume measurements must be accompanied by the symbol m^3 to ensure correct interpretation of the calculation (e.g. $2.500 \times 5.000 = 12.5$ m^2, which indicates surface area; and $2.500 \times 5.000 \times 0.150 = 1.875$ m^3, which indicates volume).

The area of an object or surface is determined by the number of square metres that the shape contains and can be calculated on a range of shapes that are both irregular and regular in size. The area measurement of a hectare (ha) is used for land subdivision because of the larger areas involved, but when measuring or describing building blocks the unit of measurement is square metres (m^2). Area measurement in square metres is used in building construction to determine, for example:

- floor area of a building, used as a means of describing the size of the building
- roof area to determine the quantity of sheet roofing
- area of a building block to determine minimum and/or maximum coverage for building regulations.

FROM EXPERIENCE

It is important to have a thorough knowledge of how surface areas and volumes are calculated as plumbers are required to problem-solve strategies across a range of areas.

Calculating the surface area is also used to obtain the number of bricks required to construct a wall, or for the number of pavers required for a path or driveway.

The area of plane figures and rectangles is found by multiplying the length by its width ($L \times W$). Examples 5.5–5.12 are typical area measurements for regular shapes.

EXAMPLE 5.5

A building with a flat roof measures 14.600 m × 9.100 m. Calculate the area of roof sheets required, allowing for 10% waste when cutting.

Area of the roof = 14.600 m × 9.100 m
= 132.860 m^2
Add 10% for cutting waste
= 132.860 m^2 × 0.10 (10%) = 13.286 m^2
= 132.860 m^2 + 13.286 m^2
= 146.146 m^2 of sheet roofing required

EXAMPLE 5.6

Determine the number of roofing sheets required for Example 5.5, if each sheet is 3.600 m long with a 0.760 m wide effective cover.

Area to be covered = 146.146 m^2
Area of one sheet 3.600 m × 0.760 m (effective cover) = 2.736 m^2
Therefore 146.146 ÷ 2.736 = 53.41 = 54 sheets

Note: The actual width of the roof sheet will be wider than the 'effective cover'. This allowance is made as roof sheets will overlap when installing.

EXAMPLE 5.7

A wall 5.400 m long and 2.400 m in height is to be covered with metal wall cladding. Calculate the net wall area if the wall has a window 2.400 m × 1.200 m and a door 2.100 m × 0.900 m.

Area of wall 5.400 m × 2.400 m = 12.960 m^2
Deductions for:
Area of window 2.400 m × 1.200 m
= 2.880 m^2
Area of door 2.100 m × 0.900 m
= 1.890 m^2
Total deductions (window and door) = 4.770 m^2
Net wall area 12.960 m – 4.770 m = 8.190 m^2 of metal sheeting required

EXAMPLE 5.8

The gable roof of a building 16.750 m long with a rafter length of 3.600 m is to be covered with Custom Orb® roof sheeting. Calculate the area of the gable roof and the number of full-length sheets for each side, if the effective cover of one sheet is 760 mm (0.760 m).

Area of each side of the roof is 16.750 m × 3.600 m = 60.300 m². As a gable roof has two sloping sides the area calculated will be doubled to allow for roof sheets to cover both sides. Total area of the roof will then be: 120.600 m².

The total number of sheets required will be:

Number of sheets per side × 2
= 16.750 m ÷ 0.760 m (effective cover)
= 22.039 (say 23) × 2 = 46

Therefore, 46 sheets will be required to cover the entire roof.

EXAMPLE 5.9

Calculate the area of roofing for a triangular-shaped carport with a base length of 8.950 m and a perpendicular length of 10.250 m.

Area of a triangle = ½ base × perpendicular length
therefore (8.950 ÷ 2) × 10.250 (dividing the base by 2 will give half of its length)
= 45.869 m² roofing

Area measurement for irregular shapes in the next set of examples are for shapes with straight sides, such as irregular quadrilaterals (four sides) or irregular polygons (five or more sides).

Irregular figures may be subdivided into triangles, parallelograms or rectangles. To find the area of the whole figure, calculate the individual areas of the subdivided shapes and add all the parts together.

In some cases, areas may be determined by considering the figure as a rectangle from which parts have been removed. For example, if a quadrilateral has only two parallel sides, its area may be found provided the length of the two parallel sides and the perpendicular distance between them is known. The formula to calculate the area of this figure is:

½ the sum of the parallel sides × perpendicular distance between parallel sides

Typical examples of calculations involving irregular shapes are provided in Examples 5.10–5.12.

EXAMPLE 5.10

Find the area of Figure 5.31.

$$\text{Area} = \left(\frac{20.000 + 36.000}{2}\right) \times 10.000$$

= 28.000 × 10.000
Total area = 280 m²

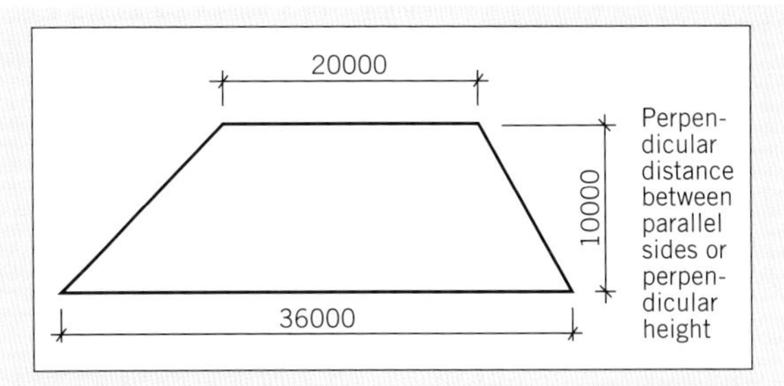

FIGURE 5.31 Trapezoid

EXAMPLE 5.11

Find the area of the block of land shown in Figure 5.32.

$$\text{Area} = \left(\frac{90.000 + 96.000}{2}\right) \times 46.000$$

= 93.000 × 46.000
Total area = 4278 m² of land

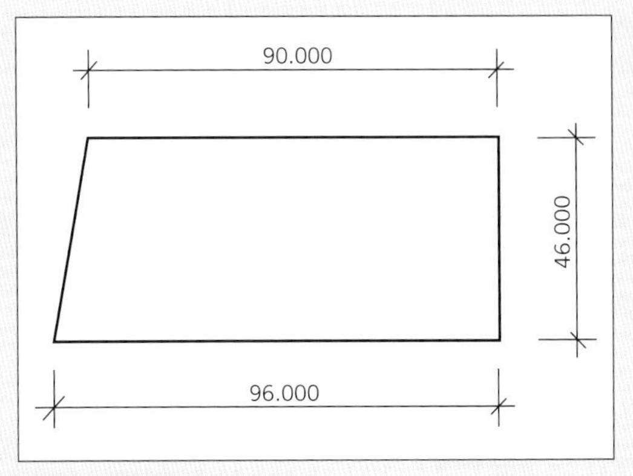

FIGURE 5.32 Quadrilateral

EXAMPLE 5.12

Calculate the area of geotextile matting required for a transpiration bed with the dimensions shown in Figure 5.33. Add 15% for waste to the total area calculated.

Area of triangle A = 14.400 ÷ 2 × 4.600
= 33.120 m²

Area of triangle B = 14.400÷ 2 × 1.600
= 11.520 m²

Total area of matting required

33.120 + 11.520 = 44.640 m²

Add 15% for waste:

44.640 m² × 0.15(15%) = 6.696 m²
Total matting required = 51.336 m²

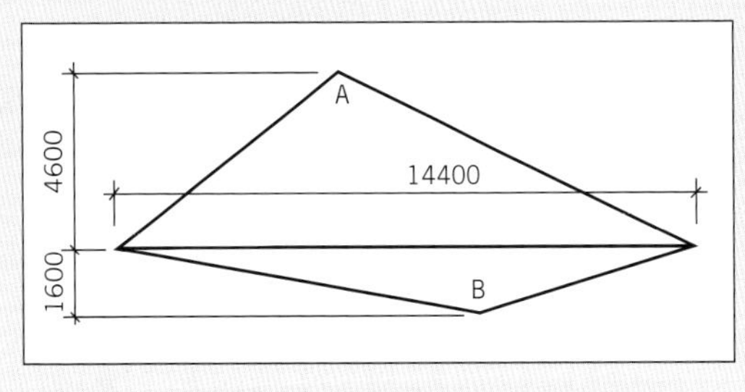

FIGURE 5.33 Polygon-shaped transpiration bed

Volume measurement

Volume is measured in cubic metres (m^3) and is found by multiplying the length by the width by the depth (or thickness).

Calculations of volume in the industry are used to determine, for example:

- volume of soil excavated from a drainage trench
- volume of soil to be removed over an area of a sloping site to provide a level area
- amount of material required as filling (e.g. crushed gravel as pipe support and embedment)
- quantity of materials to be able to do a job (e.g. cubic metres of sand to backfill a trench in a road)
- volume of concrete for an access chamber or stormwater pit.

Typical examples of calculations of quantities are provided in Examples 5.13 to 5.15.

EXAMPLE 5.13

A trench is to be excavated in soil on a level site for a water main to a depth of 600 mm from the top of the ground. The trench is to be 18.500 m long by 0.450 m in width. Calculate the cubic metres of soil to be excavated.

Volume = length × width × depth

Amount of soil to be excavated

= 18.500 × 0.450 × 0.600

= 4.995 m^3 of soil removed

EXAMPLE 5.14

Calculate the volume of sand required to backfill the trench in Example 5.14, if the depth of the backfill is 450 mm.

Volume = length × width × depth

Amount of sand required

= 18.500 × 0.450 × 0.450

= 3.746 m^3

EXAMPLE 5.15

Calculate the volume of concrete for a slab-on-ground for an LPG tank base. The base is 8.600 m long by 4.200 m wide and is 200 mm thick.

Amount of concrete required

= 8.600 m × 4.200 m × 0.200 m

= 7.224 m^3

When calculating the volume of concrete, a percentage is added to the total to allow for irregular forms, varying thickness, and losses caused by spilling when transporting and depositing the concrete. Typical waste allowances would be 10% for concrete poured on the ground and 5% for concrete placed into forms.

Calculation of solid shapes

This section outlines the various calculation methods and characteristics of common three-dimensional objects found in the construction industry. It highlights the surface or pattern development techniques required in the construction industry, which will be discussed further in Chapter 9.

Prisms

These are solid objects with two ends formed by straight-sided figures, which are identical and parallel to one another (see Figure 5.34). The sides of the prisms become *parallelograms*. The ends may be formed by common plane geometric shapes (e.g. square, rectangle, triangle, pentagon, hexagon or octagon).

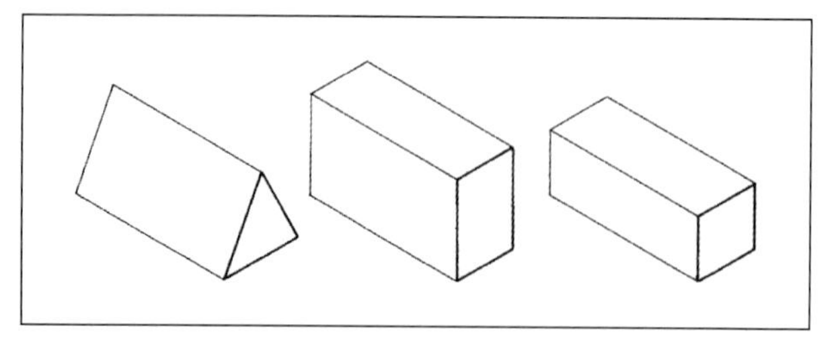

FIGURE 5.34 The prism

Cylinders

These have their ends formed by circles of equal diameter. The ends are parallel and joined by a uniformly curved surface (see Figure 5.35).

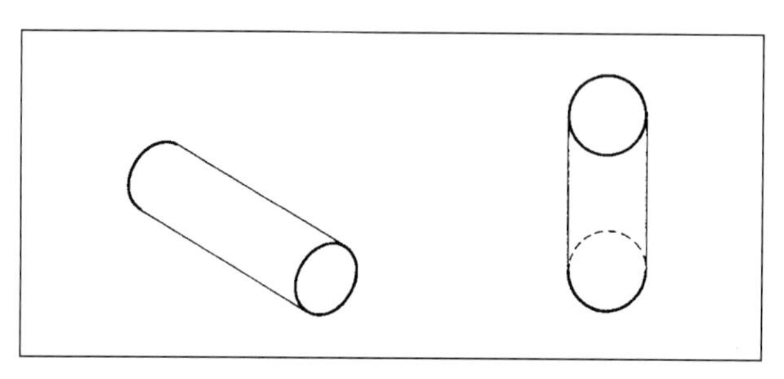

FIGURE 5.35 The cylinder

Cones

These have a circular base and a uniformly curved surface, which tapers to a point called the *apex* (see Figure 5.36).

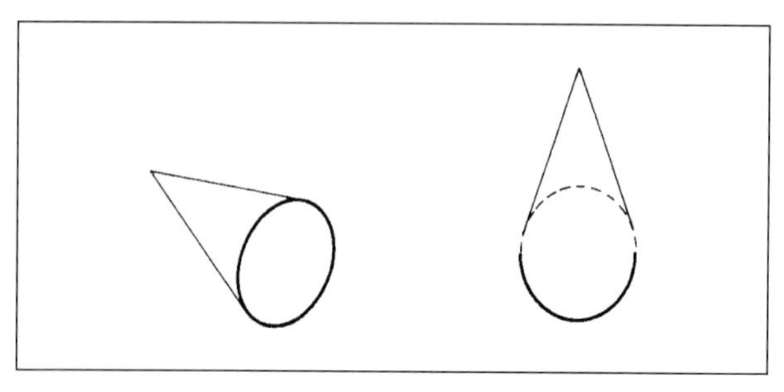

FIGURE 5.36 The cone

Pyramids

These are solid shapes with a base consisting of a straight-sided figure (square) and triangular sides, which terminate at the apex (see Figure 5.37).

FIGURE 5.37 The pyramid

Surface development and area of solid shapes

When the surface of a solid object requires measuring, or a true shape is required to create a template, the simplest way to provide an accurate detail is to develop the surface. This requires the true shape of all sides to be laid out flat as a continuous surface.

Drawing the detail in a two-dimensional view is not the only way to determine the amount of material required to form the surface shape. For example, calculation of the surface area would be a more effective method. The idea of laying the shape of each surface out flat still applies, but it's the calculation of each area added together that provides the information required (see **Figure 5.38**). The calculation of these surfaces is required when a plumber needs to order roof sheets for a conical- or pyramidal-style roof surface, or to work out how much sheet metal is required for a prism-like member; or a tank maker needs to work out how much Colorbond® Custom Orb® is required to create a water storage tank.

EXAMPLE 5.16

Calculate the surface area of a rectangular prism with a length of 5.540 m, a width of 1.250 m and a height of 850 mm.

$$\text{Area} = (\text{area of base} \times 2) + (\text{area of one side} \times 2) + (\text{area of one end} \times 2)$$
$$= [(5.540 \times 1.250) \times 2] + [(5.540 \times 0.850) \times 2] + [(1.250 \times 0.850) \times 2]$$
$$= 13.850 + 9.418 + 2.125 = 25.393\ \text{m}^2$$

EXAMPLE 5.17

Calculate the surface area of a cone having an inclined length of 1.800 m and a base radius of 600 mm.

$$\text{Area} = (\pi \times r \times \text{length of incline}) + (\pi r^2)$$
$$= (3.142 \times 0.600 \times 1.800) + (3.142 \times 0.6 \times 0.6)$$
$$= 3.393 + 1.131$$
$$= 4.524\ \text{m}^2$$

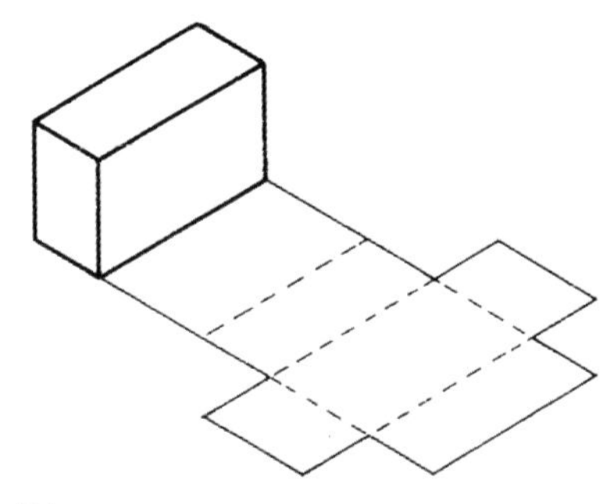

Prism
Area = (area of base × 2) + (area of 1 side × 2) + (area of 1 end × 2)
$= (L \times W \times 2) + (L \times W \times 2) + (L \times W \times 2)$

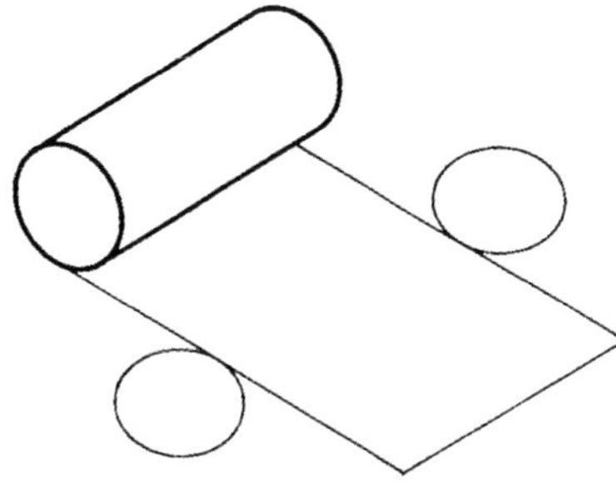

Cylinder
Area = (area of 1 end × 2) + (area of surface)
$= (\pi r^2 \times 2) + (2\pi r \times \text{height})$

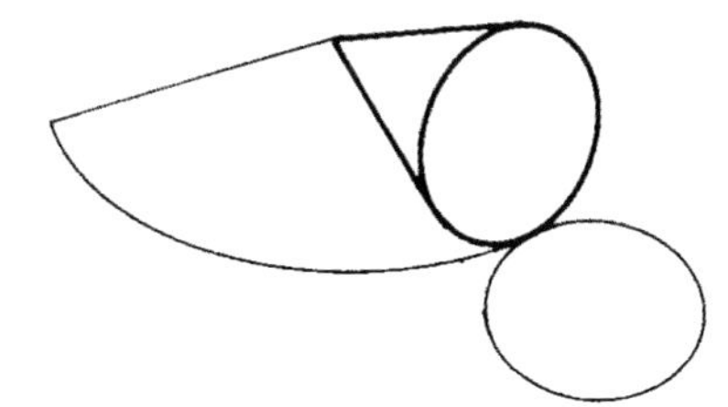

Cone
Area = (πr Length of incline) + (πr^2 base)
$= (\pi r L) + (\pi r^2)$

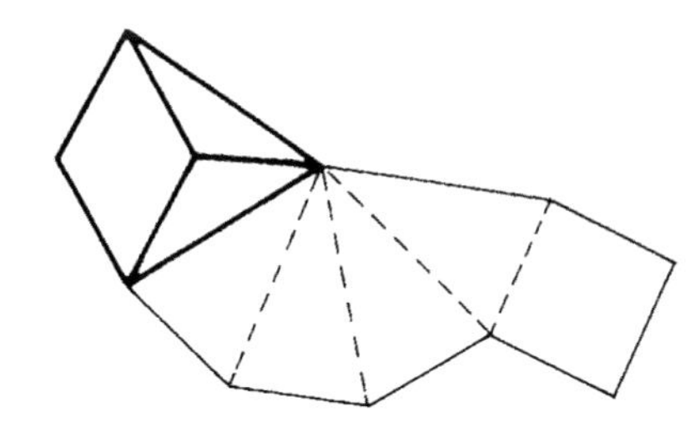

Pyramid
Area = (area of base) + [(area of 1 side) × 4]
$= (L \times W) + [(\frac{1}{2}\ \text{base} \times \text{height}) \times 4]$

FIGURE 5.38 Surface area formulae

EXAMPLE 5.18

Calculate the surface area of a cylinder with a height of 2.100 m and a base radius of 850 mm.

Surface area = (area of base × 2) + area of cylinder wall

$= (\pi r^2 \times 2) + (\pi \times r \times 2 \times \text{height})$

$= (3.142 \times 0.850 \times 0.850 \times 2) + (3.142 \times 0.850 \times 2 \times 2.100)$

Area $= 4.540 + 11.217$

$= 15.757 \text{ m}^2$

EXAMPLE 5.19

Calculate the surface area of a pyramid having a square base side of 1.200 m and an inclined perpendicular height of 1.800 m.

$$\text{Area} = \text{area of base} + \left(\frac{\text{length of base side}}{2} \times \text{inclined height} \times 4\right)$$

$$= (1.2 \times 1.2) + \left(\frac{1.2}{2} \times 1.8 \times 4\right)$$

$= 1.440 + 4.320$
$= 5.760 \text{ m}^2$

(The inclined perpendicular height is the distance from the centre of the base to the apex of the pyramid, up to the sloping face of the pyramid.)

Calculation of percentages

A percentage is a common way of expressing a fraction (e.g. ¼, ½ or ¾ as a part of the whole; i.e. 100), and written as a number followed by a percentage symbol (e.g. 25%, 50% or 75%).

Percentages are commonly used in building applications to calculate the fall of drainage pipework by using the grade as a percentage or determining wastage allowance for materials. The percentage must always be a proportion of the whole, which is 100%.

The simplest way to determine a percentage is to write down the fraction you want converted, then multiply it by 100 over 1, as illustrated in Examples 5.20–5.24.

EXAMPLE 5.20

$\frac{1}{2} \times \frac{100}{1}$ (now multiply the numerators together, then the denominators)

$= (1 \times 100)$ over $(2 \times 1) = \frac{100}{2}$

(now divide 100 by 2) = 50

Therefore, $\frac{1}{2}$ is equal to 50%.

EXAMPLE 5.21

Convert ³⁄₈ to a percentage of 100.

$$= \frac{3}{8} \times \frac{100}{1} = \frac{300}{8} = 37.5\%.$$

EXAMPLE 5.22

Convert ⁵⁄₁₆ to a percentage of 100:

$$= \frac{5}{16} \times \frac{100}{1} = \frac{500}{16} = 31.25\%.$$

This process may also be used in reverse to convert a percentage to a fraction, as follows.

EXAMPLE 5.23

This process may also be used in reverse to convert a percentage to a fraction.

$$45\% = \frac{45}{100}$$

This fraction should always be broken down to its simplest form (i.e. divide the top and bottom figures by a number common to both; e.g. 5)

$$= \frac{9}{20}$$

Percentages may also be converted to decimals by determining their value in relation to 1, as follows.

Note: Always round off decimals to three decimal places.

EXAMPLE 5.24

Percentages may also be converted to decimals by determining their value in relation to 1, as follows.

$45\% = 45 \div 100$

Move the decimal point, which automatically would be after 45 (i.e. 45.0) back to the left by the number of zeros in 100 (i.e. 2) = 0.450.

Note: Always round off decimals to three decimal places.

LEARNING TASK 5.5

OBTAIN MEASUREMENTS AND PERFORM CALCULATIONS

In the room you are currently in, determine the following using the appropriate units of measurement:

- the area of the ceiling
- the area of the walls, excluding doors and windows
- the volume of the floor, assuming it is concrete slab and has a thickness of 150 mm.

Calculate material quantities

Calculation of material quantities is a requirement for all projects undertaken in the building industry.

The principles of calculating material quantities are the same for a metal toolbox as they are for a multistorey building development. A method must be adopted that allows for the planning and calculation of material quantities in sequential order, in order to prevent mistakes or exclude materials (see Figure 5.39).

FIGURE 5.39 Plan for the process

Material quantities may be taken from plans, drawings and specifications or directly from the worksite.

All measurements that have been identified should be recorded and stored in an appropriate manner to ensure they can be obtained for later use where required.

GREEN TIP

Ordering and using the correct amount of materials on site will not only save money but can also avoid excess material being wasted and sent to landfill.

Material units

It is essential not only to identify the various materials required, but also to determine the shape, size and form in which they are available (see Figure 5.40).

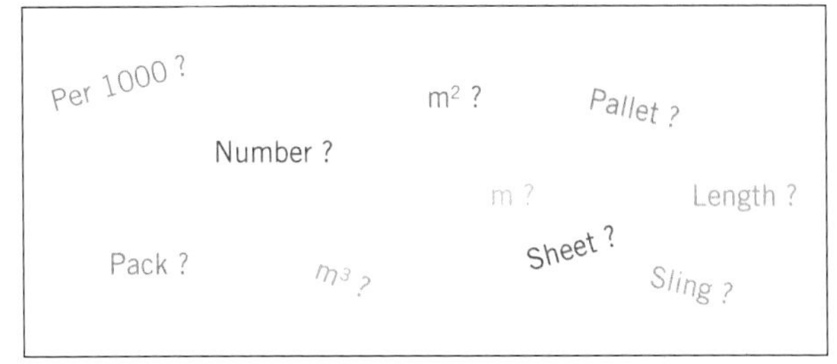

FIGURE 5.40 Which unit to use

When ordering materials for purchase or delivery, it is important to identify in what quantity the item is sold. 'Quantities' is a term used in the building industry for the calculation of materials required for a task; for example:

- Standard roof gutter is available in set lengths.
- Sheet material may be purchased by the sheet, or in packs of various quantities.
- Large quantities of pipe may be purchased by the 'sling'.
- Smaller quantities of pipe may be purchased by the metre or 'length'.
- Roofing screws are usually calculated by the number, by the thousand or by the box.
- Bedding gravel, sand or concrete is calculated by the cubic metre (m^3).

Generally, these materials are priced or costed using the same units of measurement and it may be more economical and convenient to purchase larger quantities of materials, where storage permits.

Calculating material quantities

Determining quantities may include the cost to supply materials, or a total figure that would be the total cost of supply and fixing of the materials.

Many plumbers will quote jobs on a 'per metre' basis or 'per point' basis:

- Drainage may be quoted based on the size and average depth of a pipe (e.g. 30 m of DN100 PVC-U pipe at 1 m deep) at a calculated dollar cost per metre. This would include the cost of supplying and installing the pipe complete with excavation, embedment, backfill, removal of excess soil and returning the surface and surrounds to its original condition.
- Water services may be quoted on a 'per point' basis. This would include the supply and installation of the pipe, fittings and fixings. Extra costs are then added for items such as authorities' fees, approvals and inspections, non-standard valves and prime cost (PC) items.

Typical examples of calculations of quantities are provided in Example 5.25.

EXAMPLE 5.25

Using the following information, calculate the quantities required and the cost of materials to install quadrant gutter and Klip-Lok® roof sheets to the flat roof on the building shown in Figure 5.41.

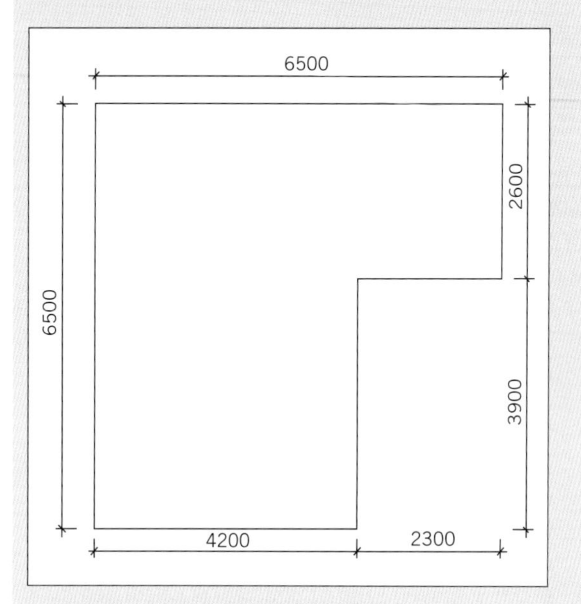

FIGURE 5.41 Irregular-shaped building

Quantities:

- 115 mm quadrant gutter
- 410 mm Klip-Lok® roof sheets.

Material costs:

- 115 mm quad gutter @ $8.80 per metre
- 410 mm Klip-Lok® roof sheets @ $13.40 per square metre.

1 Calculate the lineal metres of quad gutter by totalling the perimeter measurements of the building:

Perimeter = 6.500 + 2.600 + 2.300 + 3.900 + 4.200 + 6.500
= 26.000 m of quad gutter

2 Calculate the cost of the gutter:

Cost = 26 × $8.80
= $228.80

3 Calculate the area of roof sheeting by separating the roof into two sections:

Area = length × width
Section (a): 6.500 m × 2.600 m = 16.9 m^2
Section (b): 4.200 m × 3.900 m = 16.38 m^2
Add sections together:
16.9 m + 16.38 m = 33.28 m^2

4 Calculate the cost of the roof sheeting.

Cost = 33.28 m^2 × $13.40 = $445.95
(with no allowance for waste)

Before ordering any materials, it is advisable to double-check measurements on the site itself, particularly when ordering items such as roof sheets. Do not scale from the plans as they may differ slightly.

LEARNING TASK 5.6

CALCULATE MATERIAL QUANTITIES

1 Determine the volume in cubic metres of embedment material required for a trench that is 20 m long and 0.3 m wide, with an embedment depth of 0.2 m.
2 Allow for 10% wastage in addition to the volume required.
3 From a manufacturer's website, calculate a cost based on 10 mm aggregate (blue metal) or similar material for the volume required.

Clean up

Once all calculations have been quantified, it is essential that the work area is cleared in accordance with workplace procedures and that all materials are disposed of in accordance with local regulatory requirements, being reused or recycled where appropriate.

Tools and equipment used in reading plans and calculating quantities must be checked, cleaned and maintained in accordance with manufacturers' requirements and appropriately stored in accordance with workplace procedures. Documents that have been accessed must be completed in accordance with workplace and regulatory requirements.

REFERENCES AND FURTHER READING

Acknowledgement

Reproduction of the following resource list references from DET, TAFE NSW C&T Division (Karl Dunkel, Program Manager, Housing and Furniture) and the Product Advisory Committee is acknowledged and appreciated.

Texts

Australian Vocational Training Scheme (1996), *Student notes Module 7974C – Introduction to plan reading and maintenance*, Australian Vocational Training Scheme.

Liebing, R.W. (1990), *Architectural working drawings*, Wiley, New York.

Major, S.P. (1995), *Architectural woodwork: Details for construction*, Van Nostrand Reinhold, New York.

Web-based resources

Regulations/Codes/Laws

NSW Environment Protection Authority: **http://www.epa.nsw.gov.au**

Standards Australia: **http://www.standards.org.au**

Resource tools and VET links

NSW Education Standards Authority – Construction: **https://educationstandards.nsw.edu.au/wps/portal/nesa/11-12/stage-6-learning-areas/vet/construction-syllabus**

Training.gov.au: **http://training.gov.au**

Industry organisation sites

Australian Institute of Quantity Surveyors: **http://www.aiqs.com.au**

TAFE NSW resources

Training and Education Support (TES), Industry Skills Unit Orange/Granville 68 South St Granville NSW 2142 Ph: (02) 9846 8126 Fax: (02) 9846 8148 (Resource list and order forms)

GET IT RIGHT

The example below demonstrates an incorrect practice that can be performed when calculating volumes. Identify the incorrect method and provide reasoning for your answer.

Calculate the volume of concrete in cubic metres required for a slab measuring 4.6 m long by 2.5 m wide and 150 mm thick.

Volume = length × width × depth

= 4.6 m × 2.5 m × 150 mm

= 1725 m^3

Why is there so much concrete?

WORKSHEET 1

To be completed by teachers	
Satisfactory	☐
Not satisfactory	☐

Student name: ____________________

Enrolment year: ____________________

Class code: ____________________

Competency name/Number: ____________________

Task: Review the sections 'Identify types of plans and drawings and their functions' and 'Identify commonly used scales, symbols and abbreviations' in this chapter and answer the following questions.

1. What are plans and drawings used for in the building industry?

____

____

2. Name the key users of plans and drawings.

____

____

____

____

____

____

____

____

3. What are working drawings commonly referred to as?

____

____

4. List two items that would be shown on a floor plan.

____

____

5. What is the purpose of a section elevation?

____

____

6. List three details commonly found on a sewerage service diagram.

____

____

____

7. State the appropriate scales used to create working drawings in the building industry for the following views:

Site plans ________________ Elevations ________________

Details ________________ Sections ________________

8. State the basic information found in a plan title block for a residential building.

____

WORKSHEET 2

To be completed by teachers	
Satisfactory	☐
Not satisfactory	☐

Student name: ____________________

Enrolment year: ____________________

Class code: ____________________

Competency name/Number: ____________________

Task: Review the sections 'Identify types of plans and drawings and their functions' and 'Identify commonly used scales, symbols and abbreviations' in this chapter. Then, using the plan in Figure 5.7, answer the following questions.

1. What is the name of the view shown in Figure 5.7?

   ____

2. How many rooms on the ground floor plan require plumbing services (e.g. water, waste and gas) to be installed?

   ____

3. Make a list of the seven rooms that require plumbing services.

   ____

   ____

   ____

   ____

   ____

   ____

   ____

4. How many sanitary fixtures are shown in the en suite off bedroom 1?

   ____

   ____

5. What are the internal dimensions, excluding the shower, of en suite 1?

   ____

6. Compile a list of the plumbing fixtures required to be connected to the water supply on the ground and first floor. List each room and its fixtures separately.

7. What is the scale of this plan?

8. What do the symbols (A)— —(A) represent on the drawing? Why are they used?

WORKSHEET 3

To be completed by teachers	
Satisfactory	☐
Not satisfactory	☐

Student name: ____________________

Enrolment year: ____________________

Class code: ____________________

Competency name/Number: ____________________

Task: Review the section 'Locate and identify key features on a services plan' in this chapter. Then, using the plan in Figure 5.23, answer the following questions.

1. What is the 'RL' of the datum point?

   ____

2. What is the length of the finished floor level (FFL) of the building?

   ____

3. What does the abbreviation 'RL' stand for?

   ____

4. What does the abbreviation 'IL' stand for?

   ____

5. Where is the sewer access chamber located? State the orientation.

   ____

6. What is the IL and the depth of the sewer access chamber?

   ____

7. How many metres is the council building line set in from the street alignment?

   ____

8. What is the total area of the building block?

   ____

9. What is the length of the west side boundary?

   ____

10. What is the width of the front property boundary?

    ____

11. What is the approximate fall along the east side boundary?

    ____

12. What material is the sewer main constructed of, and what size is it?

____

13. What material is the stormwater main constructed of, and what size is it?

____

WORKSHEET 4

To be completed by teachers

Satisfactory ☐

Not satisfactory ☐

Student name: ______________________________

Enrolment year: ______________________________

Class code: ______________________________

Competency name/Number: ______________________________

Task: Review the sections 'Read and interpret job specifications' and 'Obtain measurements and perform calculations' in this chapter and answer the following questions.

1. What is the purpose of a specification?

____

2. List six items that may be listed on a hydraulics specification.

____

____

____

____

____

____

3. What is the unit of measurement used to measure how much pressure concrete can withstand?

____

4. State the two units of measure that tradespeople use on site in relation to dimensions and measurements.

____

____

5. Name the two measuring tools that most tradespeople carry on the site.

____

____

6. Describe the uses of each of the measuring tools above.

____

____

____

____

To be completed by teachers	
Satisfactory	☐
Not satisfactory	☐

Student name: ______________________________

Enrolment year: ______________________________

Class code: ______________________________

Competency name/Number: ______________________________

Task: Review the section 'Obtain measurements and perform calculations' up to and including 'Volume measurement' in this chapter and answer the following questions.

1. Using the conversion examples shown, convert the measurements in the exercise from millimetres to metres and metres to millimetres respectively.

Conversion examples

100s of mm in m	10s of mm in m	Individual mm in m
1000 mm = 1.0 m	90 mm = 0.09 m	9 mm = 0.009 m
900 mm = 0.9 m	80 mm = 0.08 m	8 mm = 0.008 m
800 mm = 0.8 m	70 mm = 0.07 m	7 mm = 0.007 m
700 mm = 0.7 m	60 mm = 0.06 m	6 mm = 0.006 m
600 mm = 0.6 m	50 mm = 0.05 m	5 mm = 0.005 m
500 mm = 0.5 m	45 mm = 0.045 m	4 mm = 0.004 m
400 mm = 0.4 m	40 mm = 0.04 m	3 mm = 0.003 m
300 mm = 0.3 m	35 mm = 0.035 m	2 mm = 0.002 m
200 mm = 0.2 m	30 mm = 0.03 m	1 mm = 0.001 m
100 mm = 0.1 m	25 mm = 0.025 m	0.5 mm = 0.0005 m

Exercise

Convert from millimetres to metres		Convert from metres to millimetres	
745 mm	=	6.0 m	=
107 250 mm	=	536.45 m	=
50 248 mm	=	27.01 m	=
3 mm	=	0.052 m	=
67 mm	=	54.209 m	=
128 mm	=	0.002 m	=
7002 mm	=	9.6 m	=
22 045 mm	=	11.08 m	=
556 mm	=	457.02 m	=
33 333 mm	=	3.44 m	=

2. Add the following measurements to find the total in metres.

5.35 + 0.345 + 11.5 = ____________________

27.467 + 0.004 + 3.32 = ____________________

3. Subtract the following measurements to find the total in metres.

   7.005 – 0.456 = ____________

   345.450 – 15.01 = ____________

4. Multiply the following measurements to find the total in square metres.

   16.7 × 5.433 = ____________

   45.00 × 0.055 = ____________

5. Divide the following measurements.

   15.00 ÷ 0.5 = ____________

   72.40 ÷ 6.2 = ____________

6. Calculate the area of a square with sides 6 m long.

   ____

7. Calculate the perimeter in metres of a square with sides 2.750 m long.

   ____

8. Calculate the area in square metres for a rectangle 5.35 m long by 4.75 m wide.

   ____

9. Calculate the perimeter in metres for a rectangle with sides 17.5 m long by 6.25 m wide.

   ____

10. Calculate the area in square metres for a parallelogram 3.75 m long and with a perpendicular height of 1.2 m.

    ____

11. Calculate the perimeter in metres for a parallelogram 5.520 m long with inclined sides 2.750 m long.

    ____

12. Calculate the area in square metres for a triangle with a base 16.6 m long and a perpendicular height of 3.2 m.

    ____

13. Calculate the perimeter in metres for a triangle with a base 4.5 m long and equal-length sides of 3.650 m long.

    ____

14. Calculate the circumference in metres for a circle with a radius of 1.5 m.

    ____

15. Calculate the area in square metres for a circle with a radius of 2.55 m.

___

16. Calculate the volume in cubic metres for a cube with sides 2.5 m long.

___

17. Calculate the surface area of a cube with sides 3.2 m long.

___

18. Calculate the volume in cubic metres for a rectangular prism with a length of 2.8 m, width of 1.75 m and height of 0.85 m.

___

19. Calculate the surface area in square metres for a rectangular prism with a length of 2.4 m, width of 1.2 m and height of 0.9 m.

___

20. Calculate the volume in cubic metres for a cylinder with a base diameter of 1800 mm and a perpendicular height of 3.6 m.

___

WORKSHEET 6

To be completed by teachers	
Satisfactory	☐
Not satisfactory	☐

Student name: ______________________

Enrolment year: ______________________

Class code: ______________________

Competency name/Number: ______________________

Task: Review the section 'Obtain measurements and perform calculations' in this chapter and answer the following questions.

1. Calculate the surface area of a rectangular prism with a length of 5.75 m, a width of 1.385 m and a height of 890 mm.

   Area = (area of wide side × 2) + (area of narrow side × 2) + (area of one end × 2)

   ____

2. Calculate the surface area of a cone with an inclined length of 2.580 m and a base radius of 1200 mm.

   Area = (πr × length of incline) + πr^2

   ____

3. Calculate the surface area of a cylinder with a height of 2.45 m and a base radius of 850 mm.

   Area = ($\pi r^2 \times 2$) + ($2\pi r$ × height)

   ____

4. Calculate the surface area of a pyramid with a square base side of 1.550 m and an inclined perpendicular height of 2.7 m.

   $$\text{Area} = \text{area of base} + \left(\frac{\text{length of base side}}{2} \times \text{inclined height} \times 4\right)$$

   ____

WORKSHEETS 5

WORKSHEET 7

To be completed by teachers	
Satisfactory	☐
Not satisfactory	☐

Student name: ______________________

Enrolment year: ______________________

Class code: ______________________

Competency name/Number: ______________________

WORKSHEETS

5

Task: Review the section 'Obtain measurements and perform calculations' up to and including 'Calculate material quantities' in this chapter and answer the following questions.

1. Convert the following percentages to fractions.

   75% = 7% =

   35% = 140% =

   ____

2. Convert the following fractions to percentages.

   $\frac{1}{40}=$ $\frac{1}{60}=$

   ____

3. Convert the following percentages to decimals.

   60% = 22.5% =

   ____

4. Surface area

   The flat surface of a skillion roof is 24.0 m long with a rafter length of 12.0 m. If a 12.0 m length of corrugated iron roof sheeting covers a width (effective cover) of 762 mm, how many sheets are required to cover the roof?

   ____

5. If corrugated iron costs $1.75 per m², how much would it cost to sheet the roof when each actual sheet width is 820 mm?

   ____

6. How many Colorbond® sheet metal panels of 2.4 m long × 1.2 m high would be required to build a fence 32.0 m long by 1.2 m high?

   ____

7. If each Colorbond® panel costs $3.85 per m², how much will it cost to buy the required panels for the fence?

   ____

8. Volume

   Calculate the volume of soil to be removed from a trench for a sewer pipe. The trench is 25.0 m long by 800 mm deep by 450 mm wide.

   ____

9. Calculate the volume of crushed rock a drainer will require to lay a bed 150 mm deep for the length of the trench.

   ____

10. If a cubic metre of crushed rock bedding costs $28.00, how much will the crushed rock cost for the bed under the sewer pipe?

    ____

PART 3

USING MATERIALS AND TOOLS

Part 3 of this textbook deals with the safe use of materials and tools for new workers that are undertaking training and education within the plumbing services sector. It will help you build and enhance the required knowledge for the safe handling and storage requirements of plumbing materials, and the safe use of plumbing hand and power tools on the worksite.

Part 3 is based on two units of competency, as described in Chapters 6 and 7. The key elements for the following competencies are as follows:

6 Handle and store plumbing materials
- Prepare for work.
- Identify hazard and risk-control information and measures.
- Handle, sort and stack materials.
- Store and transport materials.
- Clean up.

7 Use plumbing hand and power tools
- Identify hand and power tools.
- Select and use appropriate hand tools.
- Select and use appropriate power tools.
- Clean up work area.

Work through Part 3 and engage with your teacher and peers to prepare for the safe use of the required materials and tools that are essential for undertaking tasks on the worksite.

The learning outcomes for each chapter are a good indicator of what you will be required to know and perform, what you will need to understand, and how you will apply the knowledge gained on completion of each chapter. Teachers and students should discuss the knowledge and evidence requirements for the practical components of each unit of competency before undertaking any activities.

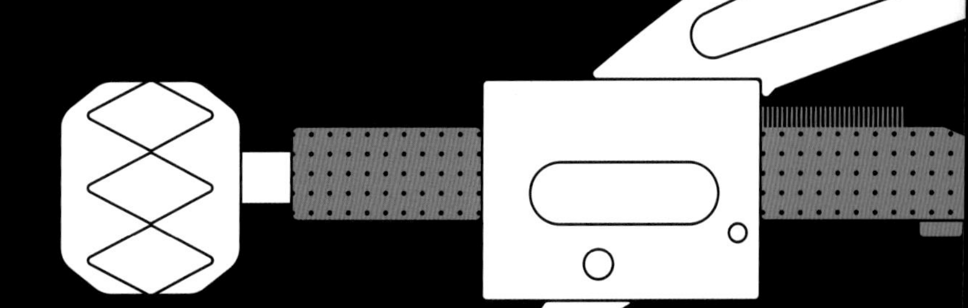

HANDLE AND STORE PLUMBING MATERIALS

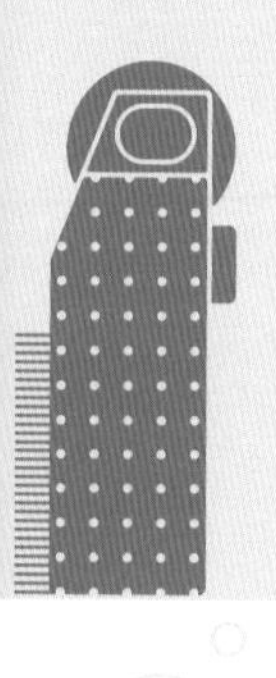

This chapter addresses the following key elements for the competency 'Handle and store plumbing materials':

- Prepare for work.
- Identify hazard and risk-control information and measures.
- Handle, sort and stack materials.
- Store and transport materials.
- Clean up.

It introduces the WHS and environmental requirements associated with handling and storing materials. It also addresses the identification and reporting of hazards and risk control associated with handling and storage of materials, and material identification, classification, selection, sorting, storage and transportation.

Prepare for work

Consideration should always be given to the safe handling and storage of all plumbing materials, tools and equipment. The most common areas where plumbers undertake these processes are:

- on a worksite
- on or within a work vehicle
- at a workshop premises.

Given the broad scope of the plumbing trade and the various situations in which plumbers may find themselves, it is important to have a basic understanding of the wide range of available materials and to become aware of their safe handling and storage.

Understanding how materials react with one another is essential not only for the safety of workers on site but also for the people who will be eventually using the plumbing system.

GREEN TIP

Choosing products that are water efficient can reduce the impact of water shortages and improve the viability of our current water supply systems.

FIGURE 6.1 Approved headwear and PPE

WHS requirements

Before you can start on site, it is important that you meet your local or state or territory WHS regulatory requirements. Hazards must be identified, assessed, controlled and reviewed to ensure a safe workplace. This can be achieved through understanding the administration and work processes that are undertaken on site, knowing how and what materials are to be used by accessing Safety Data Sheets (SDS) and Safe Work Method Statements (SWMS), and correctly selecting personal protective equipment (PPE).

Personal protective equipment

PPE is essential when working in hazardous situations and should always be used to reduce, remove or lower the risk of accident and injury. On many larger worksites, PPE must always be worn as a mandatory and safety requirement. Handling materials will usually require a person to use some form of PPE to reduce the risk of injury or illness. All PPE that is selected must be used in accordance with the job requirements and must be fitted correctly to ensure it functions as it was designed.

The following PPE items are the most appropriate for handling plumbing materials:

- *Hardhats or peaked caps* are suitable for use in construction environments where there is a risk of an object falling onto workers, such as on a commercial or multistorey site, where heavy pipework is installed overhead, or when working outdoors.
- *Safety goggles or face shields* are suitable when handling granular, powdered or chemical materials and when using power to grind or drill.
- *Earmuffs or ear plugs* are suitable when working in a noisy environment, such as near excavation machinery, or when cutting material with portable power tools.
- *Dust masks or respirators* are suitable when handling cement, or working with hazardous materials such as asbestos.
- *Leather or rubber gloves* are suitable when handling sharp-edged materials or corrosive substances such as acid.
- *Safety boots* are suitable when handling heavy or sharp materials, or when the work area contains materials stored on the floor surface.
- *Overalls or protective clothing* are suitable for handling hazardous materials such as acid or sharp edges, or to prevent exposure to the sun. Approved overalls should be used for asbestos removal.

Refer to Chapter 1 and relevant SDS for more details on PPE requirements and correct handling and storing of materials.

Product approval

Before a product can be installed on site it needs to be manufactured to a standard. A product in its simplest form can be manufactured from one material, as is the case with copper tube used in plumbing installations; or can contain a variety of materials that are combined in order to achieve the desired outcome, in the case of more complex products such as regulators or pumps. The material provides the basis for the product, which must be manufactured into its desired shape. Pipes have an internal and external diameter, and a gauge or wall thickness. The thicker the gauge, the stronger the pipe will be with respect to containing internal and external pressures. It will also be heavier, and generally more expensive. It is the combination of the material used and its dimensions that determines the properties of the product.

The minimum requirements for installation of plumbing services is controlled by Australian Standards through the principal documents *AS/NZS 3500 Plumbing and drainage* and *AS 5601 Gas installations*. These Standards provide comprehensive information on installation methods and requirements that consider the durability of a product in its application, reflecting its material properties and manufactured shape. Using the correct materials and installation methods can be achieved by referring to Standards that are relevant to the manufacture of products. For example, *AS/NZS 4020 Testing of products for use in contact with drinking water* sets out the requirements for products involved in water supply systems.

- AS/NZS 3500 Plumbing and drainage
- AS 5601 Gas installations
- AS/NZS 4020 Testing of products for use in contact with drinking water

Manufacturers for the plumbing industry require certification of their products for them to be recognised by industry regulators; in most cases this is signified by the WaterMark trademark branded on the product.

Plumbers need to be aware that the use of uncertified products may result in the installation failing inspection or not meeting regulatory authority requirements. They also need to be sure that the products are handled, stored and installed in accordance with the specifications set down in the Standards or as per approved manufacturer's installation guidelines. This process is designed to ensure that consumers can be confident that their service meets minimum agreed requirements.

Quality assurance requirements

Construction is a material-based process, and plumbing involves the use of a wide variety of materials to provide all the services covered by the trade. A lot of consideration goes into the selection of materials, as products must perform to quality assurance requirements and design-standard testing procedures if the manufacturer is to be successful. The WaterMark Certification Scheme ensures plumbing products are fit for purpose and are authorised for use in plumbing and drainage installations. The scheme is nationally consistent and products and materials that have been certified must display an approved WaterMark stamp.

In addition to certification, there are schemes that rate products according to their performance and water consumption. The Water Efficiency Labelling Standards (WELS) scheme rates products for water and energy efficiency and uses a six-star rating system. The six-star label is designed to assist people to make informed decisions when purchasing energy-efficient products.

Technological advancement in materials science and engineering is continually changing the choices of products that can be used, and their method of installation. The type of material selected will also impact on how it is handled and stored, as many plumbing materials are classed as hazardous.

Tool and equipment maintenance

All tools and equipment used for handling and transporting materials around the workplace should be checked to ensure that they are in proper working order and maintained before and after use. It is essential that lifting and moving equipment be in a safe working condition to prevent loss of time and possible injury to the operator. All equipment should be stored away in a secured area when not in use to prevent accidental damage or theft.

Never use PPE as an excuse to bypass correct WHS procedures. PPE should always be considered to be your last line of defence.

LEARNING TASK 6.1

PREPARE FOR WORK

A common hazardous material that would be used in the workplace or within the practical workshop is 'red primer' for PVC-U piping. Download the SDS for this product from the manufacturer's website and provide short answers to the following questions:

1. What are the hazards?
2. What are the prevention measures?
3. What are the response precautions?
4. What are the handling requirements?
5. What are the storage requirements?
6. What are the first aid measures?
7. What PPE should be worn?

Identify hazards and risk-control information and measures

Plumbing is a broad-based trade that covers many areas and business structures, from sole traders with one apprentice to large companies that undertake multi-storey construction.

Hazards are present on all worksites whether they are large or small, and may be dealt with in different ways or via different administrative processes. The apprentice should not feel pressured into working in any unsafe or hazardous situation and should feel free to express their views when it comes to a safe working environment. The employer has an obligation to protect their workers and the public from any unsafe condition that may arise at any worksite.

It is essential that safe work practices are followed in all situations, and on all worksites.

If any unsafe situation arises, it is important that it is reported to your supervisor urgently.

Identifying and reporting hazards

All persons on a worksite are responsible for the identification and reporting of hazards, and for putting processes in place that reduce or control risks. Moving, shifting and lifting materials can be one of the most hazardous operations for workers, and safe work processes and reporting systems are required in consultation with the employer or supervisor.

A means of providing a reporting system is to have standardised hazard report forms, from state or territory safety authorities and/or associated union bodies, readily available to the workforce. Hazard report forms may also be produced on site in consultation with workers and employers to suit specific requirements. Workers should complete a form and give it to their immediate supervisor as soon as a potential hazard is identified. All workplaces should have a hazard reporting system as it allows control measures to be put in place to reduce or remove the hazard at the earliest possible moment.

FROM EXPERIENCE

Identifying hazards on the worksite is a constant requirement and is every worker's responsibility. Communicating effectively with supervisors as issues arise can reduce hazards and create safe and harmonious workplaces.

Workplace inspections

Regular inspections of the workplace can determine what hazards exist and provide an opportunity to assess risks that may be present. Inspections can take place daily, weekly or monthly, depending on the requirements of the company or worksite. Standardised forms or checklists may be used and can be produced locally or sourced from relevant state or territory regulatory authorities when conducting inspections. Reports and recommendations from these inspections will allow hazard control measures to be reviewed and altered, as hazards will often change over the duration of the project.

At workplaces, permanent safety committees are normally established with representatives from management and the workforce. These safety committees are formed as required by the *Work Health and Safety Act 2011* and are responsible for carrying out WHS workplace inspections.

Controlling risks

Controlling the risks arising from hazards offers the greatest area of opportunity for reducing injury and illness in the workplace. Using a hierarchy of control (see **Figure 6.2**) in conjunction with safety procedures will assist employers and workers to engage in controlling risks and may form part of a SWMS.

The following information and procedures should be prepared and adhered to on every job site:

- Design or redesign the manual handling task using a hierarchy of control (also discussed in Chapter 1) to eliminate or control the risk.
- Undertake formal training in safe handling techniques.
- Use mechanical aids wherever possible to assist with moving, shifting or lifting, by using rollers, trolleys, block and tackle, pallet trucks or trolleys (see **Figure 6.3**), cranes or forklifts.

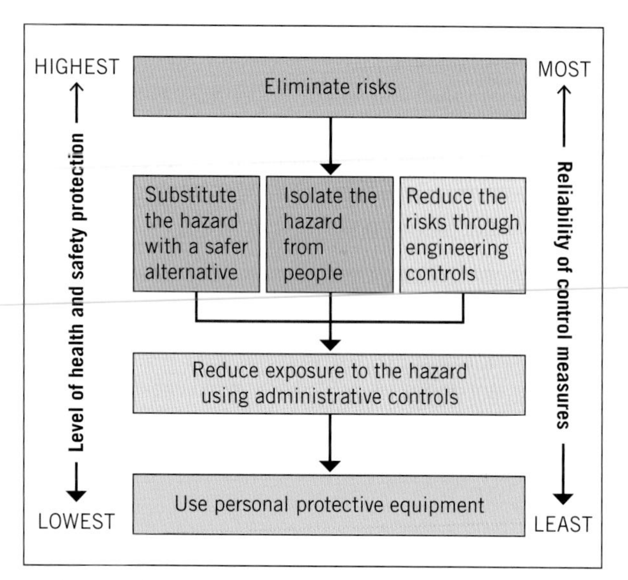

FIGURE 6.2 Hierarchy of control

FIGURE 6.3 Hand pallet trolley – 2500 kg capacity

- Have the materials delivered as close to the work area as possible to reduce the extent of the move, shift or lift.
- Team-lift heavy or awkward objects and materials.
- Follow safe lifting procedures by keeping the load close to your body; using your legs rather than your back to lift; assessing the load before lifting; turning your feet and not your waist; and placing the object or material down carefully and slowly.

Reference should be made to relevant state or territory safety authorities Codes of Practice for manual handling; also see Chapter 1.

FROM EXPERIENCE

Carrying bulky or hazardous materials up flights of stairs or in awkward situations can cause injuries. When handling and storing materials, you must ensure tasks are well planned and organised, and use teamwork where possible to allow worksites to run smoothly and efficiently.

Emergency procedures

Procedures for emergencies, fires and accidents are required to be followed when dealing with these situations. Every workplace is required to have a procedure in place and trained personnel to direct people to safety. For more detail see the 'Apply emergency response' section in Chapter 1.

LEARNING TASK 6.2

IDENTIFY HAZARD AND RISK CONTROL INFORMATION AND MEASURES

1. For a practical task that you are about to undertake in the practical workshop, obtain a SWMS and identify and risk-assess the hazards that you may encounter before commencing the task.
2. Identify the emergency procedures for your practical workshop and determine where the emergency assembly areas are located.

COMPLETE WORKSHEET 1

Handle, sort and stack materials

Plumbers are tasked with the installation and maintenance of many types of plumbing systems, materials and jointing techniques. We must understand how materials and products react with each other, and why the materials we handle and store every day behave the way they do. This will help us to select and install products that will best suit the application and provide us with protection from any hazards. All workplaces that store hazardous materials require an accessible SDS. An SDS provides information on safe manual handling techniques, what materials are contained in a product, any PPE required, what symptoms will occur from overexposure and what the required medical treatment is if this occurs.

Material properties and characteristics

A material's properties are a direct result of its internal structure. Every material has a variety of properties that may make it suitable for an application. The first feature is that it will be durable, which means it will last for a suitable length of time before failing, for any reason, in its normal use. If there is a range of possible materials with suitable properties and similar durability, a choice can then be made on a variety of other factors, including:

- cost of the material
- cost of manufacturing
- cost of installation

- availability of the product
- personal preference.

Sorting and stacking materials

Once the characteristics of materials are determined, plumbers can have a better understanding of how materials are best sorted, stacked and stored. Materials come in a range of sizes and shapes, and one single storage and stacking solution will not benefit all materials. It would be impractical to store delicate tapware in a bucket with no protective packaging around it, nor would it be sensible to store bags of concrete in an area that is constantly awash with water. Manufacturers provide guidelines on how their materials should be stacked and stored to prevent unnecessary damage (which is unlikely to be covered by their warranty). When material is delivered to the workplace, pay attention to how the material is stacked and stored. Materials such as roof sheets are generally placed on timber gluts and spaced evenly to prevent deformation, and large pipes are often placed on spacers and tied down with wire straps (see Figure 6.4). Stockpiles of aggregate material may be required to be stored and covered with plastic sheeting to prevent runoff into local waterways. Signage and barricades may be required around materials for temporary storage and to prevent double handling and always ensure hazardous materials are stored separately and by a competent person.

FIGURE 6.4 Pipe delivered to site

Note: spacers between pipes and witness marks placed there by the installer.

Material identification

A material is chosen for a task on the basis of its properties or characteristics. The properties of a material relate to the way it is or behaves in certain situations. Materials scientists study materials so they can develop products that meet people's needs. As there are so many materials with a vast array of properties, systems of classification have been made to organise groups that have similar properties.

How a product behaves in its application depends on a variety of factors, including:

- the materials from which it has been made
- how it has been manufactured
- how it has been installed
- the stresses it will be subjected to when it is in use.

Properties are also categorised to separate the requirements of different applications. Sometimes certain properties may not be relevant, and at other times a combination of properties may be needed. To understand how materials behave it is important to look at the properties of the material in closer detail. The Periodic Table of Elements groups elements together due to their individual properties.

Elements in the same columns of Table 6.1 have similar properties. The metals chromium, molybdenum and tungsten are all hard metals, while copper, silver and gold are all soft metals with high electrical conductivity.

Many elements in the periodic table are classified as metals. Metals generally have the following properties:

- They are solid at room temperature (except for mercury).
- They conduct heat and electricity.
- They are usually malleable and ductile.
- They usually form alloys.

Non-metals, on the other hand:

- can be solid, liquid or gaseous at room temperature
- are poor conductors of heat and electricity
- are brittle
- usually form compounds; for example, if a pure metal is required, its ore must be mined and processed to reverse the chemical reaction that made it into a compound, which is a combination of two or more elements. This is explained in Figure 6.5, which describes the make-up of methane gas.

Many non-metal elements and compounds are gases at normal temperatures; for example, oxygen, nitrogen, natural gas and acetylene. Others are liquid, such as water and petrol, and solids include carbon, silicon and plastics (see Figure 6.6).

Materials change between the different states as their temperature is raised or lowered. This is because the level of heat in a material determines how active the atoms and molecules are, and this determines if their forces of attraction are strong enough to keep them together.

Using the example of water, at temperatures below 0°C it becomes a solid. If heat is continuously applied to a block of ice, it increases in temperature until it gets to 0°C, which is when it starts to melt. Adding more heat at this point doesn't raise the temperature, and this is referred to as water's latent heat of fusion. The energy goes into breaking the intermolecular bonds, which turns the solid (ice) back into a liquid (water).

TABLE 6.1 Periodic table of elements (extract)

Hydrogen																	Helium
Lithium	Beryllium											Boron	Carbon	Nitrogen	Oxygen	Fluorine	Neon
Sodium	Magnesium											Aluminium	Silicon	Phosphorus	Sulphur	Chlorine	Argon
Potassium	Calcium	Scandium	Titanium	Vanadium	Chromium	Manganese	Iron	Cobalt	Nickel	Copper	Zinc	Gallium	Germanium	Arsenic	Selenium	Bromine	Krypton
			Zirconium		Molybdenum					Silver	Cadmium		Tin				
					Tungsten				Platinum	Gold	Mercury		Lead				

Light metals

Transition metals

Non-metals

Soft metals

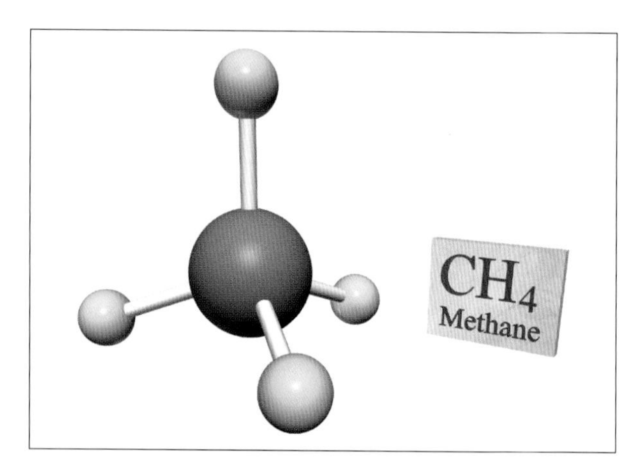

FIGURE 6.5 Hydrogen and carbon combine to form methane gas (hydrocarbon) CH_4

Source: Shutterstock.com/Antonio S.

After this phase change between states is complete, the water's temperature begins to rise until it gets to 100°C. The energy causes latent heat of vaporisation, which breaks down the water bonds and turns the liquid (water) into a gas (steam/vapour). Continued heating from this point raises the temperature of the steam.

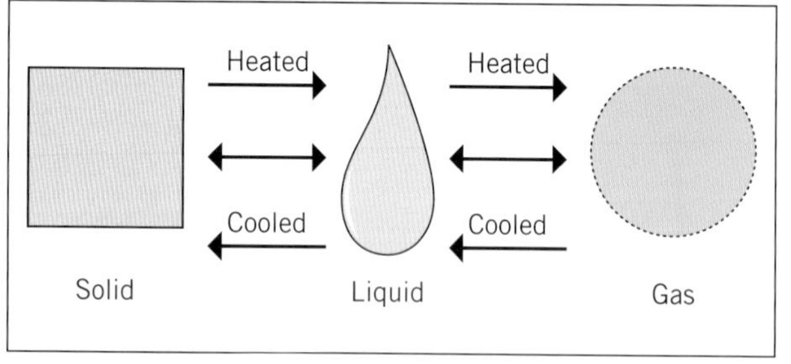

FIGURE 6.6 Three states of matter

Physical properties

These can be thought of as the material in its current state, or its characteristic properties, and include the structure of the material. Is it crystalline or molecular? Is it porous or solid? This category also includes density.

Density

A material's density is a measure of its mass and how tightly the atoms and molecules are 'packed' together. It is measured by dividing a material's mass by its volume, as shown in **Figure 6.7**.

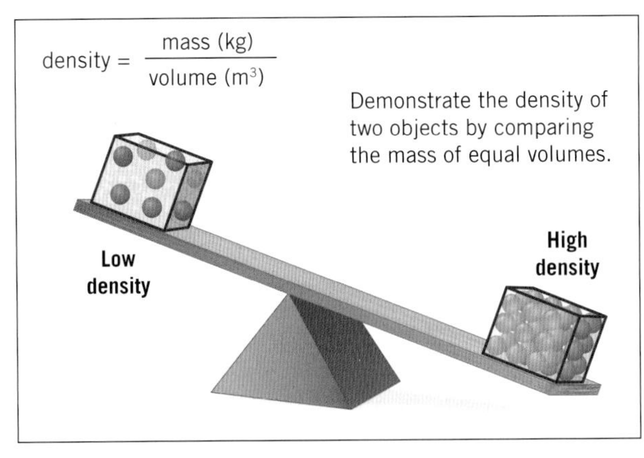

FIGURE 6.7 Density

Source: Shutterstock.com/Designua.

Density becomes an issue in construction when the load of a product needs to be supported. For example, if a roof material weighs more – or is denser – than the framing that supports it, then the framing will need to be stronger. If one pipe is denser than another, its brackets will also need to be stronger and may need to be installed closer together. Heavy materials also require greater care when it comes to lifting, so if there were a choice between similar materials or products, the less dense material or product would normally be selected.

The relative density, also referred to as specific gravity, is used to compare the density of an object in reference to another object, such as solids, liquids and gases. Water is taken as the standard measure of liquid and is given the value of '1' at 4°C, which is at its densest. Therefore, 1 L of clean water has a mass of 1 kg. A material that is twice as dense as water has a relative density of 2, and one that is half as dense as water has a relative density of 0.5.

In plumbing and gasfitting, the relative density (specific gravity) of all gases is related to air, which has a relative density of '1'.

Natural gas (NG) has a relative density of 0.6 and is therefore less dense or lighter than the surrounding air, so if it escapes it will rise or float in a room or space.

Liquefied petroleum gas (LPG) has a relative density of 1.5 and is therefore denser or heavier than the surrounding air, so if it escapes it will sink or fall to the lowest point of a room or space.

These factors must be considered when working with flammable gases to ensure that a safe environment is maintained.

Thermal properties

Thermal properties relate to how materials react to temperature change. Fluids behave in different ways and place different stresses on pipework systems. Plumbers are required to install and maintain pipework that transfers fluids and gases, and must understand the properties of what is being transferred in the pipework.

In extreme cases excess heat can cause materials to fail, which is important to be aware of when using certain jointing techniques to avoid melting materials.

Material 'heat'

When heat is added to a material, the rate at which it rises is called its specific heat capacity. This varies between materials and between the different states of the same material. To raise 1 kg of water by 1 °C would require 4.2 kJ. Specific heat capacity calculations could be used to determine the heating requirements and size required for a hot water system. The temperatures at which the 'state' changes, which are the melting and boiling points, are referred to as the latent heat of fusion and the latent heat of vaporisation, and are different for each material. This is important when it comes to heating and cooling applications, particularly on large-scale installations.

Thermal expansion

Materials and liquids expand and contract at different rates when they are heated or cooled. These rates are called expansion coefficients and can be applied to the length or volume of materials or liquids. The coefficient of linear expansion represents the amount of expansion per *metre* for every degree Celsius rise and the coefficient of volumetric expansion represents the amount of expansion per *litre* for every degree Celsius rise. The temperature variation is difference between the minimum and maximum temperatures that the material would be subjected to in a certain situation. Materials will have their own coefficient and will expand at different rates. Knowing the coefficient for a material allows the calculation of its expansion for any temperature variation. For example:

Linear expansion = original length × temperature variation × expansion coefficient

Volumetric expansion = original volume × temperature variation × expansion coefficient

These coefficients are typically very small, but are useful in situations of large temperature range, as the forces generated are very strong. Allowances can be made for the material to expand freely to prevent stress being exerted on the material or the structure. Refer to Chapter 8 for further information on expansion and coefficients.

Thermal conductivity/transfer

Thermal conductivity is how well heat moves and conducts through a material. It is relevant when looking for the transfer of energy in heat exchange systems, or for preventing heat loss or gain with insulation.

Mechanical properties

The mechanical properties of a material characterise how it will behave when force is applied to it. This is important to understand when it comes to material selection. A pipe may be able to withstand the internal pressure of the liquid inside it without bursting, but it also may need to be strong enough to support the compressive load of the soil when buried and not be damaged through contact with other materials. A range of different properties that relate to mechanical situations are described below.

Strength

The strength of a material is a measure of its ability to withstand stresses or loads. These loads can be applied in a variety of different ways.

Tensile strength

Tensile strength refers to the ability to resist stretching or pulling apart. Stretching can be applied to a material by pulling from both ends. Note the change reaction to the load (see Figure 6.8).

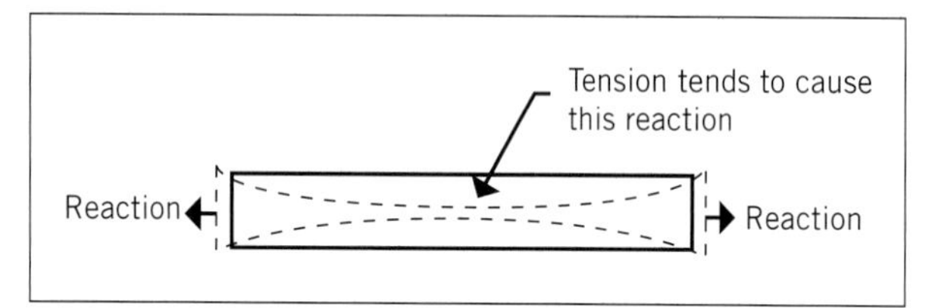

FIGURE 6.8 Tension

Note the result of bending, which creates tension on the surfaces away from the load (see Figure 6.9), and pressure from within a pipe (see Figure 6.10), which creates tension around the circumference of the pipe.

Tensile strength is most often referred to when describing a material's overall strength properties. It influences installation procedures so that materials are able to contract with temperature change without placing too much tension along the length of the material. A material should also be supported at regular intervals to prevent bending and should be

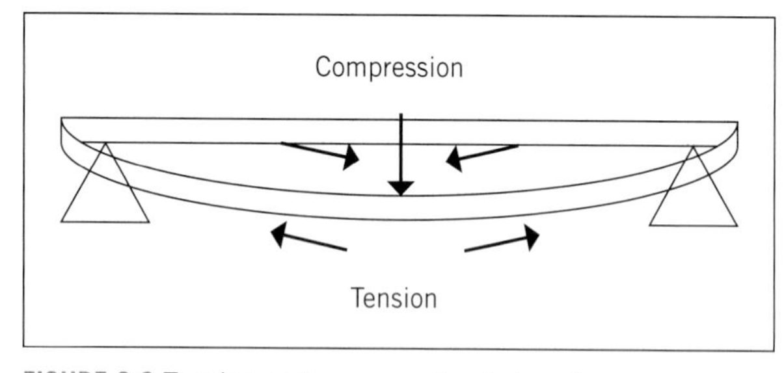

FIGURE 6.9 Tension and compression in bending

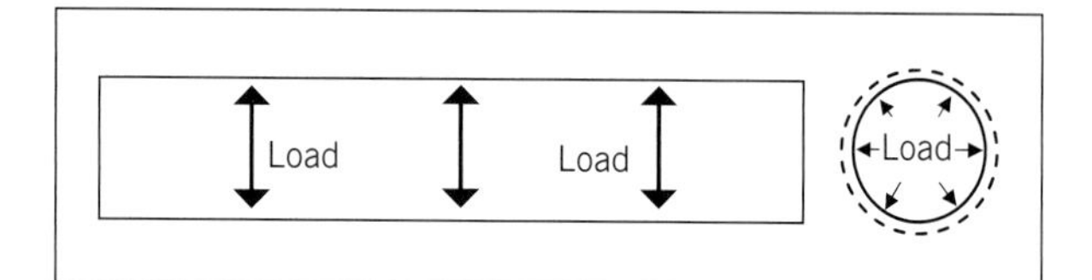

FIGURE 6.10 Tension due to pressure within pipes

strong enough to withstand the pressure of the liquids inside them.

Compressive strength

Compressive strength is the ability to resist crushing or squashing (see Figure 6.11). This is a less common situation in plumbing, but may apply if a material is bent or buried, or is not allowed to expand freely through temperature change in and around the pipe.

Shear strength

Shear strength is the ability to resist tearing. This occurs when loads are concentrated across a material in opposite directions. A suspended pipe will have shear force near its supporting brackets as the weight of the pipe and its contents pulls against the force of the bracket holding it up (see Figure 6.12).

Torsional strength

Torsional strength is the ability to resist twisting. This is a situation normally associated with screwing actions when materials are joined. Weaker materials may fail if too much force is applied (see Figure 6.13).

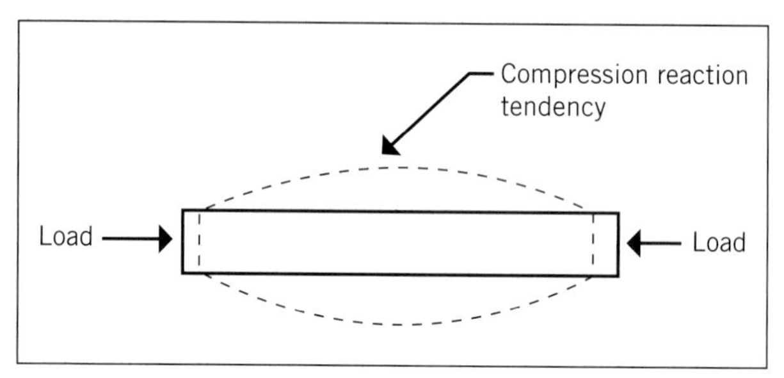

FIGURE 6.11 Compression

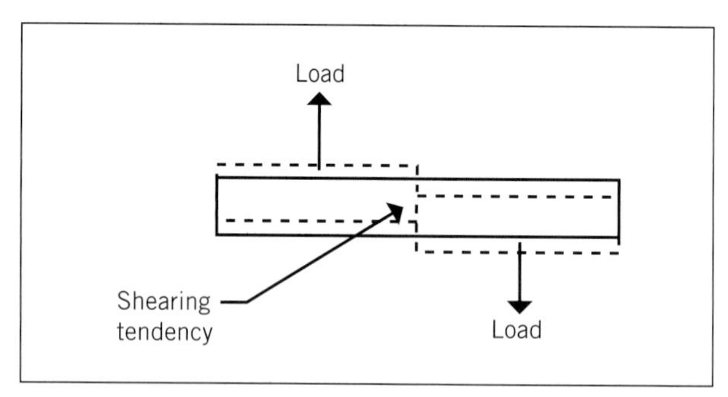

FIGURE 6.12 Shear

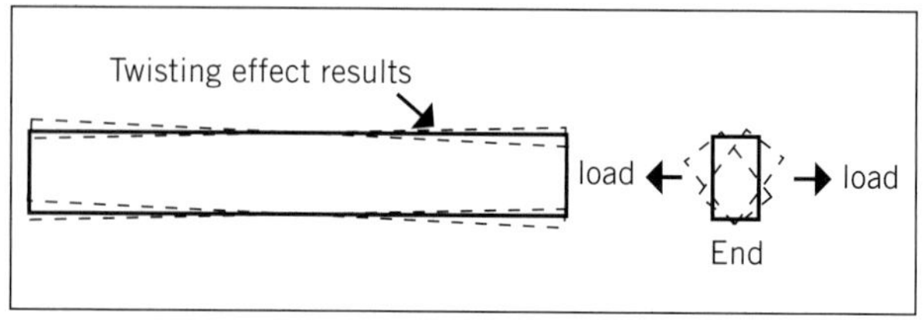

FIGURE 6.13 Torsion

Stiffness

It is important to note that all materials, regardless of their strength, deform under load.

For small loads this is not noticeable, but for larger loads this can be measured. In construction and plumbing the aim is to keep the stress on a material within its elastic limits. This means that if the load is removed the material will return to its original shape and resist deformation, which is a property referred to as the 'stiffness' of the material.

Brittleness and plastic deformation

Each material has an elastic range and, if overloaded, plastic deformation takes place. This means that the internal structure of the material has begun to break down and when the load is removed it won't return to its original shape. Brittle materials such as glass do not plastically deform and instead 'shatter' when they fail. Materials that plastically deform are called ductile, which means they can deform without being damaged, or malleable, which means they can deform under compression. This means they can be drawn into lengths or flattened out without breaking (see Table 6.2).

TABLE 6.2 Ductility and malleability of metals

Ductility		Malleability
Gold	MORE	Gold
Silver		Lead
Platinum		Silver
Iron		Copper
Nickel		Aluminium
Copper	LESS	Tin
Aluminium		Platinum
Zinc		Zinc
Tin		Iron
Lead		Nickel

Malleable or ductile materials are generally tougher than brittle ones.

Toughness

Toughness is another desirable property for a material to have. It means that the material can sustain minor damage, such as abrasions and small cracks or cuts, without seriously affecting its strength. The structure of the material distributes the load around the weakness instead of concentrating it at the point of damage, which can lead to failure.

Hardness

The hardness of a material is its resistance to indentation, marking and scratching. Hard materials are used as abrasives for grinding and cutting softer ones. Softer materials may need protection from harder ones in situations such as backfilling a trench over pipework.

Work hardening and annealing

Working or deforming metals at low temperatures can increase their strength and hardness (see Figure 6.14). This distortion of the crystal structure puts tension in it, and dramatically reduces its ductility. This may be desirable if a material is too weak in its 'natural' form for an application. Ductility can be returned, and strength and hardness reduced, by annealing the metal, which is a process of heating it above its recrystallisation temperature, which is below its melting point. This allows the crystals to realign, releasing the inbuilt tension.

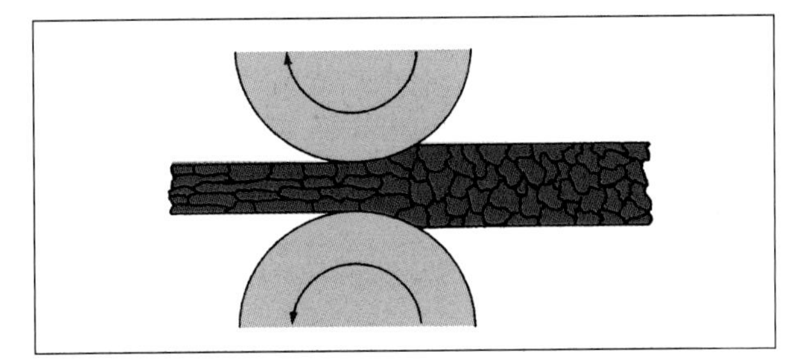

FIGURE 6.14 Work hardening

Chemical properties

Materials that react chemically are generally found in an impure form and must be processed in order to make them useable for engineering purposes. It is important that in their application they don't react with their environment and lose the properties with which they have been manufactured.

Atmospheric corrosion of metals

Metals may corrode or oxidise slowly through direct contact with oxygen in the atmosphere. This provides a film of oxide that coats the surface of the material and in some cases provides a protective layer that prevents further oxidisation; such as the dull surface that may be found on an old piece of aluminium. In other situations, the surface oxides allow oxygen to penetrate or dissolve readily, allowing further corrosion to take place. This process may be accelerated in certain environments where pollutants such as acids are present in the air or mixed with rain. Coastal environments also provide a corrosive atmosphere as the salt air reacts with the metal oxides, allowing them to be washed away. The reactivity of the different metals is represented in the Noble (or Galvanic) Scale shown in Table 6.3.

From this table, aluminium is on the more active end of the scale, and, as the oxide coating forms, the aluminium will be protected from further atmospheric corrosion. Iron is in the opposite situation, as its oxides readily flake and break off, exposing new metal below. Steels, which are iron/ferrous alloys, can be coated with zinc and other materials to protect them from corrosion.

TABLE 6.3 The Noble Scale

More active (anode) – Less active (cathode)	Gold
	Platinum
	Silver
	Copper
	Lead
	Tin
	Nickel
	Iron
	Zinc
	Aluminium
	Magnesium

Electrolytic corrosion of metals

Electrolytic corrosion is also influenced by the Noble Scale. When dissimilar metals are connected by water, a battery-type situation is set up. The more active metal, known as the anode, loses electrons and corrodes to the less active cathode. When using dissimilar metals in wet environments, they should be insulated – or, if water flows from one to the other as in the case of roofs, it should flow from the anode onto the cathode. Sacrificial anodes are used in storage hot water systems to reduce the amount of corrosion within the cylinder.

Polymers

Polymers are not corroded in the same way that metals are. It is generally the case that ultraviolet (UV) radiation can break down the bonds in polymers over time, and, if they are to be exposed, they require stabilisers. The wide range of polymers provides a variety of reactions to different chemicals, with some being resistant to acids and alkalis but not to organic solvents.

Ceramics

Ceramics are generally quite corrosion resistant, as they are products of naturally formed compounds. They have already reacted to enter a stable state, but they may be attacked by acids if they are not protected by a thin, glass-like layer referred to as glazing.

Material classification

This section takes a closer look at the different categories in Table 6.1 by identifying useful sub-categories and adding other classifications known as composites and natural materials.

Metals

Metals have been commonly used in plumbing for many centuries. The word 'plumbing' comes from the Latin word for lead, *plumbum* (periodic table symbol Pb), which was the material that early pipes were made from. The person who worked with lead was first known as *Plumbarius*, and later this was shortened to *Plumber*. Table 6.1 classifies many elements as metallic, and while there is considerable diversity within the group (see Figure 6.15), the properties of metal are generally described as being:

- good conductors of heat and electricity
- dense solids
- strong, stiff and ductile.

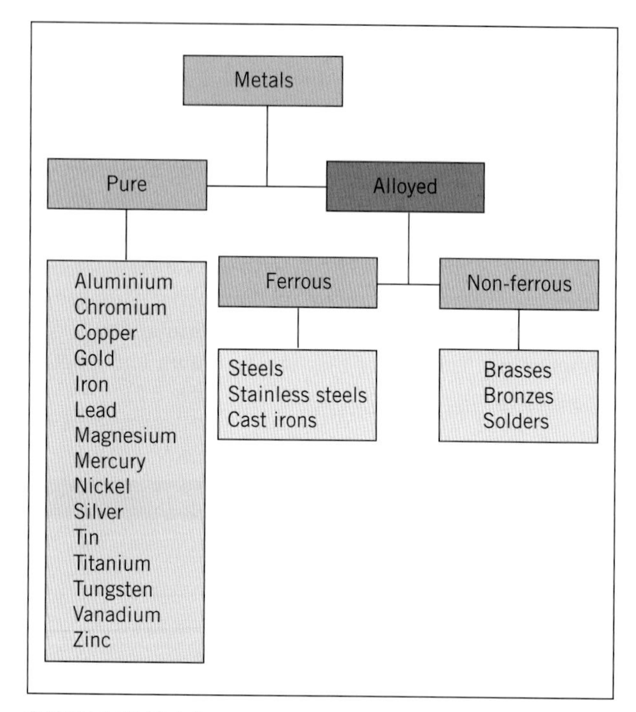

FIGURE 6.15 Metals

In practice it is difficult to obtain completely pure metals, and it is common practice for metals to be 'alloyed' to improve their properties.

Alloys

Alloys are formed in a molten state by adding other metals or non-metals to a form a homogeneous base metal. These additives dissolve and change the crystalline structure, resulting in a material that still behaves like metal, but has new properties that make it more useful in different applications.

For example, a change from just 0.1% to 0.5% carbon in plain carbon steel can change the material from being soft and ductile to being strong and tough.

Alloys are further broken down into ferrous and non-ferrous categories. Ferrous materials contain iron-based products and include various types of steel and iron.

Polymers

Polymers are the base materials for what is generally termed 'plastics' (see Figure 6.16), and are extremely

TABLE 6.4 Properties of pure metals

Metal	Specific gravity	Melting point °C	Recrystallisation temp °C	Coefficient of expansion (x 10^{-6} m)	Thermal conductivity	Electrical conductivity	Stiffness (YM GPa)	Tensile strength MPa	Hardness (Mohs)
Aluminium	2.7	660	150	25	190	37	70	45	2.75
Chromium	7.2	1900		5	87	7	280	350	8.5
Copper	9.0	1090	200	16	390	58	130	220	3
Gold	19.3	1060		14	310	45	78	120	2.5
Iron	7.9	1540	450	12	78	10	210	540	4
Lead	11.4	330	20	29	37	5	16	30	1.5
Magnesium	1.7	650	150	8	160	23	45	150	2.5
Mercury	13.5	−40			8	1			
Nickel	8.9	1450	600	13	89	14	200	317	4
Silver	10.5	960	200	19	418	60	83	140	2.5
Tin	7.3	230	20	20	63	8.5	50	220	1.5
Titanium	4.5	1670		9	23	2	116	900	6
Tungsten	19.3	3410	1200	4	180	19	411	980	7.5
Vanadium	5.8	1900		8	30	5	128	750	7
Zinc	7.1	420		30	112	16	110	280	2.5

Note: Values are indicative.

common in plumbing and other industries. They are the preferred choice for water supply to new residential properties as they are cost effective and easy to install. They are non-metallic compounds based mainly on carbon and silicon.

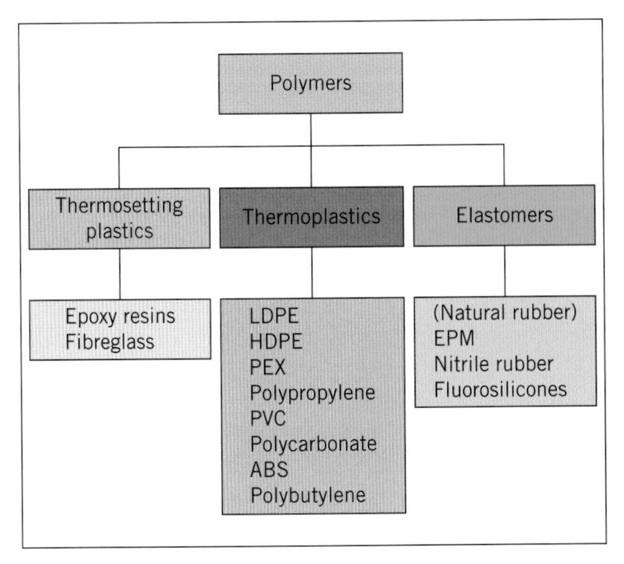

FIGURE 6.16 Polymers

The large chains of molecules can combine in different ways to produce several different types of plastics called thermosetting plastics, thermoplastics and elastomers.

- *Thermosetting* plastics are quite hard and brittle, and they cannot be softened by heat.
- *Thermoplastics* are less brittle and can be deformed without breaking, but they don't readily return to their original shape. Thermoplastics can be softened by heat and reworked into different shapes.
- *Elastomers* can be stretched and will return to their original shape.

The reason for these quite different properties is the amount of 'cross-linking' between the polymer chains. Cross-linking provides a network within the polymer structure that resists deformation. Thermosetting plastics are heavily cross-linked, making them stiff and brittle. Elastomers are lightly cross-linked, allowing them in some cases to be stretched up to five times their original length. Thermoplastics don't have cross-links and therefore, when they are deformed past their elastic limit, there is no mechanism for them to return to their original shape.

The amount of polymerisation (i.e. the types of molecules connected to the sides of the monomers and the existence and extent of cross-linking) influences the properties of the material. In addition to these factors, additives that control colour, UV stability and fire resistance can be included in the mixture.

In relation to metals, polymers have low densities, low heat and electricity conductivity, high thermal expansion, and low strength and stiffness, and can stretch further.

Composites

Composite materials are a combination of materials that, when formed, create a new material with complementary or superior properties. The individual base materials by themselves may possess certain properties that may make that material desirable. Creating combinations usually involves trying to increase strength and hardness, which can be achieved in several ways.

Mixing

Composite materials can be combined and bonded by chemical reactions by use of water or a binding compound. The component materials are distributed randomly but evenly throughout the mixture to disperse the enhanced properties. Concrete and mortar are common examples of mixing a combination of materials to achieve a desired outcome. Alloys are similar to composites; however, the mixing takes place above their boiling point.

Multilayer composite piping

Multilayer composite pipes (see Figure 6.17) use a layering process where an aluminium (AL) tube is reinforced with an internal and external cross-linked polyethylene (PEX) layer. The three layers are combined using an adhesive or bonding process during manufacture of the material, which is marketed as a PEX/AL/PEX system.

FIGURE 6.17 Multilayer composite piping

Common usage of metals

There are various metals that are commonly used in the plumbing industry for several different purposes. The following metals are chosen due to their availability, quality and ease of use.

Aluminium

In the plumbing industry aluminium can found in roof sheeting, flashing and drainage applications due to its low-density and corrosion-resistant properties. It is often pre-painted, or used in reflective foil insulation products.

Multilayer composite piping is increasingly being used for water, gas and heating applications. The aluminium in this application provides strength and rigidity when compared to PEX piping, and is impermeable to gas. Multilayer composite piping is available in a range of colours to suit colour-coding requirements for different services.

Copper

Copper's durability and workability make it a popular material for plumbers, where it is mainly used in pipe form for water, gas and sanitary applications. In sheet form it is occasionally used for roofing in corrosive environments, or where durability is an important requirement. Its thermal conductivity makes it ideal for heat exchangers such as in refrigeration, air-conditioning, boilers and solar collectors.

In installation, consideration needs to be given to copper's high rate of expansion, and the reduction of tensile strength that takes place when annealing of the pipe occurs during the soldering or brazing of fittings and the formation of bends.

Brass, a copper alloy, is used for fittings and valves and may often be plated.

Lead

Lead is a heavy, dense material that rapidly forms a protective oxide layer that makes it corrosion resistant. However, this layer can be slowly dissolved by water, and unfortunately lead is a toxic substance that accumulates in the body and can cause damage to the nervous system and kidneys, as well as blood and brain disorders. Lead affects young children and it can also affect pregnancy outcomes. It is thought that the use of lead pipes for water use in Ancient Roman times gave rise to widespread dementia.

For these reasons lead is generally being removed from the environment. Petrol and paint are now lead free, and soft solder for existing joints on water supply must be less than 0.1% lead by weight. It is not advisable to use lead in roofing applications, especially when the water is collected for domestic use. Working with lead is an obvious work hazard and guidelines on SDS should be followed.

Used for roof flashings, lead comes in rolls of different width and thickness that can be cut to size and readily shaped for the application.

Zinc

In its pure form zinc can be used in roofing applications. It is not commonly used as a roof covering as it is not strong enough to support itself in sheet form. It can be laid on a platform structure in a batten roll method, or used as an alternative to lead as a flashing material. Zinc is also used as corrosion protection through zinc coatings or plating on ferrous metals – this process is often called galvanising.

Ferrous alloys

In plumbing installations, mild steels have the greatest use due to their lower cost, malleability and availability. Galvanised mild steel piping is no longer used for drinking water applications or below ground due to corrosion. Black mild steel (BMS) is used for pipes in mechanical services applications.

Steel is also used extensively in structural situations such as brackets and fixings, and in gratings and covers for pits and chambers.

The most common types of ferrous alloys found in the roof plumbing industry are:

- galvanised sheet steel
- Zincalume®
- Colorbond®.

Galvanised sheet steel

Galvanised sheet steel comprises a mild steel sheet of metal that has a pure zinc coating, applied by a continuous hot-dipping process, to protect it against corrosion.

For many years it was the best available material for roof and wall sheeting installations. It can also be used for making:

- barge trims
- downpipes
- ductwork
- flashings for roofs
- rainwater tanks
- ridge capping
- sheet metal
- spouting.

Zincalume®

Zincalume® sheets have a metallic coating, that consists of a mixture of zinc, aluminium and a small portion of silicon, applied to a mild steel base sheet. A continuous hot-dipping process is used to apply the coating. Zincalume® provides a more traditional look with its metallic finish and offers two to four times the corrosion resistance of galvanised steel.

Colorbond®

Colorbond® sheets are manufactured using a similar process to Zincalume®, with the addition of a special process in which a painted finish is baked onto the surface. Many colours are available to suit all types of buildings and customer requirements.

Zincalume® and Colorbond® are widely used for roof and wall sheeting installations. They can also be used for fabricating flashings for roofs, spouting, downpipes, barge trims, ridge capping and rainwater tanks.

Alloy steels

Alloy steels contain greater quantities of different alloys than plain carbon steels and are made for a wide variety of reasons, which include improving strength, hardness and wearing properties. Elements that may be used with steel to enhance properties include aluminium, chromium, cobalt, copper, molybdenum, nickel, niobium, phosphorus, titanium, tungsten and vanadium.

One of the more common alloy steels that a plumber may use is stainless steel, of which the main alloy is chromium, with concentrations ranging from 10% to 25%. Stainless steels are corrosion resistant for the same reasons that non-metals are. A very thin, hard layer of chromium oxide forms on the surface of the material, protecting it from further attack. The different chromium concentrations give different grades of protection, and the other alloys influence mechanical properties, with tensile strengths ranging from around 400 MPa to 800 MPa.

Stainless steel is used in a variety of products where corrosion resistance is required. The most common applications are kitchen sinks, laundry tubs and commercial kitchens, but it also has applications in piping systems and roofing products, and is useful in a variety of joining systems.

Cast irons

Cast irons are formed by casting the molten mixture into a finished shape; for pipes this is achieved in a spinning mould, with the centrifugal force keeping the liquid to the outside until it solidifies. This economical method of manufacture has been used for a long time, and produces a product quite different from steel, with cast iron being generally more brittle, harder and corrosion resistant.

Cast iron has been commonly used for casting baths and, in its pipe form, for gas and water conduction. It was also used extensively for sanitary plumbing stackwork. It is difficult to work due to its brittle nature and weight (mass).

Grey and ductile irons

Grey irons have been, and still are, used for sanitary piping systems and cast products such as baths, pit covers and grates. The increased toughness of ductile irons has seen them replace grey cast-iron water mains. Ductile iron fittings are commonly used in conjunction with large water main systems, using either PVC-M or PVC-O pipe and rubber ring joint systems.

Polymers

Polymers in the plumbing industry are used in a range of versatile and increasingly popular products. They are used extensively and generally come under three categories: thermoplastics, thermosetting plastics and elastomers.

Thermoplastics

These plastics make up most of the polymers used in the plumbing trade and have progressively replaced metal and ceramic piping systems in lower-temperature applications. Being designed for certain applications, the different types of thermoplastics have specific mechanical and chemical properties.

Polyethylene

Also known as polythene, this is the simplest polymer chain, being based on the ethylene monomer. Despite its simplicity, it comes in several forms, depending on the extent of polymerisation, the length of the molecular chains, and the extent of branching – all of which can be controlled to manage properties. High-density polyethylene (HDPE) and medium-density polyethylene (MDPE) have the simplest form with minimal branches, allowing the molecules to be packed tightly together. This makes it more rigid and temperature-resistant than the progressively low-density polyethylene (LDPE), making it suitable for sanitary and heated-water applications. The LDPE pipes are generally only used for irrigation applications.

Another form of polyethylene is made by cross-linking bonds in the polymer structure to produce PE-X (cross-linked polyethylene). Cross linking can make the material more elastomeric in nature and therefore capable of handling heated-water applications.

Polyvinyl chloride (PVC)

PVC is based on another simple monomer and is one of the most extensively used plastics, with uses in both sanitary and water applications in the plumbing industry (see Figure 6.18). PVC is naturally rigid and has a tensile strength of around 50–70 MPa. It is quite often softened with plasticisers, which led the original product to be called unplasticised PVC (or PVC-U).

More recently, the products named modified PVC (PVC-M) and oriented PVC (PVC-O) have been approved for use in both sanitary and water supply applications. Both these products increase the strength and toughness of PVC. PVC-M does this by the addition of chlorinated polyethylene in concentrations of up to 6%, while PVC-O achieves it by stretching the material at higher temperatures in the manufacturing process.

PVC can also be used in roofing products, depending on the suitability of the location geographically.

FIGURE 6.18 PVC fittings

Polypropylene

Polypropylene's methyl (CH_3) group makes this material stronger and stiffer than polyethylene, enabling its use in sanitary and heated-water applications. Polypropylene can also be combined with the ethylene monomer to produce the 'random copolymer' polypropylene-ethylene (PP-R or EPM), which improves its impact resistance but marginally reduces its strength.

Polybutylene (PB) and acrylonitrile butadiene styrene (ABS)

These two thermoplastics have more complex structures, being made from combinations of different monomers, and have properties that give them approval for use in sanitary, water and heated-water services.

Polycarbonate

Polycarbonate is a strong, tough, transparent polymer that can maintain its properties at 100°C. It is used for transparent roof sheeting (see Figure 6.19), as well as safety glass and bulletproof glass.

FIGURE 6.19 Polycarbonate skylight

Thermosetting plastics

The rigidity and brittleness of thermosetting plastics make them a less common material within the plumbing industry. Two-part epoxy resins are used as joining compounds, but it is in composite form with fibreglass that thermosets become more useful, due to their increased strength and toughness. Depending on the amount of fibre, and its orientation, tensile strength can be improved from as low as 50 MPa to in excess of 1000 MPa. These composites are called glass-reinforced plastics (GRPs) and are used for moulded products such as arrestors, separators and baths, but can also be used in sanitary, stormwater, water and heated-water piping systems.

As well as epoxies, there are also phenolic, polyester and vinyl ester thermosetting plastics.

Elastomers

Elastomers, due to their high elasticity and flexibility, are not suitable for piping systems, but have properties well suited to flashing applications, particularly when there are compatibility issues between metal alternatives. In these exposed situations they must be UV stabilised and can also have colour additives.

Elastomers are often made from chains of different monomers, such as in butadiene acrylonitrile, known as nitrile rubbers, and ethylene propylene diene monomer (M-class) (EPDM) rubber, both of which are used in flashings. Elastomers are commonly used as adhesives, seals and moulded parts of products.

Ceramics

Ceramic materials play an important role in the plumbing industry, and although there may be a decline in use in some areas, they are still a dominant item in many facets of plumbing.

Vitrified clay pipe (VCP)

Also known as earthenware pipe, VCPs were predominantly used for sanitary and stormwater drainage applications, but have now largely been replaced by polymer alternatives, which are less brittle. VCP pipes, however, are very durable and are still preferred by some designers in certain applications, and are still used in specialised areas such as sewer mains, jacking pipe systems and generally aggressive environments, due to their increased resistance to corrosive materials and excess heat.

Vitreous china

Ceramics are still dominant in fixtures such as toilets, basins and prime cost items (PC items), where their smooth, hard surfaces allow them to be easily maintained and to blend in with the tiled wall and floor finishes.

Cement

Cement is the binding agent in concrete and mortar and is made from limestone and clay that have been crushed, mixed together with water and burned at a very high temperature, where they fuse to form marble-sized 'clinkers'. The clinkers are then ground finely together, with a small percentage of materials such as calcium sulphate, fly ash or granulated iron added, to form the cement powder.

Composites

Precast concrete-based products are also common in plumbing and include large-diameter pipes for stormwater, septic tanks, water tanks, pits and kerb inlet lintels. These products are normally reinforced by steel mesh or bars, or in some cases metal fibres.

Glass and cellulose (plant) fibres are also used when manufacturing fibre-reinforced cement (FRC) products, which are lighter alternatives to the concrete products. Ceramic aggregates such as sand, quartz and fine aggregates can also be bound in thermosetting resins such as epoxy, polyester and vinyl ester to concentrations of around 15% to form similar products.

Natural aggregates

Large quantities of sand and crushed stone are used for the support and protection of buried pipes. Standards will indicate the size and type of aggregate in these applications, as well as the requirements for making cement mortar and concrete on site.

LEARNING TASK 6.3

HANDLE, SORT AND STACK MATERIALS

Identify and select three types of materials – one metal alloy, one polymer and one ceramic product commonly used in plumbing – and provide short answers to the following questions:

1 Where are the base materials found?
2 What are they used to make?
3 How should they be handled safely?
4 How should they be stored?

COMPLETE WORKSHEET 2

Store and transport materials

Hazardous materials often require specialised handling and storage solutions, and plumbers are at risk every day due to the amount of hazardous materials required to complete various plumbing tasks. Many plumbers are unaware of the potential risks of these materials; therefore, it is important to always read the SDS for the hazardous material to ensure that all work undertaken is carried out safely.

Hazardous materials

All hazardous materials should be handled, transported and stored in accordance with details provided by an SDS. These instructions are available from the point of sale and must be readily available at the worksite before materials are handled and stored (see Figure 6.20). An SDS should be read carefully to prevent the risk of incident or injury and understand the handling requirements of the material. It is important that manufactured products are not damaged or contaminated before or during installation so that they can perform in the way they were designed to. This can be achieved by storing and stacking materials to manufacturers' guidelines in conjunction with relevant Codes of Practice.

FIGURE 6.20 SDS file located near work area

Transportation and storage

Transportation of materials can take many forms, which may range from moving materials around a large building site to transporting materials on the back (or within) a work vehicle. Regardless of the mode or area of transportation, materials require appropriate handling and storage to ensure the material remains in good condition.

Manufactured products, which include sanitary fixtures, pipes, fittings and tapware, should be carefully packed in protective wrapping to prevent accidental damage during transportation. Small items should be packed in sturdy boxes surrounded by soft padding or sealed in plastic packets to prevent the loss of small components. Large items should be packed in sturdy boxes or crates placed on suitable pallets or timber gluts for ease of unloading.

When transporting copper pipes, lengths of guttering or PVC pipe, care should be taken to prevent them from sliding forwards or off the rear of the vehicle. They should be securely strapped onto roof racks designed for that vehicle, with the protruding ends fitted with caps or other soft protectors to prevent accidental injury to the public. When travelling along public roads, hi-vis long-load flags (see Figure 6.21), attached to the material at the rear of the vehicle, may be required to alert other drivers and pedestrians and to prevent injury. Materials must not protrude past a certain distance – check the regulations within your local area. An ideal way to transport lengths of copper tubing is to securely fit a 100 mm diameter PVC pipe to the racks, with screw cap ends fitted for ease of loading and unloading.

FIGURE 6.21 Hi-vis long load flag attached to pipe material

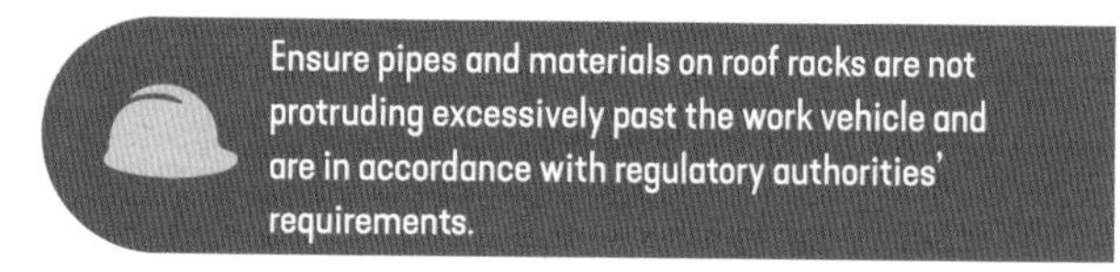

Ensure pipes and materials on roof racks are not protruding excessively past the work vehicle and are in accordance with regulatory authorities' requirements.

Smaller items loaded into the back of a vehicle should be contained in boxes, adequately restrained or contained within safety cargo netting to prevent damage from materials and products sliding around or flying out of the vehicle. Fines may apply if materials are deemed not to be safely secured on the vehicle.

Hazardous materials should be clearly marked and identified by authorised and trained personnel. Appropriate safety signage or barricades (see Figure 6.22) must be used where required to prevent unauthorised access to stored or stacked materials. Hazardous materials may also require designated storage areas (see Figure 6.23) that comply with workplace requirements and Australian Standards.

FIGURE 6.22 Pipes supported and stored near work area

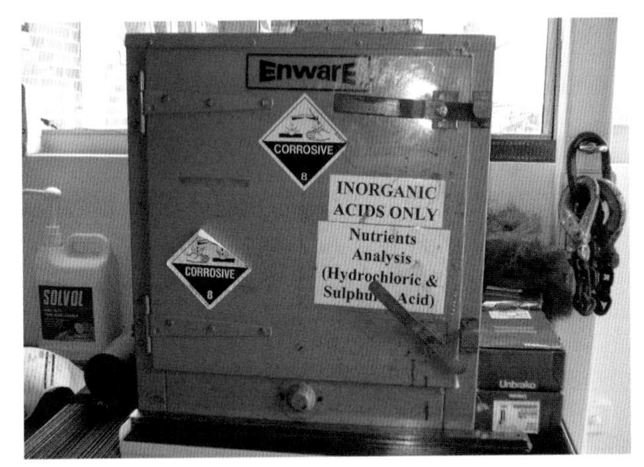

FIGURE 6.23 Dangerous goods cabinet

FIGURE 6.24 Trolley with oxyacetylene bottles

Special precautions should be taken when handling, transporting and storing acetylene, LPG and oxygen cylinders, particularly where there is a potential for the build-up of leaking gases. Gas bottles/cylinders should be transported and stored in an upright position in a specially constructed cabinet that is vapour-proof to the rest of the vehicle and is vented to the atmosphere. Where oxygen and acetylene cylinders are to be transported on the construction site, they should be transported using an approved trolley with a suitable chain to prevent the cylinders from becoming dislodged during transport (see Figure 6.24).

When storing oxidising and flammable gases on site, minimum separation distances are required in line with the relevant Australian Standard. Construction of vapour barriers or fire walls may be required to achieve separation distances in the work area.

Every work situation is different, but where materials do require storage, several common principles should be followed:

- *Minimal handling* – keeping handling to a minimum not only saves time but also reduces the risk of injury to workers and damage to materials.
- *Security* – having a secure location prevents theft and damage of products, and the associated time and expense of replacing them.
- *Maintenance of properties* – when materials are to be stored for extended periods, consideration needs to be given to their position on the property to prevent danger and inconvenience to owners.
- *Storage order* – items that are used less frequently or only needed towards the end of a project should be stored behind high-use or first-required materials. Other considerations include the following:
- Pipes and sheet materials need to be supported to prevent deformation.
- Brittle products should be stored well away from possible impacts such as access areas.
- Soft products should not be placed under other materials, as distortion or crushing may occur.
- Excessive heat can cause deterioration in some polymer products and joining compounds, so they should be stored in a cool or shaded area.
- Materials that could potentially corrode should remain in their protective packaging where possible to prevent damage.
- The storage environment should be dry, as moisture assists corrosion and can damage surface finishes.
- Moisture may also start chemical reactions in products; for example, hardening cement.
- Plastics not designed for exposed locations may begin to break down in sunlight (UV) and should be adequately protected in the original packaging or be covered with UV-resistant plastic sheeting.

- Special care should be taken with roof sheeting or flashings stored on a roof as excess weight can cause structural failure. Wind poses a danger as roof sheets can fly off and cause damage.
- Many liquid products have 'use by' dates and should not be stored beyond the time stated on the label.
- Gas cylinders should be stored separately, according to their hazard classification and their separation distance, in an area that is well ventilated.
- Hazardous materials must be handled and transported according to regulatory requirements, including appropriate signage, markings and safety precautions.

For further information relating to the safe handling of hazardous materials, refer to Chapter 1.

Delivery

Materials may be delivered to a site or workshop, and used immediately or progressively as required. On delivery, products need to be checked against the order to ensure that:

- the correct products have been delivered
- the correct quantities have been delivered
- the products are in good condition.

Making these checks means that missing or damaged products can be replaced quickly, which prevents unnecessary or complicated disputes with suppliers and clients.

Environmental factors

For bulk loose materials, either delivered to site or created through excavation, measures need to be taken to prevent material run-off due to rain or wind. Failure to keep this material within the site boundaries can also attract fines from the local council. Protection may be achieved by:

- covering the material with fabric or plastic
- wetting the material down
- locating the material away from natural stormwater courses
- diverting stormwater flow around the material piles
- using site perimeter barrier fencing designed to catch sediment
- removing excess material promptly from the site
- revegetating cleared areas as soon as possible
- storing material according to manufacturer's requirements.

GREEN TIP

Environmental regulations need to be observed when working with potentially hazardous or contaminated liquids or powders near waterways, stormwater and sewers. Contaminated materials can harm humans and animals, and potentially damage infrastructure.

LEARNING TASK 6.4

STORE AND TRANSPORT MATERIALS

Provide short answers to the following questions:

1. What are two ways that lengths of copper tubing can be transported on a vehicle?
2. How are oxygen and acetylene cylinders stored on the construction site?
3. How would oxygen and acetylene cylinders be transported in an enclosed work vehicle?
4. How would hazardous chemicals be stored in the workplace?

COMPLETE WORKSHEET 3

Clean up

Ensuring the work area is clean and clear of obstructions will provide a safe working environment and reduce hazards. Material that is no longer required on the site should be reused, recycled or removed from the site.

All tools and equipment must be thoroughly cleaned and checked for damage to ensure they are in good condition for the next time they are used. Any faulty equipment must be reported to a supervisor to prevent an accident or injuries from occurring.

At the end of each working day check that materials are not protruding into walkways, offcuts are removed from the ground, spilt liquids or powders have been cleaned up, lids are securely fitted to containers and that there is no danger of stacked material collapsing due to poor stacking methods. Signage that may have been temporarily erected when using tools and equipment must be removed. All cleaning processes must follow workplace procedures and relevant state or territory regulatory requirements. All required documentation must be completed in accordance with workplace requirements.

GREEN TIP

Materials that are left over or no longer being used should be reused appropriately or recycled.

REFERENCES AND FURTHER READING

Acknowledgement

Reproduction of the following resource list references from DET, TAFE NSW C&T Division (Karl Dunkel, Program Manager, Housing and Furniture) and the Product Advisory Committee is acknowledged and appreciated.

Texts

Flinn, R.A. & Trojan, P.K. (1981), *Engineering materials and their applications*, 2nd ed., Houghton Mifflin Company, Boston.

Graff, D.M. & Molloy, C.J.S. (1986), *Tapping group power: A practical guide to working with groups in commerce and industry*, Synergy Systems, Dromana, Victoria.

National Centre for Vocational Education Research (2001), *Skill trends in the building and construction trades*, National Centre for Vocational Education Research, Melbourne.

NSW Department of Education and Training (1999), *Construction industry: Induction & training: Workplace trainers' resources for work activity & site OHS induction and training*, NSW Department of Education and Training, Sydney.

NSW Department of Industrial Relations (1998), *Building and construction industry handbook*, NSW Department of Industrial Relations, Sydney.

Philip, M. & Bolton, B. (2002), *Technology of engineering materials*, Butterworth Heinemann, Oxford.

SAI, MP 52, Miscellaneous publication: Manual of authorisation procedures for plumbing and drainage products, Sydney.

Schlenker, B.R. (1974), *Introduction to material science, SI edition*, John Wiley & Sons, Brisbane.

WorkCover Authority of NSW (2009), *Building industry guide: WorkCover NSW*, Sydney.

Web-based resources

Regulations/Codes/Laws

SafeWork NSW Codes of Practice: **https://www.safework.nsw.gov.au/resource-library/list-of-all-codes-of-practice**

Australasian Legal Information Institute (AustLII) legislation database: **http://www.austlii.edu.au**

Safe Work Australia: **http://www.safeworkaustralia.gov.au**

WorkSafe Victoria: **https://www.worksafe.vic.gov.au/**

Resource tools and VET links

Training.gov.au: **http://training.gov.au**

NSW Education Standards Authority – Construction: **https://educationstandards.nsw.edu.au/wps/portal/nesa/11-12/stage-6-learning-areas/vet/construction-syllabus**

Industry organisations' sites

Australian Building Codes Board – WaterMark Certification Scheme: **https://www.abcb.gov.au**

CITB (SA Construction Industry Training Board): **http://www.citb.org.au**

Building Trades Group Drug & Alcohol Program: **http://www.btgda.org.au**

SafeWork NSW: **https://www.safework.nsw.gov.au**

Audiovisual resources

Short videos covering topics such as safety signs and tags, job safety analysis and dealing with chemical safety are available from the following organisations:

- Safetycare: **http://www.safetycare.com.au**
- TAFE NSW: **https://www.tafensw.edu.au.**

SafeWork NSW publications

'Hazardous manual tasks: Code of Practice', Cat. no. WC03559

'Housing industry site safety pack', Cat. no. 2977

'Personal protective equipment' – Fact sheet, Cat. no. WC05893

'Reading labels on material safety data sheets', Cat. no. 400

'Work method statements', Cat. no. 231.

TAFE NSW resources

Training and Education Support (TES), Industry Skills Unit Orange/Granville 68 South St Granville NSW 2142 Ph: (02) 9846 8126 Fax: (02) 9846 8148 (Resource list and order forms).

GET IT RIGHT

The photo shows an incorrect method of transporting oxygen and acetylene cylinders in an enclosed vehicle.

What is the correct method? Provide reasoning for your answer.

WORKSHEET 1

To be completed by teachers	
Satisfactory	☐
Not satisfactory	☐

Student name: ______________________

Enrolment year: ______________________

Class code: ______________________

Competency name/Number: ______________________

Task: Review the section 'Prepare for work' up to and including 'Identify hazards and risk-control information and measures' in this chapter and answer the following questions.

1. When should PPE be used to handle materials?

2. State the two items of PPE that must be worn when handling products that contain asbestos.

3. What should be worn to protect the hands when handling sharp-edged materials?

4. What is the purpose of the WaterMark Certification Scheme?

5. Who is responsible for reporting hazards in the workplace?

6. When and how should a hazard be recorded on site?

7. What is the benefit of a workplace inspection?

8. What does the abbreviation 'SWMS' stand for?

9. How are risks controlled in the workplace?

WORKSHEET 2

To be completed by teachers	
Satisfactory	☐
Not satisfactory	☐

Student name: ______________________

Enrolment year: ______________________

Class code: ______________________

Competency name/Number: ______________________

Task: Review the section 'Handle, sort and stack materials' and answer the following questions:

1. What does the abbreviation 'SDS' stand for?

   ____

2. List four factors that affect how materials behave.

   ____

   ____

   ____

   ____

3. List the three different states of water.

   ____

   ____

   ____

4. What is a material's density?

   ____

   ____

5. How does the density of material affect the construction process?

   ____

   ____

   ____

6. Briefly describe the meaning of a material's tensile strength.

   ____

   ____

7. Briefly describe the meaning of a material's shear strength.

   ____

   ____

8. Why is it important to know the thermal expansion of a material in linear metres?

____

____

9. What does the co-efficient of linear expansion represent?

____

____

10. What will occur if a brittle material is overloaded?

____

11. What is the term for a material that is easily stretched or flattened without breaking?

____

12. List the two types of corrosion commonly found in the plumbing industry.

____

13. What causes a metal to corrode or oxidise?

____

14. Is zinc considered to be an anode or cathode in the Noble Scale?

____

15. What is a major cause of polymer breakdown, and how does this affect the material?

____

____

To be completed by teachers	
Satisfactory	☐
Not satisfactory	☐

Student name: __________

Enrolment year: __________

Class code: __________

Competency name/Number: __________

Task: Review the section 'Material classification' up to and including 'Store and transport materials' and answer the following questions:

1. State the three main properties of metals.

2. Briefly describe how alloys are formed.

3. Explain why lead is generally being removed from the environment.

4. Identify and list two ferrous metals.

5. List two materials other than metals or polymers that may be used on site.

6. State the common name that classifies polymers.

7. How should small items be contained on the back of a work vehicle?

8. How should gas cylinders be transported in a vehicle?

9. How should gas cylinders be stored on site?

USE PLUMBING HAND AND POWER TOOLS

7

This chapter addresses the following key elements for the competency 'Use plumbing hand and power tools':

- Identify hand and power tools.
- Select and use appropriate hand tools.
- Select and use appropriate power tools.
- Clean up work area.

It introduces the identification and safe use of hand tools, power tools, pneumatic tools, plant and equipment used in the plumbing industry.

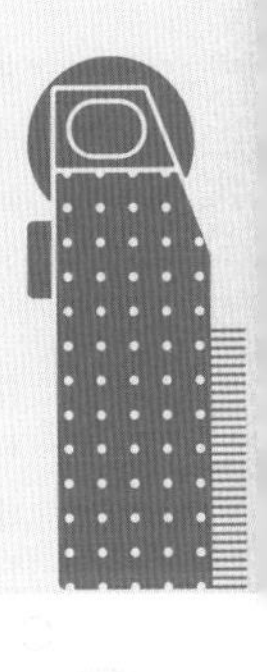

Identify hand and power tools

Tradespeople working in the plumbing services sector are required to operate and maintain a wide variety of tools and equipment. In order to be proficient in the trade, apprentices and trainees must be able to competently identify individual tools, be familiar with their component parts, understand how they operate and have the ability to maintain and repair their tools and equipment. Correct selection and safe use of a suitable tool for the task at hand will assist in producing a quality product.

Tools are designed as an extension to the human form. The actions of cutting, slicing, chiselling, forging and bending are all made easier by using specially manufactured tools. Prehistoric tools were made from raw (native) metals in the form of rocks. During the Copper Age, these native metals were transformed by hammering or slicing into usable shapes. In the Bronze Age, the natural stone was altered by changing the native ore into a metal by using a high-temperature furnace, and during the Iron Age the metal was processed to create cast iron, wrought iron and steel products.

Modern technology has developed tools and machinery that have made our lives and work easier, with the ability to create structures only dreamed of in times gone by. The modern construction worker makes use of many hand, electrical, pneumatic, fuel and battery-operated tools on projects that range from simple residential homes to the impressive landmark structures seen in Australian cities.

Although power tools are generally superior to hand tools, plumbers also need to understand how to use hand tools properly, including their safe use and regular maintenance, and an appreciation of the function of individual tools. The following sections outline some of the common tools and equipment that are used and maintained by plumbers, and the associated requirements in ensuring that they are used safely.

WHS and workplace requirements

The use of hand and power tools makes completing tasks very efficient; however, you should always use hand and power tools as they were designed and never become complacent in their use. Always read the safety instructions and refer to Safe Operating Procedures (SOP) before undertaking tasks. If you have not used the tool before, always seek proper instruction from a competent person and never assume you know how to correctly use the tool. If you have not been trained to use a tool, do not use it!

Always identify, assess and controls hazards on site and ensure all Safe Work Method Statements (SWMS) are completed and Safety Data Sheets (SDS) are referred to before tools are used with any materials. Codes of Practice may also be used for industry best practice guidelines.

Only use electrical tools that have been inspected by a competent person and have a current safety tag attached. A safety tag will not guarantee that an electrical tool is safe to use, so always inspect the tool prior to use. When using electrical equipment, you must use a residual current device (RCD), which will switch off the electrical supply if there is a fault in the electrical system. Always perform a safety test on the RCD before use and ensure that it is tagged and tested.

All guards and mechanisms must be checked and in place, and hands and feet are to be kept away from any moving parts and blades. Consider using non power tools in situations where it may be deemed hazardous or too dangerous. Quality assurance requirements must be identified, and all company compliance processes followed. Following these steps can significantly reduce injury and incidents from occurring and ensure you go home safe at the end of each workday.

Select and use appropriate hand tools

Hand tools are an essential part of a tradesperson's everyday life and they allow us to accurately measure and perform tasks. Tools come in a variety of sizes and uses, and without hand tools, no work would occur at all. Before tasks are to be undertaken it is essential that the correct personal protective equipment (PPE) is selected to ensure that work is carried out safely.

Personal protective equipment

When used correctly, PPE will reduce the risks and hazards associated with using hand and power tools. PPE should be provided to all workers by their employer to ensure that all tasks are carried out safely. All workers should receive appropriate instruction and training on the correct use and fitting of PPE to ensure it is used and protects as it was designed. PPE that does not fit or has not been fitted correctly may result in illness or injury and defeat the purpose of wearing the PPE in the first place. When using hand and power tools with materials, it is essential that you refer to an SDS for that chosen material to select appropriate PPE.

Marking out and measuring tools

Marking out is one of the main functions that is carried out by plumbers in the workplace. It is the process of taking dimensions or measurements from working drawings or hydraulic plans and transferring these details onto the work piece or job site. Marking-out tools should be kept in good condition to maintain accuracy and to ensure they work as they were designed.

Measuring implements

Measuring implements will vary depending on their application.

Steel rule

Other name: engineer's rule.

Steel rules (see Figure 7.1) can be used to mark out or determine the size of a work piece. They can be marked with a combination of metric and imperial graduations and are very accurate when used as a straight edge, for measuring pipe diameters or for scaling measurements from a plan.

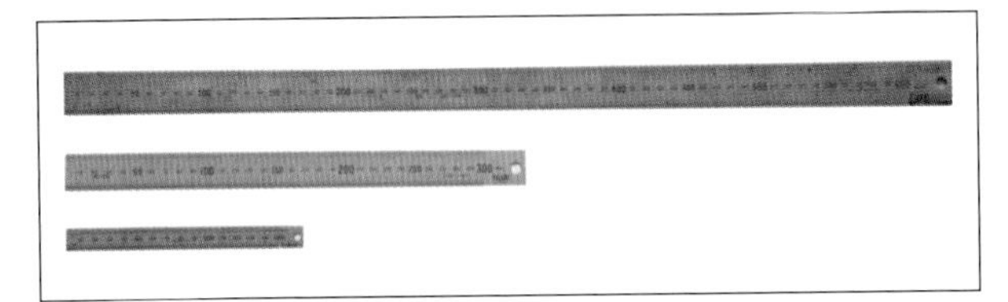

FIGURE 7.1 Steel rule in 300 mm, 600 mm and 1000 mm lengths

Four-fold rule

Other names: jointed rule, zig-zag rule or folding rule.

Constructed from timber or plastic, with brass joints, folding rules (see Figure 7.2) are used to measure the size of a work piece. They are convenient for use in confined spaces due to their compact nature when closed and are available with both metric and imperial graduations. This rule can be used to mark out gutter and downpipe angles.

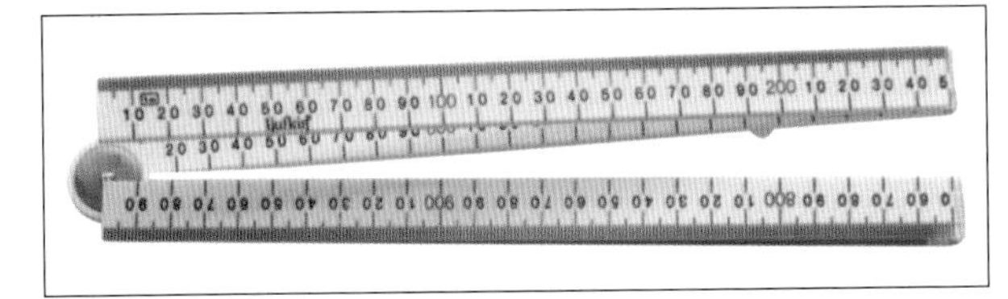

FIGURE 7.2 Folding rule

Tape measure

Other name: retractable tape measure.

The tape measure (see Figure 7.3) is constructed from a steel strip that is spring-loaded and enclosed in a plastic casing. Available with metric and/or imperial dimensions, it can be used to accurately mark pipe lengths or clipping placements, or to position fixtures according to the manufacturer's specifications.

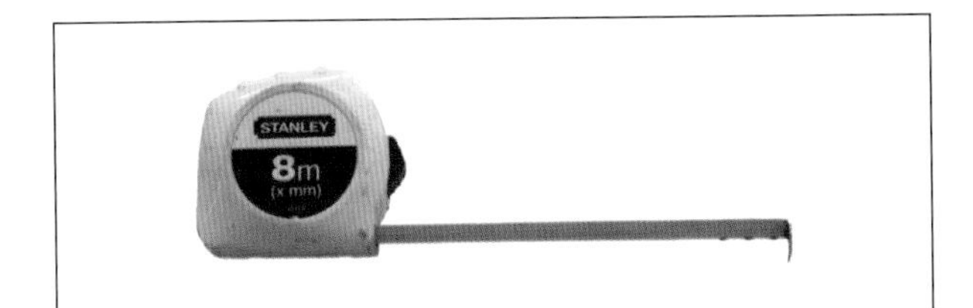

FIGURE 7.3 Tape measure

Wind-up tape measure

The wind-up tape measure (see Figure 7.4) is generally available in 10 m, 20 m and 30 m lengths. This tape measure is predominantly used for measuring larger dimensions such as calculating quantities of roofing or guttering materials, marking out drainage and stormwater, or marking out site boundaries. The tapes are constructed of fibreglass for maximum flexibility and durability, and are retracted by winding a handle.

FIGURE 7.4 Wind-up fibreglass tape measure

Setting-out and marking implements

Setting-out and marking implements can be used to mark out on a variety of materials.

Try square

Variations include the engineer's try square, mitre square and combination square (see Figure 7.5).

An engineer's try square is manufactured from steel and is used to mark a line at right angles to a straight edge or to check if a work piece is square. Mitre squares have a 45° angle on the stock (handle)

FIGURE 7.5 From left: try square, mitre square and combination square

and are used by plumbers to mark cut-outs in guttering or downpipe material. The combination square can be used as a steel rule or as a mitre square to mark 45° or 90° cuts on material.

Sliding bevel square

Other names: T-bevel or bevel gauge.

The sliding bevel square (see Figure 7.6) has an adjustable stock or handle that allows the blade to be adjusted to form any angle. The bevel is used to transfer angles for downpipes, and roof slopes angles onto rainwater flashings or sheet metal products.

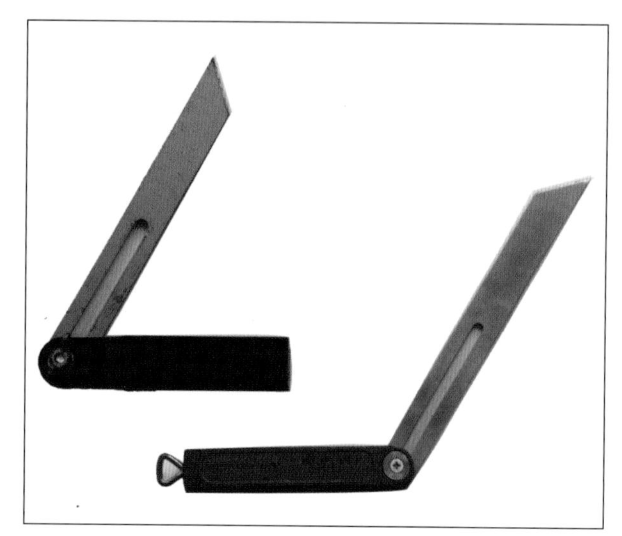

FIGURE 7.6 A sliding bevel square can be adjusted to suit any angle

Scriber

Other name: scribe.

A scriber (see Figure 7.7) is a sharpened steel shank that is used to mark lines onto sheet metal surfaces. To maintain accuracy, the point of the tool needs to be regularly sharpened on a linishing wheel.

FIGURE 7.7 Scriber

Spring divider

Other name: spring compass. Variations include the winged compass, beam compass and trammel.

Dividers and compasses (see Figure 7.8) have sharp, pointed, adjustable legs and are used to scribe arcs or circles onto sheet metal surfaces. They can also be used to 'step off' dimensions or divisions on a line or arc when creating pattern developments. A beam compass or trammel (see Figure 7.9) is used to scribe larger arcs or circles onto sheet metal material.

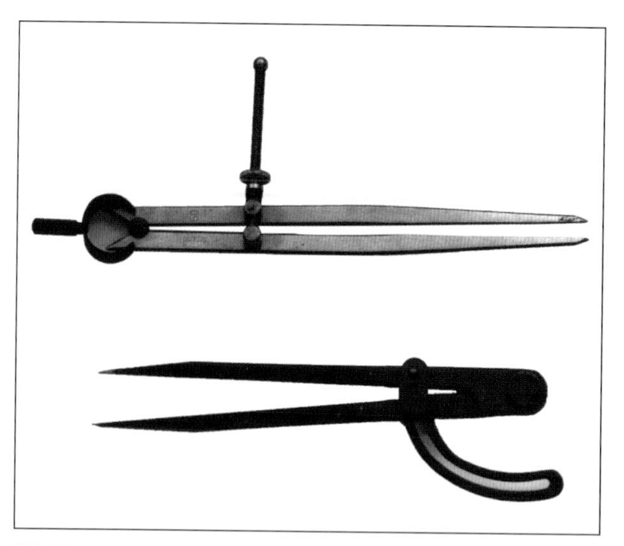

FIGURE 7.8 Spring dividers

FIGURE 7.9 Using a trammel to draw an arc onto sheet metal

Chalk line reel

Other name: flick line.

A chalk line (see Figure 7.10) is a retractable string line that is dusted in chalk. When extended and pulled taut the line is 'snapped' or 'flicked' to leave a straight, chalk impression upon the surface. This tool is commonly used by roof plumbers to mark out straight screw lines when fixing metal roofing or marking out straight runs of pipe support on concrete.

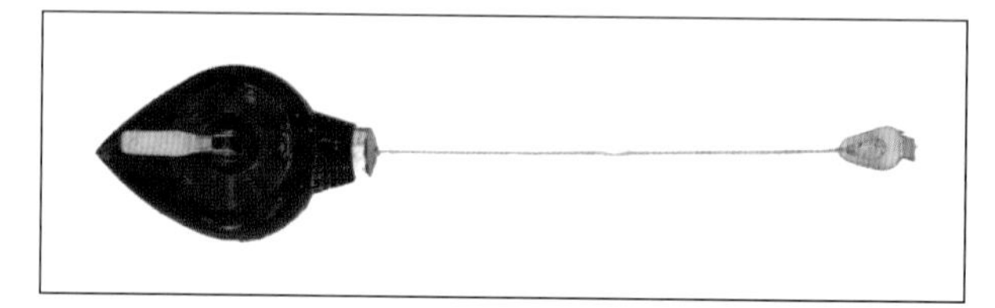

FIGURE 7.10 Chalk line reel, which can also be used as a plumb-bob

Cutting tools

Cutting tools are used to shape materials, so it is essential that these tools are regularly maintained. Cutting edges need to be kept sharp or be replaced if they become damaged.

Hacksaw

Hacksaws (see Figure 7.11) have a pistol grip handle and incorporate an adjustment screw to maintain the correct tension on the cutting blade. They can be used to cut a variety of materials through careful blade selection. Blades are measured in 'teeth per inch' (TPI) and should be fitted with the cutting teeth facing forward.

Fine-pitch blades come in two sizes: 32 TPI blades (recommended for use on sheet metal rainwater goods) and 24 TPI blades (recommended for use on copper tubing or PVC pipe).

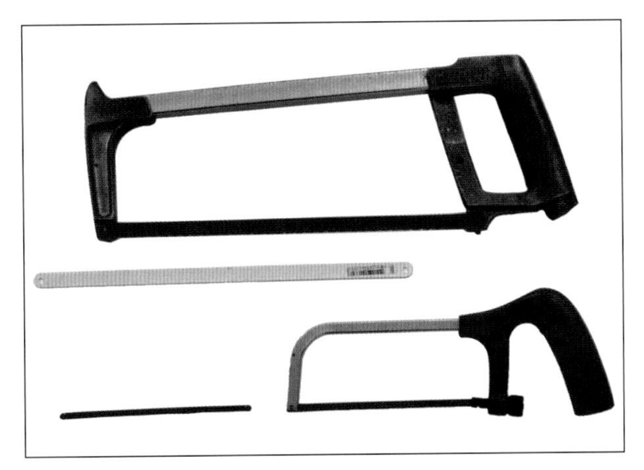

FIGURE 7.11 Hacksaw and junior hacksaw

Coarse-pitch blades also come in two sizes: 18 TPI blades (recommended for use on medium-gauge steel pipes) and 14 TPI blades (recommended for use on heavy-gauge steel pipes or PVC pipes).

The junior hacksaw (see Figure 7.11) and mini hacksaw are designed for use in confined spaces and awkward cutting positions.

Handsaw

Variations include the rip saw, cross-cut saw and panel saw.

Handsaws (see Figure 7.12) consist of a pistol-grip handle connected to a tapered steel blade with set teeth on the bottom of the blade. The main differences in the variations of the handsaw lie in the number and shape of the cutting teeth. Fine cross-cut saws can be used in the plumbing sector to cut timber or PVC pipe of 100 mm diameter or over, and it is imperative that the teeth are routinely *sharpened and set* to maintain an effective cutting edge.

FIGURE 7.12 Handsaws

Snips

Variations include straight snips, curved snips, offset snips, aviation snips, jeweller's snips and gilbows (see Figure 7.13).

Snips are used for cutting straight or curved sections in thin sheet metal rainwater products such as gutters, downpipes, flashings, cappings and corrugated iron, and for cutting banding straps found on slings of roof sheets. Rolled jaw snips have longer handles to increase leverage, and aviation snips operate using a compound action that also increases the leverage action on the jaws, therefore requiring less effort to cut. Small snips, also called jeweller's snips, are available with straight or curved blades and are useful for trimming thin sheet material and cutting small-radius curves, arcs and circles, either externally or internally. The most common types of snips used for roof plumbing are aviation snips, with the two most common being right cut (green handle) and left cut (red handle). Most right-handed plumbers tend to prefer the red-handled

FIGURE 7.13 Straight snips, aviation snips (green and red) and jeweller's snips

left-cut aviation snips, as when held in their right hand the cutting action will cut to the left, which is towards the centre of the body. In most circumstances, plumbers will be required to use both left- and right-cut aviation snips, due to having to cut in tight and awkward positions.

Bolt cutter

Bolt cutters (see Figure 7.14) operate on a leverage action and are used for heavy cutting of rods, bars and thick-gauge wire. They are mainly used in plumbing to cut reinforcement bars and steel fabric in preparation for concrete work.

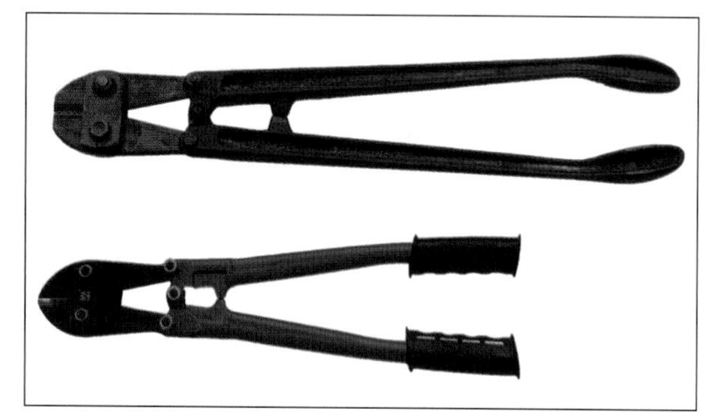

FIGURE 7.14 Bolt cutters

Retractable trimming knife

Other names: Stanley knife. Variations include the snap-blade knife.

There are numerous general-purpose trimming knives available for use by tradespeople (see Figure 7.15). All these knives have disposable, razor-sharp blades and can be used to cut a variety of materials. They are extremely dangerous if misused – always direct the cut away from your body, and retract the blade when not in use.

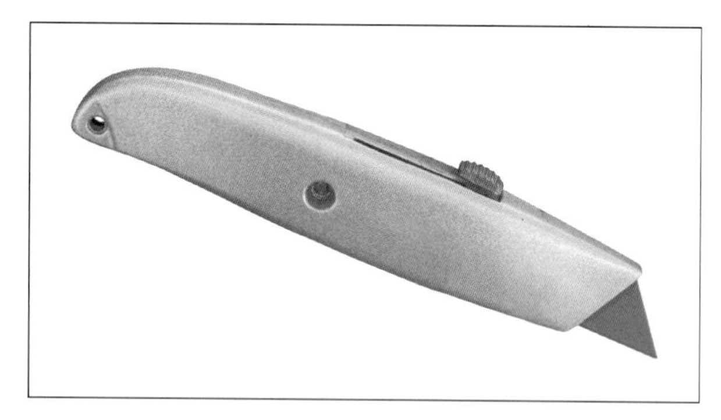

FIGURE 7.15 Retractable trimming knife

Source: Shutterstock.com/Yogamreet.

Hacking knife

Other name: chipping knife.

A hacking knife (see Figure 7.16) has a heavy steel blade with tapered sides, a sharpened cutting edge and a thickened top edge that allows the knife to withstand blows from a hammer. Hacking knives are used by plumbers to make penetrations in sheet metal products for the insertion of snips.

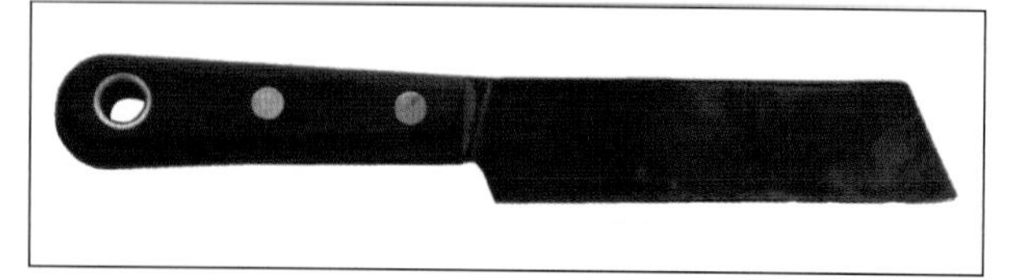

FIGURE 7.16 Hacking knife

Shave hook

Shave hooks are constructed with a choice of differently shaped heads to suit different surfaces (see Figure 7.17). These were commonly used by plumbers to clean the oxide coating from rolled lead prior to lead welding/burning process. Their use may be required in specialised heritage work to retain or recreate original aspects or features of a building from that time period.

FIGURE 7.17 Shave hooks and lead roll with oxide coating removed

Hand file

Variations include the flat file, round file, half-round file, triangular file, square file, needle file, knife file and warding file (see Figure 7.18).

Files are cutting and finishing tools that have a multitude of small, hardened-steel teeth that are used to shape or smooth flat or curved metal surfaces. Files are available in four grades of cut:

- Bastard-cut files are the roughest, coarsest type.
- Second-cut files are a medium to smooth grade.
- Mill saw files are used to sharpen snips.
- Smooth-cut files have the finest teeth and are used for smoothing or finishing off work.

Files need to be regularly cleaned with a file card (file brush) to prevent them becoming clogged with metal filings.

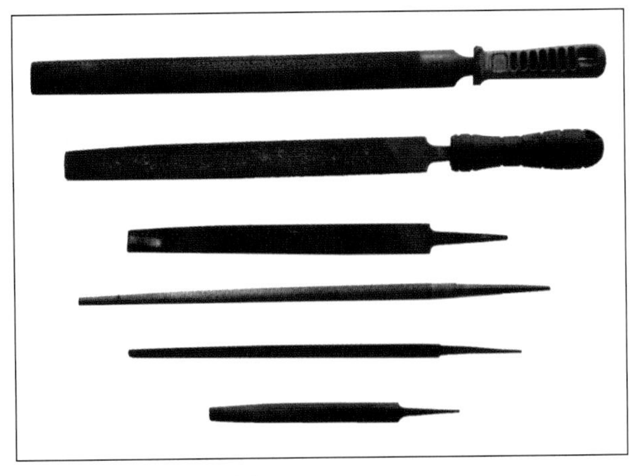

FIGURE 7.18 A range of files

Hole saw

A hole saw (see Figure 7.19) is a cylindrical saw that has specially treated cutting edges and is constructed from toughened alloy steel. Hole saws are used in conjunction with a power or battery-operated drill and can be successfully used to cut holes through most metal, plastic or timber materials. To prevent blunting during operation it is imperative that the teeth are kept well lubricated.

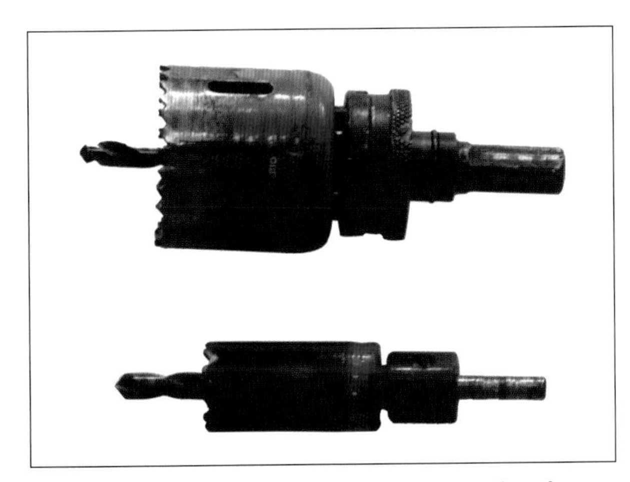

FIGURE 7.19 Hole saws are used to make penetrations for piping

Stocks and dies

Stocks and dies (see Figure 7.20) are used to produce an external thread on the surface of mild steel piping. The thread produced is a BSPT (British standard pipe thread). This is the standard thread used for plumbing fittings. The die is a round or square block of hardened steel that has a series of cutting teeth in the centre. The stock is the tool used to hold and turn the die and operates on a ratchet action. Correct lubrication of the die and pipe surface is essential to maintain a sharpened cutting edge. They are useful for connecting onto existing steel services, where regulations permit. Some plumbers will use these dies for cleaning or repairing various external threads on fittings.

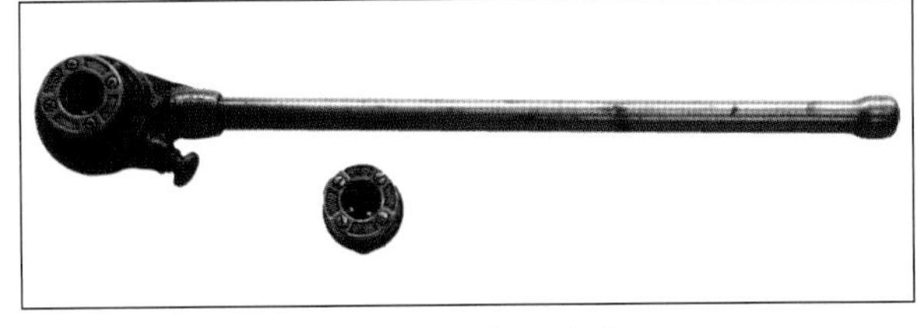

FIGURE 7.20 Ratchet action stocks and dies

Taps and dies

Taps and dies (see Figure 7.21) are the tools used to cut threads for metric or imperial nuts and bolts. They are constructed from carbon steel or high-speed steel that has been hardened and tempered. Either a tap wrench or die stock is used to hold and rotate the taps and dies during the cutting process. These are also useful for cleaning and repairing threads on bolts, nuts and threaded rod.

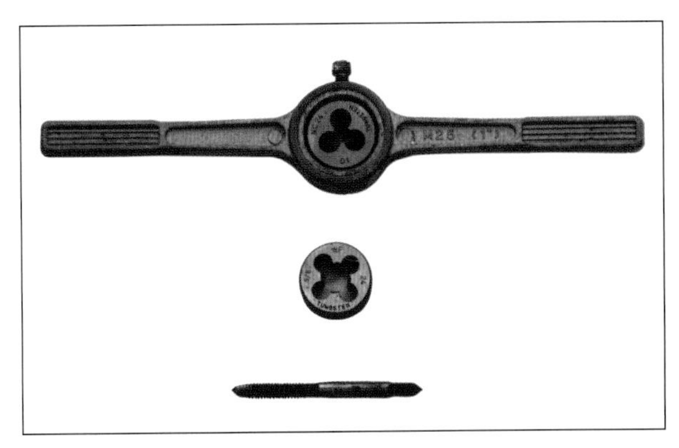

FIGURE 7.21 Taps and dies

Drill bits

Variations include the twist drill, stub drill, double-ended drill, masonry drill, spade bit and auger bit (see Figure 7.22).

FIGURE 7.22 A variety of drill bits

Twist drills are used in conjunction with electric or battery-operated power drills to form round holes in solid objects. The twist drill clears the swarf, or waste material, from deep holes via the two helical flutes that run along the length of the drill bit. Essential maintenance includes removing any swarf and periodic sharpening that involves regrinding the point so that the cutting angles are equal.

It is recommended that you 'punch' the surface of any item being drilled in order to stop the drill bit from 'running off'. A metal surface will require the use of a centre punch, whereas a masonry surface may require the use of a punch or even a pilot hole being drilled with a smaller bit.

Masonry drill bits incorporate a tungsten carbide tip and are used to drill holes in brick, stone, concrete and ceramic tiles. A spade bit is used to drill enlarged holes in timber suitable for pipework or low voltage cable penetration.

Impact tools

Impact tools are used to change the shape of a material or strike an object by using direct or indirect force.

Direct impact tools

Direct impact tools are used to hit or drive a nail or an indirect impact tool into a material.

Hammer

Variations include the claw hammer, peining hammer, mash (lump) hammer, sledgehammer and brick hammer.

A hammer is a direct impact tool and has a hardened, tempered steel head that is fastened to a timber, fibreglass or steel handle. Due to the impact action of the hammer head it is recommended that safety eyewear be worn to prevent injury from any splintering material.

Claw hammer

Claw hammers (see Figure 7.23) incorporate a weighted hammer head (225 g to 910 g) that is generally used to drive nails into timber and a split-pein or claw that is used to remove nails when dismantling work.

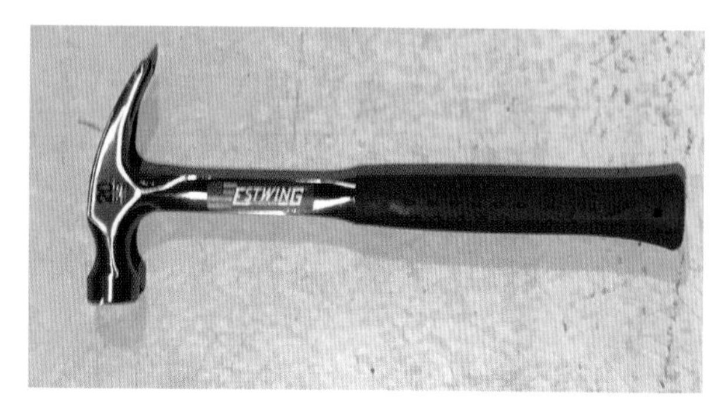

FIGURE 7.23 Claw hammer

Peining hammers

There are two main types of engineer's peining (also known as 'peen' or 'pane') hammers used in the plumbing services sector. The cross-pein or Warrington hammer (see Figure 7.24) combines a hammer head for driving nails or punches with a cross-pein that is used for dressing sheet metal work. The ball-pein hammer (see Figure 7.25) incorporates a flat-surfaced or convex hammer head that is used to strike chisels or punches with a ball-pein that is used for doming or shaping the shank of solid rivets.

FIGURE 7.24 Cross-pein hammer

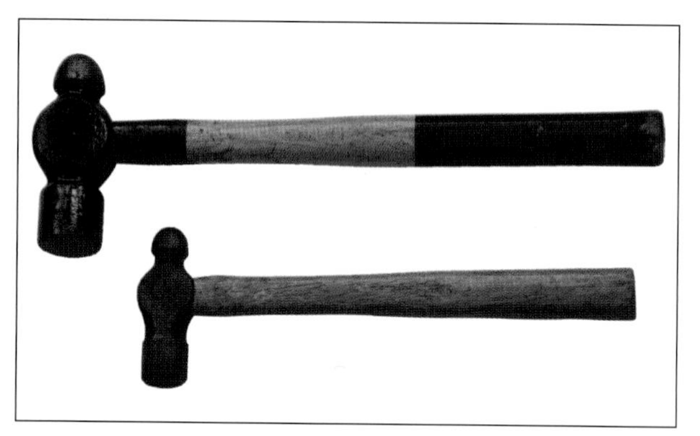

FIGURE 7.25 Ball-pein hammer

Lump hammer

A lump, mash or club hammer (see Figure 7.26) is generally used for demolition work on masonry walls or for driving brick bolsters or plugging chisels. They come with wooden, steel or fibreglass handles.

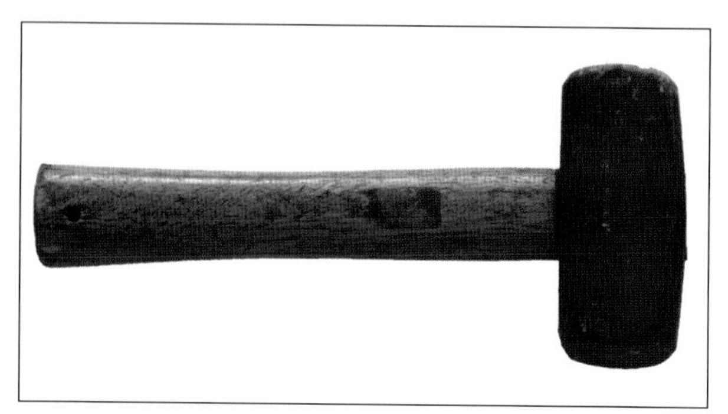

FIGURE 7.26 Lump (mash) hammer

Sledgehammer

Sledgehammers (see Figure 7.27) have a double-faced steel head with a long timber or fibreglass handle. They are used mainly in demolition work on masonry structures or to drive surveyor's pegs or barricade stakes into the ground.

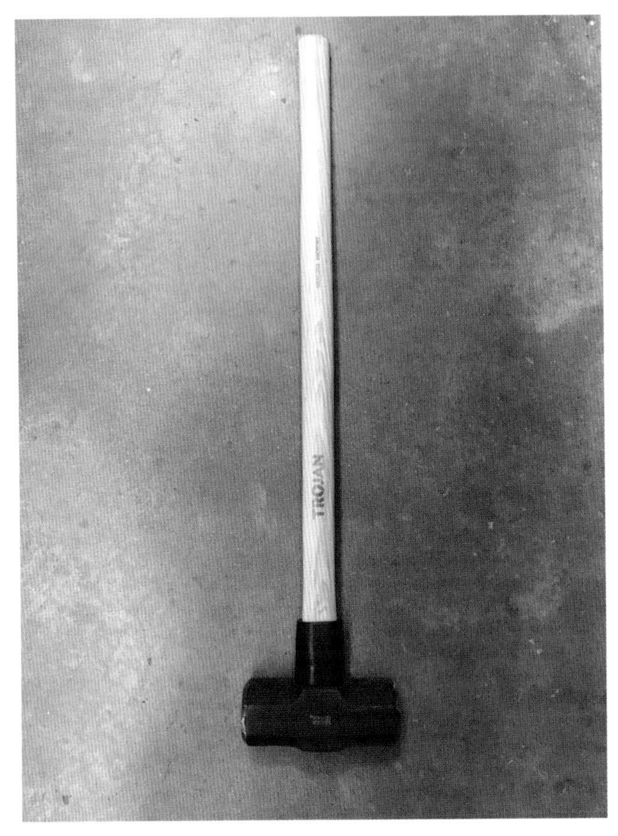

FIGURE 7.27 Sledgehammers have a long handle, making them easier to swing

Brick hammer

A brick hammer is used to drive a brick bolster or masonry chisel. Some brick hammers incorporate a scutch holder (see Figure 7.28). The scutching tool is used to prepare masonry surfaces prior to rendering.

Mallet

Other name: soft-faced hammer. Variations include the wooden mallet (see Figure 7.29), tinman's mallet, bossing mallet and bending stick.

A tinman's mallet is a light, direct impact tool used to flatten deformations in sheet metal products where a blow from a steel hammer would cause indentations in the metal surface. Other variations are used to shape or dress soft metals such as lead, copper, aluminium and zinc sheeting.

Hatchet

Other names: tomahawk or hunter's hatchet.

A hatchet (see Figure 7.30) is a heavy, direct impact tool consisting of a steel, wedge-shaped head and a curved, wooden or fibreglass handle. It is generally

FIGURE 7.28 Brick hammer and scutch comb

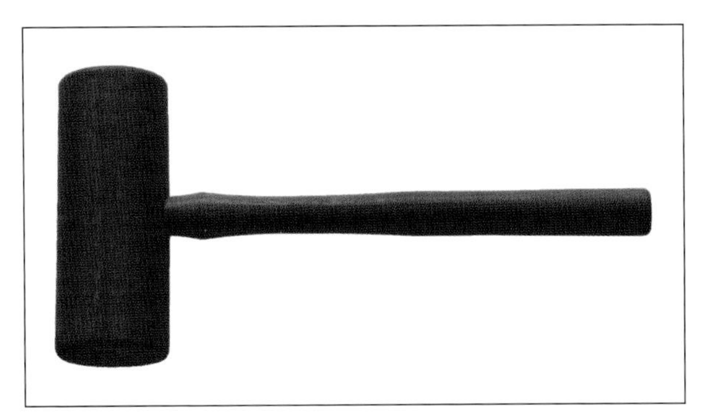

FIGURE 7.29 Wooden mallet

used to trim or shape timber and is very effective in cutting through exposed tree roots. It is imperative that the cutting edge is regularly sharpened and that safety glasses are worn during use.

Indirect impact tools

Indirect impact tools that are struck with a hammer (e.g. bolsters, cold chisels and centre punches) should always be carefully maintained. 'Mushroomed' heads and blunt tools can become quite dangerous to work with, as splinters of mushroomed ends may come flying off or excessive force on blunt tools being used may cause injury.

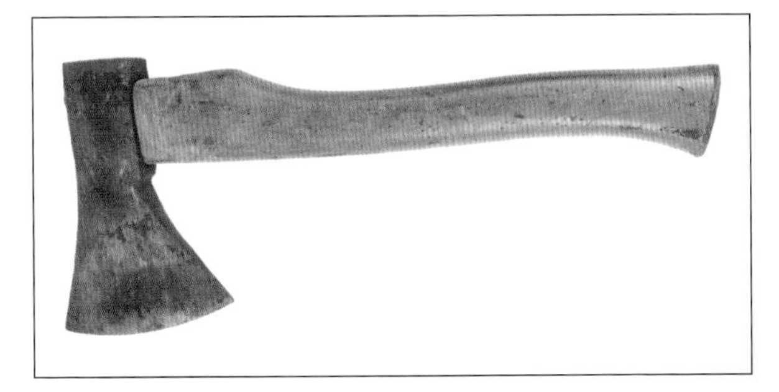

FIGURE 7.30 Hatchet

Source: Shutterstock.com/Kunertus

Pinch bar

Other names: wrecking bar, jimmy bar or jemmy bar.

A pinch bar (see Figure 7.31) consists of a round or octagonal steel shank and incorporates a flat, chisel-like end for levering and a hooked claw end that is used for de-nailing timber and for general demolition work.

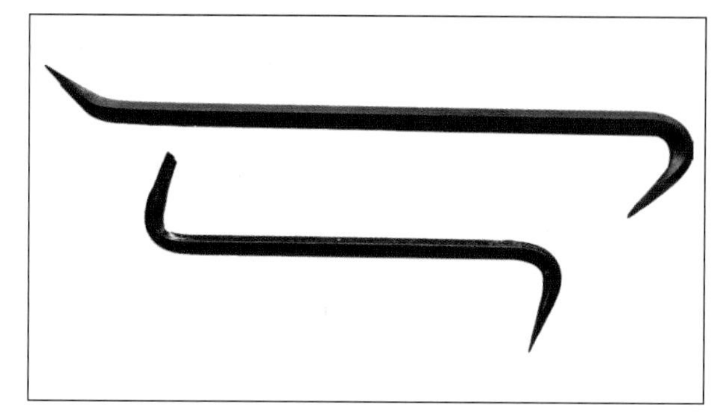

FIGURE 7.31 Pinch bar

Bolster

A bolster is a steel-bodied tool with a broad, flat face that forms a thin cutting edge ranging in length from 65 mm to 115 mm (see Figure 7.32). It is used with a mash (lump) hammer by bricklayers and stonemasons to provide a shear cut through masonry products.

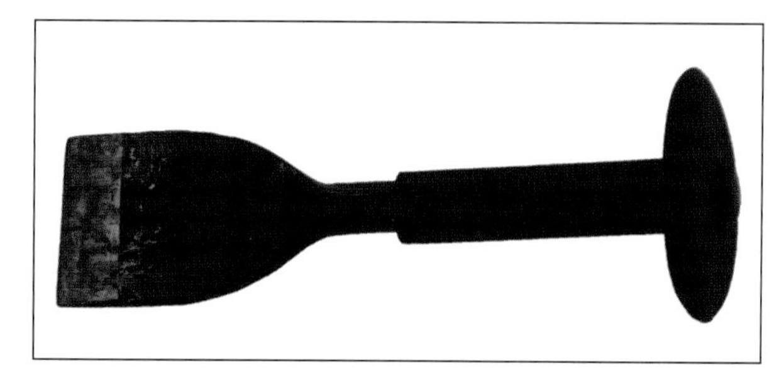

FIGURE 7.32 Bolster

Cold chisel

Variations include the cape chisel, diamond point chisel and plugging chisel.

Cold chisels (see Figure 7.33) are used to create penetrations in sheet metal, cut off rivet heads, break spot welds, remove sections of bricks or remove seized nuts. A plugging chisel is used in conjunction with a lump hammer to remove mortar joints between brickwork.

Punching tool

Variations include the centre punch, nail punch, wad punch and prick punch.

Constructed of a knurled steel shaft and angled point, centre punches are used to mark the centre point of a hole to prevent twist drills from slipping on sheet metal surfaces. A wad punch (see Figure 7.34 (a)) is used to cut discs of neoprene or leather for use as sealing washers in plumbing fixtures. A nail punch (see Figure 7.34 (b)) is used to drive a nail flush with the surface level of a piece of timber, while a prick punch (see Figure 7.34 (c)) is used by roof plumbers to create a small hole in corrugated metal roofing products to guide the point of a drill bit or screw.

FIGURE 7.33 Cold chisels: new (top), 'mushroomed' (middle) and maintained (bottom)

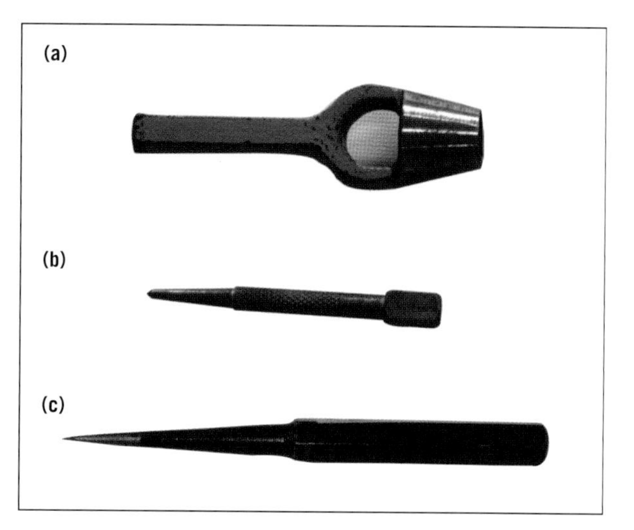

FIGURE 7.34 (a) Wad punch; (b) Nail punch; (c) Prick punch – used for roofing products

Grooving tool

A grooving tool or groover (see Figure 7.35) is used in conjunction with a hammer to fabricate a grooved seamed joint in sheet metal rainwater products such as downpipes.

Gripping, clamping and driving tools

These tools utilise simple mechanical principles to apply torque, tension and leverage to a large range of threaded fasteners or piping materials used in the plumbing trade.

Spanner

Spanners are available in several styles and in both metric and imperial sizes. Styles include single open end, double open end, podger (for erecting structural steel and replacing hot water system anodes), double

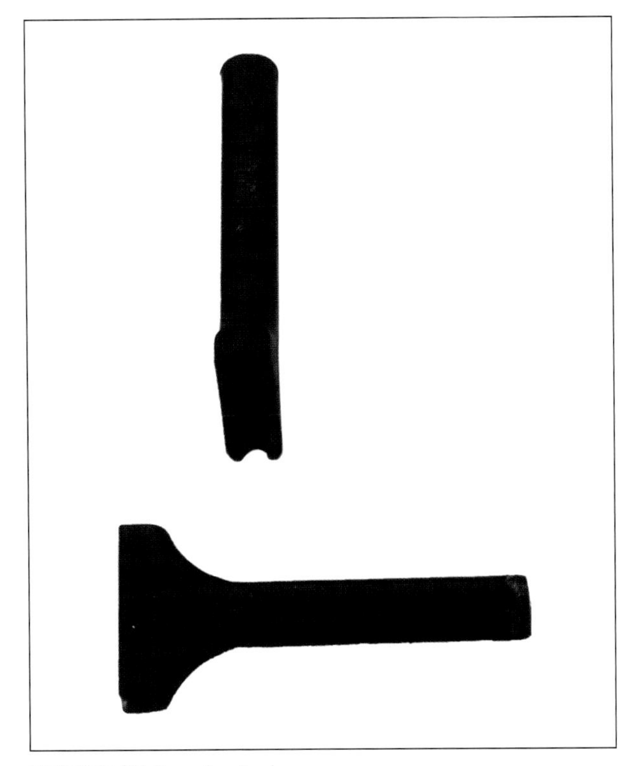

FIGURE 7.35 Grooving tool

ring end, ring and open end, and come with a variety of shaft shapes ranging from straight to almost semicircular. They can be made from drop-forged steel or chrome vanadium steel and are available in a range of socket sizes and types. Adjustable or shifting spanners are also very common. Spanners are useful for tightening or loosening nuts/bolts, kinko nuts, backnuts and tapware (see Figures 7.36–7.41).

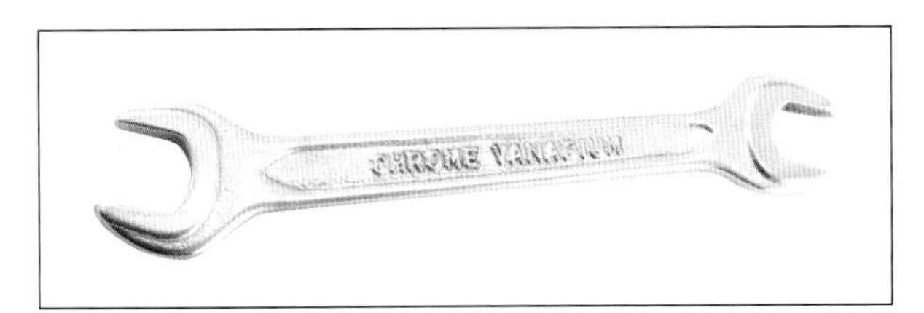

FIGURE 7.36 Chrome vanadium double open-ended spanner

Source: Shutterstock.com/Popov Nikolay.

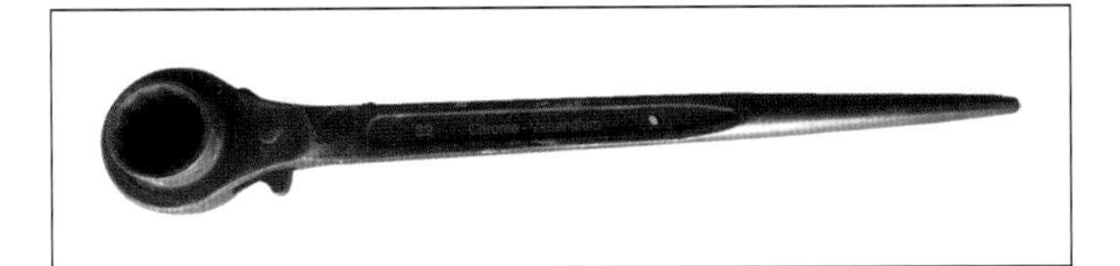

FIGURE 7.37 Podger used for structural steel

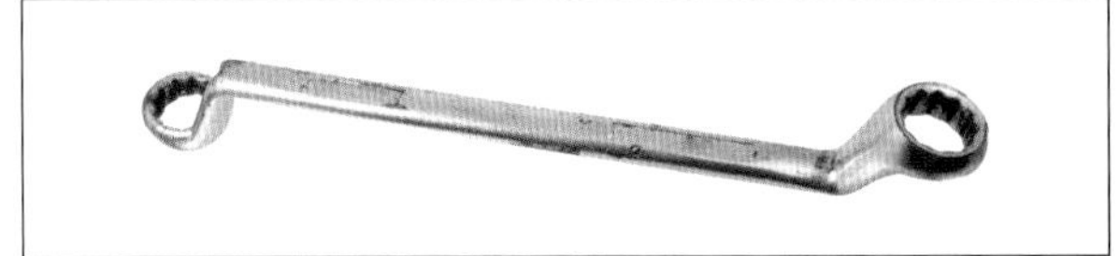

FIGURE 7.38 Double-ended ring spanner

Source: Shutterstock.com/Wongtanarak.

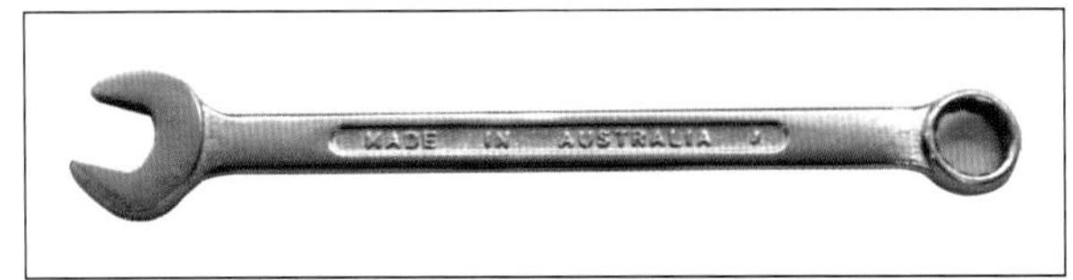

FIGURE 7.39 Ring and open-ended combination

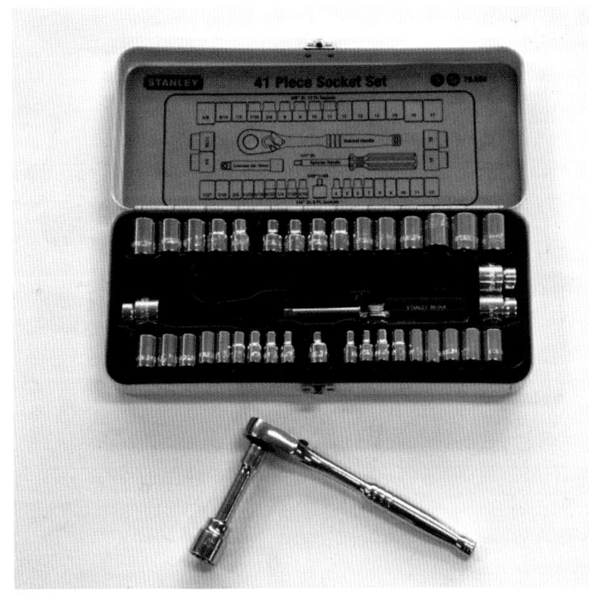

FIGURE 7.40 Square drive ratchet handle and socket set

FIGURE 7.41 Adjustable shifting spanner

Engineer's pliers

Other names: combination pliers or linesmen pliers. Variations include needle-nose pliers and circlip pliers.

Pliers (see Figures 7.42–7.44) are generally used by plumbers to grip or bend sheet metal. They can also be used for cutting, holding, twisting or stripping wire. Some pliers have specialised functions, such as circlip pliers, which are used to open or close spring-tensioned clips, or needle-nose pliers, which are useful for removing tap washers from brass recess bodies in tapware.

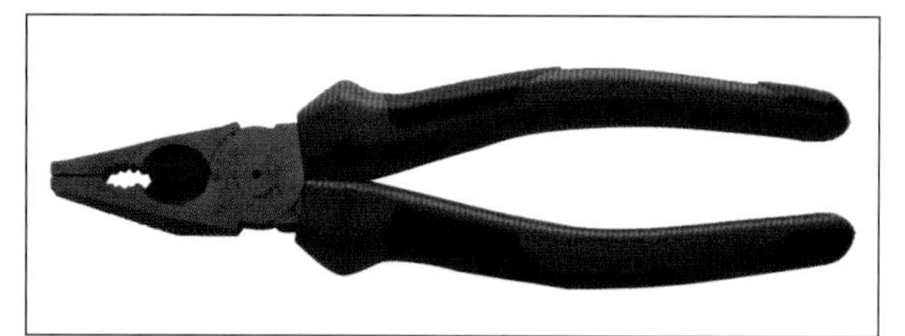

FIGURE 7.42 Combination pliers

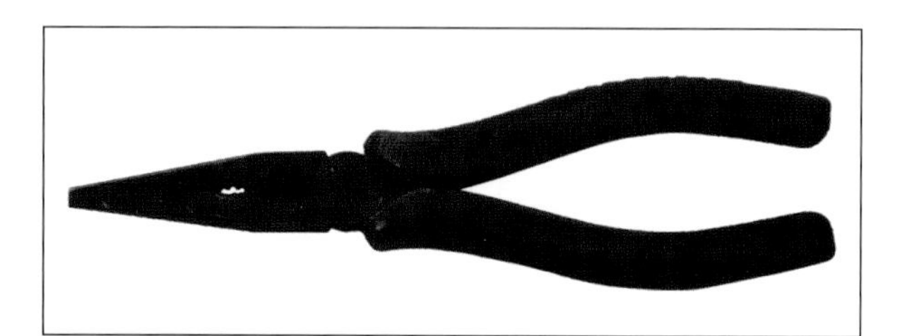

FIGURE 7.43 Needle-nose pliers

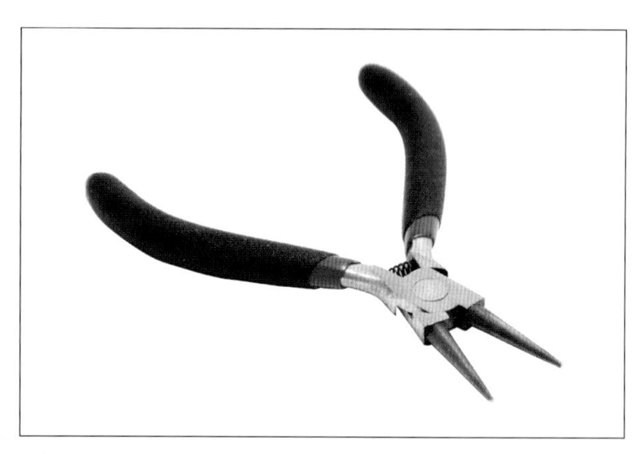

FIGURE 7.44 Straight circlip pliers

Source: Shutterstock.com/Abel Tumik.

Pincer

Pincers (see Figure 7.45) have a pair of circular jaws for holding and removing nails or fasteners and may have a small claw for levering and extracting tacks on the end of one handle.

Pipe wrench

Variations include the footprint wrench, Stillson wrench, basin wrench and multigrips (see Figures 7.46–7.49).

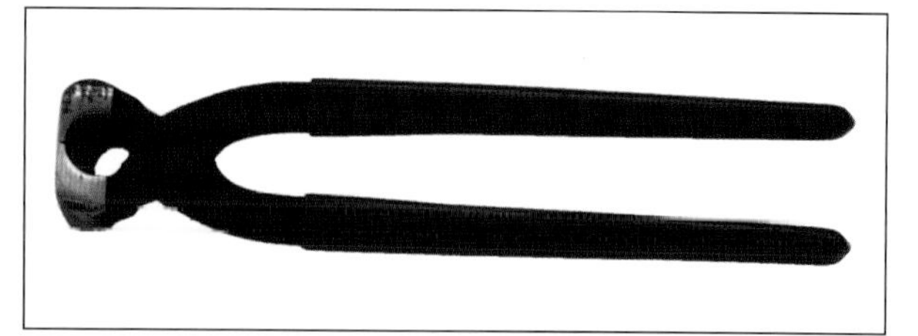

FIGURE 7.45 Pincers

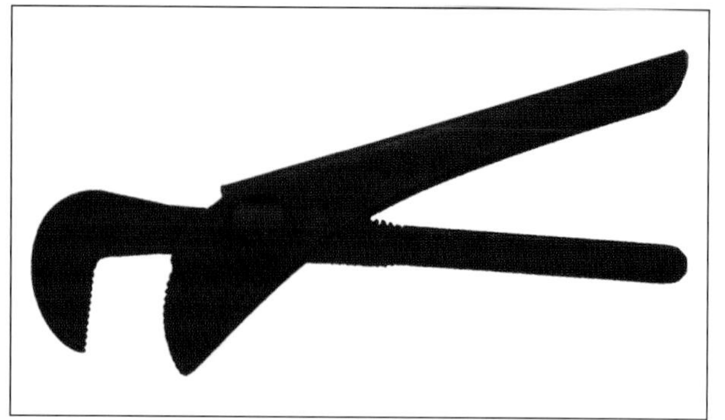

FIGURE 7.46 Footprint wrench

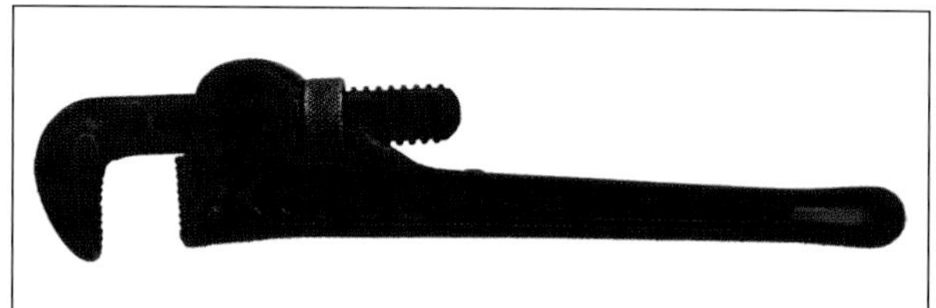

FIGURE 7.47 Stillson wrench

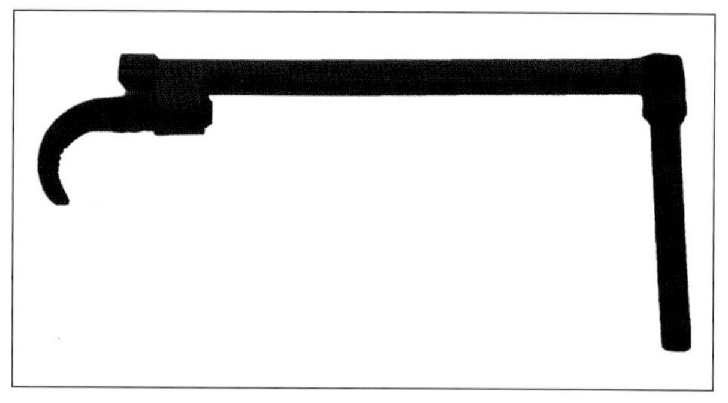

FIGURE 7.48 Basin wrench

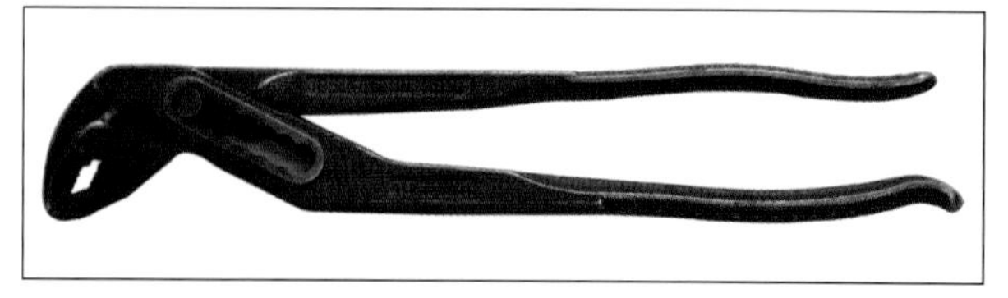

FIGURE 7.49 Multigrips

Similar to adjustable spanners, pipe wrenches are used for gripping and tightening mild steel piping or tapware backnuts. They have serrated jaws to provide grip on smooth or curved surfaces. It is important to apply force in the direction that the jaws open. It is not good practice to use footprints, Stillsons or multigrips on chrome-plated or brass fittings unless necessary, as the serrations will damage the finish. Basin wrenches have a serrated jaw at right angles to the handle and are useful in confined spaces such as working under vanity basins.

Spud wrench

Spud wrenches (see Figure 7.50) are similar to a Stillson wrench and have smooth toothless jaws and a shorter handle. They are useful for removing larger fittings, working with delicate tapware or tightening a union on a water meter without scratching or damaging the material. They are useful for tasks where a shifting spanner is 'just' not wide enough and are available in a variety of shapes and sizes to suit different applications.

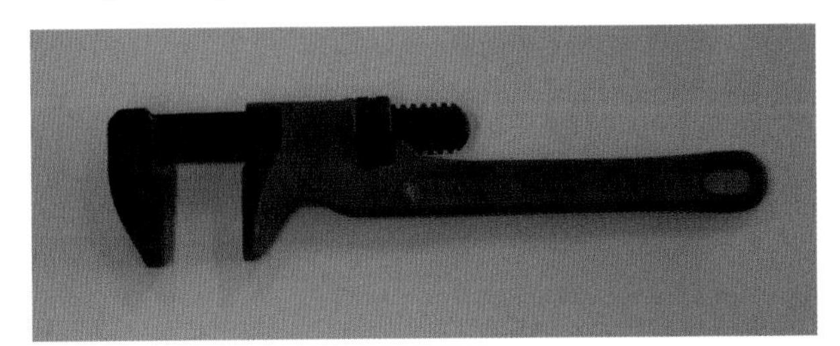

FIGURE 7.50 Spud wrench

Strap wrench

Other name: adjustable wrench strap.

Strap wrenches (see Figure 7.51) can tighten and undo a variety of different shapes where standard wrenches are not suitable or practical. They are useful when trying not to damage delicate tapware and fittings that would otherwise be damaged with serrated wrenches or for use in awkward spaces. A common use is when installing a chrome bath outlet during the fit-off stage.

Vice grips

Other name: self-locking pliers. Variations include vice-grip welding clamps, flat-nose vice grips and D-clamp vice grips.

FIGURE 7.51 Strap wrench

Vice grips are commonly used in industry to clamp sheet metal pieces together, freeing the tradesperson's hands for other activities, such as drilling. Vice grip jaws are serrated and can be clamped tight by turning an adjustment screw. The grips remain locked upon the work piece until the release lever is activated. Welding clamp or D-clamp vice grips are useful to position material when electric or fusion welding. Flat-nose vice grips have broad, flat jaws and are used by roof plumbers to fold sheet metal to desired angles (see Figure 7.52).

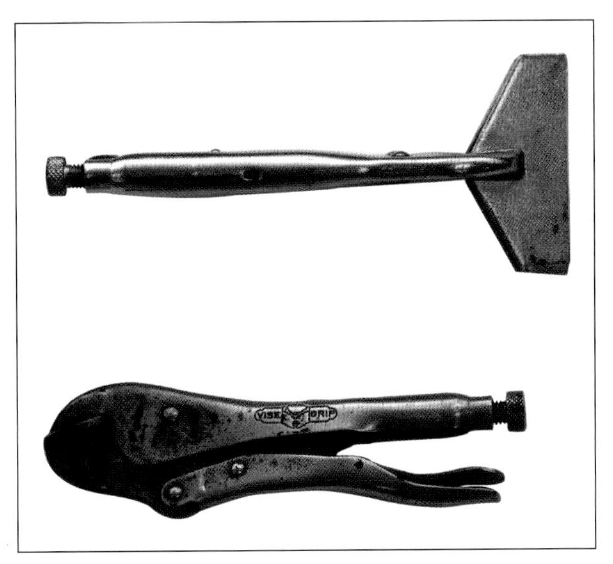

FIGURE 7.52 Vice grips and flat-nose vice grips

Vice

Variations include the swivel parallel jaw vice, pipe vice, portable pipe vice stand and chain vice.

A vice (see Figures 7.53 (a) and Figure 7.53 (b)) is a workshop tool that is firmly secured to a workbench and is used to clamp a work piece so that it can be safely drilled, filed, cut, ground or bent. Routine cleaning and lubrication of vice threads and sliding surfaces is the recommended maintenance. Portable pipe vice (see Figure 7.53 (c)) stands may also be used for multiple and large fabrication jobs to assist in faster production.

FIGURE 7.53 (a) Bench-mounted vice; (b) Bench-mounted pipe vice; (c) Portable pipe vice

Screwdriver

Variations include the stubby screwdriver, jeweller's screwdriver, offset screwdriver, insulated screwdriver and impact screwdriver.

The screwdriver (see Figure 7.54) is a tool that has its tip shaped to fit into the head of a screw. Screwdriver tips are designed to fit slotted head screws, Phillips head screws, Pozidriv screws or spline-head screws. They have modified designs to suit specific applications. The stubby screwdriver is handy in tight or confined locations, whereas the insulated screwdriver is recommended for use on electrical applications, where permitted.

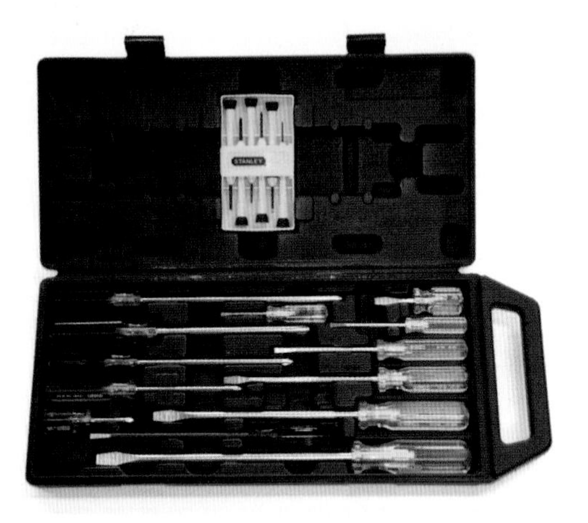

FIGURE 7.54 Screwdriver set

Allen key

Other names: hex key, Allen wrench.

An Allen key consists of an 'L'-shaped hexagonal steel shank that is used for leverage to tighten or loosen hex-head grub screws that are commonly found in tapware (see Figure 7.55). Available in both metric and imperial sizes, they may also consist of a 'T' handle set or a grip handle that has a range of sizes.

Mortar tools

Mortar tools are used to create a finished surface in concrete or rendering applications. Plumbers will often perform concreting and rendering tasks when sparging pipework for rough-ins, repairing footpaths and installing new slabs.

Screed

Screeds (see Figure 7.56) are used for levelling wet, poured concrete to provide a smooth surface. They may also be used to level landscaping materials prior to paving or brickwork. Handheld screeds range in size from 1.2 m to 3.6 m and are constructed from lightweight, strong aluminium for ease of use.

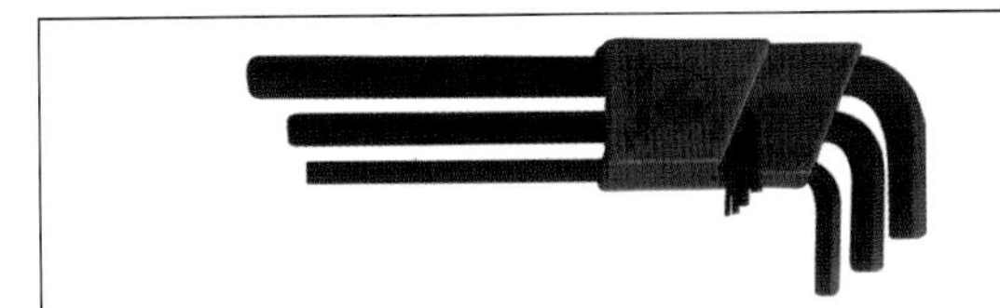

FIGURE 7.55 Allen keys

FIGURE 7.56 Aluminium screed

Float

Wood floats are used as a finishing tool for materials such as wet concrete, sand/cement render or bedding for ceramic tiling to give a smooth, textured or non-slip finish (see Figure 7.57 (a)). They are also used to apply the sand/cement render to masonry walls. Steel float finishes are usually smooth and dense to provide a hard-wearing, easy-to-clean surface (see Figure 7.57 (b)). Wood floats are also useful for floating up the surface of sand bedding prior to laying pavers, to ensure that all hollows are filled, and the surface is even.
Coving trowel/floats produce a textured finish smoother than a wood float and are often used to suit outdoor areas, or 'work in' dry colours placed on finishing concrete. They consist of a steel blade and handle, and feature a turned-up edge around the perimeter of the tool, which minimises gouging of finished concrete surfaces.

Trowel

Variations include the brick trowel, gauging trowel, pointing trowel and edging trowel.

The trowel is the traditional tool used by bricklayers to apply mortar when building with bricks or blocks. A brick trowel incorporates an angular blade with a curved edge that is hardened for cutting bricks. The multi-purpose gauging trowel is used by plumbers for general applications such as 'mudding' up pipework in wall chases (see Figure 7.58). A pointing trowel is constructed like a brick trowel but is much smaller and is used by roof tilers for pointing up ridges on terracotta and concrete tiled roofs and an edging trowel is used by a concreter to finish the edges of a concrete slab.

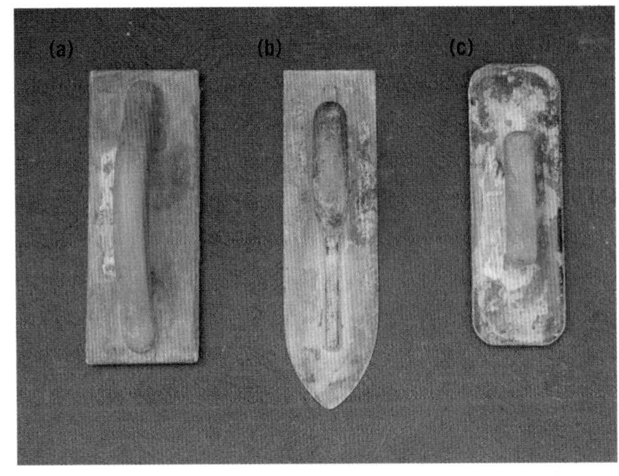

FIGURE 7.57 (a) Wood float; (b) steel trowel (c) steel float

FIGURE 7.58 A variety of trowels

Cleaning up tools

Good housekeeping and a clean work area contribute to a quality job. It is each tradesperson's responsibility to clean up their own mess and to leave a clean, dust-free and safe work environment in line with relevant state or territory WHS regulatory requirements.

Broom

Brooms are available in a variety of sizes for specific uses and may have heads containing bristles of straw or polypropylene (see Figure 7.59). Stiff-bristle brooms such as the yard broom and straw broom may be used on rough surfaces, and soft-bristle brooms may be used on smooth surfaces, as found on lined or coated floors. Broom handles are made from powder coated steel, bamboo or seasoned hardwoods that are straight-grained, strong and flexible.

FIGURE 7.59 (a) Stiff straw or millet broom; (b) Soft-bristle polypropylene floor broom; (c) Stiff yard broom of straw or polypropylene

Brooms form an essential part of the plumber's toolkit, as they may be used to finish the surface of wet concrete to provide a non-slip finish, clean large open work areas or clean up after work has been carried out. A dustpan is often used in conjunction with a broom, and these are available in a variety of materials for light- or heavy-duty applications.

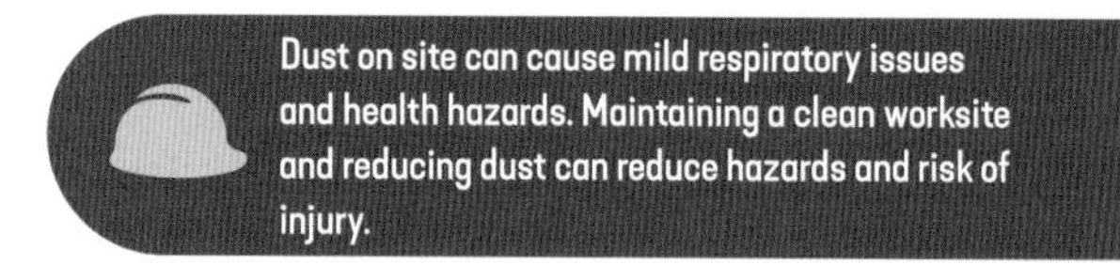

Dust on site can cause mild respiratory issues and health hazards. Maintaining a clean worksite and reducing dust can reduce hazards and risk of injury.

Pipework and metalwork joining tools and equipment

Pipework and metal working tools are a necessity in the plumbing industry. As you progress in the plumbing trade you will require more tools to complete more complex tasks. When purchasing tools, always buy the best quality that you can afford, as cheaper tools that are not fit for purpose can be detrimental to your work.

Tube cutter

Tube cutters comprise a fixed cutting wheel, tensioning handle and reamer attachment (see Figure 7.60). The cutter is rotated around the tubing and the handle is progressively tightened upon each revolution until the tube is cut. The reamer attachment is used to remove the burring created on the inside of the tube. Tube cutters are also manufactured in mini sizes and are slightly larger than the pipe itself, which allows tubes to be cut in restricted and tight areas.

FIGURE 7.60 Tube cutter

In addition, battery-operated tube cutters (see Figure 7.61) are also available for cutting tube in tight areas, and for cutting tube in quick succession where large quantities of copper tube are delivered and are required to be cut for storage or transport. Battery-operated tube cutters can cut up to 32 mm copper tube and require a 12V battery. Tube cutters are recommended for use on soft, thin-walled copper, aluminium or brass tubes.

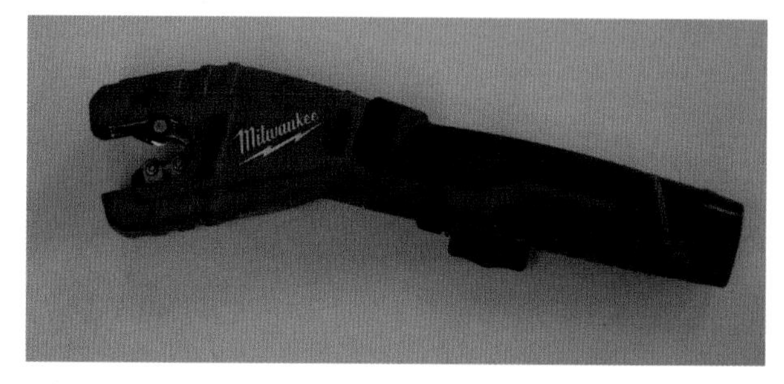

FIGURE 7.61 Battery-operated tube cutter

Pipe cutter

Pipe cutters operate in the same manner as tube cutters, but are available in single or multi-wheeled varieties and are used to cut through heavier-gauge steel or copper piping (see Figure 7.62). For earthenware drainage pipes, a clay pipe cutter (snap cutter), which

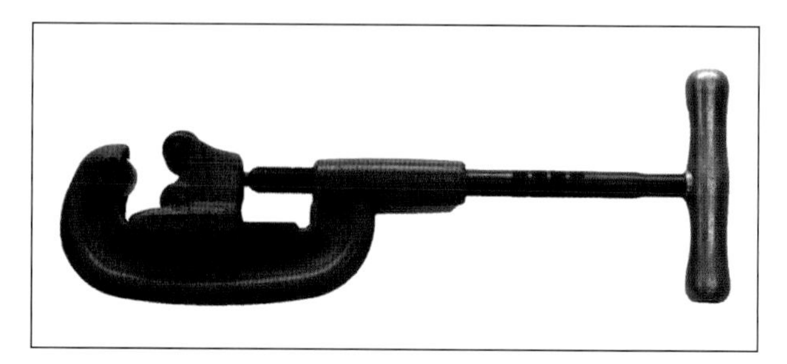

FIGURE 7.62 Pipe cutter

consists of a series of cutting wheels attached to a chain, is recommended. This tool can cut cast iron pipe and relies on compression of the wheels to cut, not a rotating or turning action.

Pipe and tube bender

Variations include the hydraulic pipe bender (see Figure 7.63), lever-type bender (see Figure 7.64) and bending springs.

The 'cold' pipe bending technique utilises a 'former' (or guide) to support the side walls of the pipe or tubing to prevent deformation occurring while under bending load. The load required to bend the pipe or tubing is created by either a hydraulic ram or a lever action. If light-gauge metal tubing is to be hand bent into a tight curve, it will kink and collapse without support. An internal or external bending spring or sand loading should be used in conjunction with annealed copper tubing to prevent deformation.

FIGURE 7.63 Hydraulic bender

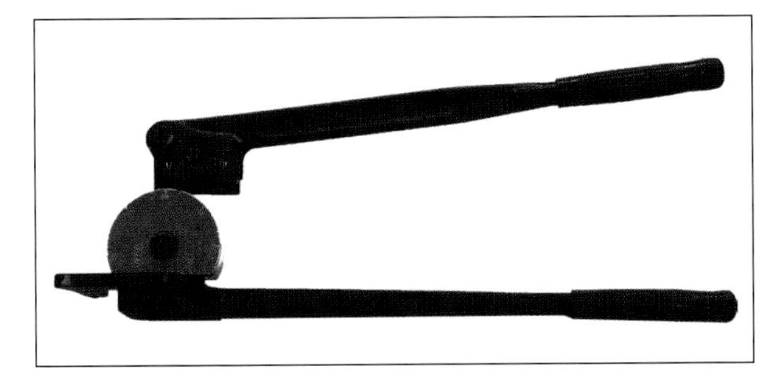

FIGURE 7.64 Lever-type tube bender

Tube expander

Pipe and tube expanders can operate with either a lever action (see Figure 7.65) or screw action (see Figure 7.66). Expanders are used to open the correctly prepared end, which allows the end of another length of tube to be inserted and brazed to create a watertight join.

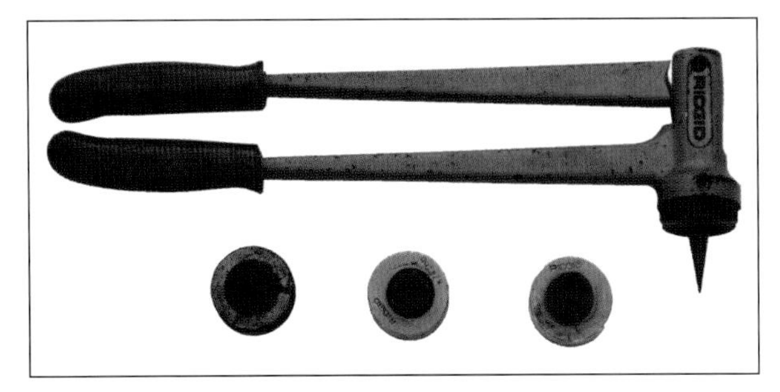

FIGURE 7.65 Lever-type tube expander

FIGURE 7.66 Screw-type expander

Flaring tool

A flaring tool (see Figure 7.67), is used to 'flare' the square-cut end of copper, brass or aluminium tubing. The tube is held in a lockable die block while a rotating, screw-fed cone creates a 45° bell shape in the end of the tubing. Flared compression fittings are used

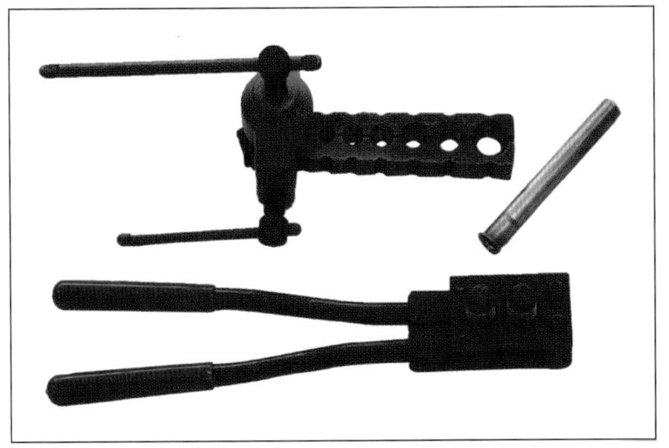

FIGURE 7.67 Flaring tool

to create a watertight or gas-tight seal in the piping system. Remember to fit a compression flare nut prior to flaring and remove the internal burr of the tube to ensure the flare sits against the conical section of the fitting. Hammer impact flaring tools are also available, which can make flaring easier in confined spaces.

Branch former

A branch-forming kit (see Figure 7.68) is used to create a T-branch in light-gauge copper tubing. The branch-forming tool is drawn out through the tubing by either a hand ratchet stock or a power-operated drill. This procedure is ideal for retrofitting or for tenancy work where alterations to pipework are required after construction is complete.

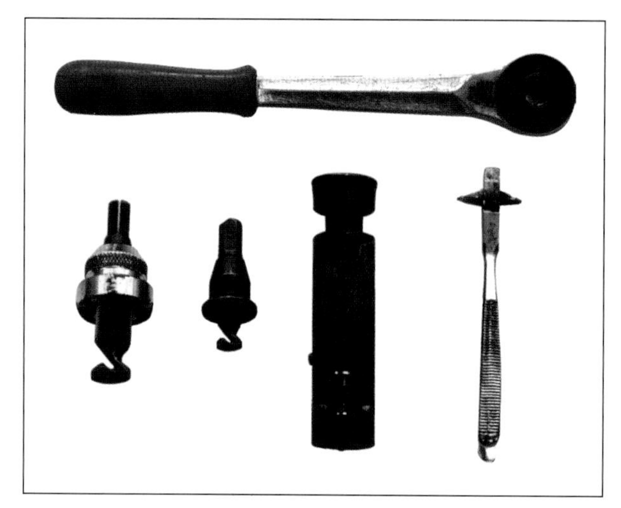

FIGURE 7.68 Manual branch-forming kit with notching tool

Croxing tool

The croxing tool (see Figure 7.69) is used to form a rippled end in thin-walled copper tube. A set of ball bearings is progressively rotated internally to form a raised ring on the external circumference of the tubing. Croxed tubes are used in conjunction with a kinco nut and olive to form a watertight compression joint that will not 'blow out' under pressure.

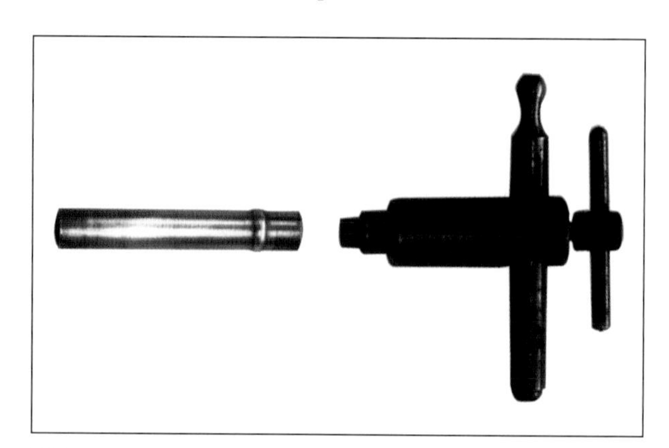

FIGURE 7.69 Croxing tool

Roll groover

The roll-grooving tool (see Figure 7.70) is either manually or electrically operated and incorporates roller bearings that form circumferential grooves externally around 25 mm to 150 mm copper tubing and light-walled steel pipe. The pipe or tubing is deformed in preparation for the accurate acceptance of roll-grooved fittings and couplings. Roll-grooved pipes and fittings are generally used for water mains and fire mains.

FIGURE 7.70 Manual roll groover

Hand riveter

Other names: pop rivet pliers (see Figure 7.71) or blind riveter.

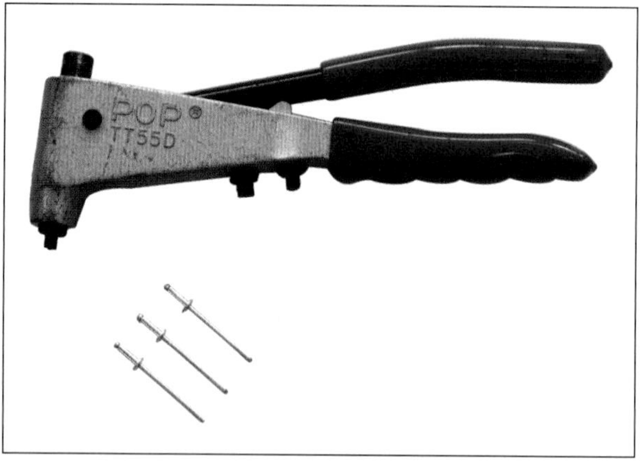

FIGURE 7.71 Pop rivet pliers

Riveters are the tools used to compress and 'set' rivets. A pop rivet gun is used to set blind, sealed or open rivets in thin-gauge sheet metal. The main advantage of a pop riveter over the more traditional forms of joining is that it can blind rivet and join two pieces of metal together when one side of the material is inaccessible.

Caulking gun

Other names: silicone gun or cartridge gun.

A caulking gun (see Figure 7.72) can be used to apply cartridge-based sealants to rainwater goods or bathrooms to create a watertight joint.

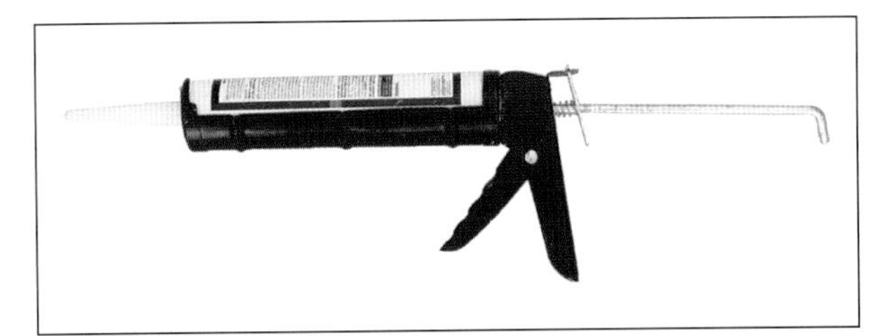

FIGURE 7.72 Caulking gun

Tap reseater

A tap reseater (see Figure 7.73) is used by plumbers to resurface brass valve seats on recessed tap bodies. The tool can be operated either manually or in conjunction with a battery drill. Reseaters are available in 12 mm cutters, which are used for recess bodies on sanitary tapware, and 19 mm cutters, which accommodate larger jumper valves for tap bodies such as exterior hose taps and loose stop valves.

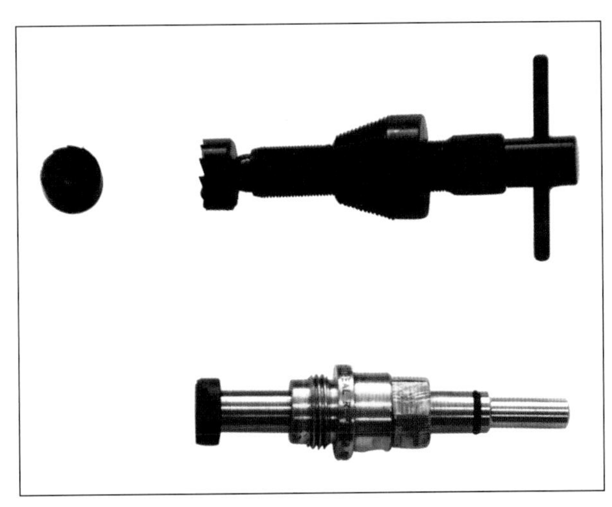

FIGURE 7.73 Tap reseating tools

Soldering iron

Other names: soldering bit.

Consisting of a weighted copper head, steel shaft and hardwood handle, the soldering iron is used as the heat source when applying soft solder to two close-fitting, galvanised steel sheet metal surfaces. Before use, a soldering iron (see Figure 7.74) must be 'tinned' and oxide surfaces removed from the soldering head, which improves its ability to hold solder (see Chapter 8's 'Preparing the iron' section and Figure 8.46). The choice of head size and weight is extremely important, and if the soldering iron is too small for the task, it will not heat the work piece sufficiently and a poor join will result. If the head is damaged or pitted from excessive use, it is easily reshaped by dressing with a hammer to the desired shape and then using a file.

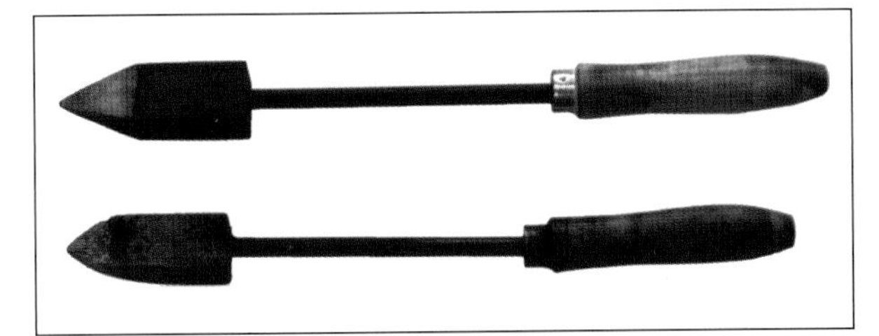

FIGURE 7.74 Soldering iron

Gas torch

Gas torches are a portable heat source that can make small jobs easier without having to get the oxyacetylene torch out. They are available to suit different requirements in the plumbing industry.

Propane torch

The propane torch (see Figure 7.75) incorporates a Bunsen burner to combust LPG (liquefied petroleum gas) that is stored under pressure in a gas cylinder. It is commonly used as a heat source for soldering copper pipework to capillary fittings or for pre-heating soldering irons.

Brazing torch

Brazing torches (see Figure 7.76) are primarily used for silver-brazing smaller diameter copper tube and provide more heat than propane torches. Until 2008, many plumbers used MAPP® gas cylinders as a fuel source for brazing torches. In 2008, the plant that manufactured MAPP® gas was shut down and as a result only substitute cylinders for MAPP® gas are now being sold. The new cylinders are designed to suit existing brazing torches, and are marketed as a MAPP® replacement gas, they contain almost 100% propylene. Some manufacturers offer brazing kits that use an equivalent sized oxygen cylinder in conjunction with the gas cylinder to provide an increased heat output.

FIGURE 7.75 Propane torch

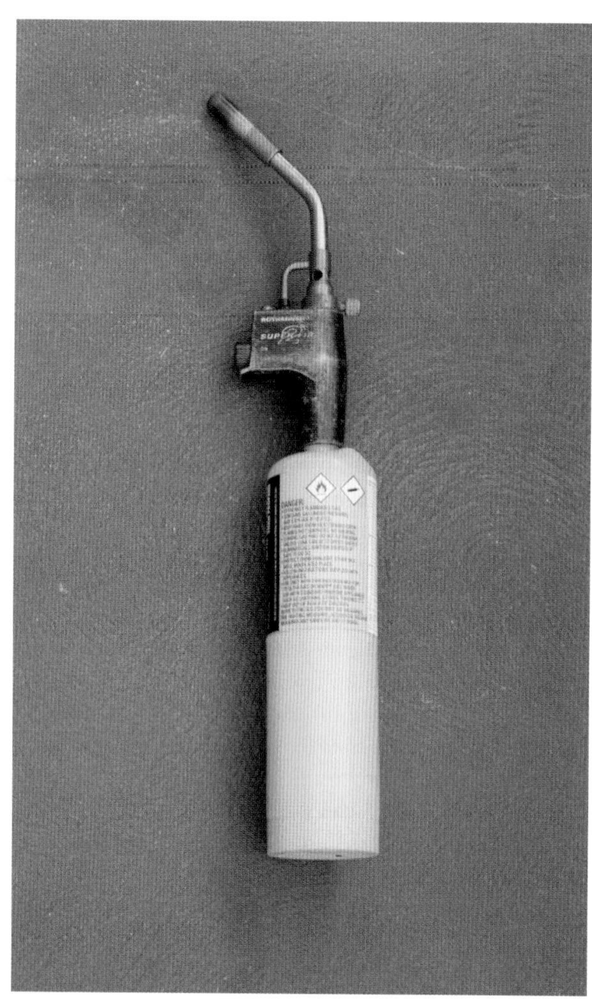

FIGURE 7.76 Brazing torch

Oxyacetylene torch

This is used by plumbers as a general heat source for silver brazing, fusion welding and oxy-cutting applications. The oxyacetylene torch incorporates a welding/brazing handpiece, a mixing tube and a welding or cutting tip, and uses a luminous flame to heat to temperatures in excess of 3000°C. Oxygen and acetylene gas is stored in pressurised cylinders and is supplied to the handpiece via a cylinder regulator that provides a constant working pressure, irrespective of varying cylinder pressures (see Figure 7.77). Chapter 10A covers oxyacetylene welding techniques in detail.

Always assess which combustible products are in the vicinity while welding to ensure the appropriate class fire extinguisher is available.

Electric welding machine

Other name: arc welder. Variations include MIG (GMAW – gas metal arc welding) and TIG (GTAW – gas tungsten arc welding).

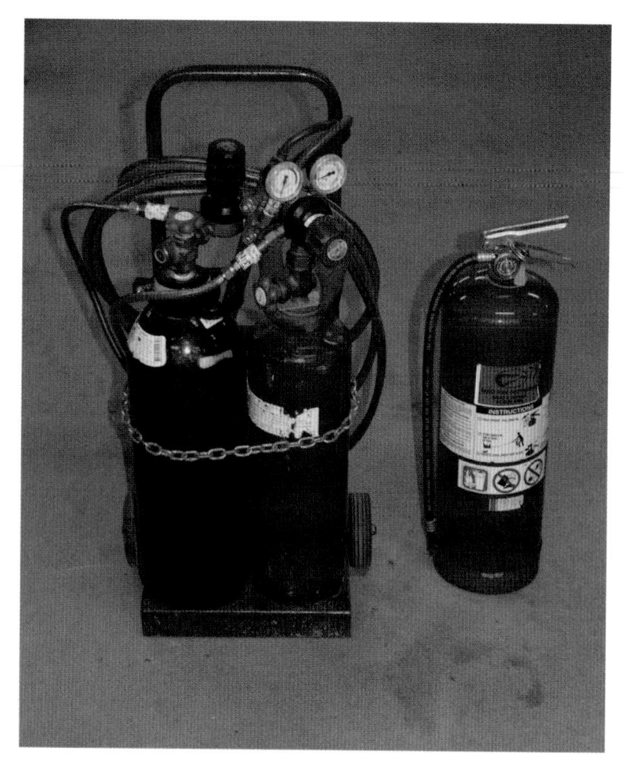

FIGURE 7.77 Oxyacetylene welding kit

The electric welding machine is used to fusion-weld metals together through the addition of an electrode filler rod. Metal arc welding creates a join by heating the material to melting point with an intense electrical discharge in the form of an electric arc (see Figure 7.78). The electrical current is regulated by a transformer, and the heat generated by the arcing of electricity melts both the filler electrode and the surface of the work piece, creating a fusion weld.

FIGURE 7.78 Metal arc welding machine

A chipping hammer and wire brush (see Figure 7.79) are used in conjunction with an electric welding machine. A wire brush is used prior to welding to prepare the surface by removing any oxides, and a chipping hammer is used on completion of the electric welding process to chip away the slag coating from an electric weld. Chapter 10B covers arc welding techniques in detail.

FIGURE 7.79 Chipping hammer and wire brush

MIG welding is a similar process to electric welding, although the filler metal is continuously fed into arc under a shield of externally supplied inert gas (gas shield) to prevent oxidisation occurring. In addition to the 'gas shielded' MIG welder, there is the 'self-shielded' (or 'gasless') MIG welder, which uses a flux-cored wire feed and produces a vapour-forming compound. Although there are no inert gas cylinders, the welding process produces a gas shield around the weld to protect against oxidation.

Tungsten inert gas welding, often called TIG welding, joins the parent metal by heat from a tungsten electrode that does not form part of the completed weld. A separate hand-held filler rod is added to the weld pool, in a similar way to flame welding. An inert gas is used to protect the weld from oxidisation.

Polymer piping joining tools

There are many manufacturers that have their own proprietary piping systems that use a range of materials and jointing techniques. Plumbers may choose materials based on price, ease of use and efficiency of installation. Despite personal preferences regarding specific materials, plumbers often come across materials that they have never used before and may be required to adapt to existing pipework. Therefore, it is advantageous to have basic knowledge of the different materials available and their approved jointing techniques.

Crimp ring assembly tool

A crimp ring assembly tool (see Figure 7.80) is used to join fittings and materials made from polyethylene (PE), polybutylene (PB) or cross-linked polyethylene (PE-X), and or composite pipes such as cross-linked polyethylene/aluminium (PE-X/AL). Many new homes and residential buildings use plastic and composite piping, and there are many plastic crimp ring pipe manufacturers in the market, all of which have their own proprietary materials and fittings. The crimping tool is used to compress a crimp ring on the pipe that has been inserted over a fitting. Crimp tools are available in either manual or battery-operated sets of jaws.

FIGURE 7.80 Crimp ring assembly tools

Compression sleeve tool

Compression sleeve tools (see Figure 7.81) are used to join PE-X piping systems. A compression ring is inserted over the pipework and the plastic pipe is expanded. The expanded section of the pipe is placed over a fitting and the compression sleeve tool is used to slide the compression ring over the expanded pipe and fitting. Compression sleeve tools are available in either manual or battery-operated forms.

FIGURE 7.81 Compression sleeve tool

Press tools

Other names: crimp tool or pipe crimper.

Press tools (see Figure 7.82) have become increasingly popular in recent years as they provide a quick and easy method for joining pipework. They are battery-operated power tools that use proprietary fittings to join copper, stainless steel and carbon steel (where permitted).

The press fitting can provide many advantages over traditional brazing methods as no heat or gas cylinders are required. Fittings contain 'o' rings at the point of insertion and, once pressed, provide a watertight or gas-tight seal. Many different types of mediums may

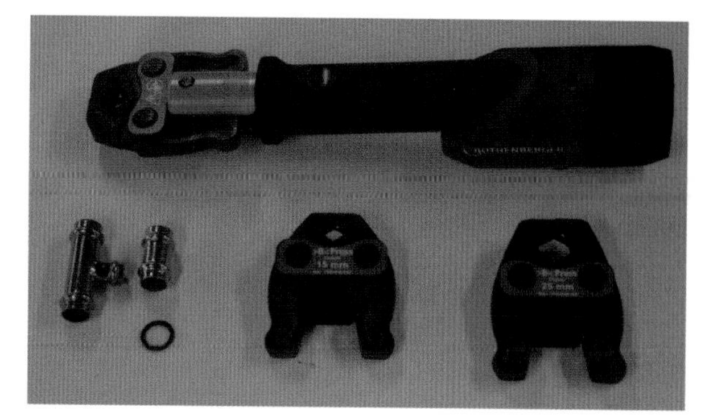
FIGURE 7.82 Press tool

be used for press fittings, and plumbers must refer to manufacturer's specifications for which fitting is suitable for a gas or water application. Manufacturers may also specify minimum distances for crimped joints near existing jointing methods. Press tools are manufactured by multiple companies and pipe fitting sizes generally range from 15 mm to 100 mm in diameter.

Levelling equipment

Levelling is used in all aspects of plumbing from calculating falls for laying of drains or determining levels for stormwater pits.

Terms associated with levelling include:

- *datum point* – any known point, line or level from which a level line may be transferred to another position; its elevation or height may be recorded and used as a permanent or temporary benchmark while carrying out a job
- *invert level* – a vertical height taken from the bottom internal surface of pipe
- level line – any horizontal line that is parallel to the surface of still water
- levelling – the determination and representation of the elevations of points on the surface of the Earth from a known datum, using a surveyor's level to measure the differences in elevation by direct or trigonometric methods
- plumb – any vertical line or surface that is exactly vertical
- *reduced level* – a vertical height or elevation taken above mean sea level.

Spirit levels

Spirit levels are generally made from aluminium and have glass or plastic vials containing a liquid with a trapped air bubble inside. The level operates on the principle that a liquid will find its own level when at rest.

There are usually at least two vials in each level: one to show whether an object is level and one to show whether an object is plumb or vertical. There are two lines marked on the glass vial, and the bubble must be centred between these lines to gain a correct reading from the spirit level (see Figure 7.83).

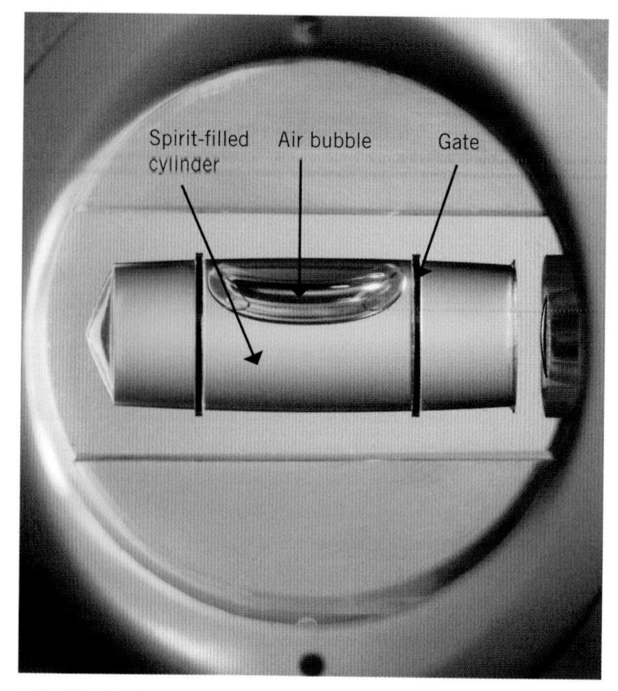

FIGURE 7.83 Reading the bubble of a spirit level

Source: Shutterstock.com/Charlie Edwards

Spirit levels are available in lengths of around 250 mm for a torpedo level (see Figure 7.84), which are commonly used when laying drains, and up to 2 m for a standard level. There are a variety of levels ranging from, but not limited to, 600 mm, 800 mm, 900 mm and 1.2 m in length. To increase the accuracy of spirit levels, a straight edge may be used in conjunction with a spirit level (see Figure 7.85) where the length to be checked is greater than the length of the level.

Remember that it is important to maintain and periodically check the accuracy of a level. This is done by simply placing the level on a flat surface and noting the position of the bubble. Then rotate the level lengthways (180°) and note the position of the bubble again. If the bubble is in the same position as the previous reading, one can assume that it is level or plumb.

Care should be taken to prevent damage to the edges of the level, and they should not be left exposed

FIGURE 7.84 Torpedo levels

FIGURE 7.85 Transferring levels using a spirit level and straight edge

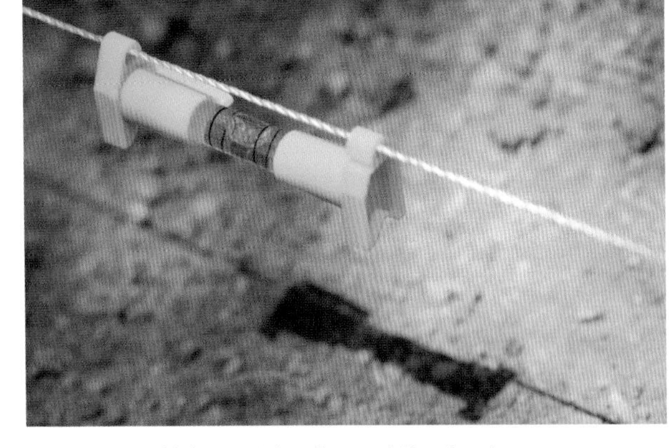

FIGURE 7.86 Using a string line and line level

Source: iStockphoto/Michael Westhoff

to the weather or stored in a manner that allows them to bow or twist.

Line level

The line level is hooked over a taut string line in the centre of its length (see Figure 7.86). It may be used to check for approximate level between two points. It is used when installing guttering or for checking site set-out lines so that measurements are made accurately.

Hydrostatic or water level

This apparatus consists of a length of clear plastic tubing filled with water until there is approximately 300 mm free of water at the top of each end. It is vital that all air bubbles are removed. When placing the tube ends side by side on two level points, the water levels should match, as water will always find its own level (see Figure 7.87). To find a level, hold one end, with the level of water on or near the desired height, and raise or lower the other end until the level of the water settles on the desired height. The other end is marked onto the wall, creating points of equal or level heights.

This level is particularly useful when there are obstructions between level points and is commonly used by plumbers when installing guttering. It is also a very economic levelling device with a high degree of accuracy, and allows levels to be transferred around corners or obstructions on building sites.

Note: Food dyes or fluorescein may be added to the water to give it an easily seen colour.

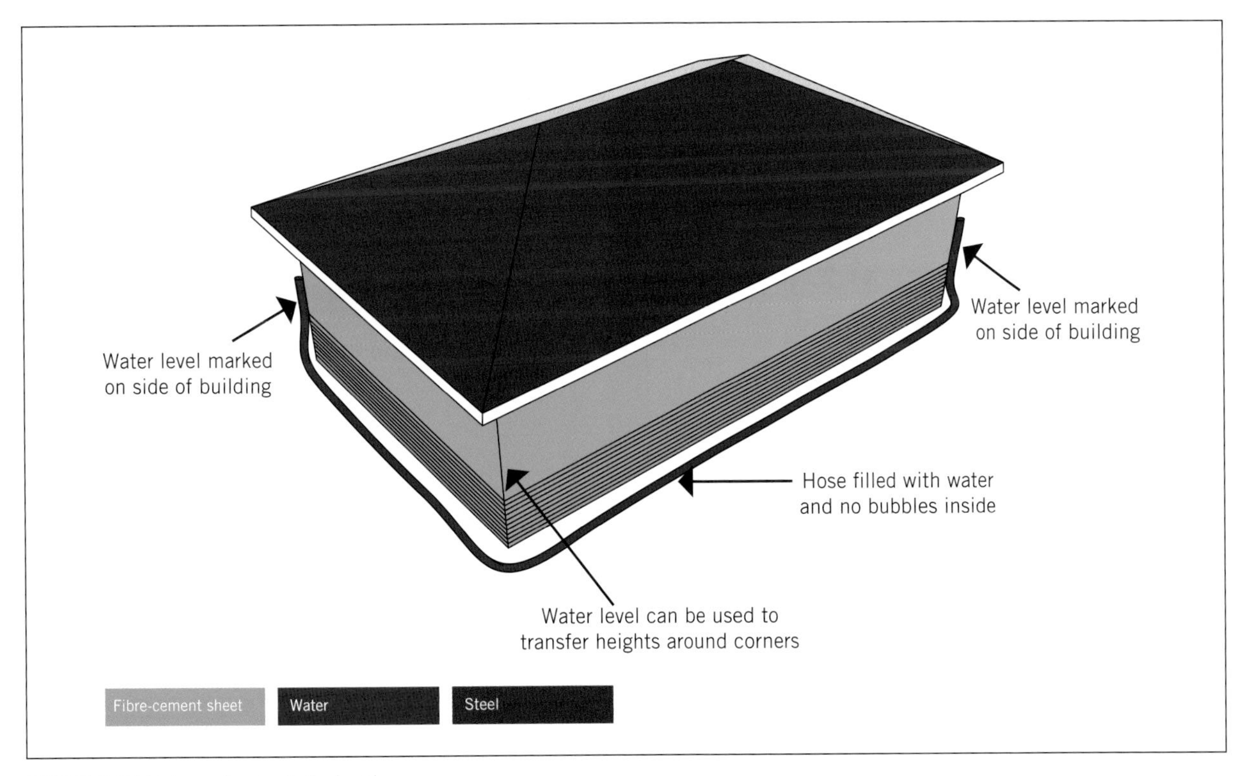

FIGURE 7.87 Using a hydrostatic level

Boning rod

Boning rods generally comprise a set of three T-shaped frames and are used to determine additional heights when two initial points – such as the start and finish of a trench – are known. The user sights between the initial frames and the intermediate boning rod to determine if it is higher or lower than the known points. Boning rods are ideal for straightening the bottoms of trenches for footings or laying pipes.

Plumb-bob

A plumb-bob is a weight attached to the end of a string line that hangs freely to line up vertically with the centre of the Earth (see Figure 7.88). It works on the principle that gravity will hold it plumb. It may be used with a straight edge or on its own, instead of a spirit level. The plumb-bob is commonly used to transfer an outlet point of guttering to the ground level below to determine the position of the stormwater drainage point.

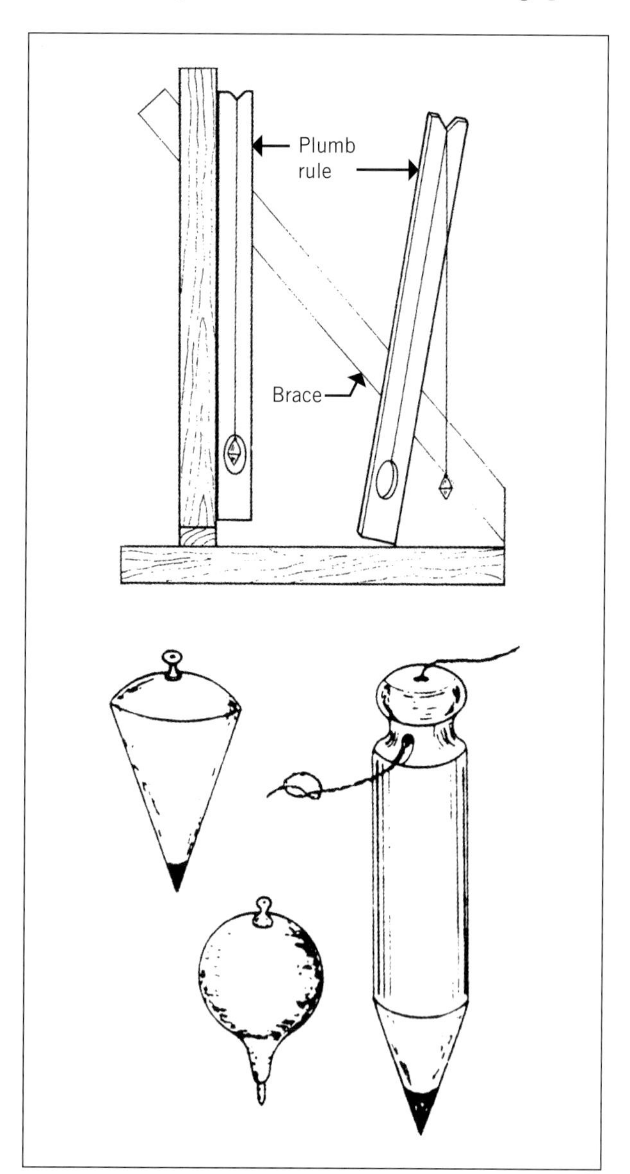

FIGURE 7.88 Plumb rule and different types of plumb-bob

Dumpy level

Other names: automatic level or levelling instrument.

The dumpy level (see Figure 7.89) is basically a telescope that gives the user a horizontal line of sight. Measurements are determined by aligning the crosshairs in the telescopic lens with the dimensions expressed on a staff. The telescopic sight rotates 360° around a vertical axis, enabling readings to be taken at different points from a fixed position. Dumpy levels are commonly used when determining the grade of drainage pipes.

Laser level

The rotating laser (see Figure 7.89) is a battery-operated electronic device that emits a laser light in a rotating 360° plane either horizontal or vertical to the Earth's surface. For site or trench grading it is used in conjunction with a receiver attached to a staff that emits an audible tone, indicating that a level line of sight has been achieved. It may also be used to transfer known levels to one side of the room to the other. The main advantage of a laser level over a dumpy level is that it allows for one-person operation.

Setting up the laser level at eye height should be avoided where possible due to the use of class lasers, which can damage or irritate the eyes. Always check the class of the laser level you are using and exercise caution.

FIGURE 7.89 Dumpy level

FIGURE 7.90 Laser level

Manual excavation tools

Tools such as crowbars, picks, shovels and mattocks (see Figure 7.91) are regularly used by plumbers in the preparation of trenches prior to the installation of drainage or water service pipes. They may also be used to clean up, loosen compacted soil, move rubbish, trim trench walls and tree roots, or break up old concrete paving or rock.

Shovelling, cleaning-out and spreading tools

After breaking compacted soils or materials into manageable sizes, shovelling or cleaning-out tools are used to move loose materials clear of the work area or to load spoil directly into a wheelbarrow or transporter for disposal.

The most suitable hand tool for this purpose is either the round-nose or square-mouth long-handled shovel. The square-mouth shovel is best suited for spreading granular materials such as sand or gravel prior to bedding pipework, while the round-nose is a universal design and performs well in materials of uniform or irregular shape and size. A solidly manufactured steel garden rake of about 500 mm in width is sometimes used to spread and level out granular filling, sand or topsoil when reinstating surfaces. Long-handled shovels allow for a more upright posture. This will prevent back strain and enable extra body weight to be applied with

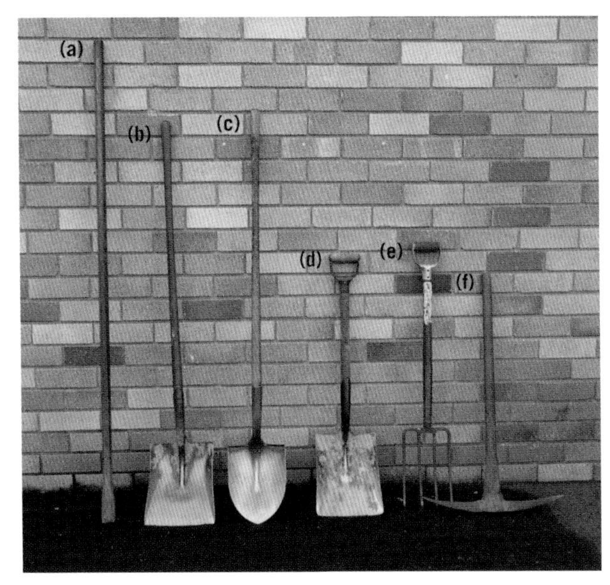

FIGURE 7.91 Hand tools for breaking, cutting and trimming: (a) Crowbar; (b) Long-handled spade; (c) Long-handled round-mouth shovel; (d) Spade; (e) Fork; (f) Pick

the assistance of the more efficient leg muscles. A short-handled, square-mouthed shovel allows more control when cleaning out loose materials from corners and edges of work.

Specialist shovels

Special purpose digging tools are generally used by plumbers when excavating for pipe installations. These digging tools include narrow-mouthed trenching shovels, post-hole shovels, long-handled round-nose shovels and D-handled spades.

General maintenance of hand tools and equipment

Hand tools and equipment make up a considerable portion of a plumber's tool kit and they should always be maintained and stored appropriately to ensure they are in good condition for when they are required.

- Impact tools such as hammers and pinch bars should be kept dry and rust free. Handles should be kept free of oil and grease to prevent slipping during use. Chisels should have any mushroom heads periodically removed to prevent splintering during use.
- Gripping or tightening tools such as spanners, pliers and wrenches must be kept rust-free, lightly oiled (or its moving parts sprayed with a water-displacing lubricant such as RP7 or WD-40) and stored in a dry, dust-free container.
- Cutting tools such as snips, knives and hacksaws must be routinely sharpened or have their cutting blade replaced. They must be kept rust-free, lightly oiled (or their moving parts sprayed) and stored in a dry, dust-free container.

- Tapes, especially metal-blade retractable tapes, should be kept dry and clean before retracting the blade, or rusting and jamming will occur. In addition, the blade should be fed back gently without allowing the hook to hit hard against the tape's body, as it may snap off.
- Four-fold rules should have the knuckles oiled and gently exercised when first used. This will prevent stiff joints from snapping.
- Concreting trowels and floats must be thoroughly cleaned, dried and stored in a toolbox after use. Do not let cement paste harden on these tools as this will allow a build-up to take place every time they are used.
- Hoses or electric leads should be loosely coiled or wrapped on a reel to prevent kinks and tangles, and should not be stored with sharp objects.
- Shovels, spades, picks and mattocks should have excess soil scraped off and then be thoroughly washed with a hose. Lightly oil exposed steel areas when these tools are not being used for extended periods of time.

LEARNING TASK 7.1

TOOL MAINTENANCE

Using sharp tools in the plumbing services industry is essential to produce an accurate, neat and professional finish. All tools require maintenance, but tools with cutting edges, blades or teeth require special attention to ensure efficient, safe operation.

Under the direction of your teacher, carry out routine tool-sharpening under supervision on the following tools:

- scribers and punches (grinding and honing)
- chisels (grinding and honing)
- hacking knife (grinding and honing)
- snips (file sharpening and oiling)
- handsaws (file sharpening and setting)
- hacksaws (replacement of blades)
- retractable trimming knife (replacement of blades).

Select and use appropriate power tools

Power tools in the construction industry play an integral role in how tasks are to be carried out, as they provide workers with the means to carry out tasks efficiently, which increases both productivity and profit. Although power tools allow tasks to be completed quickly, workers should always use power tools in the manner for which they were designed in order to prevent serious injury when in use. Adopting and carrying out the safe use of power tools in accordance with manufacturer's guidelines and safe work practices is essential to maintaining a safe workplace.

Personal protective equipment (PPE)

When used correctly, PPE will reduce the risks and hazards associated with using power tools and should be provided to all workers by their employer to ensure that all tasks are carried out safely. Before using power tools, all workers should receive appropriate instruction and training on the correct use and fitting of PPE to ensure it protects as it was designed to do. PPE that does not fit or has not been fitted correctly may result in illness or injury and defeat the purpose of wearing the PPE in the first place. When using power tools with materials, it is essential that you refer to an SDS for that chosen material to select appropriate PPE.

Electrical safety and requirements

The use of power tools requires electricity for the tools to operate, which is generally supplied through an extension or power lead. The extension lead allows the worker flexibility to move around a construction site, but this may become a vulnerable link for the worker and increases the risk of electric shock. The first line of defence against electrical shock from extension leads or power tools is the installation of a residual current device (RCD). This unit is either fixed to the power board or is a portable unit that plugs into a power point. This device senses the smallest differential in supply and demand and isolates the power in milliseconds. All power from a supply source, whether a temporary builder's service or a permanent installation, should pass through one of these units installed immediately adjacent to the power source. In addition to an RCD, always use a power point tester (see Figure 7.92) to check for correct wiring and faults before plugging in power tools and extension leads.

As an additional measure, users of electrically powered tools should wear heavy, rubber-soled shoes, which will give added protection against electrocution.

The regulations of Safe Work Australia and WHS state or territory regulators require portable electrical equipment and leads to be regularly inspected, tested and tagged by a competent person who has undergone accredited training. The purpose of inspection is to identify any defects in the leads or equipment that are

FIGURE 7.92 Power point tester

potential hazards, and to reject or advise on repair or replacement where possible. The competent person must attach colour-coded tags to the equipment as evidence that the inspection and approval has been carried out.

The colour-coded tags contain:

- the name of the person who carried out testing
- the date of testing
- the outcome of the testing
- the date when testing is to be carried out next.

Specific instructions about the duty and a summary of the regulations are contained in *AS/NZS 3760: 2010 In-service safety inspection and testing of electrical equipment* and *AS/NZS 3012: 2019 Electrical installations – Construction and demolition sites*.

7.1 STANDARDS

- AS/NZS 3760: 2010 In-service safety inspection and testing of electrical equipment
- AS/NZS 3012: 2019 Electrical installations – Construction and demolition sites

Extension leads

Extension leads in constant use and drawing near to their full current rating will generate heat and suffer a voltage drop. To minimise damage to the lead and possible hazard to users in this situation, any extension leads of a 15-amp rating must not be longer than 40 m if they have 2.5 mm^2 conductors. When in use, every lead should be unwound fully, as any heat generated cannot be exchanged efficiently and may cause damage or personal hazard to workers. Regular inspections should take place in hostile working environments.

All leads used on building and construction worksites must have plugs and sockets of clear plastic, to aid inspection for the correct termination of conductors and to identify loose or burnt terminals and wires. Moulded, non-rewireable plugs are also satisfactory.

All extension leads should be heavy-duty, be sheathed for construction, and have a rating of 10 amps (see Figure 7.93).

FIGURE 7.93 Electrical lead

Leads running from a temporary power supply or from any permanent source should be elevated clear of the ground over the distance from power source to worksite. This is achieved by using lead stands (see Figure 7.94) that have weighted or broad-based frames and vertical masts with hooks or slots that support the lead without damage. They should be constructed from or insulated with non-conductive materials.

FIGURE 7.94 Lead stands

Damage to an extension lead or to a flexible lead for a power tool or other piece of equipment means that the equipment must be taken out of service as soon as the damage is detected. Patching leads, repairing plugs or sockets, applying bandages or trying to camouflage a fault or damage to a lead is not acceptable and is an action that breaches WHS legislation of the states and territories.

Portable power tools

Before power tools are used, an employer or experienced worker who is competent should provide safe power tool instruction to new or inexperienced workers. If you are unsure on how to use a power tool, always ask a competent person for correct demonstration to ensure your safety and the safety of those around you.

It is extremely important to check the manufacturer's instruction with any power tools and to carefully note any limitations. A full risk assessment should be conducted, and a SWMS must be completed, before operating any equipment. Some 'static' machinery will require referral to a SOP and safety requirements displayed near the tool or equipment.

Battery-operated/cordless power tools

Battery-operated power tools provide plumbers with flexibility on site without compromising on performance and safety, and are increasingly popular due to their portability and affordability. Many manufacturers provide a great range of portable power tools that are interchangeable with the same rechargeable battery (see Figure 7.95). Tools can be purchased in kits or are commonly sold as individual units, without the battery, which are referred to as skins. Common battery sizes in volts (V) are 10.8V, 12V, 14.4V and 18V, although voltages can be higher for some rotary hammers. The most common power tools on the construction site use 18V batteries, as they are long lasting and provide exceptional power for a battery-operated tool. There are battery sizes that have lower and higher voltages, but they are generally specialised power tools.

FIGURE 7.95 Power tools with interchangeable battery

Basic power tools

In addition to the many tools that plumbers may carry, power tools are an essential part of every plumber's tool kit. They provide efficiency to each job as they are fast and reliable; however, they must be maintained and used in accordance with manufacturer's instructions as incorrect use may void warranties or cause safety issues for you and those around you.

Safety precautions with power tools

Portable electric-power tools can be a source of both mechanical and electrical hazards unless they are maintained and used correctly. The following list contains points of general safety and ways of avoiding damage to tools and equipment or injury to the operator:

- Keep leads neatly coiled when not in use to allow ease of future use.
- Keep leads away from oils, solvents, acids, heat and sharp objects.
- Carry leads rather than dragging them, as the insulation may wear through or be severed and expose live wires.
- Do not carry or lower tools by the lead, as the connections may come loose or break.
- Do not disconnect plugs by pulling the lead, as it may cause connection damage, cause damage to the power point, or fly back and cause injury.
- Do not leave leads partially rolled up when drawing heavy loads, as this may cause overheating and damage the wound-up section by melting the insulation.
- Do not allow the coloured cores of a lead to become visible at any position along its length or at its ends.
- Damaged leads or electrical equipment should be marked 'unsafe' or 'out of service' and be tagged out and reported to the site manager or supervisor so that maintenance can be arranged.
- Patching of damaged leads with insulation tape is not permitted. Apply a safety tag to the lead and get an electrician to repair it.
- Leads should be kept clear of the operator's feet and the machine's cutting edge.
- The power outlet switch must be turned off and the three-pin plug removed while adjustments to tools are being made. This will prevent accidents if the switch is pressed accidentally.
- If the lead is damaged while using a portable power tool, let go of the tool, turn off the power and remove the plug from the outlet. DO NOT TOUCH either the power tool or the lead until this is done.
- Protect leads with approved cable protectors when they are laid across the roadway or barrow runs.
- Flexible extension leads must be supported above any work area and passageway by using approved lead stands to provide clear access for personnel and vehicles.

- Loose pins indicate possible loose connections on leads, which may cause the plug to overheat.
- Avoid using portable power tools on damp ground, particularly if you're wearing damp boots.
- All flexible leads are required to have clear plastic plugs so that broken or exposed wires can easily be detected.
- Prevent accidental starts by never carrying the tool with your finger on the starting trigger while it is still plugged in.
- Use clamps or a vice to hold work, leaving both hands free to control the tool.
- Safety goggles must always be worn when using portable power tools.
- Keep the workplace tidy, as poor housekeeping is potentially dangerous.
- Do not wear loose clothing and secure loose long hair, as it may become entangled in moving parts of the tool.

Circular saws

A useful addition to the plumber's toolbox is the hand-held circular saw (see Figure 7.96). Saws come in a range of sizes, with the diameter of the blade giving the saw its identification. The motor of the machine is generally of the universal type, requiring a power source from either the normal domestic electricity supply or by means of a portable generator. Circular saws may also be battery operated.

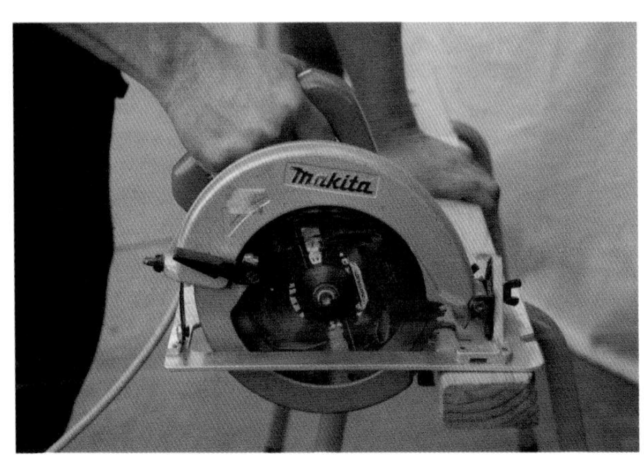

FIGURE 7.96 Circular saw

A variety of saw blades and cutting discs are available for specific tasks, and include:

- *carbide-tipped blade* – with hardened, tipped teeth that stay sharp longer
- *cross-cut blade* – designed to cut across the grain of solid timber
- *flooring blade* – good for second-hand timber where nails may be accidentally cut
- *friction blade* – for cutting corrugated iron and thin sheet metal
- *rip blade* – for cutting timber parallel with the grain.

Safety information

Below is some safety information related to the use of a circular saw. It is meant as a guide only and is not limited to the content mentioned. Refer to specific safety instructions for guidance. Always:

- warn others in the area that you are about to operate a noisy power tool
- wear approved PPE (goggles, gloves and ear protection)
- avoid wearing loose clothing and jewellery
- start the cut with the blade away from the material
- cut in a straight line to avoid kickback
- anticipate the completing cut
- locate the power cord before, during and after making the cut
- use two hands when using the saw
- allow the blade to do the work; never force the blade through the material
- support long lengths of material
- check the surrounding area is safe from flying debris
- keep bystanders clear of the work area
- keep the work area clear and uncluttered
- unplug the tool when making any adjustments (e.g. changing blades)
- check the operation of the retractable guard for the blade
- place the wide side of your boot on the material being cut (only when required and safe to do so)
- clamp the work wherever possible
- continue the blade all the way through the cut before switching the motor off.

Only:

- use a circular saw for its intended use
- use accessories designed for circular saws.

Never:

- start the saw with the blade in contact with the material being cut
- remove or fix back the safety guard.

Choosing a circular saw

The following points can help you determine what circular saw you require:

1. What size circular saw is needed?
2. What is the largest size of material you will be working with? The larger the blade diameter, the larger its cutting widths and depths.
3. How much power is needed?
4. Typically, larger blades are associated with higher-powered motors. The higher the power, the more efficiently it will handle tougher materials. More powerful motors will add additional weight, so select according to the tool's likely main use:
   - 950W to 1650W – soft to medium timbers
   - 1650W to 2400W – medium to hard timbers.

5 Key features to look for include:
   - weight and balance
   - comfortable handles and grips
   - size of the retracting guard lever
   - easy access to adjustments
   - a spindle lock
   - a laser guide generator.

Care and maintenance tips

Circular saw maintenance is essential; consider the following tips.

- Regularly check blade for missing or damaged teeth.
- Check and clear any build-up of debris within the guard.
- Check and tighten any loose nuts or bolts.
- Always check the condition and the position of the electrical cord.

Power saws

The drop saw and the compound mitre saw are other tools commonly used in the plumbing services sector to cut a variety of different materials. The operation of these tools is similar to the hand-held power saw except that power saws are mounted on a portable stand. The drop saw is very useful for cutting steel brackets and threaded rod. A compound mitre saw (see Figure 7.97) has a rotating base plate that can be adjusted to cut various angles, and a sliding mechanism that assists in cutting. Compound mitre saws are mainly used by plumbers for cutting noggins at the rough-in stage.

FIGURE 7.97 Compound mitre saw

Maintenance of power saws

Power saw maintenance is important to ensure a long service life, so consider the following tips to ensure smooth and safe operation:

- Blades can be sharpened or touched up for many reuses. Tungsten-tipped blades should be sent to a saw doctor for special sharpening.
- Dust the body and around the motor housing so that a build-up doesn't inhibit the saw's operation.
- Check for worn carbon brushes and replace them as required.
- Store saws in a dry, dust-free box when not in use to ensure long life.

Jig saw

The portable jig saw (see Figure 7.98) is used to cut concave and convex shapes in thin materials. It becomes useful on site due to its portability where access to a band saw is not possible. Jig saws may be electric or battery operated.

FIGURE 7.98 Hand-held jig saw

These saws are available in a wide range of sizes and styles, from light-duty single-speed to heavy-duty variable-speed. They may be used to cut curved shapes or along straight lines in materials such as plywood, particleboard and thin gauge metal. Care should be taken when cutting brittle material, as the blade cuts on the up stroke, which may cause surface chipping of the material. Jig saws are commonly used by plumbers to cut out timber shelving in kitchen and laundry fit outs.

Safety and maintenance of jig saws

Jig saws have an exposed blade when cutting, so consider the following points to ensure smooth and safe operation:

- Replace the blades if excessive pressure must be applied when cutting, or if burning of the material occurs.
- Never put your fingers or hands under the job while cutting.
- Make sure that sufficient clearance is available under the job being cut.
- Drill a hole in the material first so that the blade may be inserted. Do not use the blade to start a cut in hard materials, as this may shatter the blade. Plunge cutting is permissible only in soft materials.
- Ensure that air vents in the machine casing are not blocked, as this will cause overheating.
- Check, and use, the appropriate blade for each specific application.

Reciprocating saw

Another saw with a similar action to the jig saw is the reciprocating saw (see Figure 7.99). This is designed for two-hand operation to cut steel pipe, steel plate, timber and floor panels, or for awkward cutting positions. Reciprocating saws may be electric or battery operated.

FIGURE 7.99 Reciprocating saw

Electric and pneumatic drills

Drills cover a range of tools that provide rotation for boring, cutting and fastening operations. Most drills are either electrically operated (240V) or use rechargeable batteries; however, they can also be operated by air or by flexible drive from a petrol motor.

Drills are classified according to their size, use, power source, speed, the type of boring or cutting 'bit' they drive and the material they bore into. Sizes range from a palm-sized pistol-grip through to fixed machine tools for workshop use. Some drills deliver medium- to high-frequency impact or hammer forces, as well as rotating for boring into concrete, masonry or rock. Manufacturers' information charts will assist in selecting the right drill for a specific task.

Common power drill types are as follows:

- *Pistol-grip drills* – can be used one-handed for small-diameter holes, such as pre-drilling for riveting. May also be available with a side handle and used with two hands for larger holes and heavy drilling.
- *Hammer drills* (see Figure 7.100) – use a tungsten carbide-tipped bit for drilling holes in masonry materials. The impact as well as the rotation allow for easy drilling into solid materials.
- *Pneumatic drills* – used for very heavy-duty work, as found on large building sites or mines. They may be used to drill holes in concrete for reinforcing steel starter bars, into solid rock to receive long rock bolts, or even underwater where the use of electricity is not possible. They are powered by compressed air.
- *Power screwdrivers* – operate on similar principles to drills, and some drills may be used as screwdrivers due to their variable speed and reverse functions.
- *Bench drills, also called pedestal drills* (see Figure 7.101) – a workshop tool that is permanently fixed to a workbench or stand. The material to be drilled is secured in a fully adjustable vice attachment, permitting more accurate drilling. These drills are most advantageous for repetitive work.

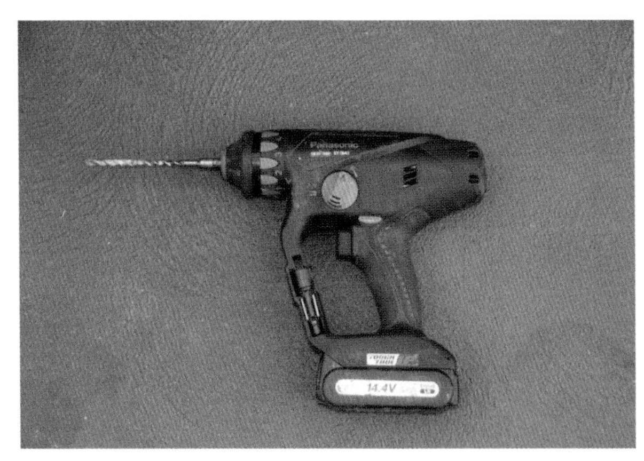

FIGURE 7.100 Rotary hammer drill

FIGURE 7.101 Bench drill

Battery-operated/cordless drills

An alternative to the electric drill is the battery-operated drill (see Figure 7.102). Most plumbers will have battery-operated drills that have the necessary range of operation for most work activities. Combination hammer drill/drivers are preferred by plumbers. Roofers often choose impact drivers for screwing off roofs where electrical leads pose access and safety issues. Do your research before purchasing a battery drill by consulting with experienced tradespeople.

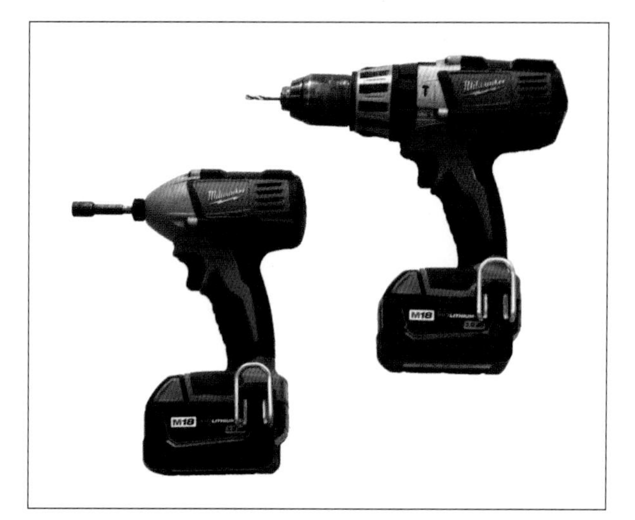

FIGURE 7.102 Battery-operated impact driver and hammer drill/driver

Common battery-operated drill types are as follows:

- *Hammer drill/driver* – this is commonly used by plumbers as it offers versatility: it has the ability to drill holes into timber or steel, drive fasteners and drill into masonry on the hammer setting. Settings are adjustable between three modes, and drills have multiple speed settings
- *Impact driver* – used primarily for fasteners due to its high-torque abilities, the impact driver uses a hex bit for attachment, with a variety of bits available to suit most fastener applications. It is used extensively in roof sheeting and cladding of buildings and for general construction work.

Safety and maintenance of power drills

Power drills increase the efficiency of work on site. Tool maintenance is essential, and the following points should be considered.

- Any electrical problem, including the replacement of leads, must be dealt with by a licensed electrician or competent person.
- Internal greasing of gears and internal cleaning should be carried out by a qualified service person.
- Brushes should be replaced as necessary and checked for contact.
- Pneumatic hoses, when not in use, should be neatly coiled, with the ends capped to prevent foreign matter from entering. Store hoses off the floor, away from acids, oils, solvents and sharp objects.

Electric and pneumatic jackhammers

Electric demolition hammers (see Figure 7.103) are efficient tools for demolition of brickwork, concrete and hard-packed clay. They are also available with a rotary hammer drill facility on smaller models for boring holes into rock or reinforced concrete. Electric hammers are available in different sizes according to their weight. They may use moil points, chisels or spade attachments and/or 13 mm to 38 mm diameter drill bits.

FIGURE 7.103 Electric demolition hammer

Pneumatic jackhammers are available to carry out similar work to that done by the smaller demolition hammers. They are used for chipping and drilling. Small points, chisels and drill bits ranging from 12 mm to 25 mm may be used with a 7 kg hammer.

The attachments for jackhammers generally fall into three different categories:

- *Spading hammers* – these are used for light demolition and spading of clay and shale. Infill points, chisels and clay spades may be used with a 13 kg hammer.
- *Rock drills* – these are used to drill small to large holes into rock or concrete with drill steels up to 1.85 m long. These are available in sizes of 17 kg and 25 kg.
- *Air breakers* – these are used for rock excavation, concrete demolition, clay digging and bitumen cutting (see Figure 7.104). Moil points, chisels and clay spades can be used with hammers in sizes of 28 kg, 37 kg and 42 kg. They are all available with silencers.

Safety and maintenance of jackhammers

Jackhammers are an essential tool during demolition and excavation work; however, if not used correctly hazards can arise. Consider the following points to reduce the risk of injury:

- Safety goggles and earmuffs should be used, as well as a silencer, to cut excessive noise.

FIGURE 7.104 Air breaker used for heavy work

- Moil points, chisels and other attachments need grinding and hardening on a regular basis to allow efficient operation.
- Air hoses should be loosely coiled when not in use, with ends capped to prevent foreign matter from entering.

Angle grinders

The primary function of an angle grinder (see Figure 7.105) is to grind material such as metal. By adding other accessories, you can cut metal and masonry, and remove rust and paint. Angle grinders can be easily misused, and it is imperative that you select the correct one for the job at hand.

FIGURE 7.105 Angle grinder with safety guard

Ratings

Angle grinders are rated by the following:

- Disc diameter – the main disc sizes for angle grinders are 100 mm, 115 mm, 125 mm and 230 mm. The larger the disc diameter, the greater the cutting depth.
- Power – typical power ranges for angle grinders are 550 W to 2300 W.
- Speed – the smaller angle grinders run at approximately 10 000 revolutions per minute (rpm), whereas the larger 230 mm angle grinders run at approximately 6500 rpm.

Angle grinder accessories

The following accessories can be fitted to an angle grinder:

- cutting and grinding discs for cutting and grinding metal and masonry
- wire cup wheels for removing rust
- sanding discs for removing paint
- diamond discs for cutting masonry
- cutting stands for greater stability and safety when cutting
- wall chasing attachments for creating channels in concrete and masonry walls for pipes to be embedded (note that it is preferable to use a wall chasing machine and suitable vacuum for this process).

Safety tips

The following tips will help to increase safety at the worksite:

- Warn others in the work area that you are about to operate the tool.
- Always use the side handle supplied with the tool.
- Never use your angle grinder without the guard in place.
- Ensure that you use the correct clothing and safety equipment (safety glasses, ear protection, gloves, dust mask and possibly a respirator).
- Check all grinding and cutting discs for cracks prior to use. If a crack is visible, do not use the tool as the disc may disintegrate and cause injury from projectiles.
- The accessory should suit the tool being used. Check the rpm rating on the accessory and the tool.
- The grinder should be laid on its back and away from any obstructions after it is switched off, as the disc takes some time to stop turning because of its high rpm.
- Regular cleaning and dusting of the air vents around the motor housing is essential to prevent blockage and to stop the motor from overheating.

- Always keep the leads clear of the cutting/grinding area and never wrap the leads around your arms, in case of kickback.
- Check and replace the carbon brushes as required. If you are unsure how to do this, have it done by an approved service person.
- Never use an angle grinder on a metal roof, as sparks will create swarf and potentially corrode the roof.
- Ensure there is no flammable material in the near vicinity when using an angle grinder.

Electric shears and nibblers

Electric shears (see Figure 7.106) or 'hand shears' are useful for cutting straight lines or curves in flat sheet metal. They operate using a scissor-cutting action similar to aviation snips and can accurately cut large sheet metal sections.

FIGURE 7.106 Electric shears

Electric nibblers (see Figure 7.107) are used for awkward cuts in light gauge sheet metal rainwater goods and for cutting corrugated roof sheeting. The nibbler punches a slot in the metal, which creates a large amount of swarf (i.e. small pieces of metal) that must be removed on completion of cutting to prevent damage to the finished surface.

Safety procedures for shears and nibblers

Shears and nibblers are generally used on the roof, which can be a hazard, so to manage risks always:

- use an RCD when working with power tools on metal roofs
- check leads for damage and protect them from sharp edges
- wear appropriate PPE.

FIGURE 7.107 Electric nibblers

Concrete and masonry cutting and drilling equipment

Plumbers now have access to a greater number of tools and equipment when cutting and drilling concrete and masonry. Although these tools are designed to make the job quicker and easier, there are also some inherent dangers associated with the use of them. The following information should be viewed as a guide only.

Diamond-tipped, concrete and masonry cutting and drilling equipment is used for:

- chasing channels in concrete, brickwork or other building masonry surfaces to allow for the inclusion of conduits, cables, pipes or flashing
- wall sawing in concrete, brick or other masonry walls to provide for ducts or pipes, or to remove sections of existing walls
- core drilling circular holes in reinforced or precast concrete, bitumen surfaces, panels for tilt-up structures, brick and other structural materials that are used for electrical, plumbing, heating, sewer and sprinkler installations.

Other applications include drilling holes for anchor bolts or lifting rods.

People using concrete and masonry cutting and drilling equipment face a wide range of hazards, such as silica dust, toxic exhaust fumes, saw kickback, blade fracture, falling walls, electrocution, vibration, noise, slips, falls, and injuries from manual handling.

Most at risk are operators of hand-held concrete and masonry saws (see Figure 7.108). This equipment is more prone than fixed saws to the violent forces unleashed when a saw blade jams inside a cut. These forces – commonly referred to as kickback, push-back or pull-in – are difficult and sometimes impossible to control, and place the operator at risk of serious and potentially fatal injury from an out-of-control concrete saw.

FIGURE 7.108 Demolition saw

Information, instruction, training and supervision are essential in all concrete and masonry cutting and drilling operations. When using concrete or masonry cutting or drilling equipment:

- *always* warn others in the immediate work area that you are about to operate the machine
- *always* follow the manufacturer's instructions for safe use
- *always* use the correct blade size recommended by the manufacturer; oversized blades are dangerous
- *never* remove the guards
- *never* work off ladders or items such as milk crates, chairs, buckets or steel drums; use a scaffold or work platform if the work cannot be safely reached from the ground
- *never* hold a hand-held saw or drill higher than shoulder height
- *never* use a hand-held saw for inverted cutting or drilling.

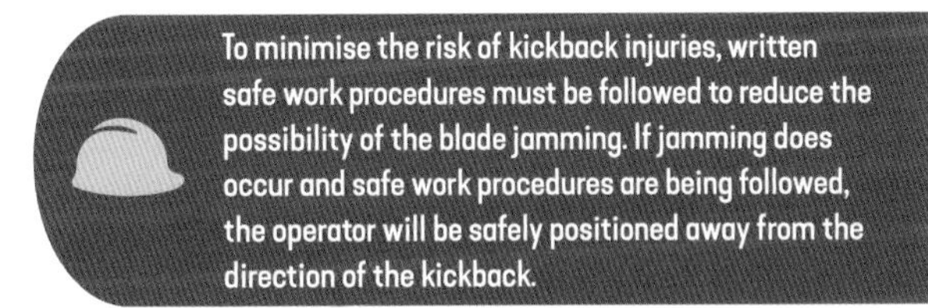

To minimise the risk of kickback injuries, written safe work procedures must be followed to reduce the possibility of the blade jamming. If jamming does occur and safe work procedures are being followed, the operator will be safely positioned away from the direction of the kickback.

Cutting and drilling equipment, especially saw blade discs and drill bits, should be removed from machines and stored where they will not be damaged between uses.

Risks and hazards

The risk of hazards on site is always high, so it is important to identify hazards before you start work or before they occur. When a hazard is identified you must assess the level of risk that is associated with the hazard. Once assessed, use the hierarchy of control (see Chapter 1) to remove or reduce the hazard and reduce the risk. Always review control measures on a regular basis.

Who is at risk?

The people most at risk of injury or harm from concrete and masonry cutting and drilling are those operating the equipment and anyone who is nearby.

Operators who occasionally use hired concrete cutting equipment, particularly hand-held or 'quick-cut' saws or unfamiliar machines, are at greatest risk because of their likely inexperience and often inadequate safety information, instruction, training and supervision.

People using hired concrete cutting equipment are also less likely to have a competent person standing by in case of a hazardous situation arising.

Operators are least at risk when employers, in consultation with workers, have developed an effective workplace safety policy covering concrete and masonry cutting and drilling operations, and safe work procedures for each type of concrete and masonry cutting equipment and job.

What are the risks?

The most likely risks of injury or harm come from, and are not limited to:

- flying saw fragments
- saw kickback, push-back or pull-in
- out-of-control or free-running cutting machines
- falling concrete and masonry
- inadequate scaffolding
- noise
- electrocution
- hazardous dusts from dry cutting and drilling, such as silica dust, contributing to lung disease
- slips, trips and falls
- manual handling or strain injuries
- vibration damage to circulation, nerves and joints
- suffocation or poisoning from hazardous fumes or gases emitted by petrol motors and other equipment or damaged gas supply services.

What are the hazards?

Some hazards are common to all concrete and masonry cutting and drilling operations; however, there are hazards specific to individual types of equipment, such as:

- *Kickback, push-back or pull-in* – these are potentially violent forces that occur suddenly and can be difficult to control. Employers and hire equipment suppliers must ensure that operators have information and training on safe work practices.
- *Obstructions or resistance in the material being cut* – these can cause sudden kick-back, push-back or pull-in movements of the saw.
- *Crooked or off-line cuts* – these can cause the saw to bite or pinch, resulting in kickback, push-back or pull-in reactions. (These reactions are also most likely to occur with hand-held saws.)

- *Pinched cuts* – these are caused when the object being cut moves, resulting in the cutting groove tightening on the saw blade, thus increasing the risk of kickback.
- *Blunt cutting edges* – these are caused by using a saw blade or drill bit with the wrong diamond cutting bond.
- *Unsafe grip, stance or stop-start procedures for hand-held saws* – these can cause the saw to swing out of control and come into contact with the operator, or strike objects that may cause the saw to fall and run free on the ground.
- *Worn, misshapen, cracked or damaged saw blades, or the wrong type of blade* – these can cause the blade to wobble, vibrate, shatter, or fragment and fly off. Guarding on most concrete and masonry equipment is designed to protect the operator from flying blade fragments, but not others in the workplace.
- *Worn blade shaft* – incorrectly fitted blades or the wrong type of blade for the job can cause wear on the central shaft, causing even new blades to shudder and resulting in early wearing and risk of shatter.
- *Wrong-size blades* – these are blades that are either too large or too small or are the wrong type for the cutting machine or size and shape of the concrete or masonry item being cut. For example, a small-diameter blade used to cut a thick slab may not penetrate sufficiently, which increases the risk of kickback or blade-shatter should the blade strike resistance.
- *Insufficient flow of coolant water* – this can cause overheating and expansion of both metal and masonry, resulting in poor performance, jamming, severe blade damage and projectile hazards.
- *Hand-held saw cutting above shoulder or below knee-height* – this can reduce operator control and increase the risk of kickback, push-back or pull-in injury.
- *Toxic fumes* – without adequate ventilation, petrol motor emissions containing carbon monoxide and other toxic gases can build up to hazardous levels, which increases the risk of asphyxiation.
- *Insufficient guarding* – guarding on some concrete or masonry saws is more effective than on others. Part of a safe work procedure should be to ensure that the manufacturer's recommended guarding is fitted to such saws. Removing guarding can greatly increase injury risk.
- *Electric wires, gas or water pipes* – exposing services, especially in existing structures, can put the operator at risk of slipping, electrocution, exposure to gases, or explosion.
- *Power cords* – when attached to electric-powered cutting equipment and other machinery, these may be cut or damaged. Pools of water coolant and slurry could cause electrocution due to an immersed cord.
- *Uneven or unstable surfaces* – these can increase the likelihood that the operator may trip or stumble, causing an unexpected movement of the blade and resulting in kickback.
- *Wet, slippery floors* – coolant water and slurry on floors can cause slips and falls.
- *Obstructions in access ways* – blocks of masonry and bricks in areas where the operator and others must stand, work or move can cause trips and falls.
- *Vibration* – whole body, hand or arm vibration caused by prolonged use of cutting or drilling equipment can cause nerve, circulatory and joint damage.
- *Working alone* – this can be hazardous because of the potential need for assistance in the event of an emergency situation or injury.
- *Noise* – excessive noise from concrete cutting and drilling is a workplace hazard, and appropriate hearing protection must be worn.
- *Hazardous dusts* – these are emitted by cutting and drilling operations or equipment that does not use water for cooling cutting parts and capturing dust. Concrete dust may carry high levels of crystalline silica dust, and repeated exposure can cause silicosis, which is a scarring and stiffening of the lungs. The effects are irreversible, invariably resulting in death. Coarser rock particles can cause short-term throat irritation and bronchitis.

Refer to Chapter 1 of the text for more information on the exposure of hazardous dusts such as crystalline silica. Relevant Codes of Practice in relation to safe work practices with masonry material can also be sought by visiting Safe Work Australia's website (http://www.safeworkaustralia.gov.au) or relevant state and territory safety regulators' websites.

Workshop equipment

Most fabrication of materials is performed on site; however, depending on what type of work is undertaken, you may be required from time to time to produce or replicate items using workshop equipment.

The following plant and equipment may be utilised to assist in completing jobs on the construction site.

Electric pipe-threading machine

Variations include power dies and power vice.

This power-driven equipment secures and rotates heavy-gauge steel piping. Several attachments can be added to manipulate the pipe. The quick-action dies are used to create an external thread on the pipe, the pipe cutter is used to trim piping to the required length, and the reamer attachment can be used to remove any internal burring. Pipe-threading machines (see Figure 7.109) are commonly used on galvanised and black mild steel pipes for fire sprinkler and fire hydrant service

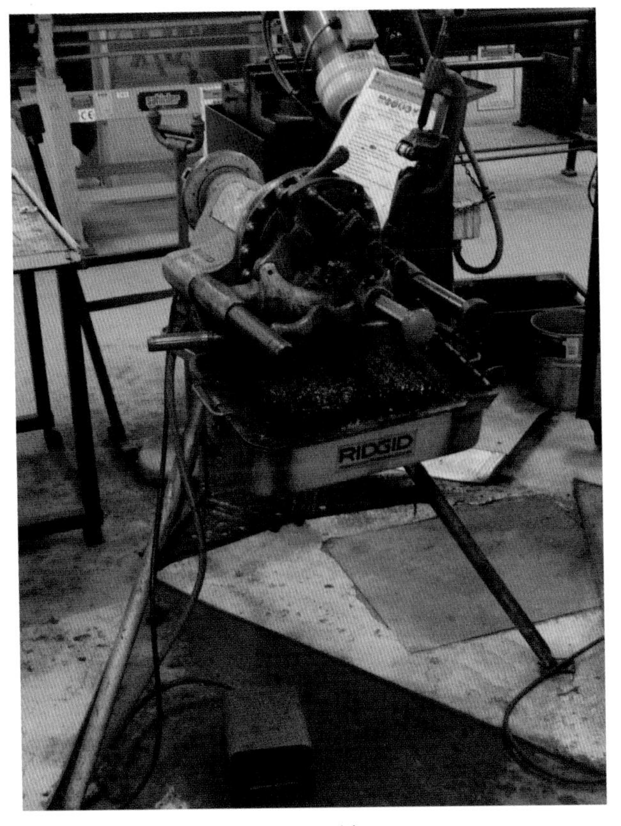

FIGURE 7.109 Pipe-threading machine

pipes. Always prepare and follow a SWMS and observe the SOP, wear appropriate PPE and have the equipment regularly serviced by authorised personnel.

Sheet metal folders

Variations include the hand folder, panbrake folder, magna bender and bench folder.

Folders are used in sheet metal fabrication work to bend flat sheet metal to set angles. The hand folder has a one-piece clamping blade and is used for basic folding. A panbrake folder (see Figure 7.110) has its top blade constructed in removable sections permitting the construction of more intricate folds. The magna bender uses an electromagnet to lock the metal in position during the folding process. The locking plates are interchangeable and come in a variety of lengths, making it the most versatile machine for folding intricate shapes. Bench folders are generally used to create small folds in sheet metal such as safety edges and grooved seams.

Apprentices or trainees should complete a SWMS prior to using a folder and observe the SOP attached to, or near the machine.

When using sheet metal folders, ensure locking plates are clamped by the user only to avoid pinch injuries.

FIGURE 7.110 Panbrake folder

Roll formers

Other name: rolling machine.

Manually operated rollers are used to curve light-gauge sheet metal to form cylindrical shapes. Two adjustable rollers are moved up or down and a third roller is locked into place to create the required radius curve in the sheet metal (see Figure 7.111).

FIGURE 7.111 Roll-forming machine

Guillotines

Guillotines can be manually, electrically or hydraulically operated. The hardened steel cutting blade uses a shearing action to slice sheet metal to the required size (see Figure 7.112). Manually operated 'treadle' guillotines require the user to physically press down upon the cutting lever to cause the blade to shear through the sheet metal.

FIGURE 7.112 Sheet metal guillotine

Safety and maintenance of guillotines

Consider the following points before using a guillotine.

- Due to the HIGH risk of injury you should not use a guillotine unless you have received appropriate training.
- Apprentices/trainees and first-time users should complete a SWMS prior to using a guillotine and all users should carefully observe and follow the SOP, which should be attached near to, or onto, the machine.
- Identify all hazards and associated risks prior to operating.
- Ensure that the area is clear of other personnel prior to operation.
- Ensure that fixed guards are in place to prevent hands or other parts of the body from entering the cutting space.
- Use the correct lifting procedures when handling sheet metal.
- Cut only within the specified capacity of the machine.
- Keep working parts well lubricated and free of rust and dirt.
- Immediately shut down and isolate equipment if faults are located.

Specialist plumbing equipment

Plumbers are required to carry specialist equipment due to most pipework being hidden or difficult to access. Therefore, it is important to have a thorough understanding of how this equipment is operated and maintained.

Drain-cleaning equipment

Drain-cleaning equipment is utilised by plumbing contractors/workers to remove blockages caused by tree roots, grease/fat, silt or other foreign objects in sewerage, stormwater and trade-waste service pipes. Small handheld augers can also be used in situations such as blocked basins connected to floor wastes in sanitary plumbing systems. Variations include the rotary drain cleaner and high-pressure water-jetter.

Rotary drain cleaning equipment

A rotary drain-cleaning machine (see Figure 7.113), commonly termed an 'electric eel', consists of an electrically operated motor that rotates or spins a flexible steel cable that can have a variety of different cutting heads attached. The cable is fed into the pipeline until it cuts its way through the blockage. Due to the nature of the work, it is vital that the correct metal studded safety gloves are worn, and an RCD is installed in the electrical supply to prevent electrocution. Even though these machines are extremely robust and reliable, it is recommended that they be routinely serviced by qualified personnel.

FIGURE 7.113 Drain-cleaning machine

Source: Shutterstock.com/pusit nimnakorn.

Water jetter

There is also a large variety of water-jetting machines available to clear blockages in pipework. They range from the domestic trolley-mounted machines, which can be wheeled directly to the job, to large truck-mounted water-jetters that are used for cleaning authorities' mains (see Figure 7.114). A water-jetting set-up generally has an engine that drives a pump to

FIGURE 7.114 Truck-mounted water-jetting machine

generate high pressure. A hose reel is connected from the pump to a secondary remote mini reel, which is then used to clear the blockage with use of an appropriate integral turbine cutting nozzle. The cutting nozzle utilises the extremely high-pressure water to dislodge or cut the blockage within the pipeline. There are different types of heads available, depending on what is causing the blockage.

FROM EXPERIENCE

When using a high-pressure water jetter, workers should always determine appropriate hand signals for isolation of the machine. Good communication skills and teamwork are essential between workers to ensure no injuries occur.

Extreme care should be taken when using this machine. It is essential that a SWMS be prepared and observed by any user of this equipment.

GREEN TIP

When using drain-cleaning equipment, environmental regulations need to be observed to ensure sewage does not enter stormwater pipes or drains.

Closed circuit television (CCTV) inspection systems

A CCTV pipeline inspection system (generally called a 'sewer camera') incorporates a mini camera attached to a push rod cable (see Figure 7.115). The cable is fed into the piping system and relays images to a portable monitor at surface level. The operator can survey the images to determine the condition of the pipeline and pinpoint any faults for rectification. Images and videos can be downloaded onto USB sticks and provided to the customer to help facilitate necessary repair work and price estimation.

FIGURE 7.115 Sewer camera (CCTV)

FROM EXPERIENCE

Being able to identify an approximate location of damaged pipework can prevent unnecessary excavation work. Embracing new technologies such as CCTV allows tasks to be carried out efficiently and effectively at a competitive price.

Pipe and cable locators

Pipe and cable locators are a precision instrument that is used for determining the position of buried pipes, cables or wires (see Figure 7.116). The device is battery operated and consists of a transmitter and a receiver. The transmitter generates an electrical impulse that is applied to the pipe or cable to be traced. When the receiver detects the signal from the electrical impulse it emits an audible tone, enabling the operator to trace the path of the buried pipe or cable.

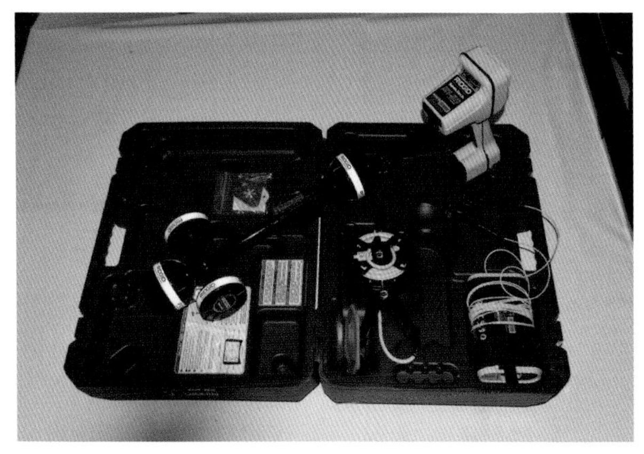

FIGURE 7.116 (a) Pipe and cable locating equipment; (b) Locating a pipe

LEARNING TASK 7.2

SELECT AND USE APPROPRIATE POWER TOOLS

Undertake safe power tool instruction with a teacher or instructor in relation to the following basic portable electric-power tools:

- pedestal drill
- jig saw and reciprocating saw
- grinder
- cordless drill
- masonry drill
- nibbler and shears.

Note: For safety and compliance with WHS requirements, it is recommended that power-tool instruction be carried out with an experienced worker who is competent. A SWMS should be completed, the SOP read by the users both before and after use, and appropriate PPE worn.

COMPLETE WORKSHEET 3

Basic plant and equipment

Basic plant and equipment may not always be required on the worksite and plumbers will often keep their equipment in a safe location such as a workshop or home garage, as it would take up a considerable amount of room in the work vehicle. When equipment is required for a task, it is loaded onto the work vehicle and used on the worksite.

Wheelbarrows

Wheelbarrows are extremely common for carrying material on building sites. They are used to carry both wet and dry material to places that heavy and large-capacity machines are unable to reach. The single, central-wheel type usually consists of a frame constructed of tubular steel with a bolt-on heavy steel tray (see Figure 7.117). Wheelbarrows are sometimes galvanised, with strong bracing at points of stress. They have a range of capacities, varying from 80 L to 120 L for the commercial market.

Vibration rammer

A vibration rammer (see Figure 7.118), also referred to as a wacker or wacker packer, is a petrol-engine rammer that uses a vibrating plate and tamping action to ram earth. The rammer is used for compaction when backfilling excavated trenches on the construction site to minimise subsidence by increasing the density and strength of the soil. Variations also include plate compactors that can be used on different soils, including fine aggregates such as sand.

FIGURE 7.117 Standard wheelbarrow

FIGURE 7.118 Vibration rammer

Concrete mixers

Concrete mixers are commonly used to mix concrete for small jobs such as bases for rainwater tanks and path repairs. They are useful for mixing sand/cement mixes to patch or render masonry walls after pipe rough-ins have been completed.

Tilting mixers

The smaller-sized mixers required for building sites are usually tilting mixers (see Figure 7.119). They are powered by an electric motor, a petrol engine or a diesel engine. Electric mixers are mounted on a steel chassis with the power unit at one end. The whole steel mixing drum and drive is mounted on a steel trunnion, which is tipped by a hand-operated lever or wheel.

FIGURE 7.119 Tilting drum mixer

Inclined drum mixer

The inclined drum mixer is conical-shaped and is fitted to mixing trucks (see Figure 7.120). The drum has only one opening and the mixing blades are arranged so that the drum will discharge when the rotation is reversed.

FIGURE 7.120 Inclined drum mixer with a capacity of 3 m^3 to 6 m^3

Routine maintenance of mixers

Consider the following maintenance points to ensure concrete mixers remain in good condition:

- Check electrical connections for proper and secure attachment. No bare wires or coloured cables should be visible at any connection point.
- Make sure all electrical connections are waterproof.
- If petrol motors are used, check oil levels as required, and clean, adjust and replace spark plugs as required.
- Cleaning should take place as soon as possible after final use of the mixer. On very hot days this may need to be carried out as a quick wash between mixes.
- Use a stiff brush and water to clean off remaining slurry from the outside and inside of the barrel.
- If the slurry or leftover mix has started to harden inside the barrel and on the mixing blades, pieces of broken brick or dry blue metal may be placed in the barrel with water, which allows the mixing action to dislodge any stubborn remains.

Note: Do not allow leftover slurry in the barrel to completely set prior to cleaning, as this will build up with future use and prevent effective mixing action.

Site generators

A site generator is normally a petrol-powered, air-cooled portable unit used to supply 240V power to operate electric-powered tools (see Figure 7.121). It becomes an essential unit on new, large housing estates where mains power is sometimes not yet available, and in country or isolated areas where access to electricity is not possible. Most sites will normally have a power supply, as temporary power and water are generally the first services made available on a worksite.

FIGURE 7.121 Portable site generator

Source: Shutterstock.com/Milos Stojanovic.

Most generating units are compact enough to fit into the back of a utility. Very large diesel-powered generating units for operating heavy equipment with high current demand are available and are mounted on the back of a truck or purpose-made trailer.

General capacity

Always determine what power requirements you need before hiring or purchasing a generator. The following outputs should be used as guide:

- 2kVA: maximum output is 1600W/6 amps (sufficient to operate two powered tools).
- 5kVA: maximum output is 4000W/16 amps (useful to operate up to five powered tools).
- 8kVA: maximum output is 6800W/26 amps (useful to operate light machinery).

Routine maintenance of generators

Consider the following maintenance points to ensure generators remain in good condition:

- Check the oil level before use each day when using a four-stroke engine.
- Check the petrol level each day and make sure the correct oil–petrol mix is used on a two-stroke engine.

- If the generator fails to start, check the spark plug. Clean and reset the plug or replace it as required.
- Clean the air filter regularly to prevent a build-up of oil choking the airflow.
- Replace the pull start cord when it becomes frayed or breaks.
- Check power point outlets for wear or damage.
- Make sure that equipment is cleaned after use and before storage.

Site compressors

A site compressor is usually portable and is used to operate nail guns, paint-spraying equipment, needle guns for paint removal, small sand-blasting guns, small air tools and to supply air for respiratory masks via an air purifier (see Figure 7.122).

FIGURE 7.122 Portable site compressor

Small compressors are generally powered by electricity, while larger compressors are powered by either petrol or diesel fuel. The air capacity and flow rate of the compressor should be based on individual requirements and the tools that are to be used. Each tool will have different flow requirements, and flow rates are measured in litres per second (L/sec) or cubic feet per minute (cfm).

Trailer-mounted site compressors are also available for use with a breaker or rock drill on demolition work and for breaking up firm soil (see Figure 7.123).

FIGURE 7.123 Trailer-mounted site compressor

Source: iStock.com/philipimage.

Larger units are also available to operate a range of air tools and sand blasters. Jobs include large excavation, breaking concrete and demolition; they can also be used as standby power sources in case of factory compressor breakdown or power restrictions.

Routine maintenance of compressors

Consider the following maintenance points to ensure compressors remain in good condition:

- An operating compressor unit should be set in a clear and level area, where cool air that's free of dust is available.
- Prior to starting, oil, water and fuel levels must be checked to avoid the possibility of engine wear.
- When starting up the plant, the air valves should all be opened. After the engine has been started and has had sufficient time to warm up, the air valves may be closed.
- When stopping the plant, the air in the receiver should be released first by opening the valves; then the engine itself should be stopped by allowing it to lose speed gradually.
- Pneumatic tools should be handled carefully and serviced frequently, which will contribute to trouble-free use of the plant.
- Prior to coupling up the tools to the air line and putting them into service, any moisture or dirt in the line should be blown out to prevent it entering the working parts of the tools.
- All bolts should be tightened up regularly, and the shanks of chisels and points should be the correct length and cross-section.
- Air lines should be checked to ensure all joints are secure to reduce air loss at these points.

Careful maintenance and attention to these points will lengthen the life of equipment and ensure the best working efficiency.

Industrial vacuum cleaners

Industrial vacuum cleaners (see Figure 7.124) are designed for wet and dry vacuuming of floors in factories, shopping centres and commercial premises, for asbestos removal, and for use in building sites. They are suited for the removal of surface water from flooded floors when core hole drilling, removing wetted asbestos dust and removing water from waterlogged carpets. Vacuum cleaners range from small capacity with a single motor to large capacity fitted with dual electric motors.

Note: Vacuum cleaners used to remove asbestos contaminated dust (ACD), or any other hazardous material, must be used specifically for that purpose. Bag removal/replacement and cleaning of these machines must be carried out within strict safety guidelines. Protective clothing and a respirator must be worn. Attachments with brushes should not be used as they are difficult to decontaminate. Bag removal must take place within the contaminated work area.

FIGURE 7.124 Industrial vacuum cleaner

Industrial vacuum cleaners are also available in smaller backpack models.

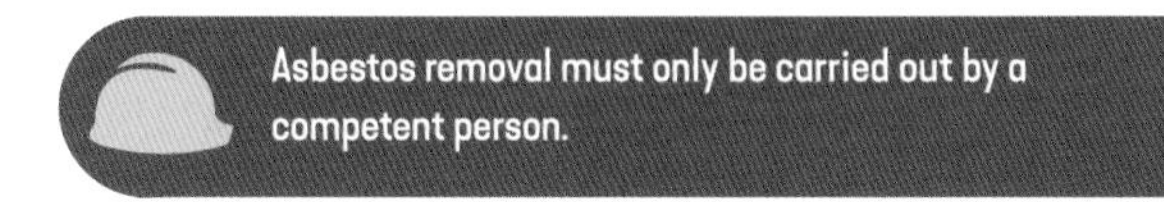

Asbestos removal must only be carried out by a competent person.

Safety and maintenance of vacuum cleaners

Consider the following maintenance points to ensure vacuum cleaners remain in good condition:

- Always check machine leads for bare wires, cracks or cuts before use. Do not use the machine if any are found.
- Ensure that the lead is always behind the machine when operating.
- Never use extension leads.
- Never stretch the lead to its limit and use the closest power point to proceed.
- Switch the machine off at the power point before unplugging.
- Never remove the plug from the power point by jerking the cord.
- Always empty the bag after use.
- Clean filters regularly.
- Always wipe the machine over after use with a clean damp cloth and thoroughly dry it prior to storage.

Elevated work platforms

Elevated work platforms (EWPs) include cherry pickers, scissor lifts and boom lifts, and are all designed to allow easy access to elevated work areas. They may be used for pipe installation, gutter installation, tree lopping, factory maintenance, changing light globes/fittings in factories, or accessing hard-to-reach places.

Cherry pickers

Cherry pickers are powered by diesel engines or batteries and have controls that are mounted in the platform basket. They range in lift height depending on the size of the machine, and are generally driven by a competent person with relevant accreditation. Trailer-mounted models are also available and have hydraulic outriggers for stability (see Figure 7.125).

FIGURE 7.125 Trailer-mounted cherry picker

Scissor lifts

Scissor lifts (see Figure 7.126) are self-propelled and are powered by mains electricity, battery or petrol. They lift vertically only and reach a lower height than a boom lift, but can generally take more weight.

Boom lifts

Boom lifts (see Figure 7.127) offer more manoeuvrability and can work over or around obstacles. They come in a variety of sizes and outreach options.

Man lifts

Man lifts are petrol-powered hydraulic lifts for one person and light equipment. They have a safe working height of up to 9 m.

Safety and maintenance of EWPs

EWPs offer greater access for tradespeople to complete projects safely and efficiently. Safety regulations must be followed when using EWPs, which includes wearing a harness while in operation and ensuring that you have received appropriate training and accreditation.

FIGURE 7.126 Petrol-powered scissor lift

Source: Shutterstock.com/Brynteg

FIGURE 7.127 Boom lift

Consider the following maintenance points to ensure EWPs remain in good condition:

- All petrol or diesel engines must have oil levels checked regularly.
- A safety harness must be worn at all times when working in the basket or on the platform.
- Platforms must not be used near overhead electrical cables.
- The operator of the trailer lift/cherry picker must have a Certificate of Competency for the safe operation of the device/s.

Basic ladders and trestles

Safe Work Australia and state or territory regulatory authorities are responsible for the implementation of regulations relating to ladders and trestles. WHS state or territory regulator inspectors have the authority to enforce safety regulations made under the WHS legislation of their jurisdiction.

Ladder types

Ladders are constructed from aluminium, reinforced plastic, fibreglass or timber. Selection of the correct ladder for the job is an important safety consideration (see Figures 7.128–7.132), as falls from ladders contribute greatly to general workplace accidents. Selection is assisted by the following information being permanently marked or labelled in a prominent position on the ladder:

- the name of the manufacturer
- the duty rating (i.e. INDUSTRIAL USE or DOMESTIC USE) and the load rating in kilograms
- the working length of the ladder (closed and maximum working lengths for extension types)
- electrical hazard warning: DO NOT USE WHERE ELECTRICAL HAZARDS EXIST (metal types)
- on single and double-sided stepladders: TO BE USED IN A FULLY OPEN POSITION.

When working with ladders there are some important things to consider:

- Ladders must be in good condition, free from splits, knots, and broken or loose rungs.
- *The 1:4 base:height ratio rule with ladders must be followed*. For example, the foot of a 4 m ladder should be at least 1 m away from the wall against which the ladder is leaning. Make sure the top of the ladder extends at least 1 m above the landing or platform (see Figure 7.133).
- The ladder should be securely fixed at the top and bottom and footed securely on a firm level foundation (see Figure 7.134).
- Never put ladders in front of doorways, or closer than specified by state or territory safety regulatory authorities to bare electrical conductors. An electrical current can jump from a conductor to an aluminium ladder without direct contact.

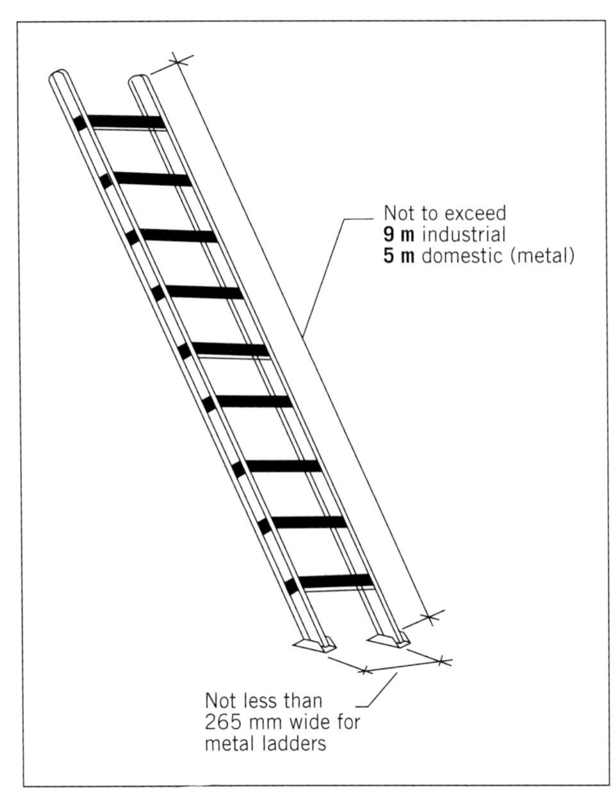

FIGURE 7.128 Single ladder

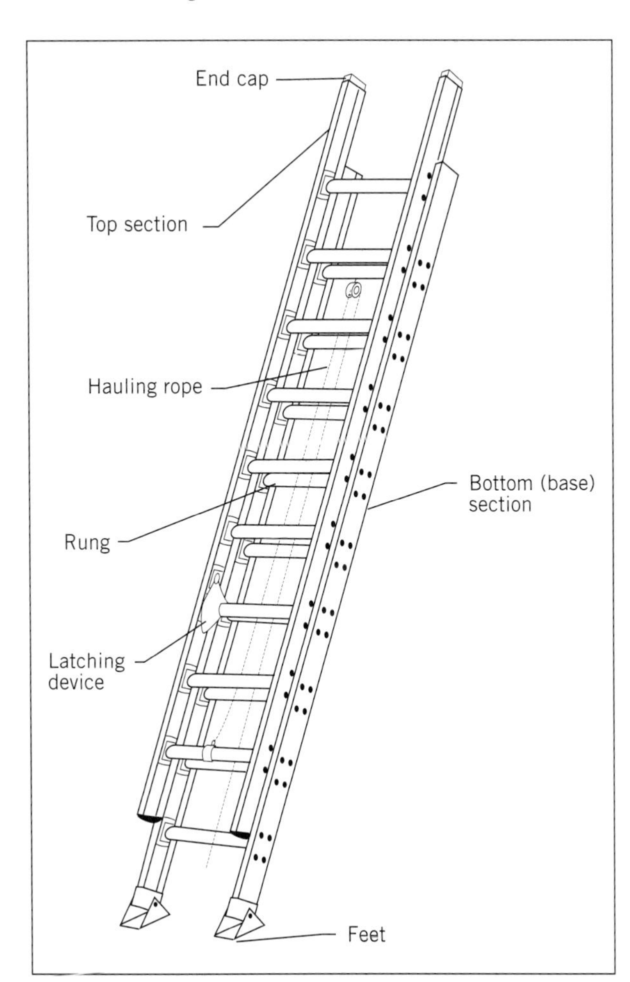

FIGURE 7.129 Extension ladder, maximum length 15 m

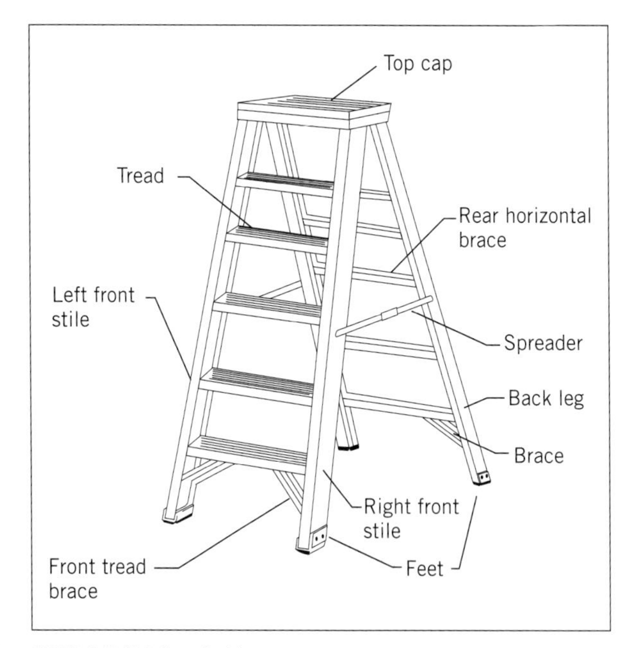

FIGURE 7.130 Stepladder

FIGURE 7.131 Dual-purpose step/extension

- When working with or on electrical equipment, use only fibreglass ladders. Do not use metal or wire-reinforced ladders when working near exposed power lines.
- One person at a time should be on a ladder, and tools should be pulled up with a rope. Workers ascending, or descending should always face the ladder.
- Two ladders must never be joined to form a longer ladder.
- Ladders should not be placed against a window.

FIGURE 7.132 Trestles used to create a working platform

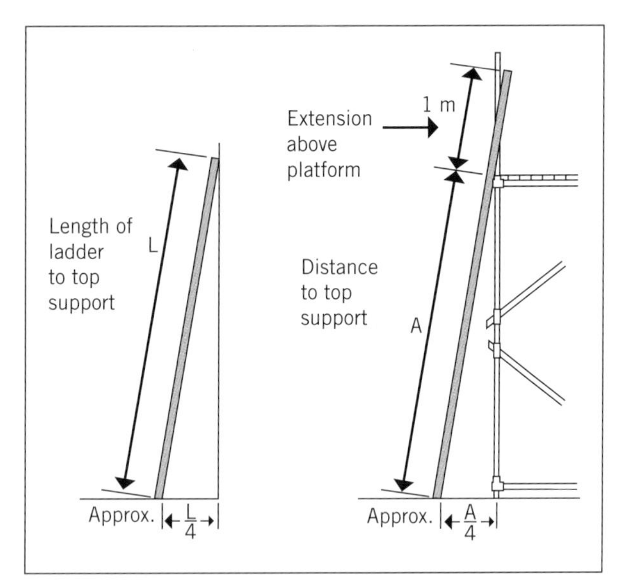

FIGURE 7.133 Base-to-height ratio rule

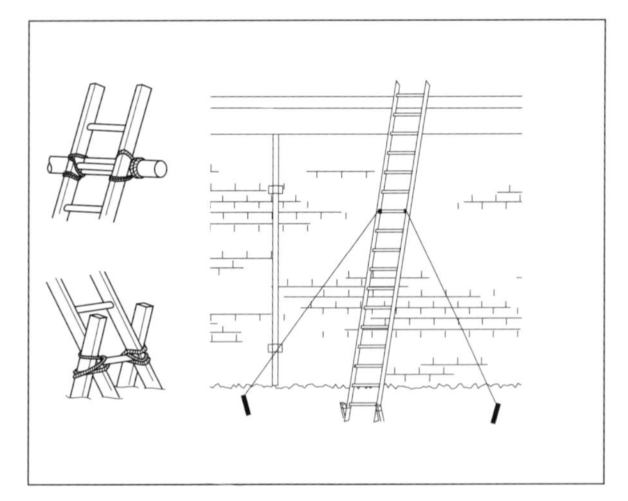

FIGURE 7.134 Method of securing ladders

- Timber ladders should never be painted, as this could cover faults in the timber.
- Before using a ladder, check wind conditions.

Clean up work area

Hand and power tools make tasks very efficient, but they can produce waste such as swarf, offcuts and wood shavings. Good housekeeping while working with hand and power tools is essential to prevent accidents and injuries from occurring. After certain tasks or at the end of each workday, ensure the work area is clean and tidy, and stack and store any materials for later use. Always recycle materials where applicable in an appropriate manner, observing sustainability principles and concepts during work processes.

Tools and equipment

All tools and equipment have different maintenance requirements based on manufacturer's specifications and must be thoroughly cleaned and checked for damage to ensure they are in good condition for the next time they are used. Any faulty equipment must be reported to a supervisor to prevent an accident or injuries from occurring. Once tools and equipment have been checked, they should be located and stored in a safe suitable location in line with workplace procedures.

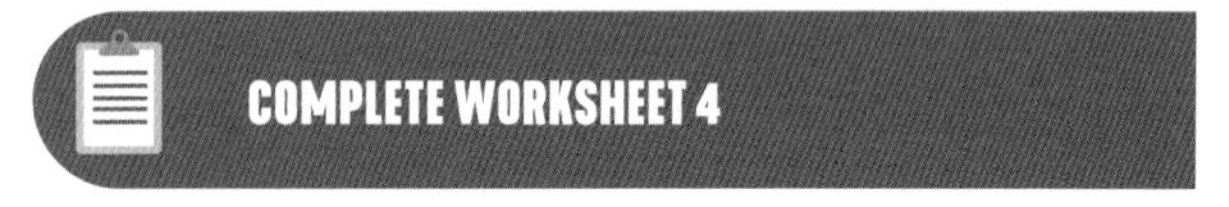

REFERENCES AND FURTHER READING

Acknowledgement

Reproduction of the following resource list references from DET, TAFE NSW C&T Division (Karl Dunkel – Program Manager – Housing and Furniture) and the Product Advisory Committee, is acknowledged and appreciated.

Texts

Holloway, D. (1994), *Made simple: Woodworking joints*, Harlaxton

Jackson, A. & Day, D. (1984), *The complete book of tools and how to use them*, Treasure Press, London

Metal and Engineering Industry (1994), *Training manual: Hand and power tools*, ACTRAC Products Ltd, Victoria

Noll, T. (1997), *The encyclopedia of joints and jointmaking*, RD Press, Sydney

Web-based resources

Resource tools and VET links

Training.gov.au: **http://training.gov.au**

NSW Education Standards Authority – Construction: **https://educationstandards.nsw.edu.au/wps/portal/nesa/11-12/stage-6-learning-areas/vet/construction-syllabus**

Industry organisations' sites

Infolink: **http://www.infolink.com.au**

Just Tools: **http://www.justtools.com.au**

Audiovisual resources

Short videos covering topics such as basic levelling, dangers of compressed air, and falls in the workplace are available from Safetycare: **http://www.safetycare.com.au**.

TAFE NSW resources

Training and Education Support (TES), Industry skills unit Orange/Granville 68 South St Granville NSW 2142
Ph: (02) 984 68126 Fax: (02) 9846 8148 (Resource list and order forms).

GET IT RIGHT

The photo below shows an incorrect practice that can be performed when using an angle grinder.

Identify the incorrect method and provide reasoning for your answer

WORKSHEET 1

To be completed by teachers	
Satisfactory	☐
Not satisfactory	☐

Student name: ______________________

Enrolment year: ______________________

Class code: ______________________

Competency name/Number: ______________________

Task: Review the section 'Identify hand and power tools' up to and including 'Cleaning up tools' in this chapter and answer the following questions.

1. What document must you always refer to before undertaking a task with a power tool?

____

2. Before using electrical equipment, what must you use and connect to, and why?

____

____

3. Who should PPE be provided to when working on site?

____

____

4. What is the main difference between a tape measure and wind-up tape measure?

____

5. Briefly describe two tasks that can be carried out in the plumbing sector using a wind-up tape measure.

____

____

6. What type of file would you recommend for sharpening snips?

____

7. What is a sliding bevel used for?

____

8. List two ways a chalk line reel could be used.

____

____

9. Name three different types of snips used in the plumbing services sector.

10. State the direction the cutting teeth on a hacksaw blade should be facing.

11. List three different hacksaw blade sizes in 'teeth per inch' (TPI) and list the recommended materials to be cut for each of the blade sizes.

12. Name four different types of hammers that are used in general plumbing activities.

13. For what purpose would you commonly use a centre punch?

14. Describe the term 'mushroomed head' when referring to indirect impact tools and state the potential danger of this condition and the steps taken to prevent any danger.

15. State what type of wrench would be suitable for tightening delicate tapware.

16. Name the tool that would be used to tighten kinko nuts underneath a vanity in a restricted position.

17. Why is it NOT recommended to use a footprint wrench to tighten a chrome-plated bath spout into position?

18. Name the tool that would be used to tighten or loosen grubscrews on tapware.

___

19. Where would an edging trowel be used in the construction industry?

___

20. Name the type of broom that you would recommend for cleaning up a driveway that has had a delivery of blue metal temporarily stored upon it.

___

WORKSHEET 2

To be completed by teachers

Satisfactory ☐

Not satisfactory ☐

WORKSHEETS 7

Student name: ______________________

Enrolment year: ______________________

Class code: ______________________

Competency name/Number: ______________________

Task: Review the section 'Pipework and metalwork joining tools and equipment' up to and including 'General maintenance of hand tools and equipment' in this chapter and answer the following questions.

1. What is a reamer attachment on a tube cutter used for?

____

2. What is the difference between a tube cutter and pipe cutter?

____

____

3. State two different methods used to prevent annealed tube from deforming under load during hand bending.

____

____

4. What is the main advantage of branch forming compared to installing a prefabricated T-piece?

____

5. Name a situation in the plumbing sector where a croxing tool would be used.

____

____

____

6. State what a tap reseater would be used for, and name one situation where this tool would be used.

____

____

7. What are brazing torches primarily used for?

____

____

8. What is the purpose of the regulator that is fitted to oxygen and acetylene welding cylinders?

____

9. List two types of crimp ring assembly tools.

____

____

10. Describe how you would check the accuracy of a spirit level.

____

11. Define the term 'datum point'.

____

____

12. What is the main advantage of using a laser level?

____

13. Describe the main benefit of using a water level compared to other types of levels.

____

14. State the main advantage of using a long-handled shovel.

____

WORKSHEET 3

To be completed by teachers	
Satisfactory	☐
Not satisfactory	☐

Student name: ______________________________

Enrolment year: ______________________________

Class code: ______________________________

Competency name/Number: ______________________________

Task: Review the section 'Select and use appropriate power tools' up to and including 'Specialist plumbing equipment' in this chapter and answer the following questions.

1. When should a residual current device (RCD) be used on the construction site?

   ____

2. State one item of clothing that power tool operators should wear to give them additional protection against electrocution.

   ____

3. Safe Work Australia requires that all power tools and leads are tagged and tested by a competent person. What is the purpose of this examination?

   ____

4. When used on construction sites, why must plugs and sockets of electrical leads and power tools be made of clear plastic?

   ____

5. In relation to safety with power tools, why should leads not be left partially rolled up during use?

   ____

6. What is the benefit of battery-operated/cordless power tools?

   ____

   ____

7. Portable electric-power tools can be a source of both mechanical and electrical hazards unless they are maintained and used correctly. List four general ways of avoiding damage to tools and equipment or injury to the operator.

   ____

   ____

   ____

   ____

8. What is a rip blade on a circular saw used for?

   ____

9. List two materials that can be cut using a reciprocating saw.

10. What type of power drill would be recommended for fixing dyna-bolts into a concrete slab?

11. What type of drill would be used to fasten hex screws into roof sheeting?

12. State six possible risks that may occur when operating concrete and masonry cutting and drilling equipment.

13. Describe the difference between a hand folder and a panbrake folder.

14. State one advantage of using CCTV inspection systems on an existing drainage system.

15. State the main differences between a rotary drain-cleaning machine and a water-jetter.

16. Describe the operating principle of a pipe and cable locator.

To be completed by teachers

Satisfactory ☐

Not satisfactory ☐

Student name: ______________________

Enrolment year: ______________________

Class code: ______________________

Competency name/Number: ______________________

Task: Review the section 'Basic plant and equipment' in this chapter and answer the following questions.

1. Describe the recommended method of removing leftover slurry from the inside of a tilting drum concrete mixer.

____

2. State why a portable site generator is an essential piece of equipment when working on a new housing estate.

____

3. What is the correct procedure that must be followed to prevent damage to an air-operated appliance prior to coupling it to a hose from a compressor?

____

4. List three types of elevated work platforms (EWP).

____

____

____

5. State what safety requirements must be met before working on an EWP.

____

6. In your state or territory, which safety authority implements the safety regulations for the use of ladders and trestles?

____

7. Draw a simple sketch to explain the base-to-height ratio rule for ladder use.

8. State how ladders should be secured to prevent them from slipping or blowing over.

____

9. What type of ladders are recommended to be used within close proximity of bare electrical conductors?

____

10. Describe how you would ascend and descend a ladder.

____

11. State why timber ladders should never be painted.

____

PART 4

SITE WORKS

Part 4 of this textbook deals with the types of work that may be carried out on site for new workers that are undertaking training and education within the plumbing services sector. It will help you build and enhance the required knowledge for cutting and joining sheet metal, marking out materials, and welding using oxyacetylene and arc-welding processes on the worksite.

Part 4 is based on four units of competency, as described in Chapters 8 to 10B. The key elements for the following competencies are as follows:

8 Cut and join sheet metal
- Prepare for work.
- Identify joining requirements.
- Cut and join sheet metal.
- Clean up.

9 Mark out materials
- Prepare for work.
- Determine job requirements.
- Mark out the job.
- Clean up.

10A Weld using oxyacetylene equipment
- Prepare for work.
- Prepare materials and welding equipment.
- Perform welding.
- Clean up.

10B Weld using manual metal arc welding equipment
- Prepare for work.
- Identify welding requirements.
- Prepare materials and equipment for welding.
- Weld items.
- Clean up.

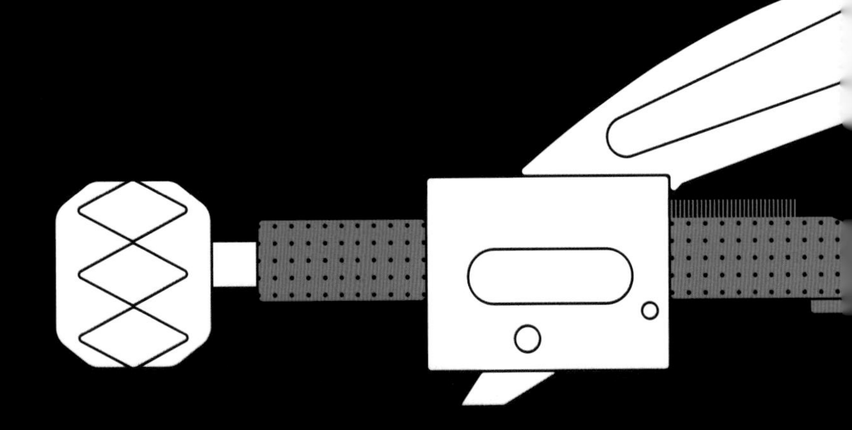

Work through Part 4 and engage with your teacher and peers to prepare for the different types of site works that may be carried out on the worksite.

The learning outcomes for each chapter are a good indicator of what you will be required to know and perform, what you will need to understand, and how you will apply the knowledge gained on completion of each chapter. Teachers and students should discuss the knowledge and evidence requirements for the practical components of each unit of competency before undertaking any activities.

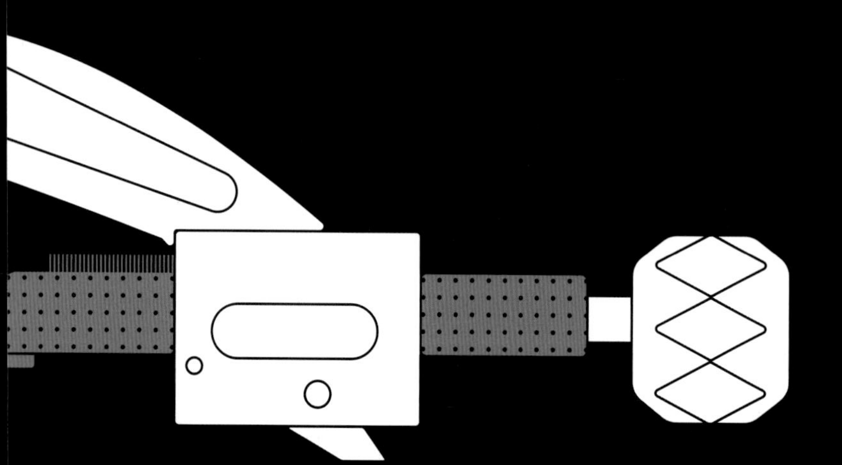

8

CUT AND JOIN SHEET METAL

This chapter addresses the following key elements for the competency 'Cut and join sheet metal':

- Prepare for work.
- Identify joining requirements.
- Cut and join sheet metal.
- Clean up.

It introduces the skills and knowledge required to cut and join sheet metal using various cutting and joining techniques. It also addresses the fabrication, installation, repair functions, fastening methods and sealing techniques required for the prevention of leaks caused by capillary action and thermal expansion.

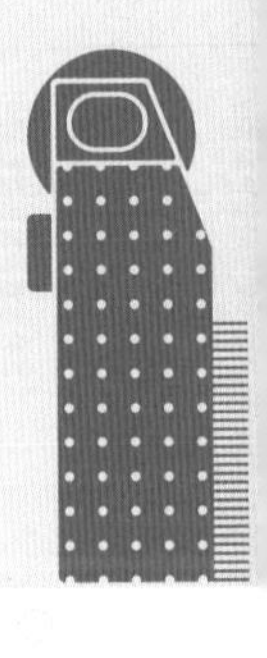

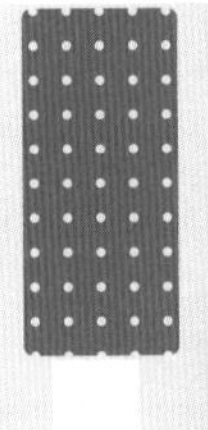

Prepare for work

Working with sheet metal has a long history in the plumbing trade. The variety of available materials has changed dramatically over the years, with most of the materials used in the past being mild steel, galvanised mild steel, lead and copper. Technology has now provided us with many different suitable alternatives that may be used alongside traditional sheet metal products.

Sheet metal items such as roofing, guttering, downpipes (spouting) and flashings are manufactured in specialist factories that either provide materials directly to contractors or deliver to plumbing suppliers to fill specific orders for specific projects. Most items that are now specified for roof work and flashings are Colorbond® or Zincalume®, as they are easy to fabricate and have a higher resistance to corrosion.

Skills in working with sheet metal have also changed dramatically as plumbers were previously required to be highly skilled when working with, and joining, sheet lead, sheet copper and other forms of sheet metal. These materials were joined mostly with the assistance of soft soldering. Plumbers' skills are now predominantly required to weatherproof buildings with Colorbond® or Zincalume®, using silicone and pop rivets as the joining medium.

Sheet metal has a long history of use as a material for a wide range of applications. It has been used successfully over hundreds of years in products ranging from basic tools, weapons and artwork to modern buildings, cars and aeroplanes, and to simple household items such as refrigerators, hot water heaters (see **Figure 8.1**), stoves and kitchen sinks (see **Figure 8.2**).

FIGURE 8.1 Hot water heater with a sheet metal outer casing

Source: Image supplied by Dux Hot Water.

FIGURE 8.2 Stainless steel kitchen sink

Source: Shutterstock.com/Joe Gough.

Sheet metal working is a highly skilled craft and good hand skills and safe work methods are still required to perform 'heritage' roof work, which is generally undertaken by a specialist roof plumber. These skills have been used to create finely crafted metal roofs such as those found on the Queen Victoria Building in Sydney (see **Figure 8.3**) or Flinders Street Station in Melbourne. Although the nature of sheet metal installations has changed dramatically over the years, thorough knowledge of all facets of sheet metal installation is essential.

FIGURE 8.3 Queen Victoria Building with copper-domed cupola

Source: iStock.com/Sergeo_Syd.

PART 4

8

In order to complete the job of cutting and joining sheet metal in an efficient, safe and professional manner, it is essential to plan and prepare work activities. This includes obtaining information from building plans, drawings or specifications and becoming familiar with workplace quality assurance requirements. In addition, WHS and environmental requirements associated with cutting and joining sheet metal must be considered and adhered to, and meet the requirements of state and territory regulations.

It is essential to have knowledge of the materials being used for the work in order to select compatible materials to join them using the most suitable and effective methods for production of a durable and high-quality product.

When planning and preparing for tasks, consideration should be given to the sequencing of activities and scheduling work with other tradespeople, which will minimise the risk of workplace accidents, save time and money and lead to the production of a quality job. Planning and preparation are essential to ensure the job is completed correctly and in accordance with plans and specifications. It is important that specifications are read as it would be a costly error installing Zincalume® gutter when the specifications asked for Colorbond®.

FROM EXPERIENCE

Most sheet metal tasks take place on a roof, and access by an elevated work platform (EWP) may be required. Always plan and organise access to prevent delays and to ensure tasks are achieved in the most efficient way possible.

Hand and power tools required

Before tools and equipment are used, they should always be checked for serviceability to ensure they remain in good condition. Always check the work area is prepared and ready for all sheet metal tasks.

Plumbing hand and power tools used for sheet metal fabrication may include the following:

- battery-operated drill/drivers and impact drivers
- clamps to hold and bend materials – locking pliers, vice grips, duckbill vice grips and multigrips
- drill bits
- files – bastard, second cut and smoothing
- grooving tool, to either complete or repair grooved seam joints
- hacking knives, cold chisels, prick punches and centre punches
- hacksaws – 32 TPI blades for sheet metal
- hammers of various types
- hole saws
- marking tools such as scribers and dividers (see Table 9.1 in Chapter 9 for compatible markers)
- nibblers or shears for cutting metal sheeting
- riveting tools
- screw guns
- screwdrivers
- sealant guns for silicone and adhesives
- sliding bevel
- soldering equipment, including conventionally heated traditional, gas or electrically heated irons
- spanners – shifting, tube and fixed (some spanners can be purchased as specialised tools from roofing manufacturers)
- squares (both engineer's try squares and combination squares)
- steel rules and tape measures
- tin snips of various types (aviation snips, straight or curved snips, rolled jaw snips and jeweller's snips)
- wooden mallets for dressing sheet metal to reduce marking on the materials from the hammering/dressing action.

When working outdoors, care should be taken to prevent rain falling onto tools. Tools should always be dry and sprayed with a suitable multi-purpose lubricating spray, which displaces water and prevents corrosion.

Always check for existing services underneath roof sheets before penetrating with sharp tools to avoid the risk of damaged services or electrocution. To minimise the risk of injury when using power tools and equipment, care should be exercised to ensure that tools are being used in the way they were designed.

Equipment and PPE

Plumbers are required to be proficient in the use of many tools and equipment, and at times specialised plumbing equipment may be required. It can include the following:

- an LPG cylinder and torch, or a smaller MAPP replacement gas cylinder and torch head for times when the LPG cylinder runs out or to access difficult-to-reach areas
- mechanical folding equipment (see Figures 8.4 and 8.5) (for further information, refer to Chapter 7 – Figures 7.110, 7.111 and 7.112)

FIGURE 8.4 Bench folder

FIGURE 8.5 Hand-operated folder

- rollers
- anvils (see Figure 8.6)

FIGURE 8.6 Anvils

- bench-mounted plasma cutters
- guillotines (see Figure 8.7).

FIGURE 8.7 Sheet metal guillotine

Correct selection and proper use of tools is crucial to the quality of the finished product and to the safety of the operator.

Personal protective equipment

Personal protective equipment (PPE) should be worn by all workers when undertaking sheet metal tasks to assist in carrying out work safely. Appropriate instruction and training on the correct use and fitting of PPE should always be given by the employer. PPE that does not fit or has not been fitted correctly may result in illness or injury and defeat the purpose of wearing the PPE in the first place. When using hand and power tools with sheet metal, it is required that you refer to a Safety Data Sheet (SDS) to select appropriate PPE.

Always be mindful of using gas torches near existing buildings as the heat may cause ignition to existing buildings. Flammable gases must also be used in a well-ventilated area with appropriate PPE and must be stored in a safe manner in accordance with regulatory requirements.

LEARNING TASK 8.1

PREPARE FOR WORK

1. Discuss the preparation requirements for a sheet metal task that would be undertaken in the practical workshop.
2. Which work health and safety, quality assurance and environmental requirements would need to be addressed?
3. Make a list of the tools, equipment and PPE that would be required to successfully complete the task.

Identify joining requirements

Before undertaking sheet metal tasks, it is important that the correct material and joining processes are identified and checked for compliance against the plans and specifications. The compatibility of materials and sealants need to be taken into consideration to minimise or prevent corrosion when joining sheet metal materials.

Sheet metal

Sheet metal is an integral part of the plumbing and roofing industry and can be defined as thin sheets of metal ranging from approximately 0.15 mm to approximately 6 mm thick, which are used in the manufacture of ductwork, flashings, roof products, aircraft and automobiles.

Plumbers generally work with thin, light-gauge material with a high tensile strength as it is easily fabricated on site.

Sheet metal gauges

The sheet metal gauge indicates the standard thickness of sheet metal. For most materials, as the gauge number increases, the material thickness decreases. Metal gauges are measured in both imperial and metric units (see Table 8.1).

TABLE 8.1 Sheet metal gauge conversion chart for common metal gauges (indicative only)

Imperial gauge	Metric sheet (mm)
10	3.0
12	2.5
14	2.0
16	1.5
18	1.2
20	0.9
22	0.7
24	0.6
26	0.5

Types of sheet metal

Sheet metal is defined as a ferrous or non-ferrous material formed into thin, flat pieces, which is able to be cut and bent into various shapes to create an infinite number of objects. Thicknesses vary considerably from very thin pieces (referred to as leaf or foil) to metal plate or solid metal (up to 6 mm thickness).

It is important to note that when selecting sheet metal for various applications, the compatibility of the different materials must be considered. When two dissimilar metals are combined there is an increased risk of electrolysis, which can accelerate corrosion of materials (see Table 6.3 in Chapter 6).

Ferrous metal

Ferrous sheet metals are materials containing iron and include:

- mild steel
- stainless steel
- Zincalume®
- Colorbond®

Ferrous metals will corrode or rust when exposed to a moist atmosphere, which is known as atmospheric corrosion. This reaction is a result of the oxygen in the air and water combining with the metal to form ferrous or iron oxide, commonly known as rust. This oxide forms on the surface then expands and eventually peels away in chunks. This exposes fresh metal underneath and, if left unchecked, will continue to corrode until the metal finally disintegrates.

Another characteristic of a ferrous metal is its ability to be attracted to a magnet, so if a magnet does not stick, it is unlikely to be a ferrous metal.

Zincalume® and Colorbond® are classified as ferrous metals due to the fact the materials both contain iron; however, due to the proprietary alloy coating on the steel, there is increased corrosion protection, provided that the materials are handled and installed according to manufacturer's installation guidelines.

Non-ferrous metal

Non-ferrous sheet metals are materials that do not contain iron, and include:

- copper
- copper alloys such as brass, which is a combination of copper and zinc
- lead
- aluminium
- tin
- zinc.

A characteristic of non-ferrous metals is that when oxides form on the surface, the surface becomes dull or discoloured, which provides a reasonably strong corrosion-resistant protective coating that does not flake away by itself. If this coating is removed a new coating is formed, and a certain amount of metal thickness is lost.

There has been increased development in the manufacture and use of both ferrous and non-ferrous sheet metal, ranging from the types of colours, protective coatings, cutting and sealing techniques and jointing methods.

Joining

Despite continuing developments in sheet metal technology, there are still only two basic methods of joining sheet metal. They are categorised as integral and mechanical joints.

Integral jointing

This method requires folding and manipulating the sheet metal itself to provide a strong and durable joint. These joints are known as:

- grooved seam
- countersunk grooved seam
- peined-down
- knocked-up
- Pittsburgh lock seam
- sliding cleat
- simple lap.

Most jointing techniques are manufactured by machine in a factory, with grooved seam joints being used on square, rectangular or round downpipes, using either an internal (countersunk) or external groove seam joint.

It is important to have a basic awareness of how each joint is fabricated so that suitable precautions can be taken on site to prevent the joint from breaking apart (e.g. grooved seam joints on rectangular downpipes/spouting).

Mechanical jointing

These methods result in the sheet metal being held together by fasteners, soldering or welding techniques. These types of joints are normally basic lap or butt jointed.

Mechanical joints include the simple lap joint, which is joined using:

- resistance (spot) weld
- soft-soldered joints
- silver-brazed joints
- blind or pop rivets (with silicone)
- screwed and bolted joints.

Simple lap and butt joints may be joined using:

- arc-welded joints
- tungsten inert gas (TIG) welded joints
- metal inert gas (MIG) welded joints
- oxyacetylene fusion welded joints
- bronze-welded joints.

Refer to Chapters 10A and 10B for details on the use of these methods.

Grooved seam joints

Grooved seams are commonly used as vertical or lateral joints that can be sealed or remain unsealed (see Figure 8.8).

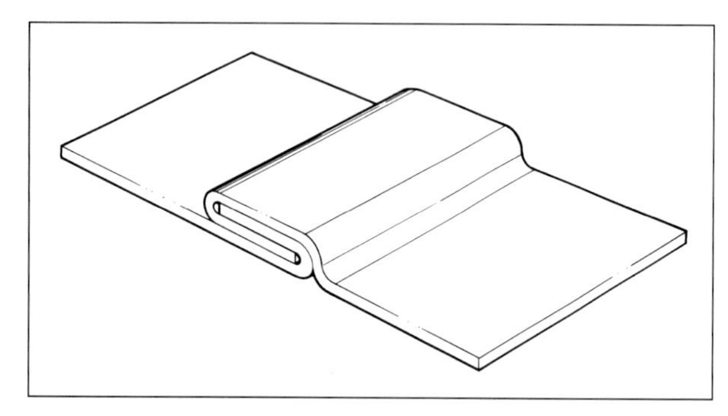

FIGURE 8.8 The grooved seam

In the construction industry they are mostly used on items that include:

- square and round downpipes
- tanks
- downpipe nozzles
- flat sheet metal roofing
- irregularly shaped transitions such as ductwork in commercial projects (see Figure 8.9).

Creating the seam

When designing a project, it is important to consider the location of the seam to allow adequate support by an anvil, a metal block, a bar or another strong flat surface. Exposure to the elements may also determine its location.

Making allowances for the joint

The size of the grooved seam is determined by the thickness of the material and the width of the grooving tool (see Figure 8.10). To determine the size of the fold for the joint, it is necessary to allow for three thicknesses of the material to be deducted from the width of the groove in the tool. The remainder is the maximum size of the fold.

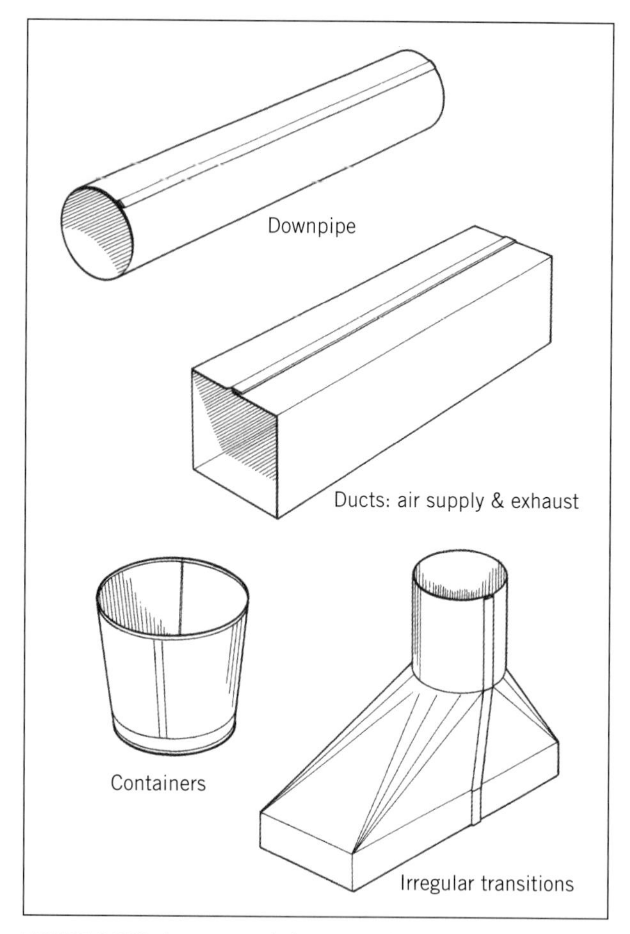

FIGURE 8.9 Various uses of the grooved seam joint

For example, a grooving tool with a groove of 6.0 mm and metal thickness of 0.5 mm will result in a fold of 4.5 mm, as follows:

6.0 mm grooving tool width – 3 × 0.5 mm material thickness = 1.5 mm

6.0 mm – 1.5 mm = 4.5 mm fold

In the plumbing industry, most joints can be made by selecting a 4.5 mm or 6 mm grooving tool, both of which are suitable for metal thicknesses of between 0.5 mm and 0.8 mm. Generally, the grooves in the chosen tool should be at least 2 mm wider than the lap folds to allow enough material to be dressed around during the final locking of the joint with the grooving tool.

To ensure sufficient material is allowed for to make the joint and complete the project, a groove seam allowance should be determined in addition to the dimensions of the product being manufactured. When marking out the project for a 6 mm grooved seam with laps of 4.5 mm, a piece of sheet metal wide enough for the dimensions of the project (base pattern), plus an allowance of three times 4.5 mm for the laps (totalling 13.5 mm) should be cut.

Half of the 13.5 mm allowance is added to each side of the dimension needed for the project to allow for the

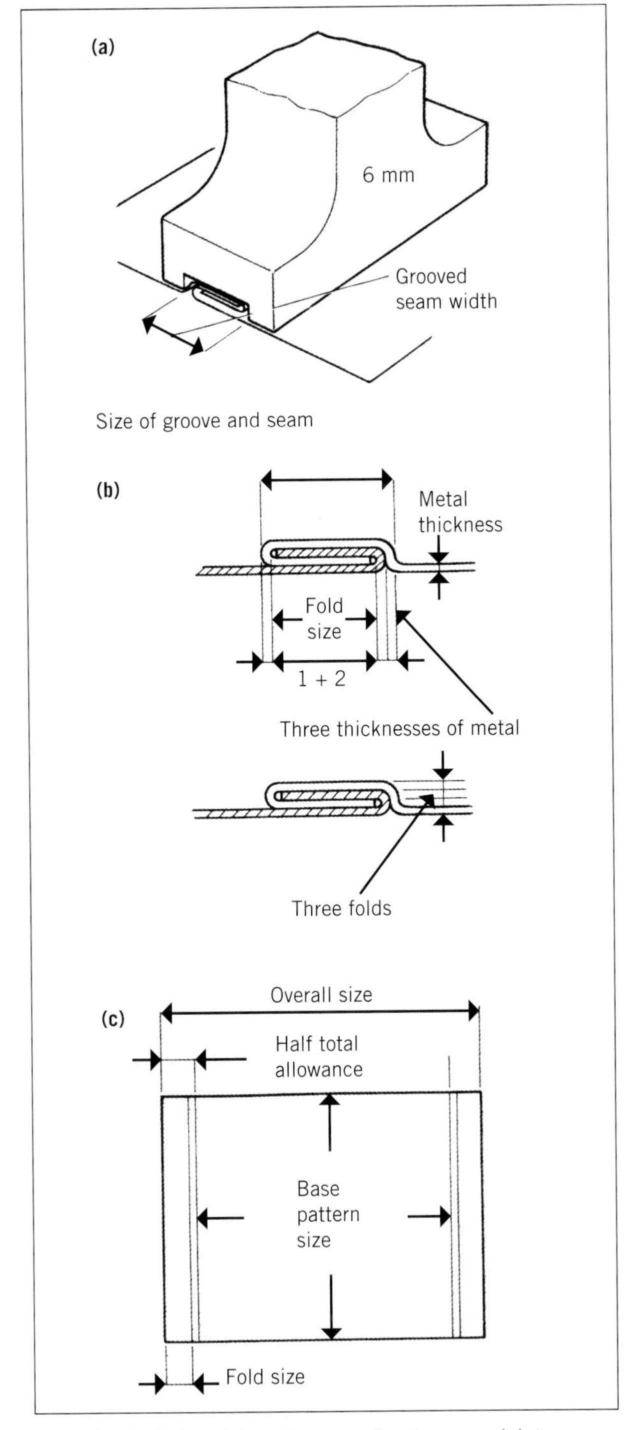

FIGURE 8.10 Determining allowance for the seam joint: (a) Size of groove and seam; (b) Thickness of seam; (c) Determining seam allowance

groove seam joint. The lap allowance required would be 6.75 mm extra on each side (see Figure 8.10(c)).

When there is a need for greater accuracy in an item using the groove seam method of joining, it is recommended that a sample piece of work be made to observe the loss of dimension in the overall size. As a guide, calculate three times the fold length plus six thicknesses of metal to avoid producing an undersized product (see Figure 8.11).

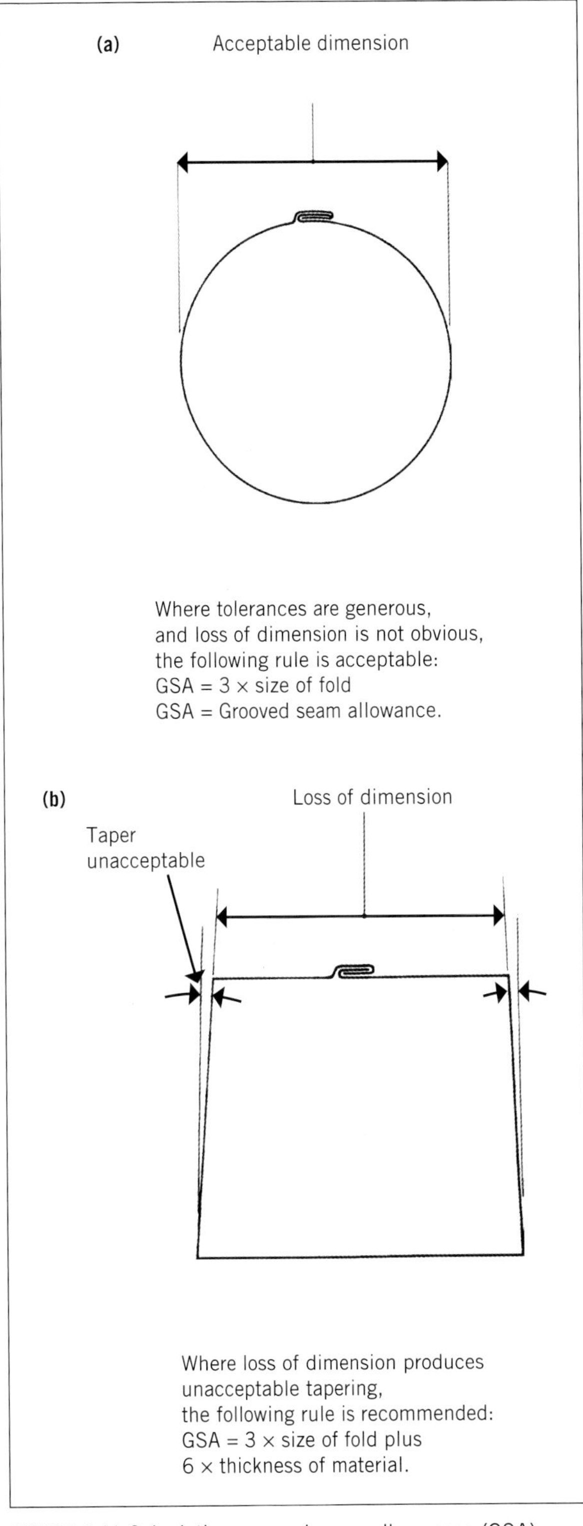

FIGURE 8.11 Calculating grooved seam allowances (GSA): (a) Acceptable dimension; (b) Loss of dimension

HOW TO

MAKE THE GROOVED SEAM JOINT

1 To make a grooved seam joint, the sheet metal is cut to shape according to the plan dimensions (see Figure 8.12), including the allowance for laps, with each end of the material being folded in a folding machine such as a hand, panbrake or bench folder.

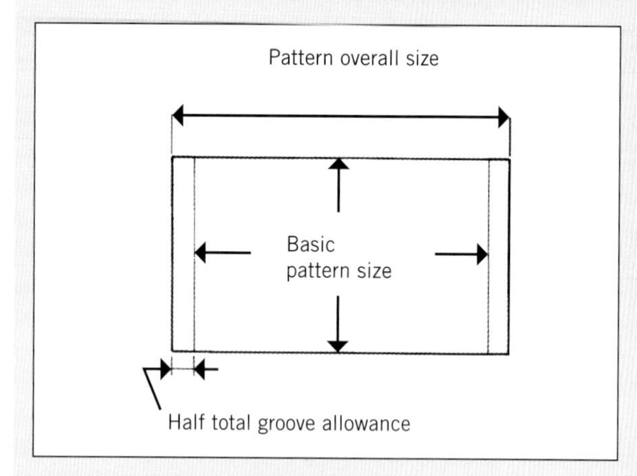

FIGURE 8.12 Overall pattern size

2 After folding the edge over on itself to slightly less than 180°, turn one of the pieces over so it faces up and turn the opposite piece down to allow them to interlock (see Figure 8.13).

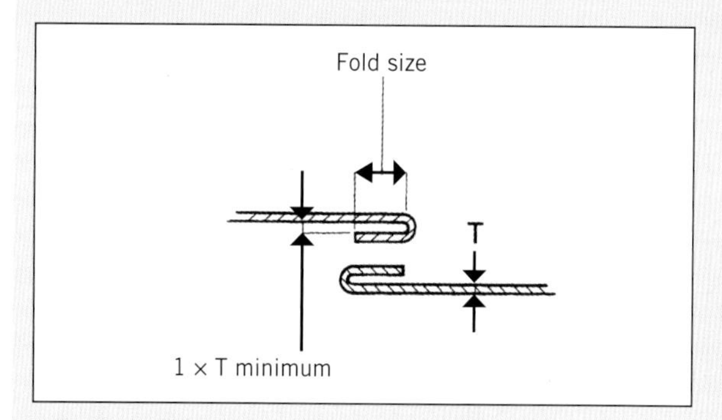

FIGURE 8.13 Edge folds – one up and one down

3 Ensuring that the laps are parallel with each other, the folded laps are interlocked and clinched at the top and then at the bottom, by applying light blows to a grooving tool with a small hammer. This process is carried out at certain intervals along the length of the seam to form the first stage of locking the joint (see Figures 8.14 and 8.15).

4 In small applications this process may be further applied to complete final locking of the joint in position (see Figure 8.16). It is advisable to centre punch the seam at intervals of approximately 50–150 mm to further lock the joint in place. This will ensure the seam does not slide or come apart during cutting and edge dressing.

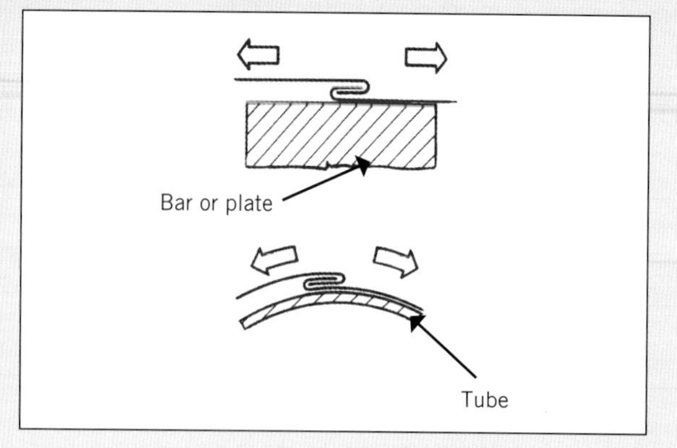

FIGURE 8.14 Support on solid base to interlock the folds

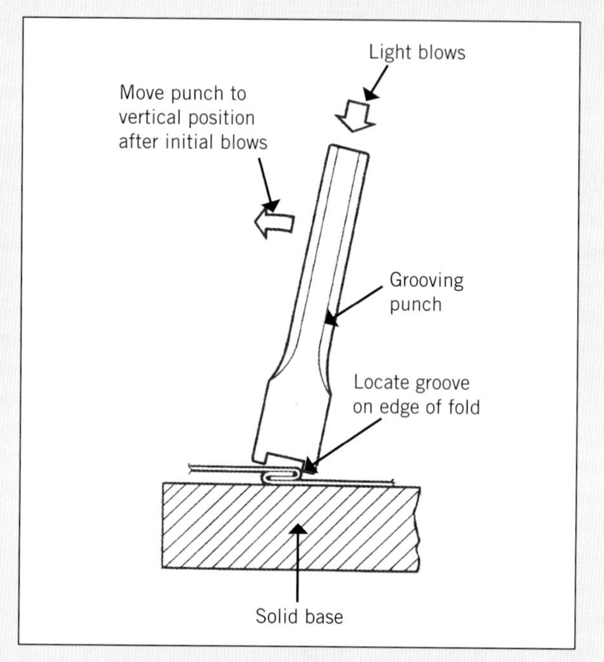

FIGURE 8.15 Clinching the groove

5 If additional sealing is required, the completed grooved seam joint may be soldered or silver brazed, depending on the materials being used. If the article is made from a material that cannot be soldered (i.e. Colorbond® or Zincalume®) the joint should be sealed with a neutral-cure silicone prior to interlocking the laps and before the first clinching procedure is undertaken. This will ensure the sealant is squeezed between the laps, resulting in a watertight joint.

6 To make the grooved seam:
 a Calculate the size of the joint fold.
 b Determine the total groove allowance.
 c Halve the total allowance.
 d Add half the total size of the joint fold to each end of the pattern.

>>

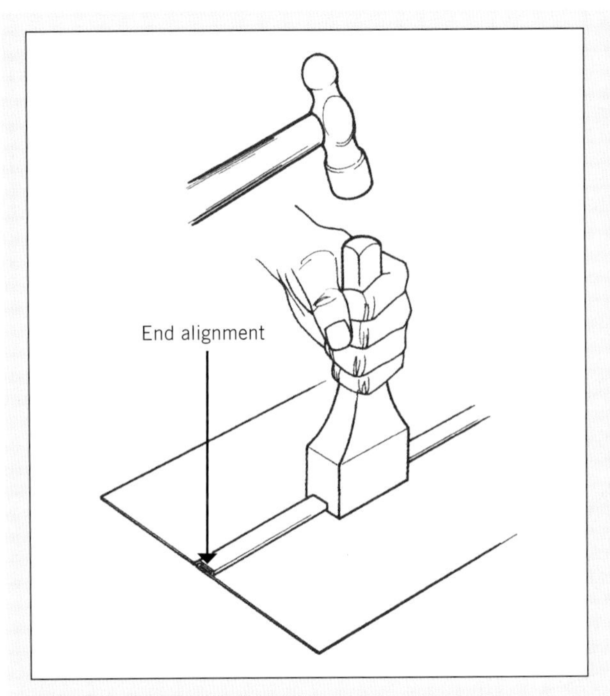

FIGURE 8.16 Grooving the joint

When using the grooved seam method to join one piece of metal to form an object such as a downpipe nozzle, it is essential to remember to make the folded edges in opposite directions so that when the rest of the job is bent or rolled up, the ends will interlock. The best technique to achieve this is to bend one side of the material in the bench folder. Once the first bend is complete, slide the material away from the bender without turning it over and reinsert it in the folder to make the second bend. The result is that the folds are now in opposite directions. Bending these folded edges should be performed before any other bends are made, although extreme care should be taken not to squash the joint folds while rolling or making other bends, as required for the job.

Countersunk grooved seam joints

This type of seam is similar to the external groove seam except the raised section of the seam is out of sight and recessed internally below the outside of the material (see Figure 8.17). The countersunk grooved seam is used where aesthetics is important as it produces a flush joint, which eliminates raised seams that may jam when slid together.

FIGURE 8.17 Rectangular downpipe/spouting with internal grooved joint

In the plumbing industry this type of joint is mostly used to:

- manufacture tanks
- create downpipes
- make downpipe nozzles.

Making the countersunk grooved seam joint

The procedure for making this type of joint is similar to external groove seam joints, up until the joint is to be locked with a grooving tool. At this stage the interlocked seam is placed on a slotted bar that is comparable to the groove in the punching tool. The seam is dressed into the slot in the bar with a mallet or dressing tool to prevent bruising or marking the metal.

Peined-down joints

The peined-down joint is generally used to join the preformed sheet metal sections of square or round tanks (see Figure 8.18). The joint is not waterproof by itself and requires sealing with an approved sealant when used on tanks. This joint is made up of flanges (see Figure 8.19), and when making allowance for laps, one section has a single flange matching the size or width of the lap for the joint and the second section has a flange that is twice the width of the joint lap, with half of the lap dressed or bent up to 90°. Each of the components is brought together and the upstanding lap is dressed over the single flange on the other section.

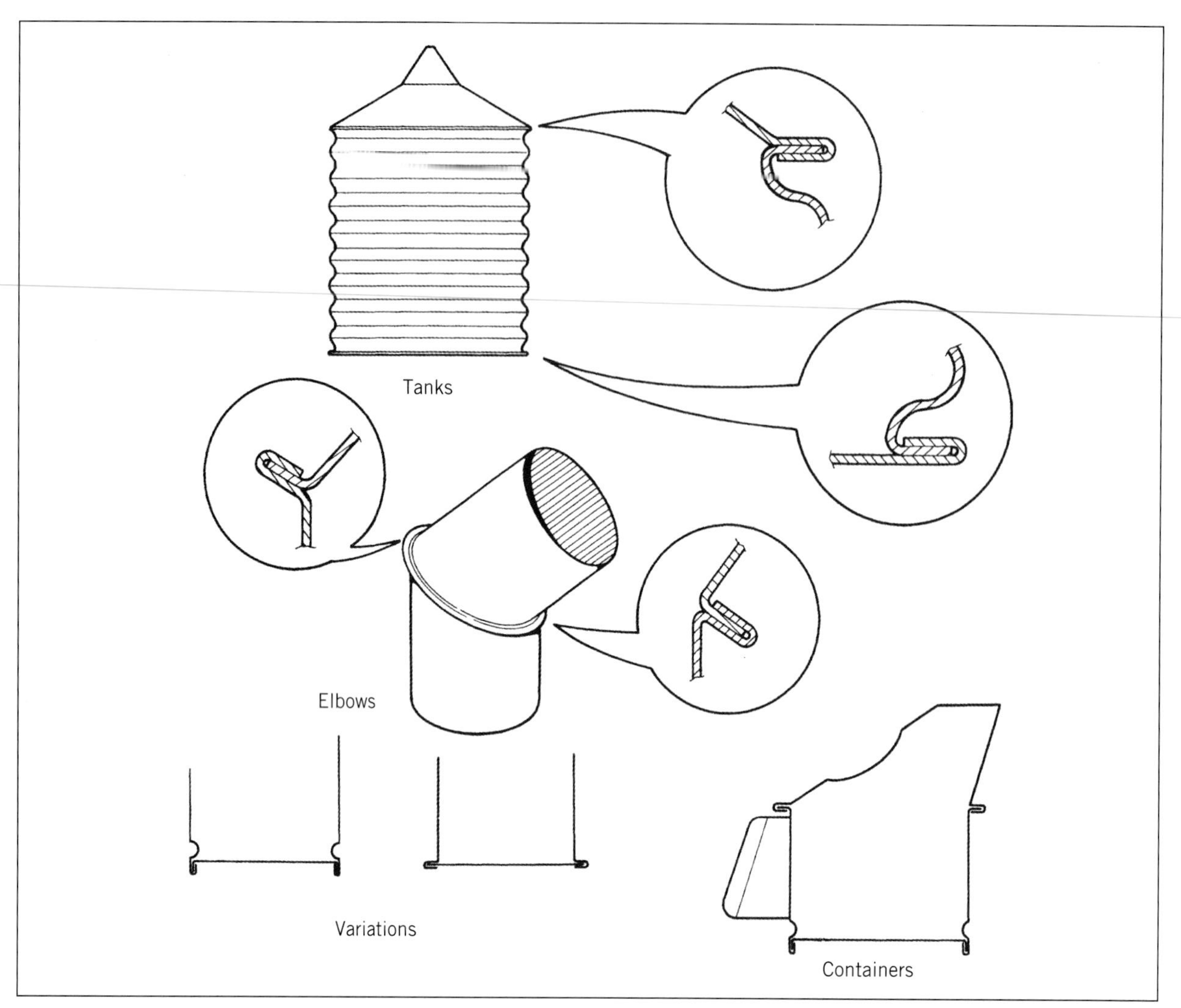

FIGURE 8.18 Peined-down joint applications

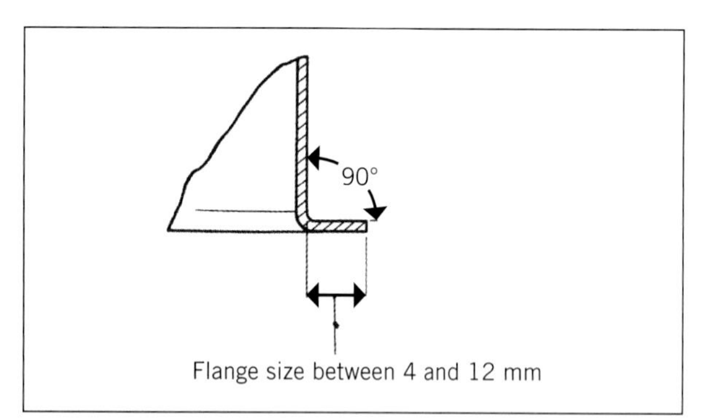

FIGURE 8.19 Flange size

HOW TO

MAKE A PEINED-DOWN JOINT

1. From the supplied plans, mark out and cut the material into shape, ensuring that adequate material is allowed for the laps.
2. Turn out the flange on the body section of the job using a hammer and a steel hand dolly (see Figure 8.27 (c), below) or a burring machine to achieve an angle of 90° (see Figure 8.20).
3. Dress and round up the cylinder. Measure the outside diameter in several places and average the measurement to ensure the cylinder is round (see Figure 8.21).

>>

>>

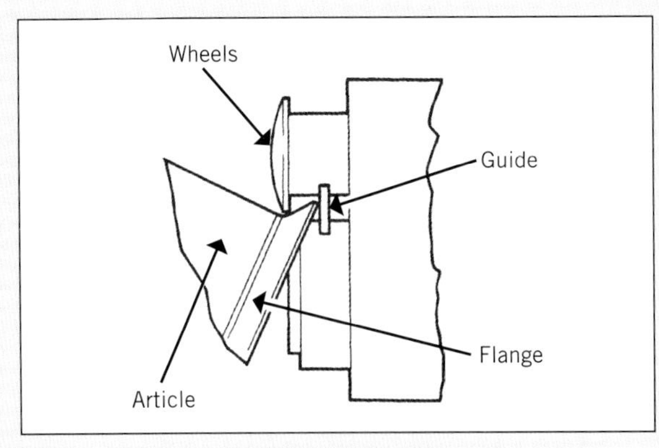

FIGURE 8.20 Turning the flange on the burring machine

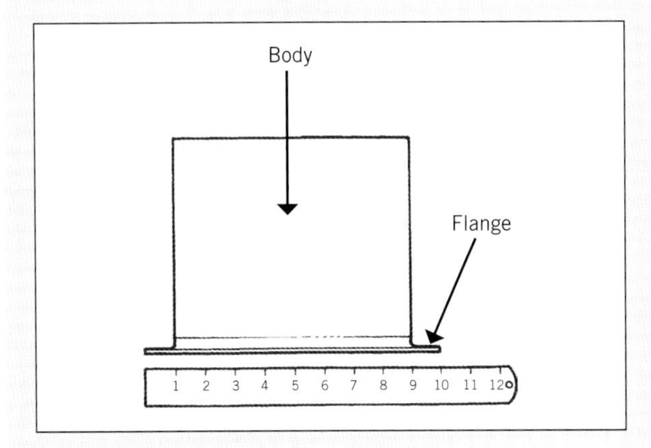

FIGURE 8.21 Measure the outside diameter

4 Cut the base to the average diameter measured previously, remembering to add two flange widths (see Figure 8.22).

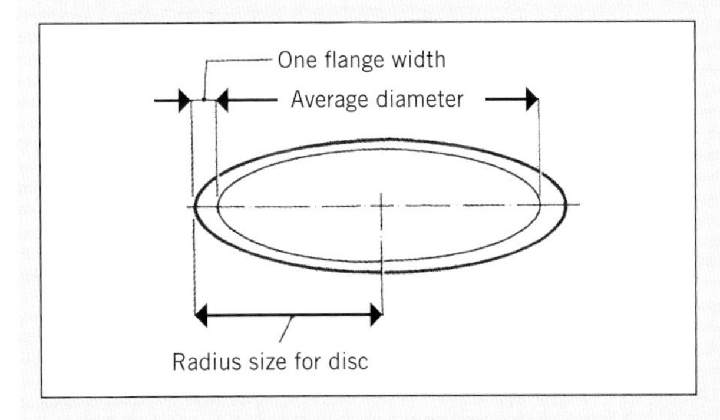

FIGURE 8.22 Disc size

5 Turn up one flange width using the burring machine (see Figure 8.23).
6 Fit the end to the body. A snug fit should allow the end to rotate without being too loose (see Figure 8.24).
7 Ensure the job is properly supported on a solid base such as an anvil or steel hand dolly. Using a cross-pein hammer, dress or pein down the flange on the end to approximately 35° all the

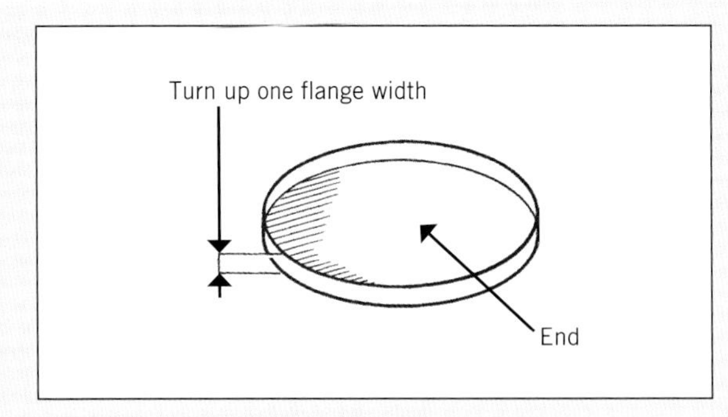

FIGURE 8.23 End flanging

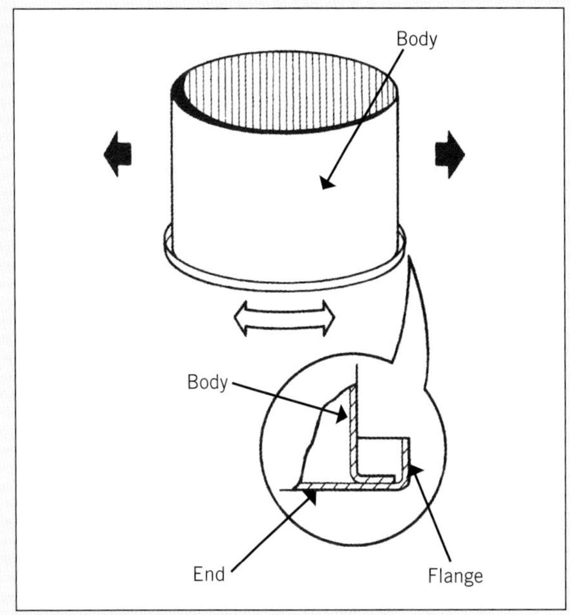

FIGURE 8.24 Fitting the end to the body

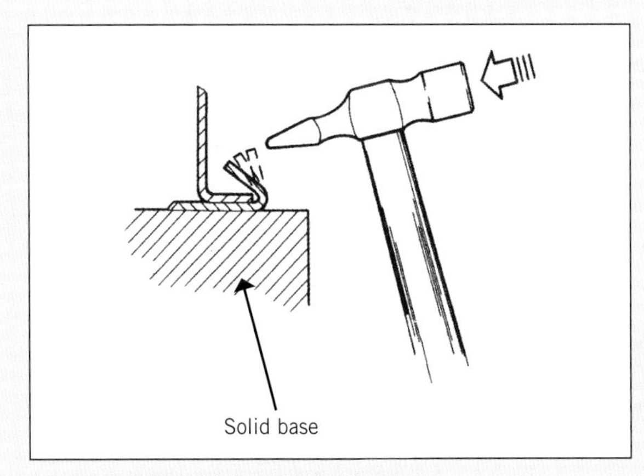

FIGURE 8.25 Dressing down the joint

way around, ensuring that the body of the article remains central in the disc (see Figure 8.25).

8 The upper lap can then be tightened by using a peining down machine (see Figure 8.26), or hand finished by tightening with a cross-pein hammer while the article is supported on a solid base.

>>

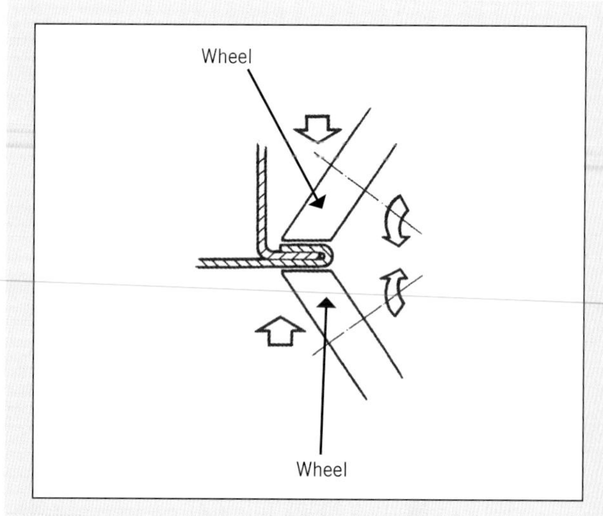

FIGURE 8.26 Tightening down the joint using a peining down machine

9 When the peining down is complete, the joint may be soldered or silver brazed if the material is compatible with these processes (e.g. galvanised mild steel or copper). If the article is manufactured from a material that cannot be soldered (i.e. Colorbond® or Zincalume®) the joint should be sealed with a neutral-cure silicone prior to the body being fitted to the base and before the first dressing procedure is undertaken. This will ensure that the sealant is squeezed between the laps, resulting in a watertight joint.

Alternative ways of locking the joint include the following:

1 Place the job in the bending machine and use the clamping pressure to complete the joint (see Figure 8.27 (a)).
2 With the article still on a solid base the joint can also be tightened and finished by hand with a cross-pein hammer and a buffer (see Figure 8.27 (b)).
3 Finish the joint by hand using a hammer and steel hand dolly (see Figure 8.27 (c)).

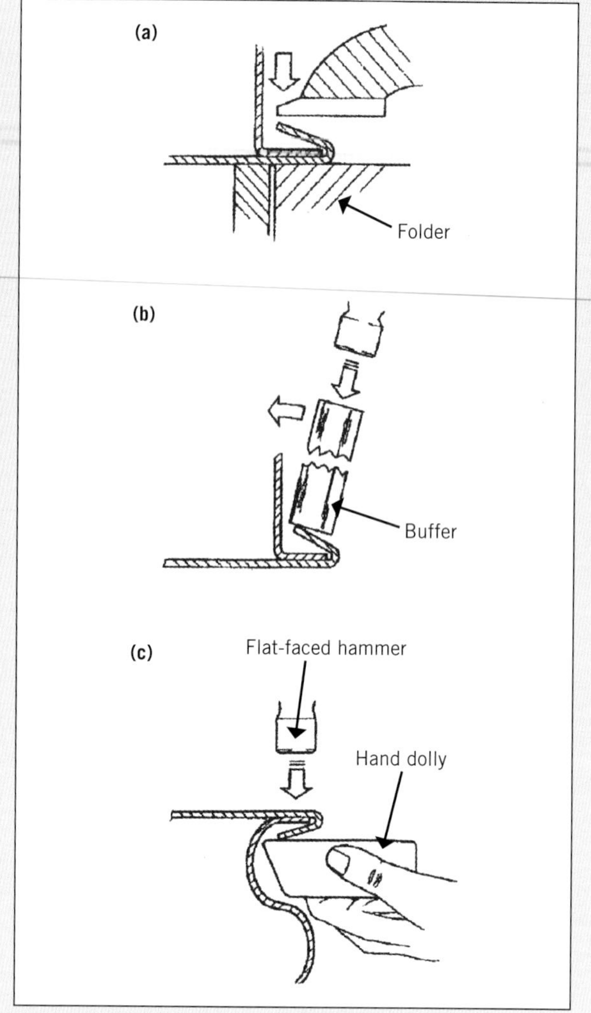

FIGURE 8.27 Alternative ways of locking the joint: (a) Using clamping pressure from a folding machine; (b) Using a hammer and buffer; (c) Using a hammer and steel hand dolly

Gloves should be worn to protect your hands against cuts and abrasions from sharp edges.

Hearing and eye protection should be worn while undertaking dressing of sheet metal as this is a noisy process and sheet material may become a projectile causing injury.

Knocked-up joints

This joint is used to attach ends to cylindrical and square projects, and is used mainly for the construction of tanks (see Figure 8.28).

The first stage of the knocked-up joint is similar to the peined-down joint, with a further step of turning up the protruding peined-down lap to the side of the project. This is known as the knocked-up stage.

The lap allowances are the same as the peined-down joint, which is equal to three times the flange size.

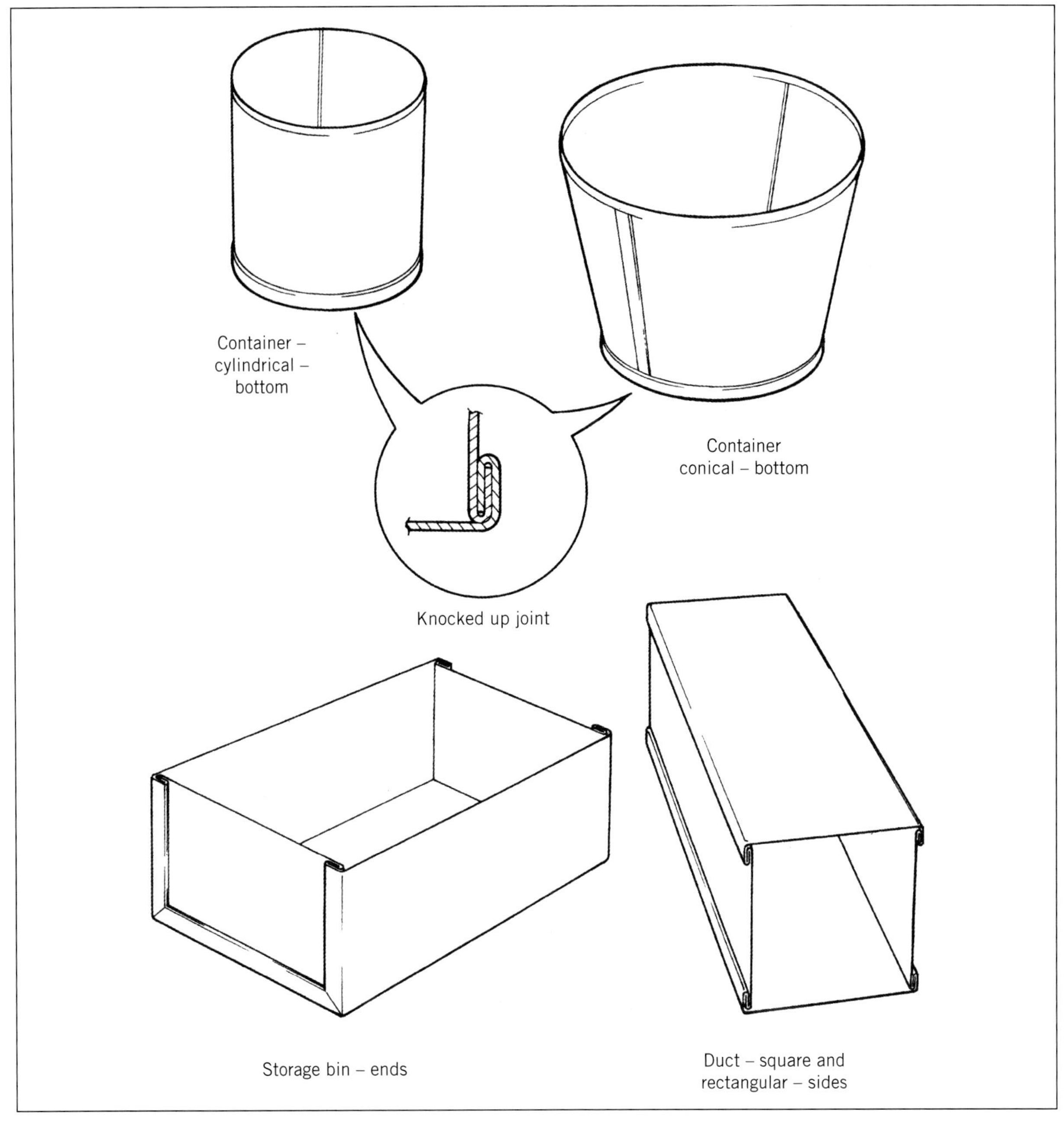

FIGURE 8.28 Various joint applications

HOW TO

MAKE A KNOCKED-UP JOINT

1. The preliminary stages of this joint construction are the same as the basic peined-down joint, but the difference is that one thickness of metal is deducted from the flange size on the end to allow movement when turning up the joint (see Figures 8.29 and 8.30).
2. The lap length on the end remains the same, as the turn-up bend is shorter. This allows room for the joint to stretch around when it is knocked up.

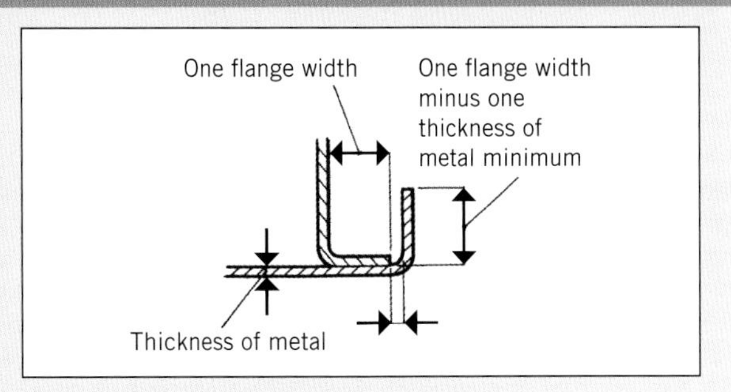

FIGURE 8.29 Size difference on end piece

>>

>>

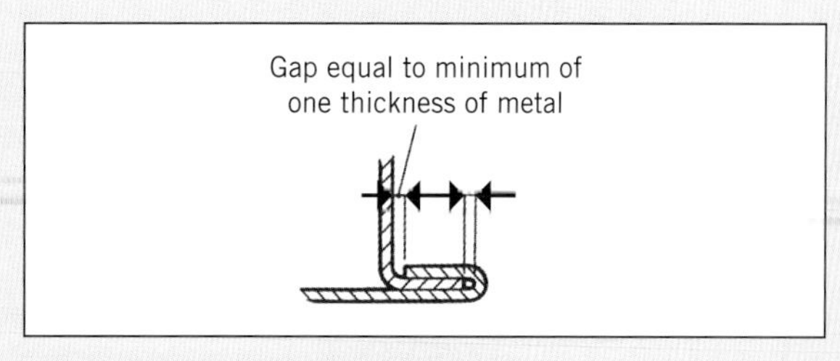

FIGURE 8.30 Preparing the joint

3 The peined-down joint is now bent through to an angle of 30°. This can be achieved with a hammer and steel hand dolly, by using folders and clamps, or with peining machines in the case of square shaped forms with straight joints (see Figure 8.31).
4 Once the peined-down joint has been bent up through 30°, the job should be positioned on a suitable solid base and the edge gradually dressed down (see Figure 8.32). The dressing down should

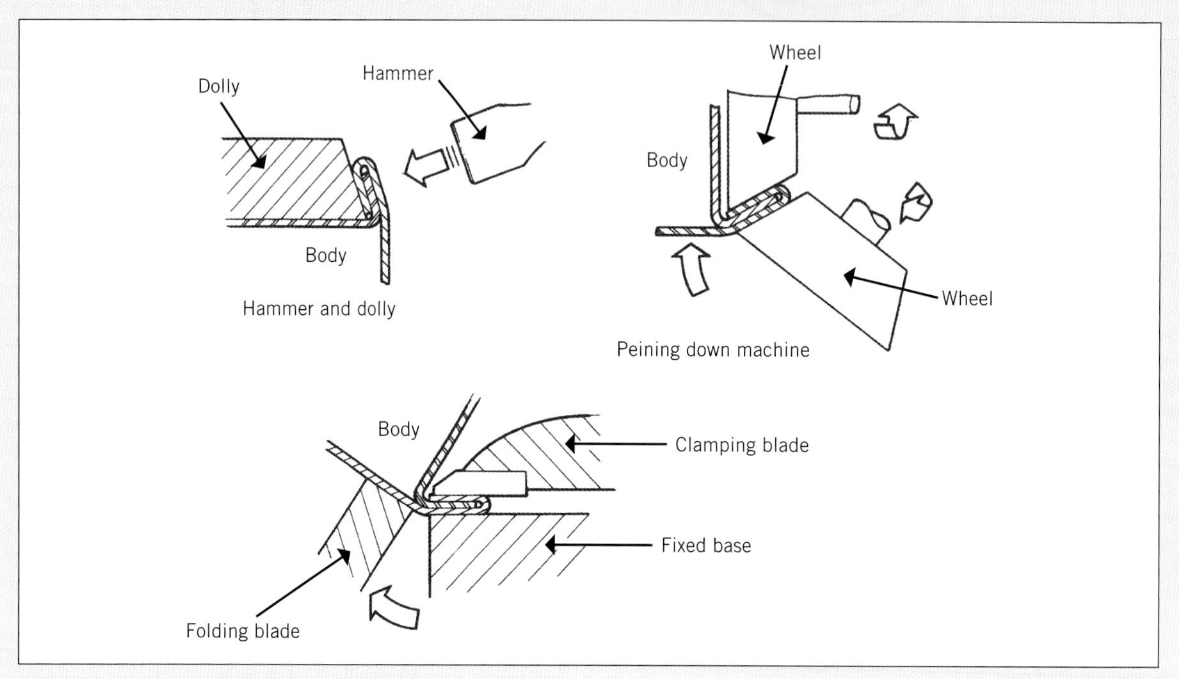

FIGURE 8.31 Peining down through 30°: (a) Using a hammer and dolly; (b) Using a peining down machine; (c) Using a clamp

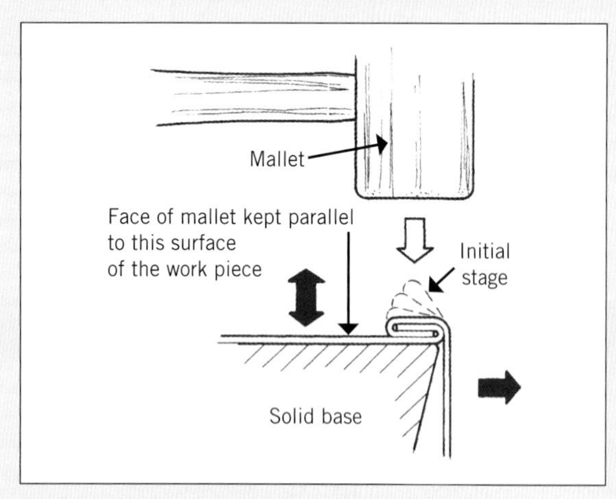

FIGURE 8.32 Knocking up the joint

be done gradually and in stages around or along the joint to allow the metal to reduce and assume the new position.
5 Some distortion may result after knocking up the joint. Although the more gradual the knocking up, the less likely this is to occur, it may be necessary to correct the base of the form by removing the distortion.
6 Correcting can be achieved by placing the base of the form on a solid steel support. Then, using a 'dubber' and a hammer, dress the edge of the joint down until it is parallel with the base (see Figure 8.33).

>>

>>

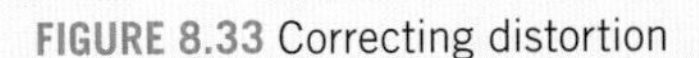

FIGURE 8.33 Correcting distortion

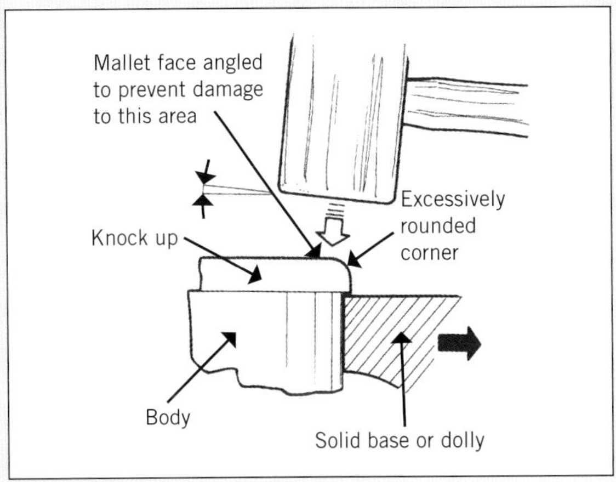

FIGURE 8.34 Reducing the size of round corners

It is also important for the aesthetics of the form to reduce the radius of excessively round edges/corners by dressing down the bottom of the knock-up (see Figure 8.34). Ensure the job is supported on a solid base during this operation.

Pittsburgh lock seam joints

The Pittsburgh seam is a corner joint system that uses a roll-forming machine called a Pittsburgh lock former. The machine is used to manufacture this joint where it is required in large quantities or larger applications, and has a series of rollers that gradually shape the lock form into the female section of the joint (see Figures 8.35 and 8.36).

FIGURE 8.35 The Pittsburgh lock-forming machine

The lap allowances for the seam will vary depending on the machine and its setting. Generally, allowances are 25–30 mm for the multiple bend section of the joint and up to 6–8 mm for the right-angle bend portion of the joint.

The Pittsburgh seam is used mainly for mechanical ventilation and air conditioning ductwork (see Figure 8.37).

Sliding cleat joints

Sliding cleat joints (see Figure 8.38) are different from other integral joining methods, as most integral joints are permanent. Cleat joints consist of a folded section forming a channel known as the cleat. The cleat interlocks with folds on the form, which allows this type of joint to be dismantled. The lap sizes vary depending on the type of cleat joint and the application or the material being used.

There are a several designs and applications of cleat joints and they are usually made on a roll-forming machine. Cleats that have a simple design can be manufactured by hand with the aid of a sheet metal folder in the workshop.

Simple lap joints

This type of joint is the most basic used for sheet metal, as its strength relies on mechanical fasteners or other techniques such as welding or soldering. Rivets or other mechanical fasteners such as screws and bolts should be used to support the joint and provide extra strength when the joint is loaded, such as during tank

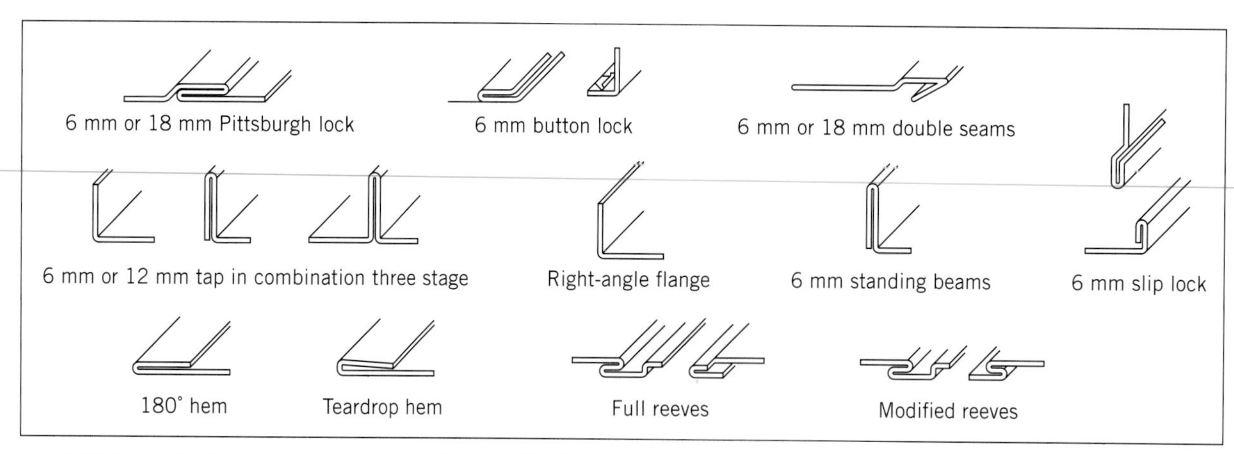

FIGURE 8.36 Joints that can be manufactured by the Pittsburgh lock-forming machine

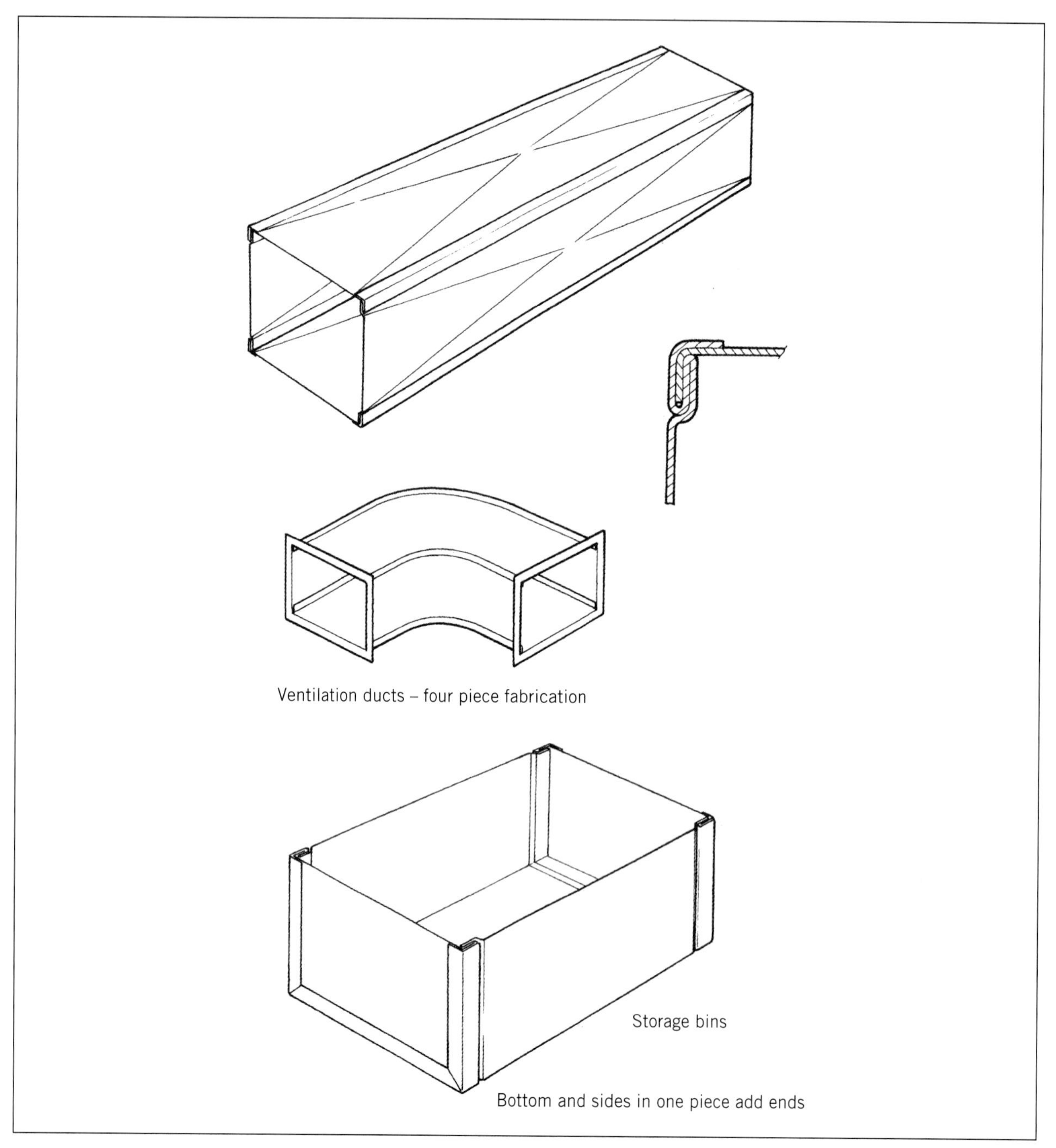

FIGURE 8.37 Uses for Pittsburgh lock joints

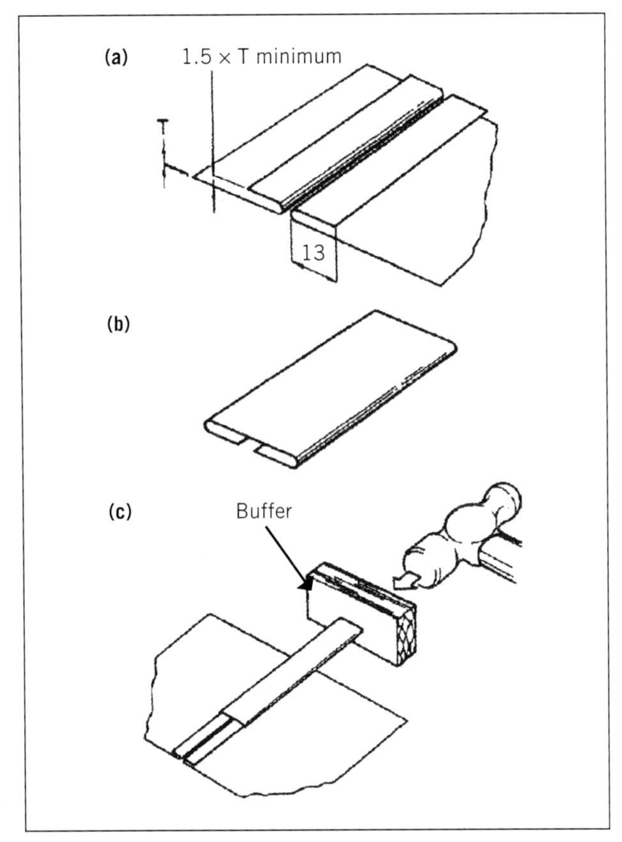

FIGURE 8.38 Driving the cleat: (a) Prepare the end folds; (b) Create the drive slip cleat; (c) Drive on the cleat

manufacture. The joint may or may not be sealed and this will depend on its exposure to the elements.

Resistance spot welding

Spot welding is one of a group of resistance welding processes in which two or more metal pieces are lap joined. It is used on sheet metal ranging from very thin to thick sections, and the process is ideal for joining a wide range of materials.

It uses specially shaped copper alloy electrodes to apply pressure while passing an electrical current through the metal being joined. Resistance to this current develops the heat between two pieces of metal, and causes them to melt and fuse together. The amount of time that the current passes through the material is an important factor and is usually determined by the type and thickness of the material. It is important to ensure that the current is not passing through for too long, as this will cause overheating and possibly burn through the work piece, resulting in an inferior weld or a completely ruined job.

Spot welding can be used to provide a strong joining method for a wide range of materials, including stainless steel, nickel alloys, aluminium alloys, Zincalume® and galvanised mild steel.

There are generally two types of spot-welding machines, which are the rocker arm and the press. The rocker arm machine has a pivoted or rocking upper electrode arm, which is activated by pneumatic power or by the operator pushing down on the pedal. The press spot welder has an upper ram that comes down in a linear motion to perform the spot weld. The portable hand-operated spot welder is the most commonly used for smaller projects and is found in fabrication workshops (see Figure 8.39).

FIGURE 8.39 Spot welder

Soft-soldered joints

Soft soldering is a method of joining materials with the use of heat in conjunction with a low-melting-point tin–lead alloy at temperatures below 420°C. This alloy is commonly known as 'soft solder' to distinguish it from hard solders and brazing alloys. Soft solders are more flexible and have a lower melting point and lower tensile strength than hard solders (see Figure 8.40). Tin has a melting point of 232°C and lead has a melting point of 327°C. A distinguishing feature of tin–lead alloys is that they have a melting point below the lead component of the alloy. Melting points can also be lowered by the addition of other elements to the alloy, such as cadmium and bismuth.

Soft soldering in the plumbing industry

Traditional soft soldering is not commonly used in industry today, as there is increased use of Zincalume®, Colorbond® and plastic products. The zinc-and-aluminium-coated steel cannot be soldered due to the aluminium in the alloy.

Although there is a major decline in use, it is still an important skill and is used to fabricate, patch and repair existing work, such as copper downpipes and flashings, and elements of heritage buildings. Soft solder can be used to join the following materials:

- black mild steel
- brass
- bronze

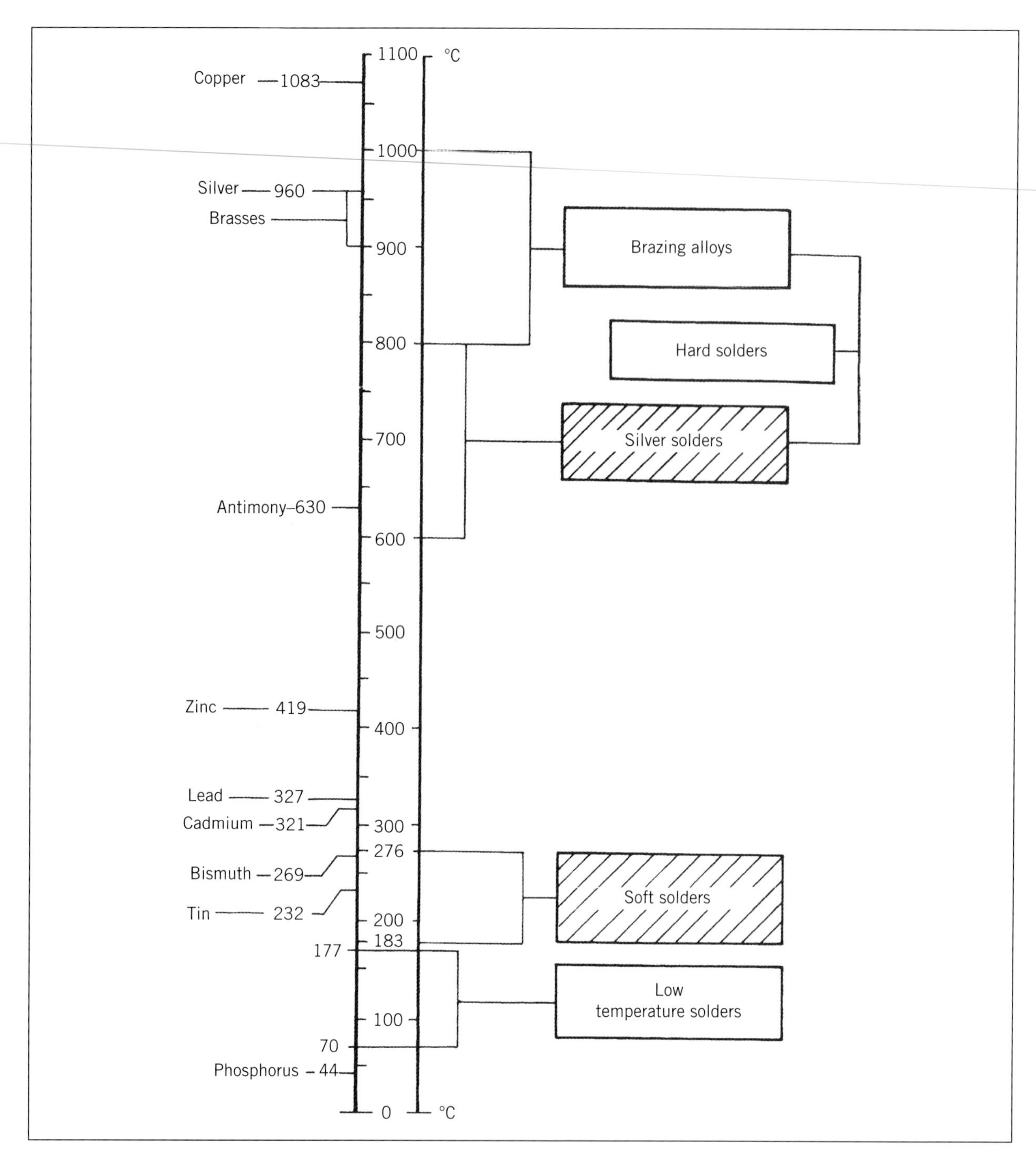

FIGURE 8.40 Melting temperatures for hard and soft solders

- copper
- galvanised mild steel
- lead
- stainless steel
- tin plate
- zinc.

Soft solder has several qualities, which include:

- *economy* – it's economical when joining a large range of sheet metal products and types
- *strength* – provided the joint has adequate laps and is of good design
- *water tightness* – it provides a water-sealed joint
- *low melting temperature* – which reduces the risk of major distortion or melting of the sheet metal being joined.

Tin–lead alloys are available in varying proportions, which is expressed as a ratio in relation to the percentage content, with the tin component always stated first. Soft solder is available commercially in the following ratios (see Figure 8.41):

- 50:50 solder, or *tinsman's* solder, is the most commonly used solder in plumbing applications.

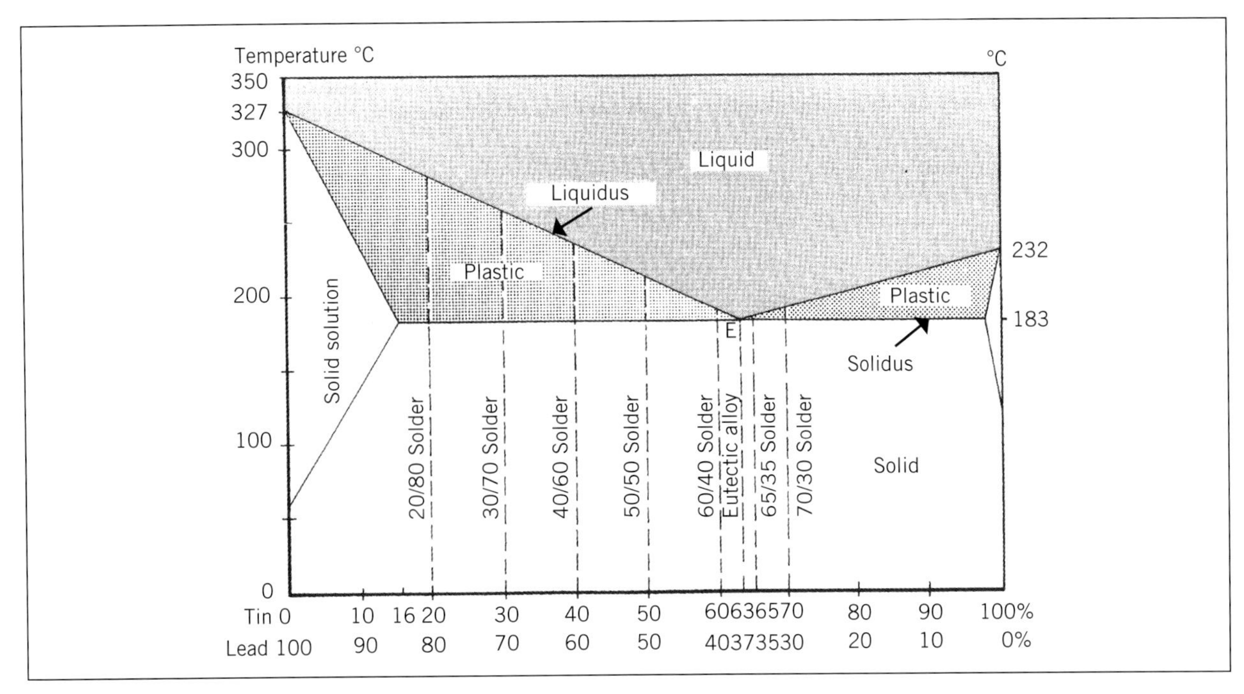

FIGURE 8.41 Melting characteristics of tin–lead solder

- 60:40 solder, or *fine* solder, is quick setting and has a higher strength than 50:50 solder.
- 40:60 solder, *coarse* solder, is generally used for loading or wiping (a process very rarely used these days). It also possesses slow-setting characteristics to extend its workability.

Soft solders are available in a variety of forms and can be purchased as sticks, bars, solid wire, core wire, pastes and creams. The pastes and creams are a combination of flux and powdered solder.

The tin:lead ratio of 63:37 forms what is known as 'eutectic' solder. At 183°C, this solder has the lowest melting point of all solders. All solders begin to melt at 183°C and are in a plastic state until the solder completely melts. The plastic range of a solder is the temperature difference between its solid and liquid states. The plastic state may also relate to a time factor, as a solder with a small plastic range in temperature will take less time to melt and less time to become solid than a solder with a larger plastic range. Solder with a larger plastic range can be more workable when wiping the excess from joints.

Note: Due to health and environmental concerns, solders containing lead cannot be used on items related to water supply. Care should also be applied to using lead flashing on roofs while collecting roof water that is to be used for human consumption. Consequently, a lead-free solder has been developed. The use of this solder may need to be considered when manufacturing tanks that will contain potable water for drinking purposes, as *AS/NZS 3500.1 Plumbing and drainage, Part 1: Water services* states that solder 'shall not contain more than 0.1% Lead by weight'.

- AS/NZS 3500.1 Plumbing and drainage, Part 1: Water services

Soldering irons

A soldering iron is a tool that is used to melt solder and fuse it to a parent metal to make a soft-soldered joint. Soldering irons are available in a range of shapes and sizes to suit the work application and the amount of heat retention required for the job (see Figure 8.42). The soldering iron has not changed a great deal over the years and has a copper head or bit, forged onto a steel shank, with an insulating hardwood handle that is protected from cracking by the fitting of a ferrule. The copper bit of the soldering iron is cubic shaped at the top to help retain the heat and has a pyramid-shaped work face to enable good contact with the work surface (see Figure 8.43).

Copper is the ideal material for soldering iron heads. The physical properties of copper suit the soldering process because of the following:

- Copper is well suited to tin–lead alloys and possesses good adhesion qualities with the solder. This in turn facilitates a pathway for the solder to the work piece.
- Copper is an exceptional conductor of heat and effectively passes heat onto the work piece.
- Copper has an ideal mass and will hold sufficient heat to carry out soldering.
- Copper is a soft metal that becomes malleable upon heating, which makes it easy to form into shape.

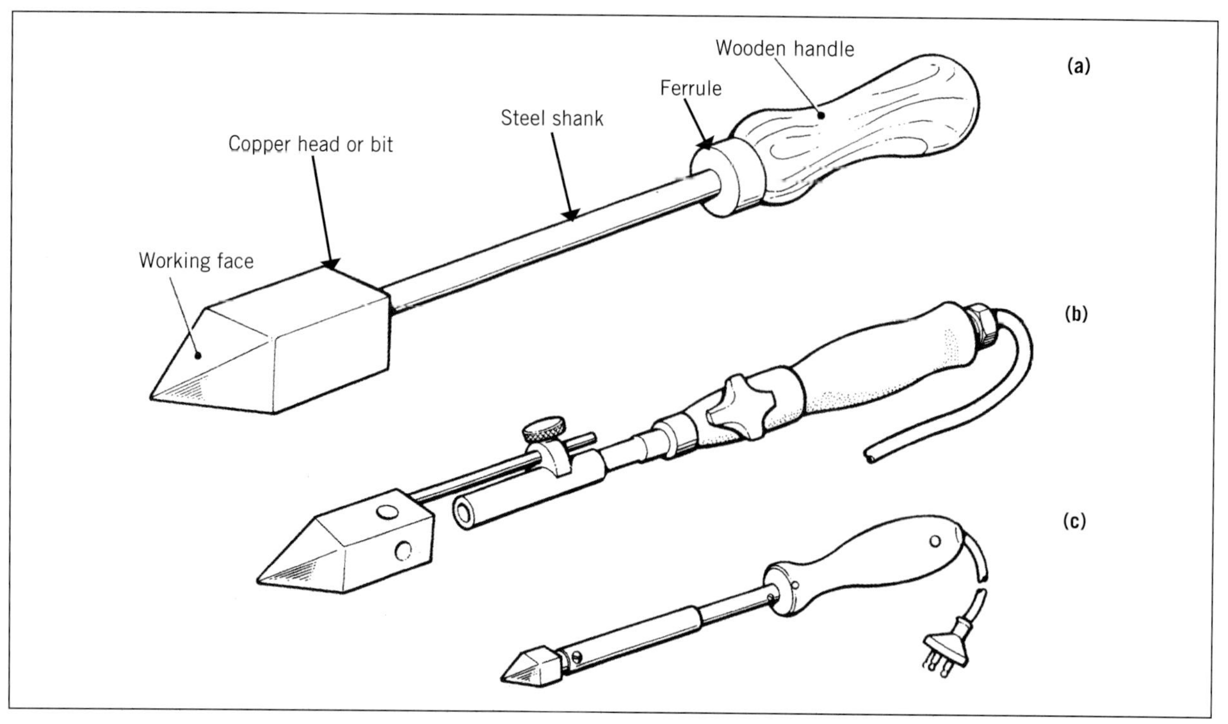

FIGURE 8.42 Typical soldering irons: (a) Traditional soldering iron; (b) Gas soldering iron; (c) Electric soldering iron

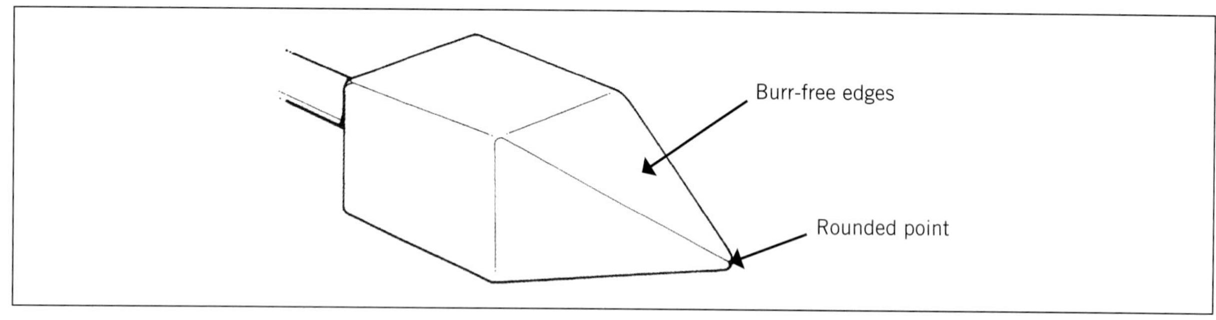

FIGURE 8.43 Soldering iron head – standard bit

Soldering irons may be categorised as follows:

- *Traditional irons* – these need to be reheated or replaced by another pre-heated iron throughout the soldering operation.
- *Gas irons* – these are heated by an attached gas heat source throughout the soldering procedure.
- *Electric irons* – these are heated via a 240V power supply.

Methods of heating the iron

There are two main options for heating soldering irons: by flame or by electric element.

Traditional soldering irons are usually heated by a gas furnace with a flame produced from a combustible fuel such as natural gas or LPG (see Figure 8.44). An alternative is a portable gas supply such as LPG in a small cylinder (see Figure 8.45).

FIGURE 8.44 Soldering iron furnace for use in a workshop

Safety precautions when heating the iron

Always keep the following in mind when heating the iron:

- Manufacturer's instructions must be followed at all times.
- Gas connections on LPG fuel cylinders have left-hand threads requiring them to be tightened in an anti-clockwise direction and loosened in a clockwise direction.
- All connections to the bottle gauges, hoses and hand pieces on portable gas equipment must be

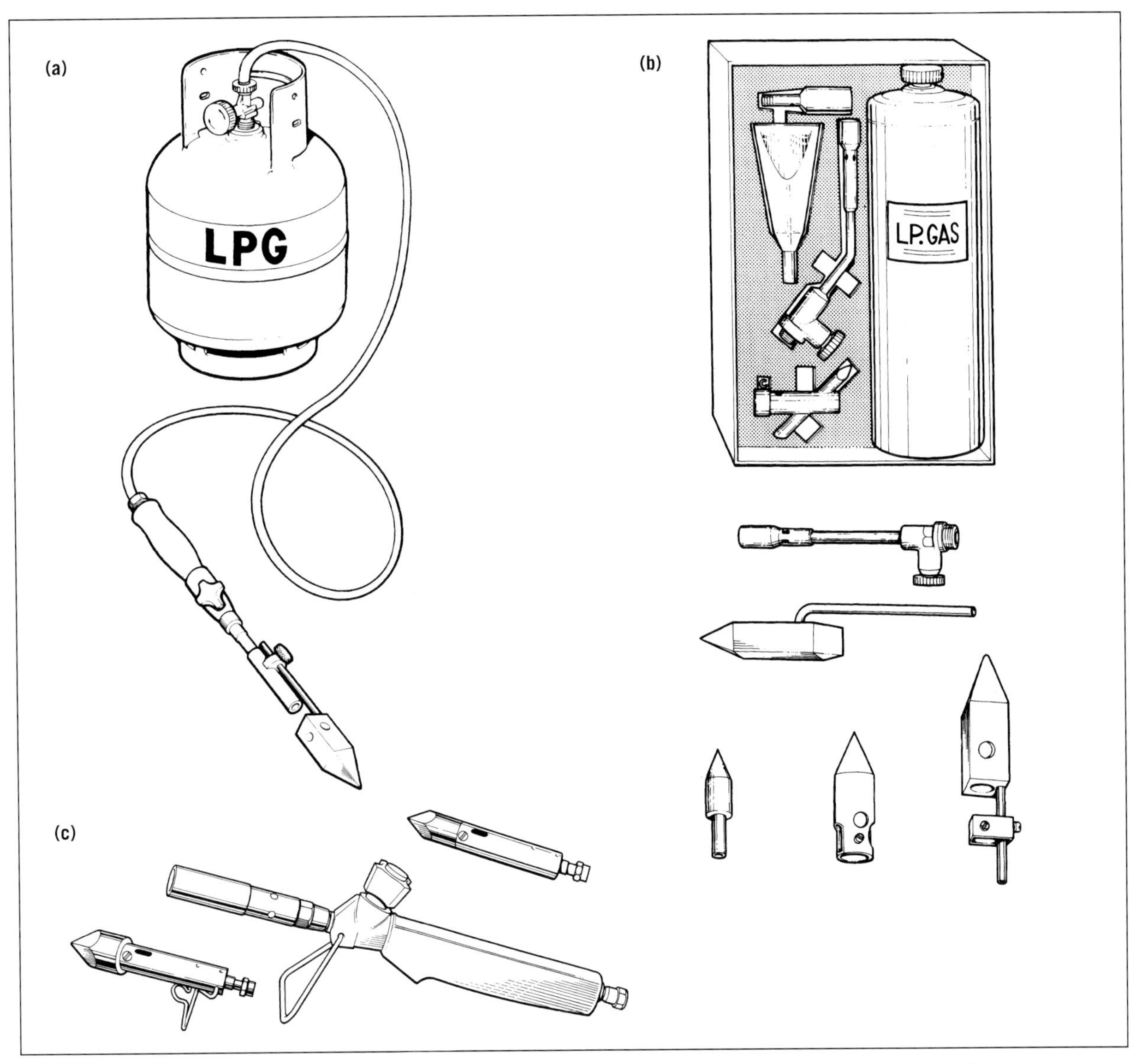

FIGURE 8.45 Alternatives for portable gas heating of soldering irons: (a) LPG-heated soldering arrangement; (b) LPG kit and attachments; (c) LPG torch, burner and bit attachments

tested with soapy water before use to ensure there are no leaks in the system.

- Do not leave valves on for extended periods before lighting, as LPG is a heavy gas and has a relative density greater than air and will accumulate in low lying areas, posing a potential fire or explosion hazard.
- Do not direct gas or flames in the direction of others. When soldering operations are completed and if the workshop is to be left unattended, all gas valves should be turned off.
- Always check leads and cables for cuts or splits when using electric irons and ensure a residual current device (RCD) is being used in conjunction with the iron.
- Ensure that leads and cables are not dragged across sharp metal edges as this can lead to exposed wires.
- Uncoil the leads and cables completely during use to prevent them from overheating and melting.
- Power points should be turned off until the iron is plugged in and before it is unplugged.

Fluxes

Fluxes are used to assist the bonding process between solder alloy and the metals being soldered. All metals will oxidise to some extent, even the most corrosion resistant. Oxides must be removed by mechanical abrasion such as cleaning copper with steel wool or emery cloth, or by using a flux, before soft solder will bond with the work piece.

Always prepare and use fluxes in a well-ventilated area and where possible use mechanical ventilation. Ensure they are stored away from heat, flames or any other potential sources of ignition. Keep them in properly marked and appropriate plastic containers in a designated storage area that is cool and dry and away from food, people, machinery and tools. Always wear PPE to protect the skin, eyes and lungs.

All inorganic fluxes are highly toxic and corrosive. Safety Data Sheets should be consulted before handling these chemicals. Brushes should always be used to apply these fluxes.

There are three primary roles that fluxes perform:

- They dissolve the oxides on the surface of the metal.
- They prevent further oxides from reforming during the soldering process.
- They lower the surface tension of the molten solder, which assists the flow of the solder and facilitates better capillary action.

There are two categories of fluxes that are used for soldering (see Tables 8.2 and 8.3):

- inorganic or corrosive fluxes (active)
- organic or non-corrosive fluxes (passive).

Inorganic fluxes

Inorganic fluxes are acidic and remove the oxides by dissolving them from the material. These fluxes are usually applied directly to the surface of the metal with a brush. Before the flux is applied, the metal surface should be cleaned of all protective oils, corrosion, greases and waxes. The cleanliness of a job determines the quality of the finished product.

Care should be taken not to apply excessive amounts of flux as it can become trapped in the laps of the joint and cause corrosion in the future or cause unsightly staining of the surfaces outside the coverage area of the solder. Inorganic fluxes should be cleaned off with the use of a wet cotton cloth.

TABLE 8.2 Table of fluxes suitable for protective coated steels

Coated steels	Coating application method during manufacture	Inorganic flux	Organic flux	Comments
Galvanised mild steel	Zinc hot dipped	Hydrochloric/muriatic acid (Raw spirits)		Dilute in water if required for new sheets
Galvabond	Zinc hot dipped	Zinc chloride (Killed spirits)		Commercial fluxes are available
Zinc anneal	Zinc hot dipped heated and rolled	Zinc chloride or killed spirits and raw spirits mixed		Commercial fluxes are available
Tin plate	Tin pure hot dipped eloctroplate	Zinc chloride (Killed spirits)	Resin	Commercial fluxes are available

TABLE 8.3 Table of fluxes for common metals when soft soldering

Metal being soldered	Type of inorganic flux	Type of organic flux	Comments
Brass: Copper/zinc alloy	Killed spirits, Sal-ammoniac	Resin, Tallow	Commercial flux available
Bronzes: —Tin —Silicon —Phosphorus	Killed spirits	Resin	Mechanical removal of oxides is necessary on surfaces exposed to a corrosive environment for long periods
Cadmium	Killed spirits	Resin	Commercial flux available
Copper	Killed spirits, Sal-ammoniac	Resin	Commercial flux available
Lead	Killed spirits	Tallow, Resin	Oxides must be scraped off regardless of flux being used
Monel: Nickel/Copper alloy	Killed spirits		Commercial flux available
Nickel	Killed spirits	Resin	Commercial flux available
Pewter: Tin/Zinc alloy	Killed spirits	Resin	Mechanical removal of oxides is recommended
Stainless steel: Chromium/Nickel content	Phosphoric acid or 50/50 Raw spirits and Killed spirits mix		Commercial flux available
Steel	Killed spirits		Residue removal is essential Neutralise
Tin	Killed spirits	Resin	Commercial flux available
Zinc	Muriatic acid		Dilute with water for new zinc surfaces

Inorganic fluxes commonly used for soft soldering

The most common inorganic fluxes used for soft soldering are:

- *hydrochloric acid* – effective as a flux on zinc and zinc-coated steel, and easily diluted when added to water, which reduces the aggressiveness of the chemical reaction on the work piece
- *zinc chloride* – best suited for use on mild steel, tin plate, copper, brass, bronze, lead and cadmium
- *phosphoric acid* – contains phosphate salts and is a thickish liquid used as a flux for stainless steel.

If acid is required to be diluted, the acid is always added to the water, as a violent reaction may occur if water is added to the acid and it may splash concentrated acid out of the container. Always refer to an SDS before use.

PPE must always be used when using inorganic fluxes as they are extremely corrosive and may cause burns and irritation to the eyes and skin and may cause severe irritation to the respiratory tract if inhaled.

Organic fluxes

Organic fluxes are chemically inactive and require the metal to be cleaned by mechanical methods to remove oxides before they are applied. Organic fluxes protect the metal surfaces from oxidisation by excluding the atmosphere during the soldering process.

When the soldering joint is complete, organic fluxes should be cleaned off with a solvent such as white spirits or methylated spirits. Cleaning the surfaces on completion means:

- a neater finish will result at the seams
- the work piece can be handled without residue
- painting of the job will be easier.

The most common organic fluxes are as follows:

- *Resin fluxes* – these amber-brown substances are extracted from the sap of pine trees. This type of flux is found in resin-core wire solders. Resin is an ideal flux for soldering copper and its alloys, such as brass and bronze, and can be used to protect tin plate, cadmium, nickel and silver during soldering.
- *Tallow fluxes* – tallow fluxes are treated animal fats. They are used on lead pewter and some brasses.
- *Commercial or proprietary fluxes* – commercial fluxes in paste form are usually purchased in various-sized containers and tubes depending on the amount required. The common base for these products is resin with other organic substances added, such as fish oil and palm oil.

Preparing the iron

Preparation of the iron is essential to a good soldering job. An iron will need to be tinned if it is new, or if the tinned surface has deteriorated due to its service. It is best to inspect the iron before use, and if the tinned surface is not shiny with an even cover of solder over the tapered faces of the bit, or it has a burnt or dull appearance, the iron should be tinned again. Tinning provides a coating over the copper bit to enable it to transfer heat to the job, hold the solder and subsequently control the flow of solder on the work piece. Tinning the iron is achieved by carrying out the following steps.

HOW TO

TIN A NEW SOLDERING IRON

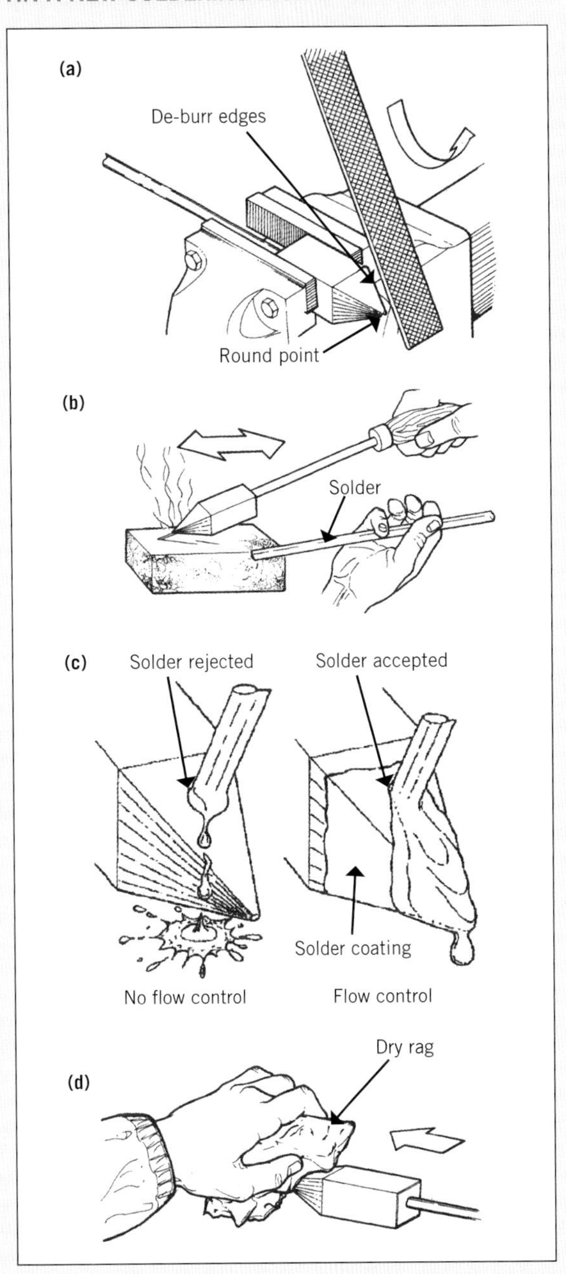

FIGURE 8.46 Tinning a new bit: (a) Heat the bit, then shape with a file; (b) Move bit back and forward and add solder; (c) Check the flow of solder on each face of the bit; (d) Remove excess flux and solder

1. Clean the new bit by filing it to remove any pitted faces and rough edges.
2. If the bit is hard to file, anneal by heating the iron until solder melts on it freely, then cool it by plunging it into cold water. Round off the point slightly with the file.

>>

>>

3 Pour a small amount of hydrochloric acid into a tray made from a scrap piece of galvanised mild steel.
4 Heat the iron to the appropriate temperature, when colours appear on the iron.
5 Test the temperature of the iron by dipping it in the pot of water; if it boils quickly and a squelching noise is heard, the iron is hot enough. It can also be tested by rubbing it on the galvanised steel tray.
6 Clean the faces of the iron in the tray of acid.
7 Place the stick of solder on each face of the iron and spread the solder over the face of the iron by rubbing it in the tray again.
8 Remove excess solder and flux by wiping with a clean cotton rag.

Note: Do not use a synthetic cloth as this will contaminate the tinned surface.

Soldering techniques

There are several items that need consideration before, during and after the soldering operations.

The following features are essential to create a successfully soldered job:

- selection and preparation of the soldering iron
- correct design and preparation of the joint
- correct selection and application of flux
- heating of the iron to the correct temperature
- correct orientation and manipulation of the soldering bit
- cleaning of the job and removal of excess flux on completion
- inspection of the completed job to ensure soundness of the joint.

Correct joint design

Simple lapped-type joints are ideally suited for joining or sealing with soft solder. There may be a need to support the joint with rivets prior to soldering. Lapped joints should be completely sweated through and well supported and close fitting to encourage the flow of solder and achieve good capillary action (see Figure 8.47).

Soldering is generally carried out as a sealant only and does not necessarily contribute to the strength of the joint; however, soldering may enhance the strength of the folded seam and, unless specified, the solder does not need to be sweated into the lap. Butt joints are unsuitable for soldering as they do not offer enough strength and will inevitably result in joint failure.

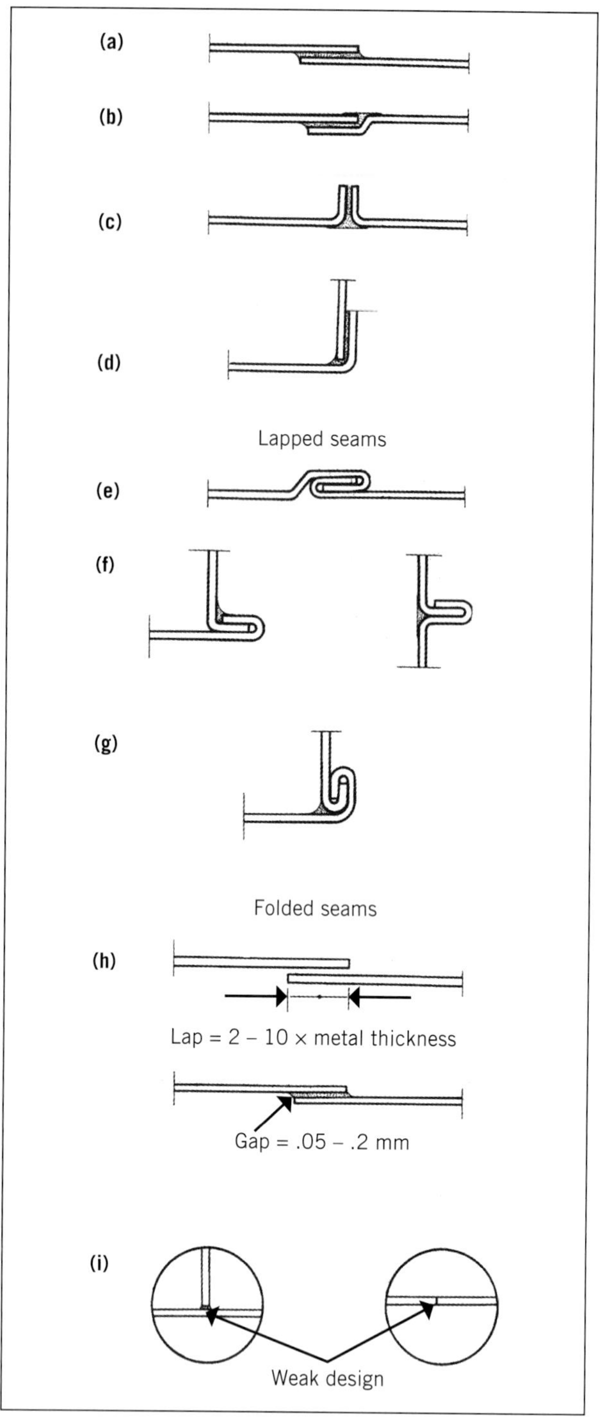

FIGURE 8.47 Correct joint design: (a) Lap; (b) Joggled lap; (c) Abutting lap; (d) Corner lap; (e) Grooved or lock seam; (f) Peined-down; (g) Knocked-up; (h) Gap and lap allowances; (i) Weak joints

Preparation of the joint

The most important aspect of any soldered joint is cleanliness. The surfaces that are to be joined should be cleaned thoroughly before they are assembled.

During manufacture and to help protect the product during storage, oils and waxes are applied to the surface of the metal. These oils and waxes must be removed by using an appropriate solvent. Surface oxides should also be removed, either with acids or mechanically using a wire brush, steel wool or abrasive cloth (see Figure 8.48).

Cleaning and inspecting the completed joint

The joint should be cleaned of excess flux and stains, and the job should be left clean and dry ready to be painted. The removal of flux will help with resistance to deterioration and prevent premature corrosion.

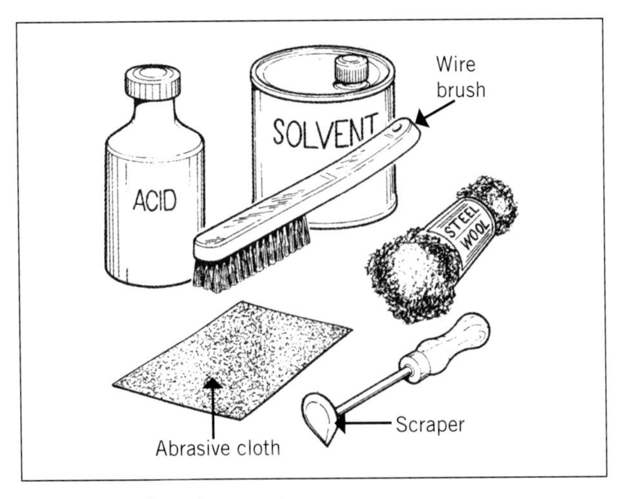

FIGURE 8.48 Cleaning agents

On completion of and after cleaning the joint, it should be inspected for soundness. A well-soldered joint should have the following characteristics:

- There should be good penetration between the laps.
- The joint should be well sealed with a neat smooth appearance.
- The solder on the upper section of the lap should be a thin layer of consistent width.

Hazards relating to soft soldering

As part of the process of safety self-regulation, anyone carrying out soft soldering should complete and follow a SWMS relating to the task, which will identify hazards, assess the risk of the hazards and manage the potential risks using a hierarchy of control.

Hazards that may be encountered in soft soldering include:

- *inhalation* – from heated fluxes, acids and fumes given off during soldering
- *heat* – from heated materials and equipment, naked flames and fluxes
- *spillage* – of corrosive liquids, such as flux
- *fire* – potential risk of fire from gas heating equipment
- *superheated flux* – overheated soldering iron dipped in flux
- *explosion* – ignited fuel gas
- *burns* – from the heat source, the soldering iron/bit, fluxes and the material being soldered.

Hazard reduction

The following procedures should be followed to ensure a safe soft-soldering process:

- Suspected gas leaks should be checked by using a soapy water solution, and identified leaks reported.
- Disconnection and reconnection of cylinders should be carried out by a competent person.
- Soldering benches should be situated in a position where good cross-ventilation is available.
- When using electric soldering irons, position away from damp or wet areas.
- Electrical equipment must have current test and tag certification.
- Gas equipment should be audited each year.
- It is recommended that low-voltage continuous-operation soldering equipment be used.
- Soldering tips should be maintained in good working order.
- Excessive gas build-up should be avoided during delayed use of the flint gun to light the gas furnace.
- Valves to a gas furnace should be turned off fully. A valve should never be turned off and then turned on again as a hot furnace could ignite gas build-up with explosive force.
- Suitable eyewash facilities should be available in the workspace prior to the commencement of the soldering.
- Inhaling fumes or vapours from acids, fluxes or gases should be avoided when soldering.
- Flux containers should not leak and should be equipped with suitable brushes.
- Protective equipment and PPE must be used to reduce the risk of burns.
- Correct storage and containers for acids and solvents should be used to minimise spillage from containers.
- Safety precautions should be adopted to prevent possible fire hazards when using gas for heating. Combustible materials must be cleared from the work area.
- Correct handling methods should be used when heating oils and liquids to reduce the risk of splashing.
- Hot irons should be cooled in an appropriate fluid and stored accordingly.
- Dip pots constitute a significant hazard and should be in good order, clean, and contain sufficient solution.
- An appropriate hot iron rest should be provided for the activity.
- Gas furnaces should not be lit using a disposable or naked flame from a match or taper.
- In the event of fire, turn off the gas supply at the isolation valve. Always know where the gas isolation valve is prior to the event of an emergency.
- An appropriate class fire extinguisher and/or fire blanket should be kept in the vicinity of the soldering area.

When lighting a gas furnace, the following procedures should be followed:

- Stand to one side of the furnace.
- See that the flint gun is operative.
- Turn on the gas slowly and ignite by means of a flint gun.
- Adjust the flame as required.
- If difficulty is encountered in using the flint gun, turn the gas off immediately, allow the gas to dissipate and repeat the process if it is safe to do so.

Riveted joints

Rivets are metal fasteners that are used to join two pieces of metal together. There are numerous rivets available to join sheet metal, but the most common types of rivets used in the plumbing industry are blind or pop rivets.

Blind or pop rivets

Blind or pop rivets are from a group of fasteners originally designed for joints that are accessible from one side only. In the plumbing trade, these rivets are used where both sides of the joint are accessible, allowing a simple joining method with an acceptable appearance. This is useful in situations where the joint is to be made in elevated areas with no solid support available, such as in roofing and guttering applications. The rivets require pre-drilled holes of the equivalent size to the diameter of the rivet. The rivets are made up of two parts, which consist of a body with a flanged portion at the top that forms the head, and a mandrel head that is pulled through the body then 'popped' or snapped off, with the tail or stem discarded (see Figure 8.49).

Blind rivets have a variety of head shapes, such as domed, flanged and countersunk, and are manufactured from various materials such as copper, stainless steel, aluminium and steel, and are also available pre-painted in a variety of colours to suit the application. Care should be taken to ensure rivets are compatible with the material that is being joined.

When marking out for rivets, they should not be set too close to the edge of the lap as the thickness left may not provide enough strength, causing the rivet to tear out.

The distance from the centre of the rivet to the edge of the metal should be at least twice the diameter of the rivet. Rivets should not be spaced closer than three times the rivet diameter, or further apart than eight times the rivet diameter or there may not be sufficient strength in the joint (see Figure 8.50).

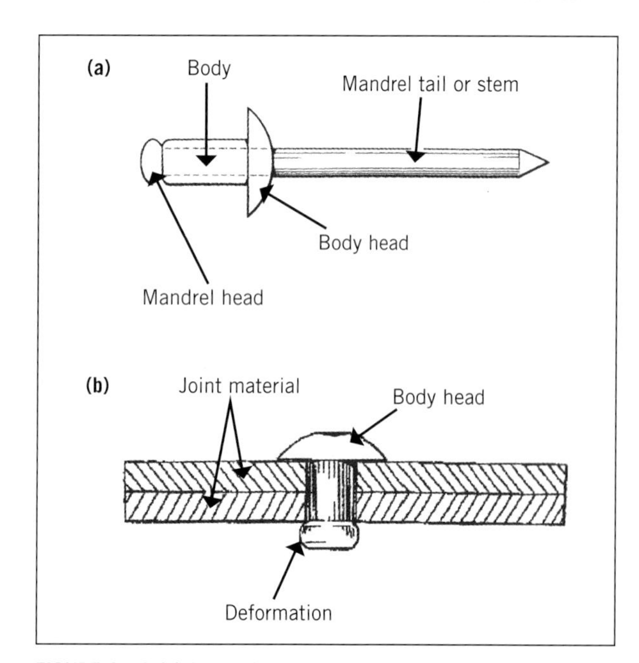

FIGURE 8.49 (a) Blind rivet; (b) Blind rivet set in place

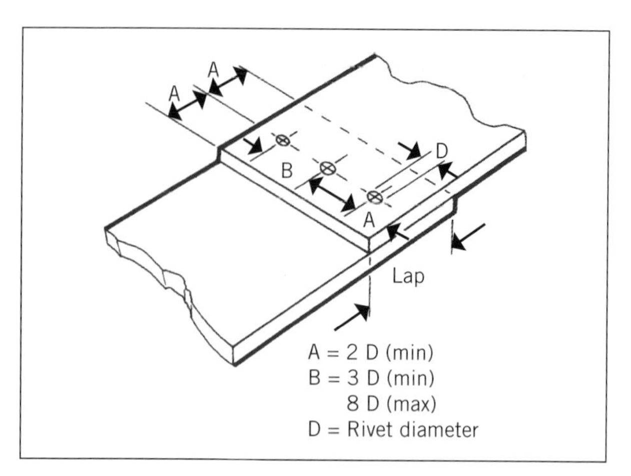

FIGURE 8.50 Location and spacing of rivets

The spacings used for general sheet metal work are 20–25 mm for the most economical results.

In the plumbing trade, rivets are normally used on lapped joints with a maximum lap width of 25 mm, with rivets spaced at a maximum of 40 mm. The spacing of the rivets can be provided on the plans or job specification, or noted in the Australian Standards and/or the manufacturer's instructions.

Tools commonly used to set the rivets

Pop rivets can be set by using pop rivet pliers (see Figure 8.51) or, as it is more commonly known, the pop rivet gun. Other examples are lazy tongs (see Figure 8.52) and lever hand tools (see Figure 7.53).

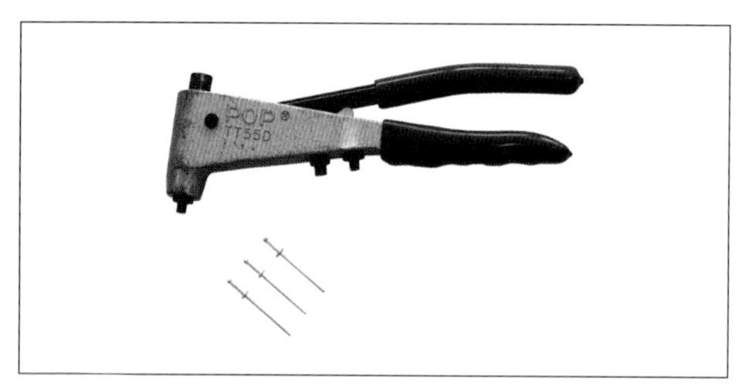

FIGURE 8.51 Pop rivet pliers

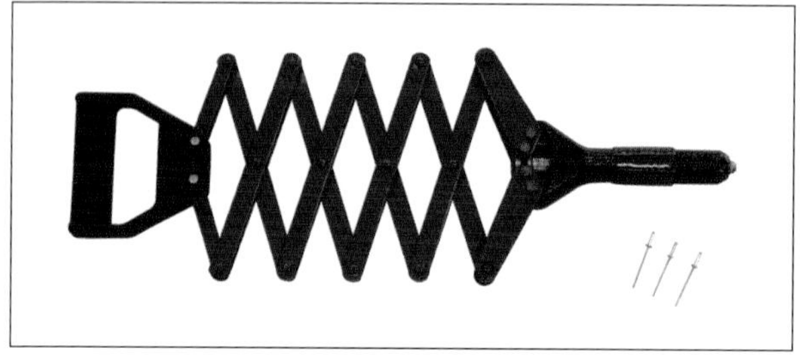

FIGURE 8.52 Lazy tongs

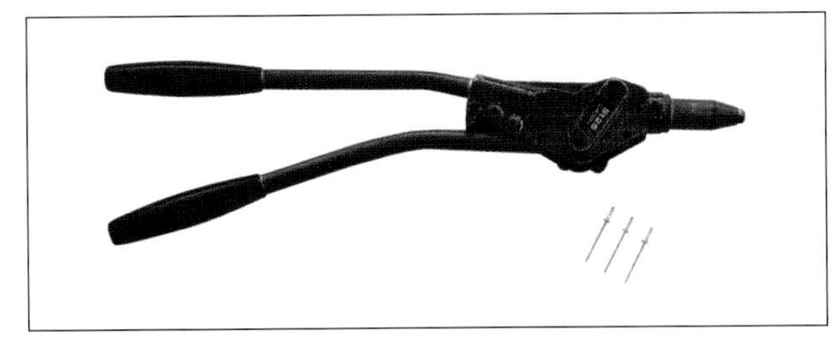

FIGURE 8.53 Lever hand tool

A faulty rivet should be removed as it will not create an effective joint (see Figure 8.54). The method of safe removal is to drill through the faulty rivet by using a drill bit of the same size as the initial drill hole (see Figure 8.55). It is recommended to add a small dollop of silicone to the area prior to resetting the new pop rivet as the hole may be slightly larger than the original hole that was drilled.

FIGURE 8.54 An example of a faulty pop rivet

FIGURE 8.55 Removing a faulty pop rivet

Threaded fasteners

Screwed and bolted joints use threaded fasteners, which offer the convenience of being able to dismantle the joint due to maintenance or the replacement of parts; they are also easily replaced if the need arises.

Bolts

Bolts (see Figure 8.56) are used as a convenient method to join materials together and are normally used with a washer and nut. The nut is attached to the bolt by rotation, tightening the nut onto the bolt. Nuts are generally tightened to a specified tension to ensure they do not loosen or break. Bolts are placed in a pre-drilled hole that is slightly larger in diameter than the bolt shank and is known as a clearance hole.

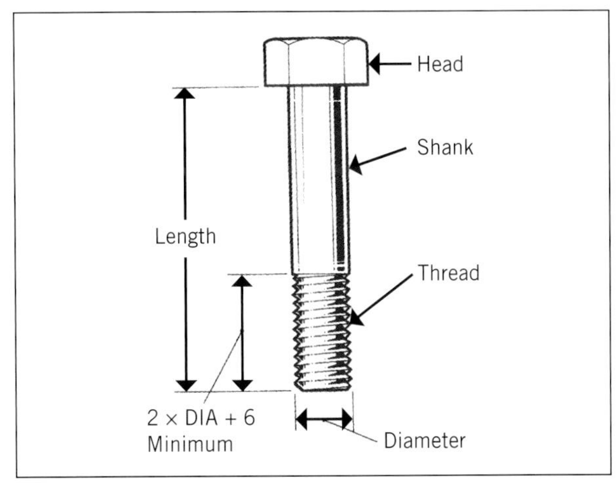

FIGURE 8.56 Bolt details

Bolts are generally made of low-tensile-strength steel, stainless steel, brass and bronze, and are available in numerous shapes, sizes and finishes. When specifying bolts, an adequate description is necessary and should include the head shape, shank length and diameter, thread form, material and finish. A bolt that is required for a job may be a steel hexagon-headed bolt with zinc plating (ZP), a 12 mm metric thread (12 mm thread size indicates 12 mm shank diameter) and a 100 mm shank length. This bolt may be described as M12 × 100, steel, hex head, ZP.

Sheet metal ranging in thickness from 1 mm upwards can be fastened by nuts and bolts. These are generally used where there is a number of materials such as sheet metal, plate, bar, angle iron or casting.

Hexagon head bolts

Hexagon bolts are the most commonly used bolt style as the hexagon head shape facilitates ease of tightening or loosening of the bolt. They are often used for heavier gauge sheet metals (see Figure 8.57).

The various types of bolts that are available are shown in Figure 8.58.

Cup head bolts

Cup head or carriage bolts are designed to fit sheet metal to timber frames such as coach bodies, hence the name. Some bolts are manufactured for specialised applications, such as the countersunk head, square neck and plow bolt, which is used as a part on machinery to hold the share sections on agricultural machines.

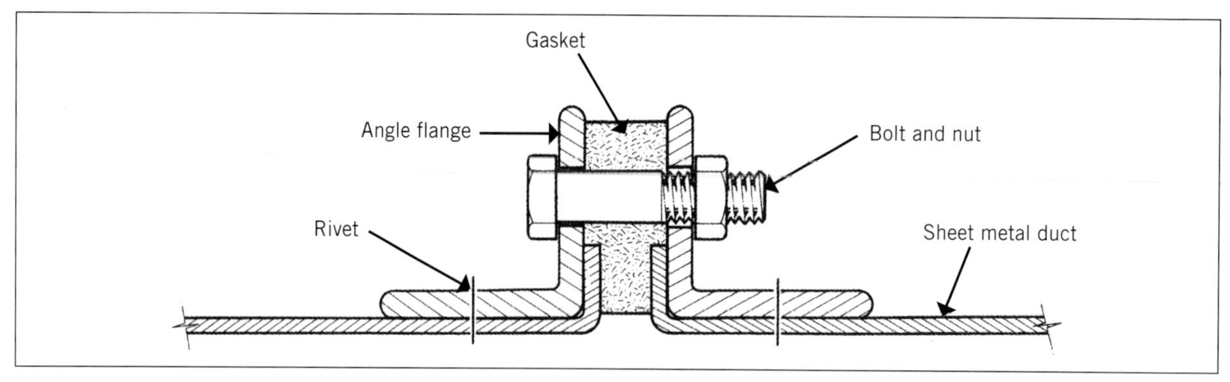

FIGURE 8.57 Bolted flange joint on an air conditioning duct

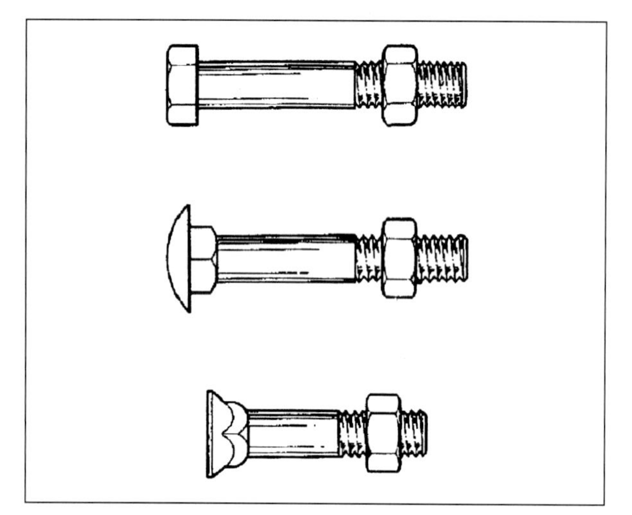

FIGURE 8.58 Various bolt types: (a) Hexagon head bolt; (b) Cup head bolt; (c) Countersunk head, square neck, plow bolt

Nuts

There is an extensive variety of nuts available for different applications (see Figure 8.59). The following nuts are commonly used in the plumbing trade. They are used for fastening gutters, roofing and pipe support systems and are attached to bolts and threaded rod (Booker rod) as required by the application.

Standard hexagon and square nuts

Standard hexagon nuts are the most common type of nut used in the plumbing trade and require tightening with a tool such as a ring spanner or adjustable shifting spanner. Square nuts are generally used for lighter-load applications than hexagon nuts. Square and hexagon nuts are adequate for most intended uses, but under some load or assembly conditions where vibration occurs, they may not be able to adequately maintain a tight joint. In these conditions lock, nyloc, captive, slotted and castle nuts may be required.

Wing nuts

Wing nuts have a rounded body with two distinct protruding wings that enable them to be fastened and loosened through rotation by hand. They are most often used in situations that require a low torque or

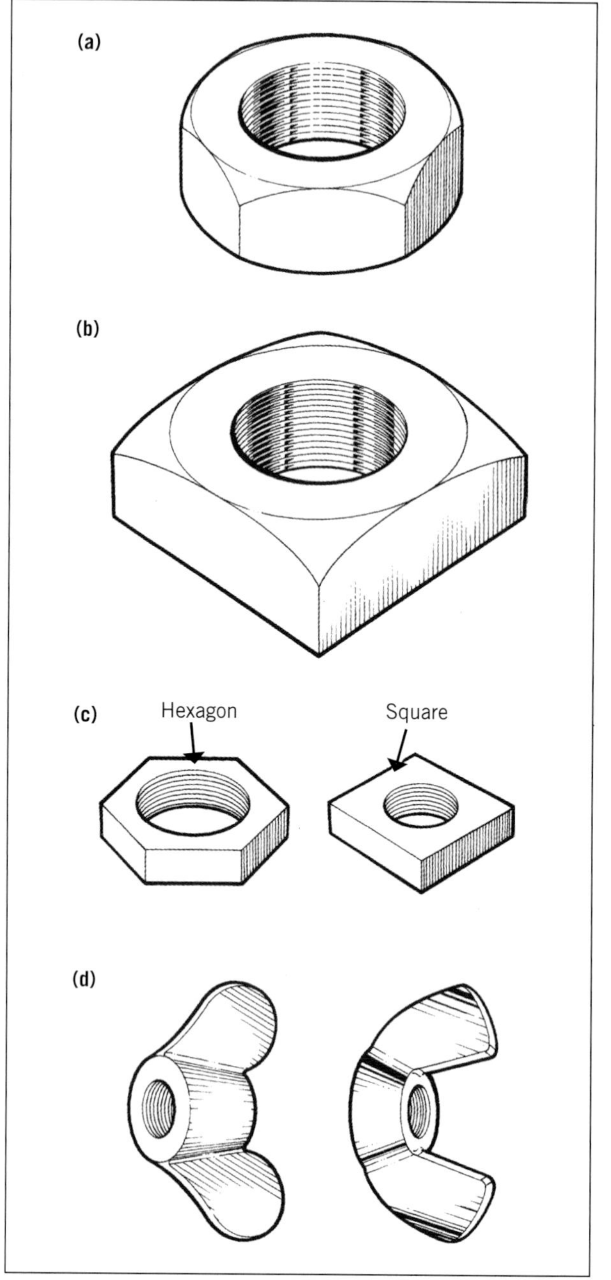

FIGURE 8.59 Various nut types: (a) Hexagon nut; (b) Square nut; (c) Pressed metal nuts; (d) Wing nuts

light loads, and are useful for applications requiring regular dismantling, such as access panels in ducting or tightening nuts in hard-to-reach places.

Screws

Screws are fastened and loosened by rotation. Some bolts have screw-type heads and may be fastened with a nut similar to a bolt. Screws have various head styles that can fasten into preformed or self-made (self-tapped) internal threads.

The head style choice of the screw (see Figure 8.60) depends on the type of driving equipment to be used, the preferred appearance and the proposed loading placed on the fasteners. Hexagon heads are used in situations that require high strength, such as the wind shear loads on roof sheeting. They can be easily tightened with mechanical devices, such as hexagon drivers in screw guns and impact drivers. Slotted screw heads may be used for some general applications, although they are not as common as they once were. Phillips-head and Pozidriv screws provide positive engagement between the driver and the screw with minimum slippage, and are generally preferred within most trades including plumbing. Although Pozidriv and Phillips-head screws are similar in appearance, there is a slight difference in the shape of the recess walls. Phillips-head recess walls are tapered and the Pozidriv recess walls are vertical. The tapered recess walls of the Phillips-head screw tend to reject the driver through vertical reaction and, without heavy end pressure being applied to the driver, engagement may be rejected, and the driver could slip. However, with the vertical recess walls of the Pozidriv screw, this vertical reaction is virtually eliminated. Therefore, the rotation of the screw can be achieved without the potential of slippage and with less downward pressure. This results in a more positive drive, hence the name Pozidriv.

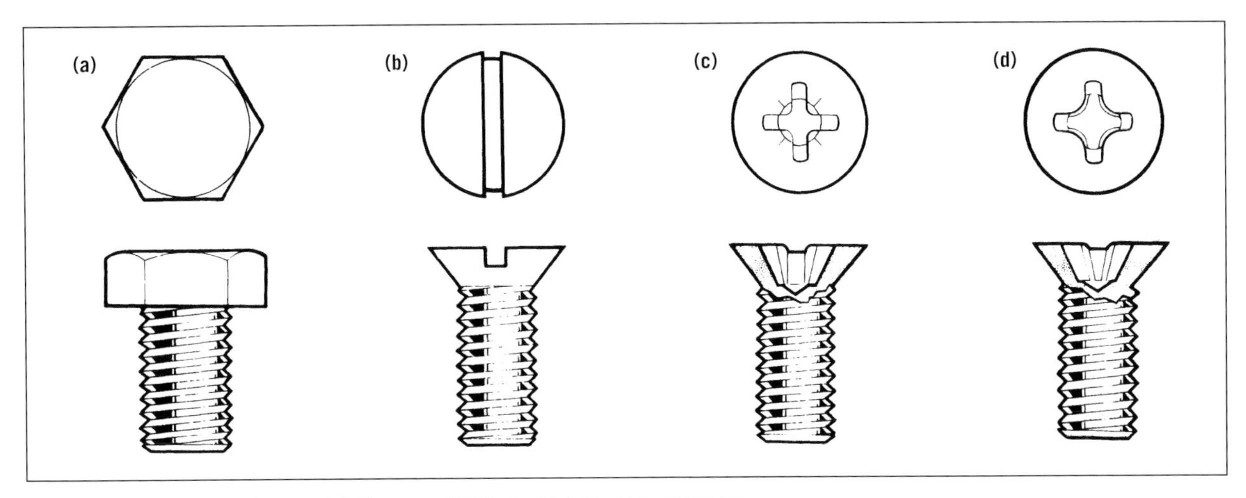

FIGURE 8.60 Screw head types (a) Hexagon (b) Slotted (c) Pozidriv (d) Phillips

There are also screws that have a combined slotted and Phillips-style head. Because of the problems associated with driver slippage, these are generally used on lighter-duty applications.

Screws may be obtained in various types, including:

- self-tapping screws
- self-piercing screws
- self-drilling screws
- machine screws.

Self-tapping screws

Self-tapping screws are made from hardened steel and they make their own thread as they are applied. They need a preformed hole in which to be screwed and form the thread, as they do not drill or punch their own hole. The threads available are either a machine thread or spaced thread (see Figure 8.61).

There is a wide variety of self-tapping screws available, which are classified as two basic types:

- thread-forming screws
- thread-cutting screws

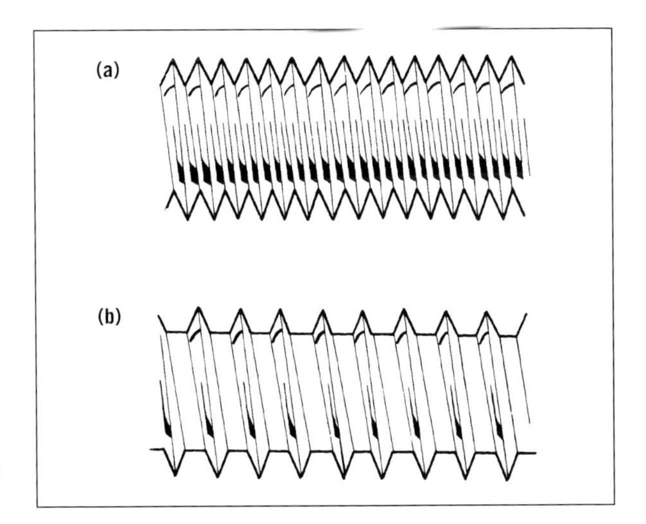

FIGURE 8.61 Thread types (a) Machine thread (b) Spaced thread

Thread-forming screws

Thread-forming screws produce a thread in the work piece by displacing the metal without actually cutting, and remove any material in the form of swarf or chips from the parent metal.

Thread-cutting screws

Thread-cutting screws have a cutting edge and chip cavities built into the screws that perform a tapping action similar to a conventional thread tap that would be used to form a thread in a metal block. The thread is cut in the same way by removing material from the parent metal. Thread-cutting screws produce swarf and cuttings. When selecting the size of the blind hole, allowances need to be made to allow for the thread to be cut into the parent metal by selecting a drill size smaller than the screw diameter specified by the manufacturer. All swarf and cuttings should be thoroughly cleaned from the work after the screws have been installed to avoid corrosion in the future. A soft bristled broom or battery-operated leaf blower will sufficiently clean the work area.

Self-piercing screws

Self-piercing screws form their own thread when used with a high-impact system such as in fastening roof and wall sheeting, flashing and framing. High-energy impact forces the point of the screw and pierces through the sheet metal. It extrudes its own pilot hole in one action and upon further downward pressure and rotation, it forms a thread in the pilot hole. The extruded hole that has been formed in the parent metal produces an increased number of threads, creating more resistance to loosening.

Self-drilling screws

Self-drilling screws have a point that forms a drill bit in order to drill their own holes (see Figure 8.62). The screw then cuts and forms its own thread in a similar way to the standard self-tapping screw. These screws are installed by battery drill or screw gun. They usually have hexagonal heads and in-built washers to prevent water or condensation leaking through the roof sheet (see Figure 8.62).

These screws are also used for fastening roof and wall sheeting, flashings and framing. The point of the screw is designed and constructed to drill the hole in the work piece, then the faster feeding thread engages and tightens the joint together. It is necessary for the unthreaded drilling portion of the screw shank to be longer than the combined material thicknesses to be fastened. Otherwise the top section of the roof sheet lapped joint will rise up the shank of the screw and strip the thread being formed by the screw before the lower section of the lap is drilled through. This will result in the lower section of the lap carrying all the load of the joint. When calculating this drill length, any air space allowances in the joint must be taken into consideration and the manufacturer's literature and specifications should be consulted.

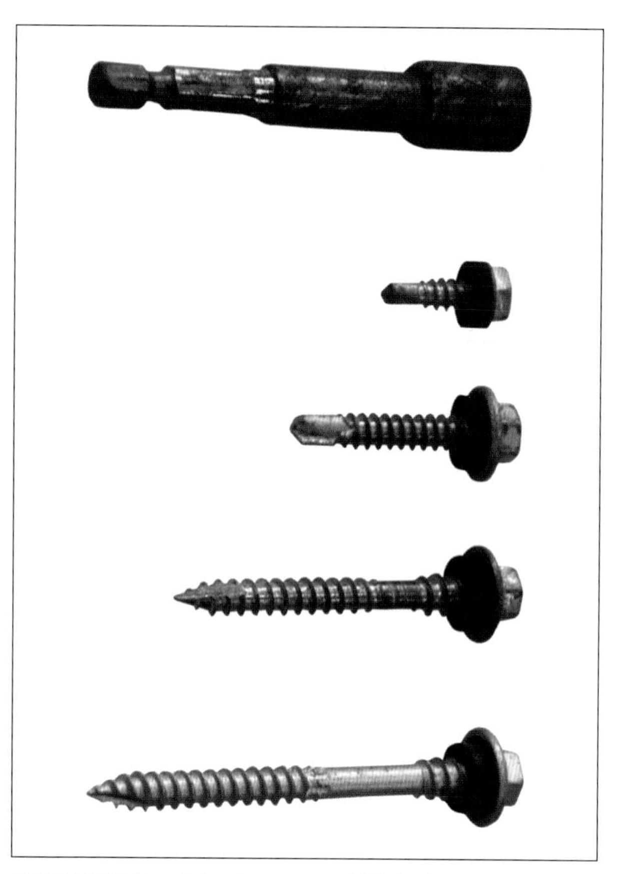

FIGURE 8.62 A variety of screws and bit for fastening

LEARNING TASK 8.2

IDENTIFY JOINING REQUIREMENTS

Identify and select a sheet metal material and joining process for a specific task. This could be a Colorbond® water tank installed on a domestic property or a copper downpipe on an architectural commercial building.

1 Describe the jointing process of the task.
2 What allowances need to be made for the jointing process?
3 What materials, tools and equipment are required?

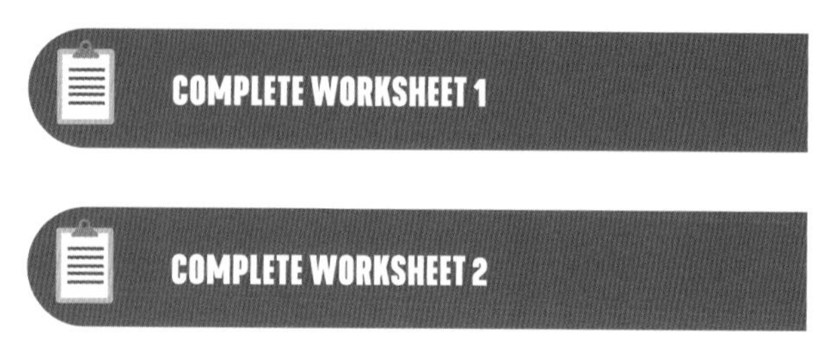

Cut and join sheet metal

In preparation for cutting and joining sheet metal, reference should always be made to plans and specifications to ensure materials are marked correctly. Selecting the correct methods and tools for cutting sheet metal to size will reduce mistakes and is essential in producing a quality product. Accuracy is achieved by marking out carefully and always checking that the edges of the sheet you are marking out are square and form 90°. From a square corner you can then mark out all your measurements.

When measurements are to be marked out, use a scribe, felt-tip pen or snips for best results. Lead or graphite pencils are not to be used when marking out certain types of sheet metal as they are an incompatible marker and may reduce the life expectancy by causing corrosion.

It is essential when working with sheet metal to measure your material twice and cut once, as this will minimise mistakes when cutting out material. If too much has been cut off, it can lead to wastage of material and can potentially cost additional money that has not been allowed for.

After marking the sheet material in accordance with plans and specifications, the sheet metal is cut to size ready for assembly.

In a workshop situation where straight cuts are required and large sheets need to be cut into smaller pieces to suit the intended application, a mechanical guillotine is usually used (see **Figure 8.63**).

FIGURE 8.63 Mechanical guillotine

It is essential to read the Safety Operating Procedure (SOP) and complete a risk assessment prior to operating a guillotine or any other workshop equipment. A Safe Work Method Statement (SWMS) should also be addressed both prior to and on completion of any activity.

All guillotines are rated according to the type and thickness of material they are capable of cutting. The manufacturer's specifications should be consulted before using the machine to ensure that the capabilities of the machine are not exceeded. When placing sheet metal in a guillotine, care should be taken that all work health and safety requirements are observed and that fingers are kept clear of the blade.

Cutting technique

When using a mechanical guillotine to cut sheet metal, cut a small nick on each side of the sheet using red handled snips on one side and green handled snips on the other. This will cause the waste side of the snips to turn down one side of the nick, so that when the sheet is placed in the guillotine it is slid in past the line to be cut and retracted until the turn-down in the nick grabs the edge of the guillotine base and stops. The marks should be cut straight and checked for final alignment by using a slight pulling action on the sheet to ensure the nick is still against the guillotine body, before the foot pedal is depressed.

Other types of power tools that can be used for cutting sheet metal include:

- power shears
- nibblers
- circular saws (fitted with metal cutting blades)
- angle grinders (fitted with metal cut-off discs).

Power shears and nibblers can be a good alternative to using hand tools as they are accurate and efficient, and when used properly will eliminate burrs on the material. However, circular saws and angle grinders can leave burred edges, which can lead to problems in the future. For these reasons, circular saws and angle grinders are not often recommended for cutting sheet metal and should only be used where necessary, as these burred edges are very sharp and can be dangerous to handle, potentially resulting in deep cuts to hands and fingers. The burred edge of steel-based materials will also corrode badly if exposed to moisture. This corrosion occurs due to the rough edge area of the cut sheet being too wide, resulting in the protective zinc coating becoming ineffective. The burrs can be removed with a file so that the edge is smooth.

Angle grinders also produce very hot chips of metal that can burn into the surface of materials such as Colorbond®, which will permanently damage the sheet. Never use an angle grinder on a roof or use carborundum cutting discs for sheet metal.

Where possible, a hacksaw with a 32 TPI blade or tin snips should be used for cutting sheet metal, as burrs are minimised with the correct cutting technique.

Examples of some sheet metal projects are shown in **Figures 8.64–8.68**.

Sealants

There are many different sealants available from a wide range of manufacturers, which are used successfully in sheet metal jointing. Sealants normally used in the plumbing industry come from the polymeric range and include natural rubber and other synthetic

FIGURE 8.64 Cutting sheet metal with curved jeweller's snips

FIGURE 8.65 A sample of completed work

FIGURE 8.66 Dressing sheet metal into shape with a wooden mallet

FIGURE 8.67 A sheet metal penetration flashing

elastomers. The primary purpose of these sealers is to form a continuous bonding layer on the surface of the opposing laps of the material to be joined. They may also act as a gap filler and joint sealer for weatherproofing or dustproofing, or to improve the appearance of the joints.

FIGURE 8.68 A sheet metal penetration flashing (different view)

It is essential to select the correct sealant to suit the application. Specifications for the job should be checked to identify the correct sealant to be used. If a sealant has not been specified, it will be necessary to check the manufacturer's product specifications in order to ensure the correct one is selected. The SDS for each sealant should be referred to prior to working with the product.

To choose the right sealant for the job, you may need to consider the desired features of the sealant, which may include:

- types of material it will be applied to and its compatibility
- the size of gap it will fill and its gap-filling ability
- its degree of mould resistance
- its waterproofing properties
- whether it can be painted
- whether it can be used indoors and/or outdoors
- whether it is solvent-based or gives off fumes when curing
- its final state when cured (i.e. solid, flexible or very flexible)
- whether it is acetic-cure or neutral-cure
- its degree of ultraviolet (UV) resistance for use in sunlight
- its level of heat and cold resistance
- its adhesion qualities
- whether it is corrosive.

Generally, the most common sealant used in the plumbing industry is a neutral-cure silicone sealant. Sealants containing acetic acid, known as acetic silicones, should not be used on zinc, Zincalume® or steel-based products, as contact with moisture during the early stages of the sealant curing will cause corrosion.

There are also other types of sealants available to suit different job situations, such as mastics to seal ductwork, water-based gap fillers suitable for exposure to the elements, and polyurethane adhesives and sealants and co-polymer sealants that can be applied to moist surfaces, which is very useful for repair work on wet surfaces.

The most common form of container used in the plumbing industry is the cartridge, which is placed in a standard, open-type hand-operated caulking gun (see Figure 8.69).

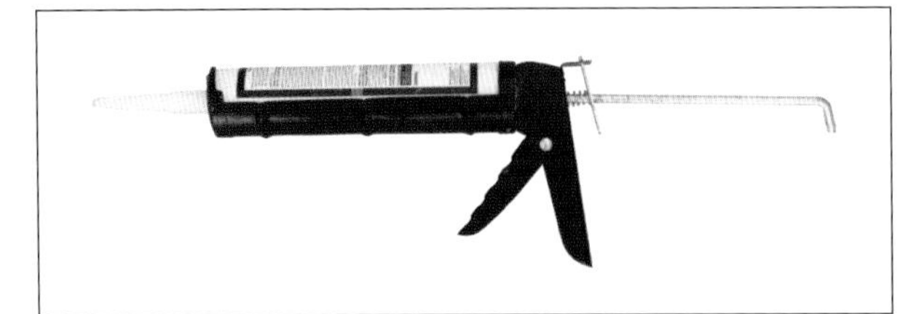

FIGURE 8.69 Caulking gun

As with all other jointing and sealing processes, cleanliness of the joint is essential to ensure proper adhesion or bonding. Joints must be clean, dry and free from contaminants, with old sealant removed when repairing existing sealed joints. An oil-free solvent such as mineral turpentine can be used to clean the surfaces of the joint prior to applying the sealant, although this product should be wiped with a clean, dry cloth before it evaporates naturally to avoid any residue being left on the joint area. Further cleaning with methylated spirits can help to remove the turpentine, although this should also be wiped off before it evaporates to avoid a residue on the joint. Always refer to an appropriate SDS for cleaning materials before use.

There are other commercial products available to clean old silicone from metal surfaces and manufacturers, specifications should be consulted, depending on the material to be cleaned. First, the old silicone should be cut away with an appropriate tool such as a scraper, retractable knife or spatula. The remover should then be applied to a thickness of approximately 5 mm over the old silicone and allowed to soak in. This will soften the old silicone, allowing it to be scraped away easily. The area should then be cleaned with methylated spirits.

When using retractable knives and sharp blades, ensure PPE is worn and all cutting performed is away from the body.

HOW TO

PREPARE A CARTRIDGE FOR THE EXTRUSION PROCESS

1 Open the gun by extracting the plunger handle all the way.
2 Place the cartridge in the gun by positioning the lower section of the cartridge over the plunger plate and then place the cartridge in the gun sleeve.
3 Ensure that the cartridge is properly engaged in the front of the cartridge gun.
4 Cut the nozzle at the desired angle with a suitable sharp knife.
5 Cut off the blank end of the thread of the cartridge.
6 Pierce the cartridge membrane or safely cut the end of the threaded cartridge and screw the nozzle on the thread of the cartridge.

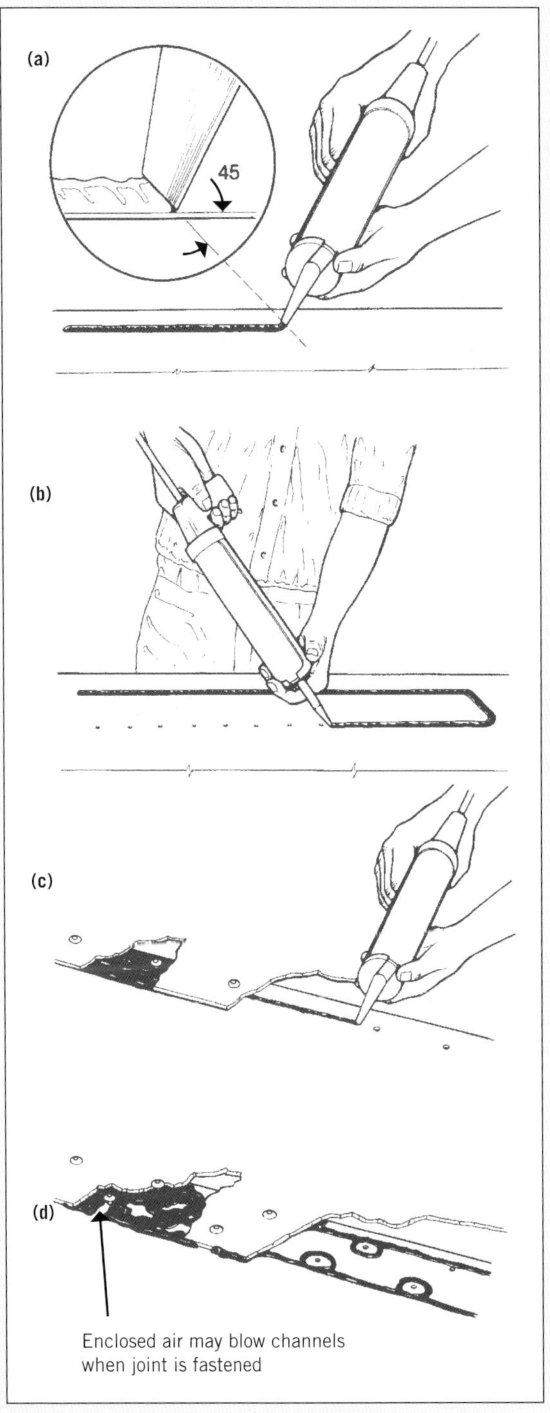

FIGURE 8.70 Correctly applying the sealant: (a) Correct angle for application; (b) Placing the sealant bead; (c) Recommended method of placing sealant; (d) This method is not recommended

>>

>>

7 Place the tip of the nozzle on the surface of the work piece (as shown in Figure 8.70 (a)) and begin to extrude the sealant.
8 Eject enough sealant to ensure that the air in the nozzle is fully expelled and wipe off any excess with a clean cloth.
9 Extrude the sealant over the surface in a consistent run where required. Regulate the quantity and width of the sealant bead as required by adjusting the speed and trigger pressure.
10 When the extrusion is complete and the sealant has been applied to the work piece, release the trigger pressure on the plunger and seal off the nozzle with a suitable cap, which will help to prevent the sealant curing in the nozzle.

Joint finishing

The silicone sealant should be applied on the same day that the surface is prepared. After extrusion and application of the silicone, the joint should be finished as soon as possible to prevent it prematurely curing, which will result in poor bonding to the surface. The maximum use time for most sealants is 10–15 minutes in humid conditions and 30 minutes in dry conditions. Care should be taken not to trap air pockets in the seal, as this may allow the seal to crack, causing it to become ineffective.

When solvents are being used, always refer to the SDS. The area should be well ventilated, and appropriate PPE used.

Final joint inspection

When any joint has been completed, and prior to the job being put into service, it is essential to inspect the work and test it for soundness. When carrying out an inspection and testing of the job, always check the following:

- All laps are properly closed with good coverage of silicone and held tight by the jointing method used.
- The correct number of mechanical fasteners have been used and are evenly spaced and in line.
- The job is complete according to the plans and specifications and all dimensions are correct.
- Gaps are sealed on all joints and the job is watertight.
- All rivets in the joints are sealed correctly and there is no possibility of water entry. If blind rivets have been used, it is important to check that they are properly sealed.
- All soldered joints are completed properly. If screws and rivets have been used in the joint, ensure they have been soldered over correctly.
- All flux residue has been removed.
- All joints that are required to be watertight, such as in a container or tank, have had a static water test performed upon them. This can be achieved by simply filling the product with water.
- Gutters and downpipes have been tested with water running through them, at a rate that is similar to or slightly more intense than when they will be in service. This can be achieved by pouring a bucket of water into the gutter. Roofing should be given a visual check, as wet roofs can be dangerous, and water should not be poured over them when workers are required to walk on them.

Capillary action

Capillary action takes place when two smooth, close-fitting surfaces draw a liquid between them. This can cause the liquid to rise against the force of gravity. This is due to the properties of water molecules, which have an ability to adhere to each other and the surfaces they come into contact with. If there is a very small space between two flat pieces of sheet metal, water can be drawn into the space and may continue moving along this space to the other side of the joint.

This may present a problem when joining metal sheeting exposed to a wet environment such as wall cladding, roof sheeting and flashings. Leaks may result due to capillary action as water is drawn into the space between the sheets, which may lead to water damage inside the building.

Where these types of joints are installed, care should be taken to properly seal the joints, or to create a larger space between the components by using a drip edge or capillary folded edge, known as a 'capillary break'.

Roof flashings and capping should be provided with anti-capillary breaks by a fold in the metal of 10 mm at an angle of 45° against flat surfaces and 10 mm at an angle of 135° against all other surfaces (see Figure 8.71).

Both pierce fixed and concealed fixed roof sheets require special attention. It is important that these sheets are lapped correctly in accordance with the manufacturer's specifications, including facing the side of the lap away from the prevailing weather (see Figure 8.72).

Expansion

Thermal expansion and contraction in sheet metal occurs due to a rise or fall in temperature of the material (see Table 8.4). Even though metals expand in all directions, most expansion and contraction in the plumbing industry is referred to as 'linear expansion'. This is expansion in a direct line along a material's length. In the case of guttering, there would be greater expansion along its length than its width, and buckling may occur if expansion is not allowed for.

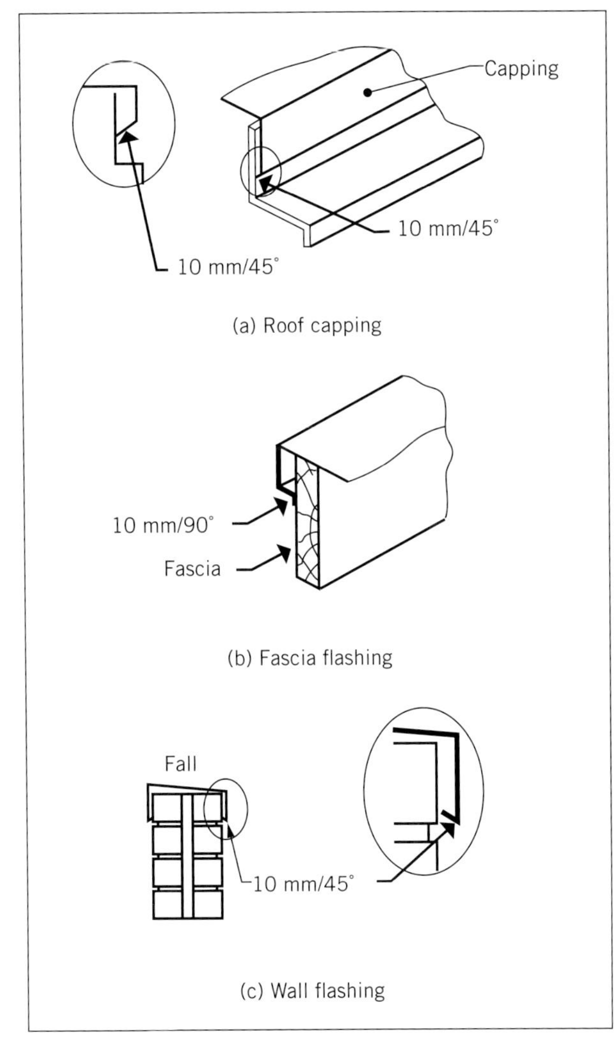

FIGURE 8.71 Anti-capillary breaks: (a) Roof capping; (b) Fascia flashing; (c) Wall flashing

TABLE 8.4 Expansion and contraction of metal sheeting

Sheet length (m)	50°C change (expansion in mm)		75°C change (expansion in mm)	
	Steel	Aluminium	Steel	Aluminium
5	2.9	5.8	4.3	8.6
10	5.7	11.4	8.6	17.2
15	8.6	17.2	12.8	25.6
20	11.4	22.8	17.1	34.2
25	14.3	28.6	21.4	42.8
30	17.1	34.2	25.7	51.4

All materials expand and contract at different rates, so a property that is known as the 'coefficient of expansion' is used to calculate expansion. When 1 m of material is raised in temperature by 1°C, it increases in size by its coefficient, or its change in length per metre. For example, the coefficient for copper is 0.0000167, which represents a 0.0000167 m increase in length per metre of material per degree rise in temperature. The coefficient of expansion of common metals used in the plumbing industry is expressed in Table 8.5.

TABLE 8.5 Coefficient of expansion

Imperial gauge	Metric sheet m
Mild steel	0.000 012
Zincalume®	0.000 010
Aluminium	0.000 024
Copper	0.000 016 7
Lead	0.000 029
Zinc	0.000 026

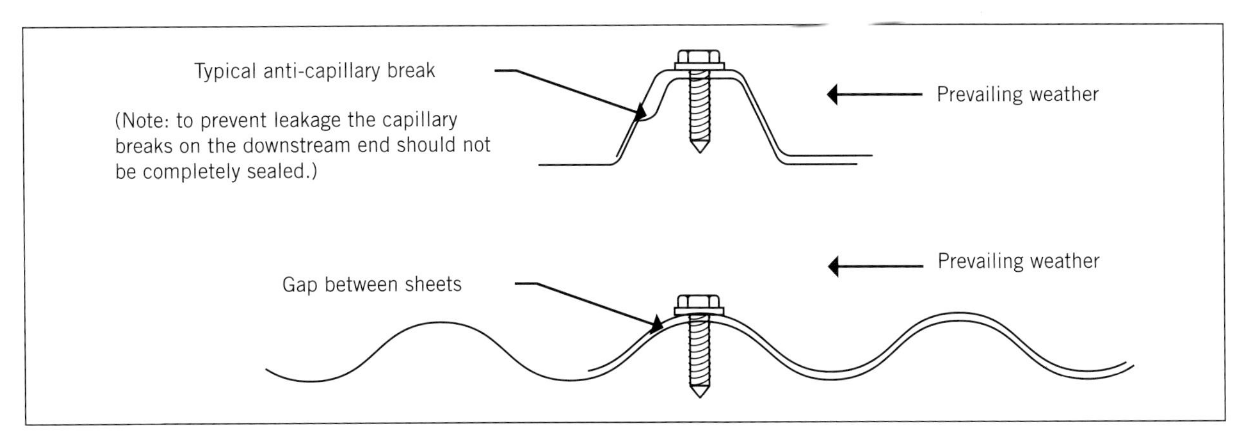

FIGURE 8.72 Side lapping: (a) Trapezoidal profile; (b) Corrugated profile

Total expansion is calculated by multiplying the length of the material by the rise or temperature differential in degrees Celsius, and the result is then multiplied by the coefficient of expansion corresponding to the material. To convert metres to millimetres, multiply the answer by one thousand, as follows:

$E = L \times T \times C \times 1000$

Where E = Expansion in millimetres

L = Length of material in metres

T = Temperature difference in degrees Celsius

C = Coefficient of linear expansion

1000 = Converting metres to millimetres

Considerable expansion may occur in a length of guttering, so provision needs to be made for expansion by using shorter lengths with expansion joints between each length (see Figure 8.73).

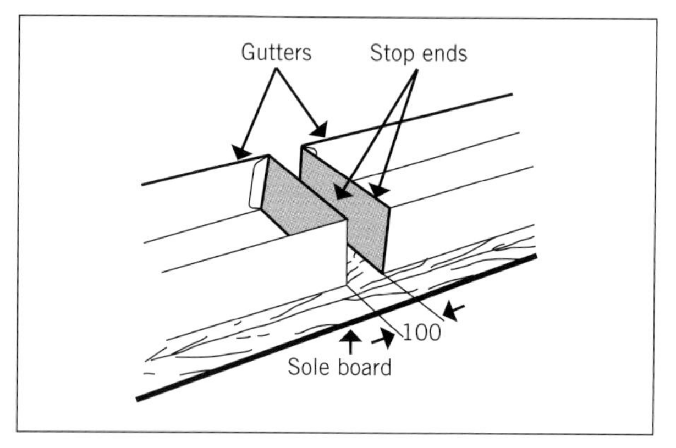

FIGURE 8.73 Allowance for expansion between lengths of box guttering

A suitable protective capping must be provided over the two stop ends that form the expansion joint in order to prevent water damage.

The movement in sheet metal that is fastened down at regular intervals, such as a roof sheet (see Figure 8.74), would normally occur from the centre towards each end of the sheet. The actual expansion specified in Table 8.4 would take place evenly along the length of sheet and would be equal at each end. The amount of expansion and contraction between the fasteners would only be a fraction of the amount shown in the table. Refer to *SA HB 39 – Installation code for metal roof and wall cladding* for further details on expansion of roof sheet metal products.

- SA HB 39 – Installation code for metal roof and wall cladding

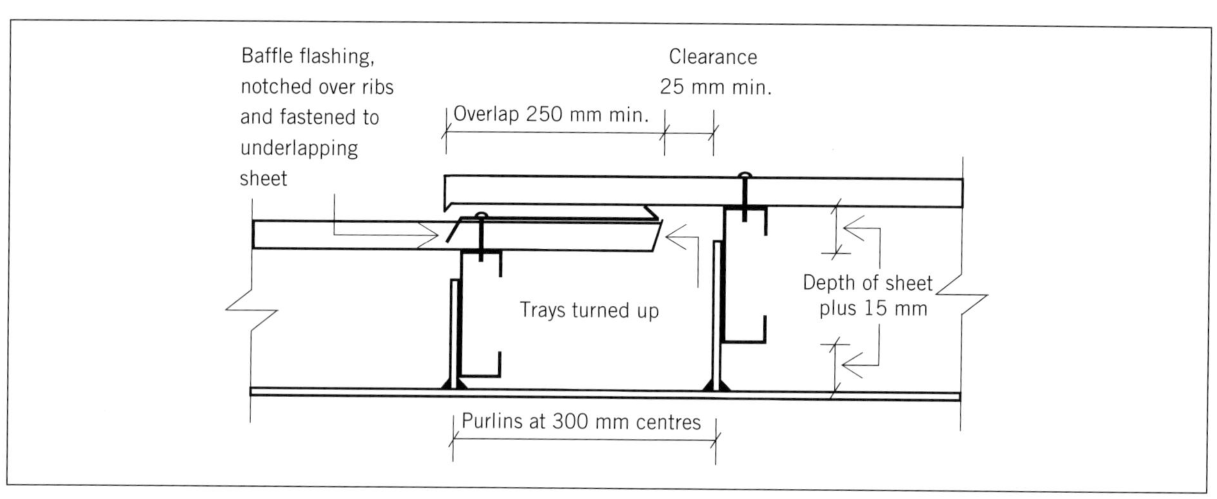

FIGURE 8.74 Note the expansion joint in the length of the roof sheeting

Sustainability principles and concepts

Sustainability principles and concepts must be applied throughout the cutting and joining process by ensuring materials are correctly marked out to minimise the amount of waste material. Any additional material may be reused where appropriate, or recycled in accordance with manufacturers' and state or territory regulatory authority requirements.

LEARNING TASK 8.3

CUT AND JOIN SHEET METAL

1 List the processes and tools required for cutting and joining a Zincalume® gutter with an approved sealant and pop rivets.

2 Select a surface preparation cleaner that would be used to remove contaminants before the joining process.

3 For the cleaner selected, refer to an applicable SDS on the handling and PPE requirements.

4 Are there are compatibility requirements that need to be taken into consideration?

COMPLETE WORKSHEET 3

Clean up

A job is not finished until the clean-up is completed. The following guidelines should be observed:

- All debris should be cleared from the worksite and placed in the designated or appropriate area.
- All leftover offcuts and materials generated during the course of the work must be sorted, and all pieces large enough for future use should be stacked and stored in a predesignated safe place.
- On completion of the job, all reusable materials should be returned to the workshop store to ensure they are not damaged.
- The smaller offcuts of sheet metal that are to be discarded should be placed in separate bins to be sold as scrap. Lead and lead alloys should be recycled or disposed of properly and never thrown in the rubbish bin, as they are damaging to the environment and are considered toxic waste.
- When handling sheet metal offcuts, suitable gloves should be worn to protect hands against cuts and abrasions.
- All debris and excess material should be disposed of or recycled in accordance with the workplace policies and any local, state or territory legislation.
- All filings and swarf should be swept up to avoid corrosion problems in the future.
- All tools and equipment should be cleaned before they are stored. Removal of all residual sealant, oils and dust from the tools is paramount, as tools are expensive and proper maintenance will prolong their service life. They should also be checked for wear, and if damaged they should be labelled and set aside so that they are not used until they are repaired.
- Maintaining and cleaning the work area is the responsibility of all personnel and should be treated as a team effort. When the work area of an individual is complete, they should then help others to clean up, rather than just leaving the workplace because they consider their part is complete.
- Ensure all required documentation is completed according to workplace requirements.

REFERENCES AND FURTHER READING

Acknowledgement

Reproduction of the following resource list references from DET, TAFE NSW C&T Division (Karl Dunkel, Program Manager, Housing and Furniture) and the Product Advisory Committee is acknowledged and appreciated.

Texts

NSW Department of Industrial Relations (1998), Building and construction industry handbook, NSW Department of Industrial Relations, Sydney

Philip, M. & Bolton, B. (2002), *Technology of engineering materials*, Butterworth & Heinemann, Oxford

Standards Australia International (2005*), MP52–2005 Miscellaneous Publication: Manual of authorization procedures for plumbing and drainage products*, Sydney

WorkCover Authority of NSW (1996), *Building industry guide*, WorkCover NSW, Sydney

Web-based resources

Regulations/Codes/Laws

Safe Work Australia: **http://www.safeworkaustralia.gov.au**

Resource tools and VET links

Training.gov.au: **http://training.gov.au**

NSW Education Standards Authority – Construction: **https://educationstandards.nsw.edu.au/wps/portal/nesa/11-12/stage-6-learning-areas/vet/construction-syllabus**

Industry organisations' sites

CITB (SA Construction Industry Training Board): **http://www.citb.org.au**

Building Trades Group Drug & Alcohol Program: **http://www.btgda.org.au**

SafeWork NSW publications

'Hazardous manual tasks: Code of Practice', Cat. no. WC03559

'Personal protective equipment' – Fact sheet, Cat. no. WC05893

'Work method statements', Cat. no. 231

TAFE NSW resources

Training and Education Support (TES), Industry Skills Unit Orange/Granville 68 South St Granville NSW 2142 Ph: (02) 9846 8126 Fax: (02) 9846 8148 (Resource list and order forms).

GET IT RIGHT

The photo below shows an incorrect practice that can be performed when cutting and joining sheet metal.

Identify the incorrect method and provide reasoning for your answer.

WORKSHEET 1

To be completed by teachers

Satisfactory ☐

Not satisfactory ☐

Student name: ______________________

Enrolment year: ______________________

Class code: ______________________

Competency name/Number: ______________________

Task: Review the section 'Prepare for work' up to and including 'Resistance spot welding' in this chapter and answer the following questions.

1. Why should hand and power tools be checked for serviceability when preparing for work?

____

2. Before using new PPE, what should be given by the employer?

____

3. Name two types of folding machines found in the workshop.

____

____

4. List two examples of sheet metal materials used in the plumbing trade.

____

____

5. Explain the term 'ferrous metal'.

____

____

____

____

6. Explain the term 'non-ferrous metal'.

____

____

____

____

7. List two ferrous metals.

____

____

8. List two non-ferrous metals.

9. List three examples of integral joints used for joining sheet metal.

10. How is the size of a groove seam joint determined when making allowances for the joint?

11. What is a knocked-up joint commonly used for?

12. Describe how the resistance spot welding process works?

WORKSHEET 2

To be completed by teachers	
Satisfactory	☐
Not satisfactory	☐

Student name: ____________________

Enrolment year: ____________________

Class code: ____________________

Competency name/Number: ____________________

Task: Review the section 'Soft-soldered joints' up to and including 'Fluxes' in this chapter and answer the following questions.

1. What are the two main metals contained within soft solder?

2. What is the most common type of soft solder used by plumbers for joining sheet metal?

3. List three types of soldering irons.

4. Why are fluxes used during the soldering process?

5. List the two categories of flux that are used for soldering.

6. What are the three primary roles that flux performs?

7. List three characteristics that need to be assessed on completion and cleaning of a soldered joint.

8. When diluting a flux containing acid, should the water be added to the acid, or the acid added to the water? Why is this?

WORKSHEET 3

To be completed by teachers	
Satisfactory	☐
Not satisfactory	☐

Student name: ______________________________

Enrolment year: ______________________________

Class code: ______________________________

Competency name/Number: ______________________________

Task: Review the section 'Riveted joints' up to and including 'Expansion' in this chapter and answer the following questions.

1. Name three tools that are commonly used to set pop rivets.

____

____

____

2. Describe the method to remove a faulty rivet.

____

____

____

3. Explain the main difference between Pozidriv and Phillips-head screw heads.

____

____

____

4. State what type of marker is to be used on sheet metal when marking out and why.

____

____

____

5. What two power tools are a good alternative to hand tools when cutting sheet metal?

____

____

6. What problem can be created if an angle grinder is used to cut sheet metal?

____

____

____

7. What features should be considered when choosing a sealant?

8. Define the term 'capillary action'.

9. How would capillary action be prevented in sheet metal installations such as roofing, flashing or wall cladding?

10. Why is it important to allow for expansion and contraction when installing sheet metal products?

11. Define the 'coefficient of expansion'.

MARK OUT MATERIALS

9

This chapter addresses the following key elements for the competency 'Mark out materials':

- Prepare for work.
- Determine job requirements.
- Mark out the job.
- Clean up.

It introduces common tools for measuring and marking out materials, and describes how to select compatible markers for commonly used materials within the plumbing trade. It also addresses basic drawing principles and pattern development methods applied to measuring and marking out projects in the plumbing industry.

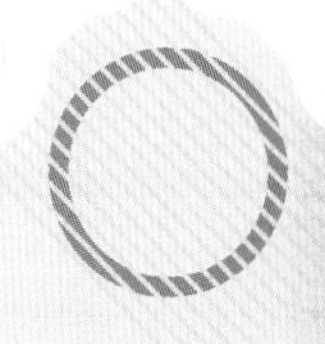

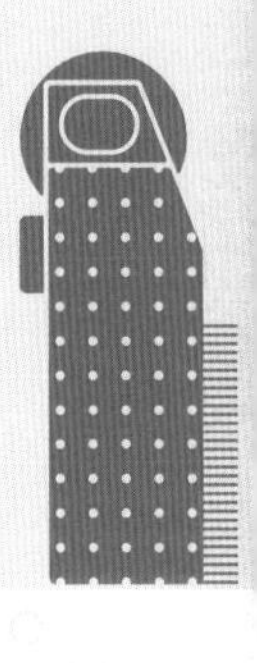

Prepare for work

Marking out materials is involved in most aspects of the plumbing trade, and is the process of selecting and preparing material, and making allowances for jointing, bending, folding, cutting and changes in direction prior to fabrication.

Marking out is an important job, and if a mistake is made at this critical stage, it could impact greatly on the rest of the job. It is important that you plan and understand each stage of the process to reduce the possibility of an error.

FROM EXPERIENCE

Problem-solving effectively and taking initiative with tasks on site are critical skills in the plumbing sector. These skills take time to develop and you should always ask for assistance from supervisors or employers to enhance these important skills.

Preparation on the job starts with accessing plans and specifications to determine the job requirements. If a practical approach is adopted during the preparation phase of your work, it will generally make the project easier and safer to complete. *Work smarter, not harder*. This rule will apply to either a workshop activity or on a building site.

Job planning

Obtaining the plans and specifications from the supervisor or builder will allow you to determine the marking out requirements for the entire job. Without them, simply no work would occur at all.

Plumbers are increasingly being required to analyse tasks and provide documentation of job break-ups. One common area where this can be seen is in work health and safety (WHS) requirements in the form of a Safe Work Method Statement (SWMS). At this stage, not only can you identify safety issues but it is also possible to identify materials and equipment that may have been left out of your original plan. Environmental and quality assurance requirements must also be identified and adhered to throughout this planning phase.

Planning involves methodically working through the job in an organised way and breaking jobs up into tasks. These tasks can then be sequenced into a logical order in conjunction with other trades on site. Clarifying your understanding of the job details with your supervisor can also reduce the likelihood of mistakes.

Working alongside other trades

As the plumbing trade is part of the building and construction industry, each trade within this industry is dependent on every other in some way. To ensure that the process of building is carried out in an efficient and uninterrupted manner, builders or project managers estimate the time that each trade will take to complete its role at each stage. Coordinators plot the estimated time and activity on a chart in a logical sequence or schedule. The company that has won the job to carry out its trade or services on this project will usually have its contract linked to keeping up with, or ahead of, this chart. (See Table 3.3 and Figure 3.3 in Chapter 3.) Should a company or trade fall behind the chart, the company at fault will usually be financially penalised. Although playing a small part within the much larger picture, each tradesperson's work, as part of the collective efforts, can have a dramatic effect upon the overall job dynamics.

Prior to marking out a new job, a plumber must be able to clearly identify any tools, materials or equipment that may have been inadvertently left out of the original plan.

FROM EXPERIENCE

Arriving at the worksite without the required materials and tools wastes valuable time and can cost money in lost wages. Planning for a job will provide you with a clear understanding of what work needs to be undertaken. This will help you to carry out the required tasks efficiently and safely.

Tools and equipment

The following hand tools are common for marking out and measuring materials on the worksite. All tools and equipment should be checked for defects prior to use and any damage reported to your site supervisor.

Vernier callipers

These callipers are used for measuring and marking material that requires tolerances of less than 1 mm. They commonly come in 150 mm, 200 mm and 300 mm lengths and can be either digital or manual read-out (see Figure 9.1). Vernier callipers can be used to check pipe sizes, bolt sizes, gauge thickness and transfer measurements.

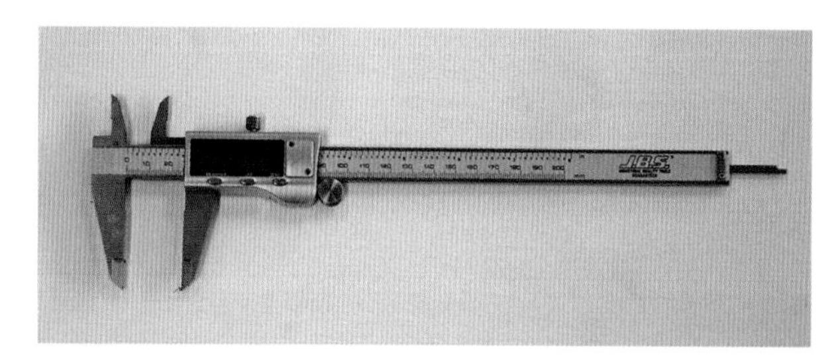

FIGURE 9.1 Vernier callipers

Graduated steel rule

Normally made of stainless steel with contrasting background in various lengths (e.g. 150 mm pocket size, 300 mm toolbox size, and 600 mm and 1000 mm workshop sizes), these rules (see Figure 9.2) are commonly used for marking out flat sheet material and can be also used as a straight edge for marking out developments.

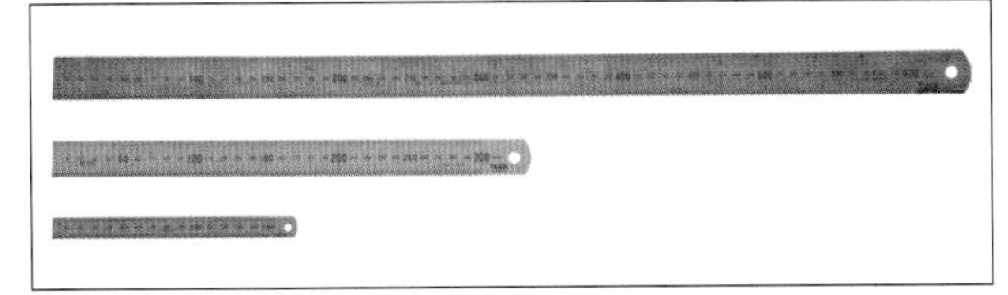

FIGURE 9.2 Steel rules

Scale rule

The scale rule is a plastic rule, which measures 150 mm or 300 mm in length and is used to scale off dimensions on plans that are not given on the plan or drawing (see Figure 9.3).

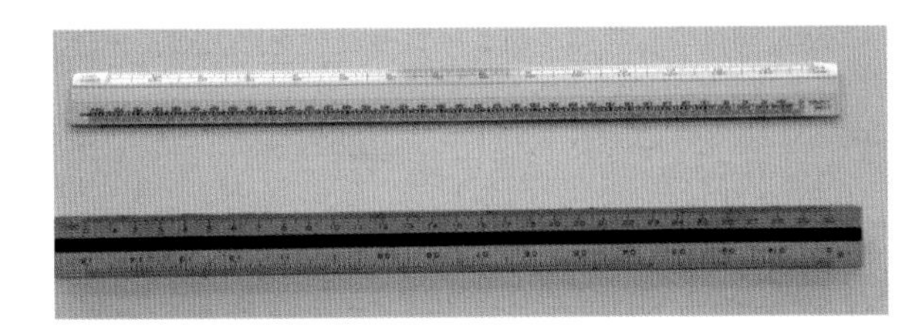

FIGURE 9.3 Scale rules

Retractable tapes

Retractable tapes have steel blades with an integral locking device. They come in various blade widths and lengths. Retractable tapes are commonly used within the building trade for medium-length measuring.

FIGURE 9.4 Wind-up tape

Source: age-fotostock/Jacek/VWPics

Wind-up tape

Wind-up tape blades are generally made from fibreglass or steel and can be up to 30 m in length (see Figure 9.4). These tapes can be used for marking out long lengths, such as pipework, drainage and roof sheet measuring and marking.

Laser distance measurer

The laser distance measurer, also referred to as a range meter, is a device that emits and returns a beam or signal at a solid object or target (see Figure 9.5). The device calculates the time the signal takes to hit the selected object and turns this time into a distance.

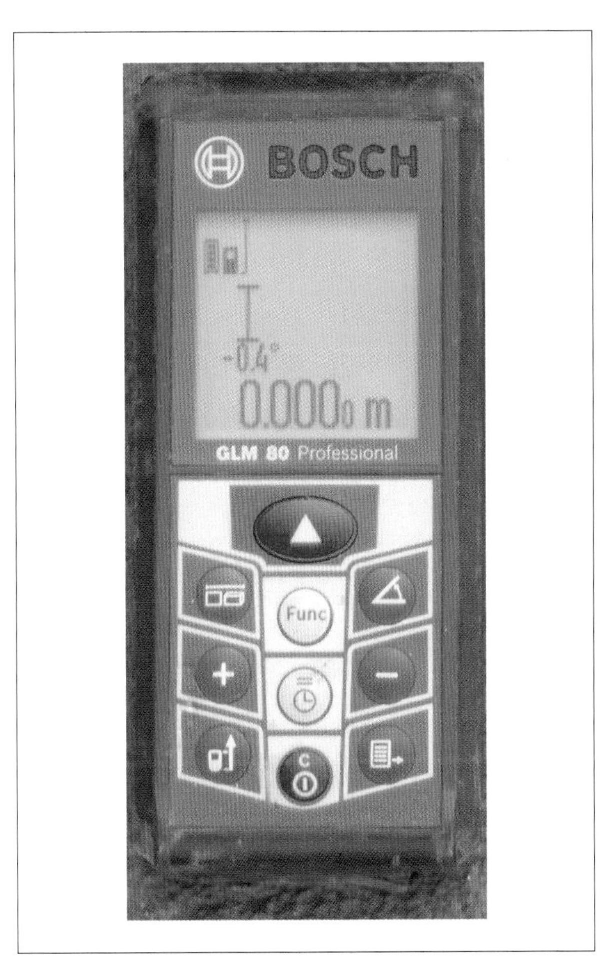

FIGURE 9.5 Laser distance measurer (range meter)

The advantage of this equipment is that it can be operated by one person, even when measuring or marking out over some distance horizontally or vertically. Most laser distance measures can perform many other tasks as well, such as area and volume calculations and recall of previous distances measured.

Laser level

The rotating laser (see Figure 9.6) is a battery-operated electronic device that emits a laser light in a rotating 360° plane either horizontal or vertical to the Earth's surface. It is commonly used to transfer known levels from one area of a building site to another, or to show a proposed building line such as a ceiling height. A staff and receiver may also be used to assist in transferring measurements.

FIGURE 9.6 Laser level

Engineer's try square

This square consists of a thin rectangular blade with a thicker stock set at 90° to the blade (see Figure 9.7). The engineer's try square can be used to check bends and folds for right angles and for squaring sheet material from the edge.

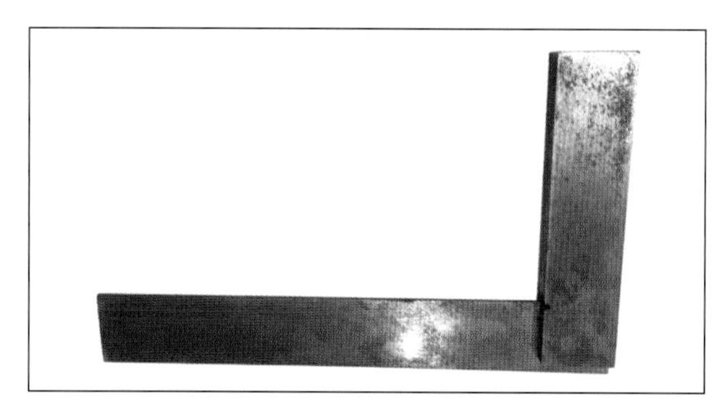

FIGURE 9.7 Engineer's try square

Builder's square

A builder's square (see Figure 9.8), also known as a flat square, has a uniform thickness to allow it to lay flat on the material for marking out, developing and checking if the corners of a room are square.

FIGURE 9.8 Builder's square

Dividers and compass

Dividers are designed for drawing circles, stepping off distances and striking an arc onto sheet metal material (see Figure 9.9). A compass is a technical drawing instrument and can perform most of the same functions as dividers; however, with the addition of a pencil or lead refills, its primary function is for drawing arcs and circles on paper.

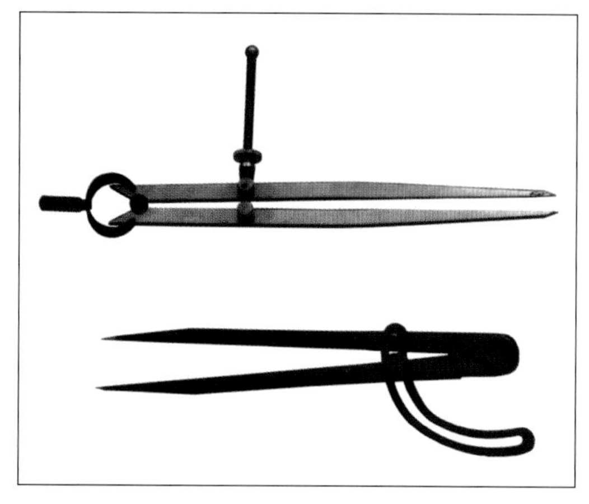

FIGURE 9.9 Dividers

Trammel

The trammel (see Figure 9.10) is used for the same purpose as dividers but has a greater reach and allows a larger radius to be drawn. It consists of two points or spikes that are adjustable along a timber or steel rail.

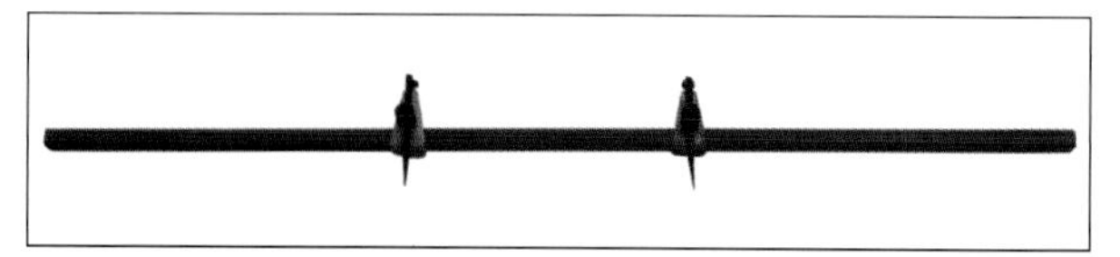

FIGURE 9.10 Trammel

Set squares

Set squares are primarily used in technical drawing and are triangular and transparent. They are very useful when used in conjunction with a technical drawing board. There are two types of set squares – the 60-30 and the 45 set square – which refer to the angles contained within the set square.

The 60-30 set square consists of 90° right angle with the remaining two angles being 60° and 30°. The 45 set square also consists of 90° right angle, with the remaining two angles both being 45° (see Figure 9.11).

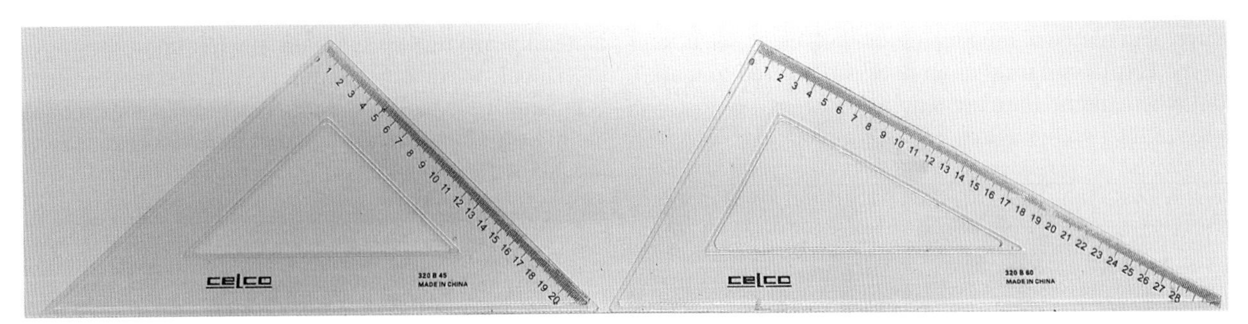

FIGURE 9.11 Set squares: 45 (left) and 60-30 (right)

The importance of understanding basic drawing principles – particularly using dividers, a compass or set squares – is that a number of different angles can be found without the use of a protractor. A circle can be divided quickly into multiples of 30°, 45° or 60° with any of these tools.

Protractor

A protractor (see Figure 9.12) is a 180°- or 360°- tool used to indicate an angle.

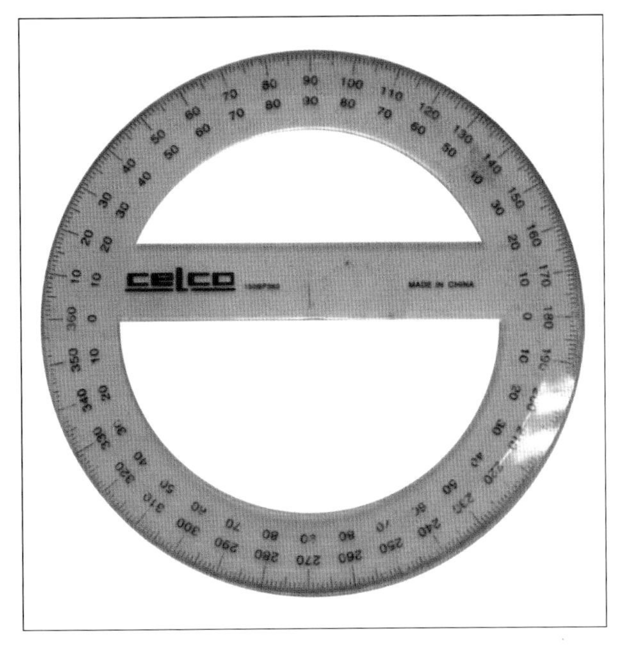

FIGURE 9.12 360° Protractor

Protractors can be made from various materials including metal, plastic or timber. The protractor can be used directly to mark off an angle on a job or used in conjunction with a sliding bevel to transfer the angular measurement.

Sliding bevel (adjustable bevel)

The sliding bevel (see Figure 9.13) is a hand tool similar to a try square, except that the blade is not fixed at the stock. The blade can be rotated around the stock until the required angle is reached. This angle is then fixed by tightening a nut or key.

FIGURE 9.13 Sliding bevel

Centre punch

A centre punch (see Figure 9.14) is a piece of hardened steel with a point at one end. The punch is hit on top with a hammer to leave an indentation or mark on the material. It can be used to locate a pivot point for dividers when striking an arc or circle, to locate a pilot hole when drilling, or to highlight marking-out lines.

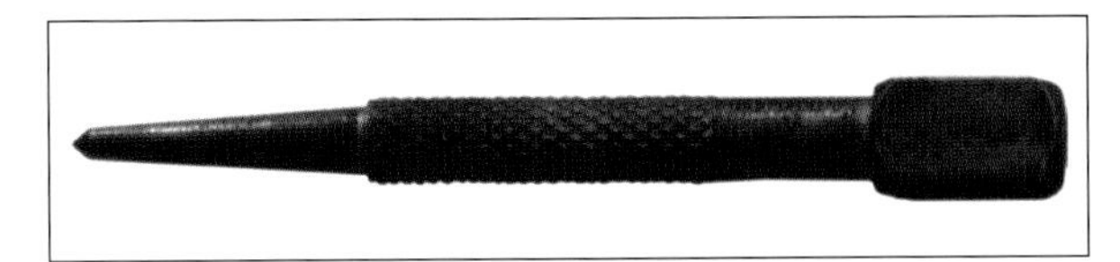

FIGURE 9.14 Centre punch

Plumb-bob

A plumb-bob is a weighted piece of metal with one end tapering in to form a point (see Figure 9.15). The other end is designed so that a string can be attached in the centre. When the plumb-bob is held by its string, the force of gravity will hold the plumb-bob perpendicular to the Earth's surface. Plumb-bobs can be used to mark out where plumb lines are required, such as where a pipe or duct will penetrate through a roof.

FIGURE 9.15 Plumb-bob

Sheet metal snips

Sheet metal snips (see Figure 9.16) can be used to help identify where cuts and folds in sheet metal are required. By making a small cut or notch in the side of the sheet, the person fabricating the job is not required to transfer marks or dimensions from one side of the sheet to the other. Sheet metal snips are also used to cut out templates where numerous copies of the same job are required.

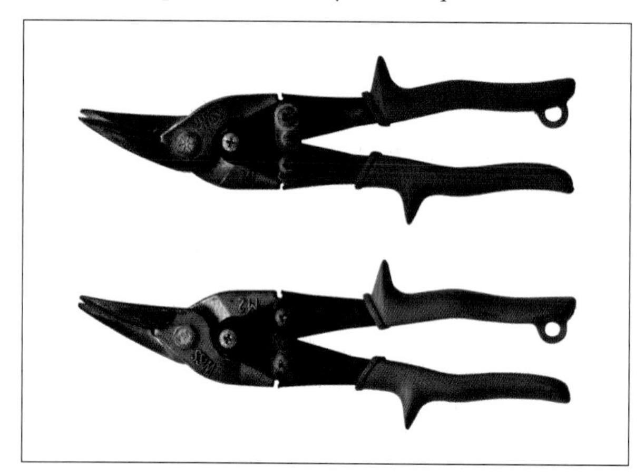

FIGURE 9.16 Sheet metal snips

Marking instruments

Material compatibility and selection is well documented, with detailed information usually found on plans, specifications and manufacturers' installation guidelines. The instrument that is used to mark out materials (see Figure 9.17) is usually left up to the individual, so correct selection is vital. If a marker that is incompatible with the material is used, it may reduce the life expectancy of the material or possibly cause fatigue of the material itself.

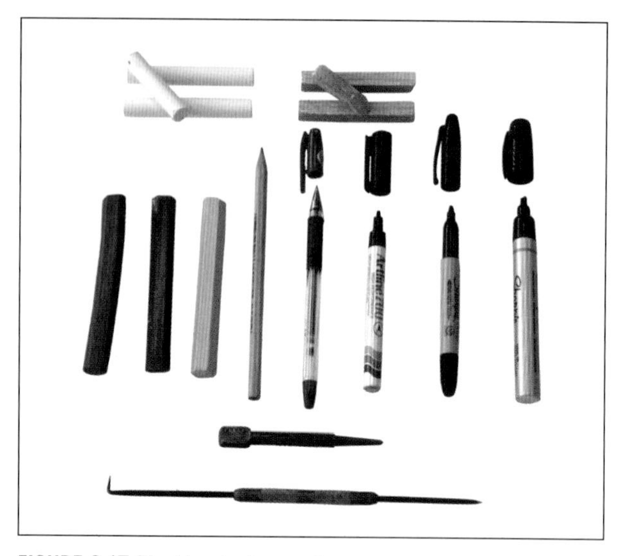

FIGURE 9.17 Marking instruments

Two examples of incorrect selection of marker and materials are as follows:

- Using a graphite clay pencil (lead pencil) to mark Zincalume® may leave permanent marks, which can lead to corrosion.
- If the centre line of thin gauge copper tube with a mechanically pulled bend has been marked or scratched with a scriber or the de-burring tool found on tube cutters, this could weaken or split the tube. This is due to the tube being put under stress during the bending process.

When deciding on what marker to use, consider the following points:

- Does the material have a smooth surface or is the surface textured?
- Does the material have a protective film applied that can be removed later?
- Will an additional protective coating be applied once the job has been installed?

Table 9.1 lists commonly used material and compatible markers.

TABLE 9.1 Plumbing materials and compatible markers

Marking instrument	Material
Ballpoint pen	Sheet metal with protective plastic, pre-painted sheet steel downpipes, and polished stainless steel
Chalk (blackboard)	Steel pipes and plates, and copper tube (subjected to heat)
Chalk and centre punch marks	Mild steel pipes and plates to be thermal cut
Chalk line	Sheet metal roofing profiles
Engineer's chalk	Steel pipes and plates
Felt-tip marker	Sheet metal and plastic pipe
Paint pen	Sheet metal, steel and plastic pipes; ideal for permanent marking of templates and patterns
Pencil (graphite clay)	Copper and plastic pipe
Scriber	Sheet metal

Personal protective equipment (PPE)

When used correctly, PPE will reduce the risks and hazards associated with marking out materials on the worksite. PPE should be provided to all workers by their employer to ensure that all tasks are carried out safely. All workers should receive appropriate instruction and training on the correct use and fitting of PPE to ensure it is used and protects as it was designed. PPE that does not fit or has not been fitted correctly may result in illness or injury and defeat the purpose of wearing the PPE in the first place.

Preparing the work area

Is the work area prepared for marking out appropriate for the job? Take a good look around prior to setting up, and consider the following:

- Are other trades likely to be working nearby, cluttering the area with their tools and equipment?
- Will there be adequate access, light and ventilation to safely carry out the work?
- Will the ergonomics be suitable? Is the workbench at a comfortable height, or will you have to crouch down while marking out?
- Is the surrounding area quiet enough to enable you to concentrate on the task or communicate with other people?

FROM EXPERIENCE

Actively communicating with other tradespeople on site helps improve your communication skills and promotes effective teamwork. Teamwork on site is essential for completing tasks and keeping on schedule.

GREEN TIP

Materials should be marked out carefully to reduce waste. Consider how unused materials could be reused, recycled or recovered, where possible.

LEARNING TASK 9.1

PREPARE FOR WORK

Provide short answers to the following questions:

1. What should first be obtained from the builder or supervisor in order to mark out materials on the worksite?
2. What document is required to be completed before work can commence?
3. State the importance of coordinating with other trades.
4. What is the advantage of using a wind-up tape rather than a retractable tape?
5. What compatible marker would be used to mark copper tube?
6. What is the benefit of preparing the work area when working with other trades?

COMPLETE WORKSHEET 1

Determine job requirements

An eye for detail is an essential component of working out job requirements. In conjunction with plans and specifications, the following points should be considered:

- Materials have been correctly selected and checked for compliance.
- Quantities and types of material have been correctly calculated.
- Appropriate pattern development methods have been considered.

Early identification of job requirements, including the need for specialised tools and materials, should be organised to meet job timelines. When organising tools and materials, allowance should be made for factors that may affect the supply of resources. A delay in sourcing material can create scheduling pressures for other tradespeople and hinder overall job progress.

Most job requirements are determined through reading plans and specifications; however, an article may be brought to a workshop for duplication. This could require specialised skills, techniques and pattern development to replicate an existing article.

FROM EXPERIENCE

When organising job requirements and ordering materials, think about the overall goal that you wish to achieve from the project. Considering a job over the long term will minimise delays in the future.

Pattern development

Having a basic awareness of what is involved in the processes of pattern development will give you a better understanding of what you should look for when ordering a particular item through plumbing and roofing suppliers, or when marking and cutting out to install the item.

Pattern development is taking a three-dimensional geometric shape, or combination of shapes, and unfolding the shape flat on the workbench so it can be reproduced for fabrication. A pattern development method is selected based on the shape and fabrication technique. When marking out sheet metal, there are three commonly used pattern development methods: parallel line development, radial line development and triangulation.

Parallel line development

Parallel line development is used for marking out shapes with parallel surfaces, such as prisms and cylinders, that can be cut at an angle. Uses include roof flashings for a cylindrical vent or square exhaust duct penetrating through a pitched roof.

Radial line development

Radial line development is used to develop shapes and sections of shapes that have tapering sides and that radiate from a common centre or apex, such as pyramids, right cones and frustums. An example of a frustum is a right cone with its pointed end cut off, like a funnel, which could be used as a tapered flashing over a round vent pipe.

Triangulation

Triangular development is a method of developing irregular shapes by dividing the object into a series of triangles without a common apex. This method is commonly used to develop square-to-round transition pieces and decorative rainwater heads.

Step-by-step development exercises

The dimensions given in the following learning exercises are based on the use of A3 paper with a 10 mm border line allowance. Each of the exercises requires some or all of the following equipment:

- 0.5 mm drawing pencil
- A3 piece of paper
- drawing board or T-square and spring clips or adhesive tape
- 60-30 set square
- compass
- dividers
- calculator.

Note: A perpendicular line is where two lines meet at a right angle.

HOW TO

RADIAL LINE DEVELOPMENT OF A RIGHT CONE

1. On the left-hand side of your A3 drawing paper in landscape orientation, about midway down, draw a horizontal line across the page. We will call this horizontal line our *base line.*
2. Come in from the left-hand side of the drawing by about a third; now draw a vertical line down perpendicularly past the horizontal line.
3. Set your compass to a radius of around 30–40 mm and draw a circle with the centre point 50 mm below the horizontal line and on the vertical line.
4. Draw a horizontal line through the centre of this circle and then divide the circle into 12 sections using your 60-30 set square. Number each section, as shown in Figure 9.19. This is called our *plan* or *base view* and is what the cone would look like if viewed from above.

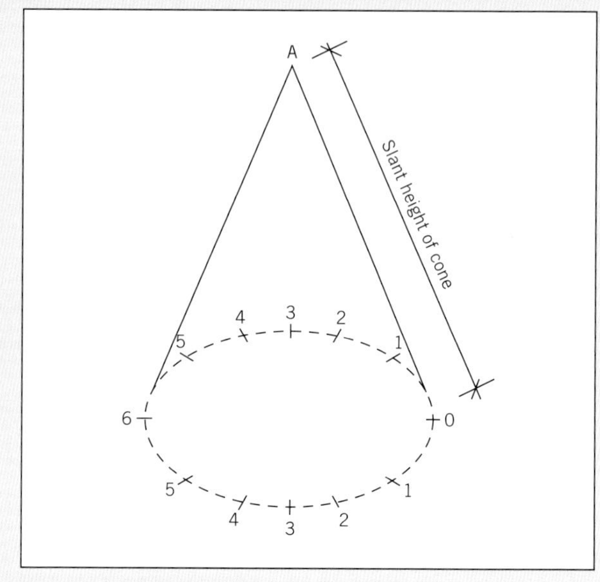

FIGURE 9.18 A right cone

>>

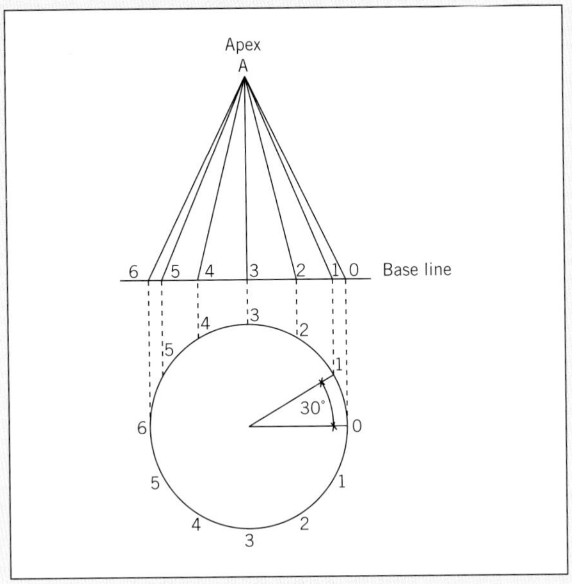

FIGURE 9.19 A right cone – plan and elevation views

5. Project the points 6 through to 0 up to the base line and measure up the perpendicular line above the base line approximately 60–70 mm. This point will become the apex of your cone.
6. Draw lines from the points off the base line to the apex to form your right cone.
   Note: The two outside lines on this cone (0 and 6) are called *true length lines*. The lines are also referred to as the *slant height* of the cone (see **Figure 9.18**). The view of the cone is commonly called the *front view* or elevation.
7. From the apex of the cone, move across the drawing to the right side about another third and draw a vertical line down. Set your compass to the slant height of the cone using a line from either 0 or 6 to the apex. Now draw an arc by placing the centre of the compass on the new vertical line at about the same height as the apex of the cone.
8. To develop the cone, we must step off around the arc the same number of intervals that we divided our plan view into – in this case 12 sections (see **Figure 9.20**). To get this measurement we can use our compass and set the distance from any one of the 12 intervals on the plan view or calculate the distance mathematically using the circumference of a circle formula $C = D \times \pi$ (i.e. diameter × 3.1416), then divide the circumference length by the number of intervals that the plan view has been divided into. As a general guide, the greater the diameter of the plan view, the greater the number of intervals will be used.
9. Start where the arc has cut through the vertical line and mark off six intervals either side of this point. Referring to the base plan, identify your points around the arc and then draw in the radial lines from these points back to the apex. This completes the development of a right cone using the radial line method (see **Figure 9.20**). Triangulation can also be used to develop this.
   Note: No lap allowance has been made for jointing. Should this be required, the lap is normally placed on the shortest edge.
   Other common variations to this drawing include:
   - the attachment of the development to the elevation by drawing the arc from the slant height or true length line
   - the drawing of only a half plan or base view, as this geometric shape is symmetrical and doesn't require a full plan.

   Note: Always use construction lines when drawing a pattern development, as this will minimise pencil indentations and will make it easier to erase mistakes. Once completed, darken the outline to provide a contrast between the construction lines.

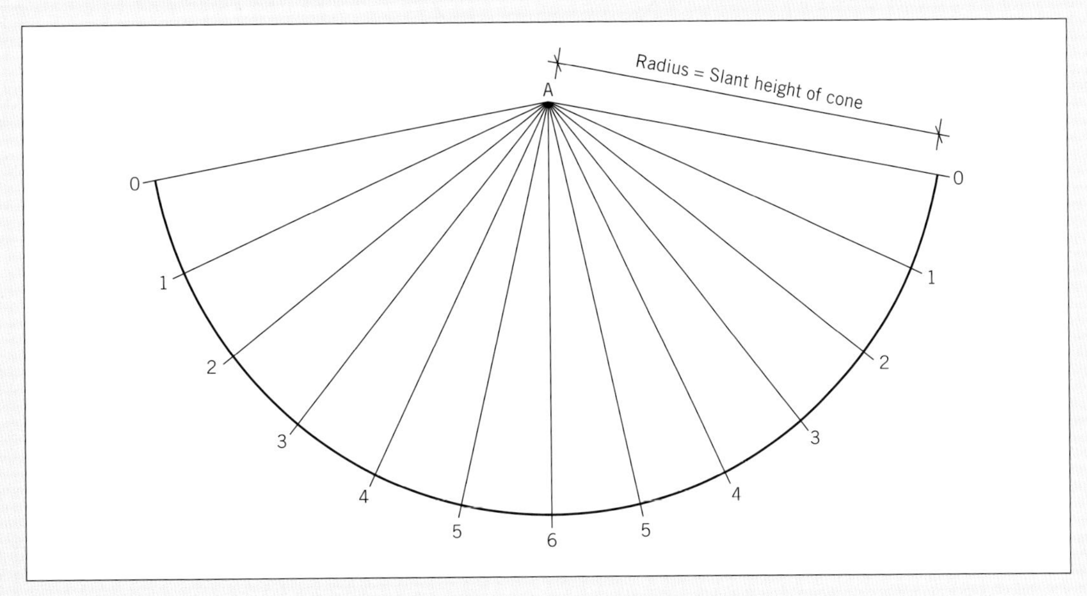

FIGURE 9.20 A right cone pattern development

HOW TO

RADIAL LINE DEVELOPMENT OF A TRUNCATED CONE

This item can be used as a weathering flashing for a round pipe penetration through a relatively flat roof (see Figure 9.21).

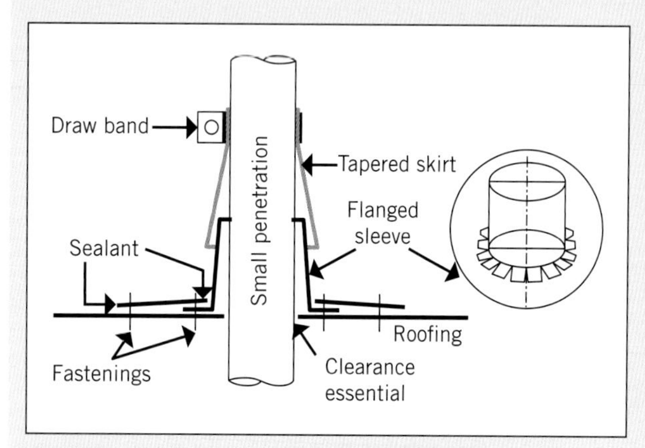

FIGURE 9.21 Small penetration with metal skirt and sleeve (note tapered skirt overlapping and weatherproofing the sleeve)

1 On the left-hand side of your A3 drawing paper in landscape orientation, about midway down, draw a horizontal line across the page. We will call this horizontal line our *base line.*
2 Come in from the left-hand side of the drawing by about a third; now draw a vertical line perpendicularly down past the horizontal line.
3 Set your compass to a radius of around 30–40 mm and draw half a circle with the centre point 20 mm below the horizontal line and on the vertical line.
4 Draw a horizontal line through the centre of this circle and then divide the circle into six sections using your 60-30 set square. Number each section as shown in Figure 9.22. This is called our *half plan* or *half base view.*
5 Project the points 6 through to 0 up to the base line and measure up the perpendicular line above the base line approximately 60–70 mm. This point will become the apex of your cone.

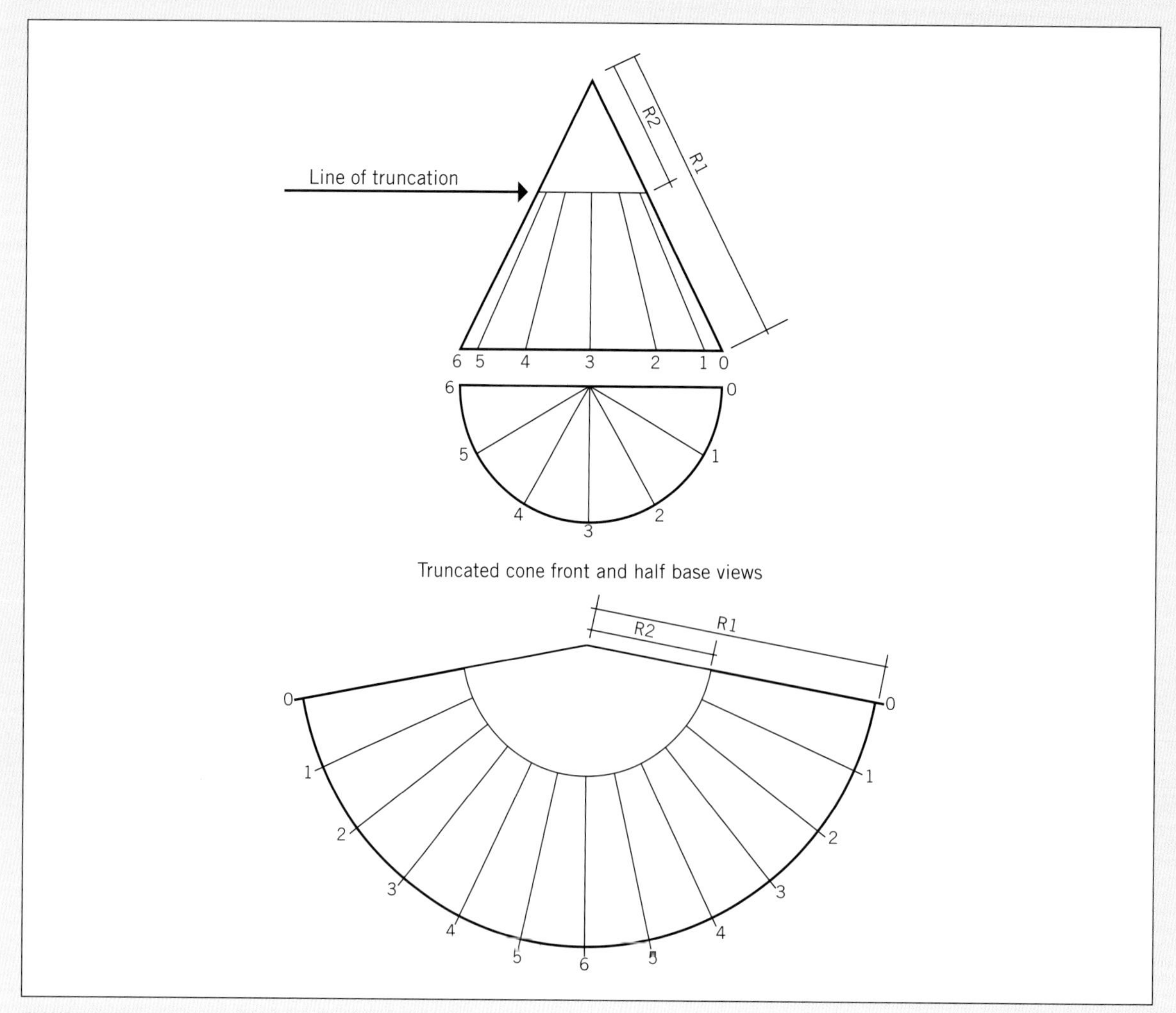

FIGURE 9.22 A truncated cone pattern development

>>

>>

6 Draw lines from the points off the base line to the apex to form your right cone.
Note: The two outside lines on this cone (0 and 6) are called *true length lines*. The lines are also referred to as the *slant height* of the cone. The view of the cone is commonly called the *front view* or *elevation*.

7 From the apex, measure down vertically approximately 30 mm and draw a horizontal line across at this point. This line can be referred to as the *line of truncation* or *horizontal cut line*. The line of truncation is an imaginary line that separates or cuts the cone at this point.

8 From the apex of the cone, move across the drawing to the right side about another third and draw a vertical line down. Set your compass to the slant height of the cone using a line from either 0 or 6 to the apex. We will call this measurement R1. Now draw an arc by placing the centre of the compass on the new vertical line at about the same height as the apex of the cone (with reference to the elevation).

9 To develop the cone, we must step off around the arc the same number of intervals that we divided our full plan view into (in this case 12 sections: six per each half plan). To get this measurement you can use the compass and set the distance from any one of the six intervals on the half plan view or calculate the distance mathematically using the circumference of a circle formula, $C = D \times \pi$ (i.e. diameter × 3.1416), then divide the circumference length by the number of intervals that the plan view has been divided into. As a general guide, the greater the diameter of the plan view, the greater the number of intervals will be used.

10 Start where the arc has cut through the vertical line and mark off six intervals either side of this point. Referring to the base plan, identify your points around the arc and then draw in the radial lines from these points back to the apex.

11 To draw the line of truncation on the development, set your compass to the distance between the apex and where the truncation line cuts point 6 or 0 in the elevation – we can call this point R2. Now carefully draw in an arc between the point 0 across to the other point 0 in the development, using the existing centre point. This completes the development of a truncated right cone or frustum using the radial line method (see Figure 9.22).
Note: no lap allowance has been made for jointing. Should this be required the lap is normally placed on the shortest edge.

A common alternative to this type of flashing is a Dektite®.

HOW TO

DEVELOPMENT OF A 30° TRUNCATED RIGHT CONE

This item can be formed and used as a weathering flashing (as part of the complete process) when weatherproofing a round pipe penetration through a sloping roof.

1 On the left-hand side of your A3 drawing paper in landscape orientation, about midway down, draw a horizontal line across the page. We will call this horizontal line our *base line*.

2 Come in from the left-hand side of the drawing by about a third; now draw a vertical line perpendicularly down past the horizontal line.

3 Set your compass to a radius of around 30–40 mm and draw half a circle, with the centre point 20 mm below the horizontal line and on the vertical line.

4 Draw a horizontal line through the centre of this circle and then divide the circle into six sections using your 60-30 set square. Number each section as shown in Figure 9.23. This is called our *half plan* or *half base view*.

5 Project the points 6 through to 0 up to the base line and measure up the perpendicular line above the base line approximately 60–70 mm. This point will become the apex of your cone.

6 Draw lines from the points off the base line to the apex to form your right cone.
Note: the two outside lines on this cone (0 and 6) are called *true length lines*. The lines are also referred to as the *slant height* of the cone. The view of the cone is commonly called the *front view* or *elevation*.

7 From the apex, measure down vertically approximately 45 mm and draw a line at 30° to the horizontal across the face of the cone. This line can be referred to as the *line of truncation* or *horizontal cut line*. The line of truncation is an imaginary line that separates or cuts the cone at this point.

8 From the apex of the cone, move across the drawing to the right side about another third and draw a vertical line down. Set your compass to the slant height of the cone using a line from either 0 or 6 to the apex. Now draw an arc by placing the centre of the compass on the new vertical line at about the same height as the apex of the cone (with reference to the elevation).

9 To develop the cone, we must step off around the arc the same number of intervals that we divided our full plan view into (in this case 12 sections, six per each half plan). To get this measurement you can use the compass and set the distance from any one of the six intervals on the half plan view or calculate the distance mathematically using the circumference of a circle formula,

>>

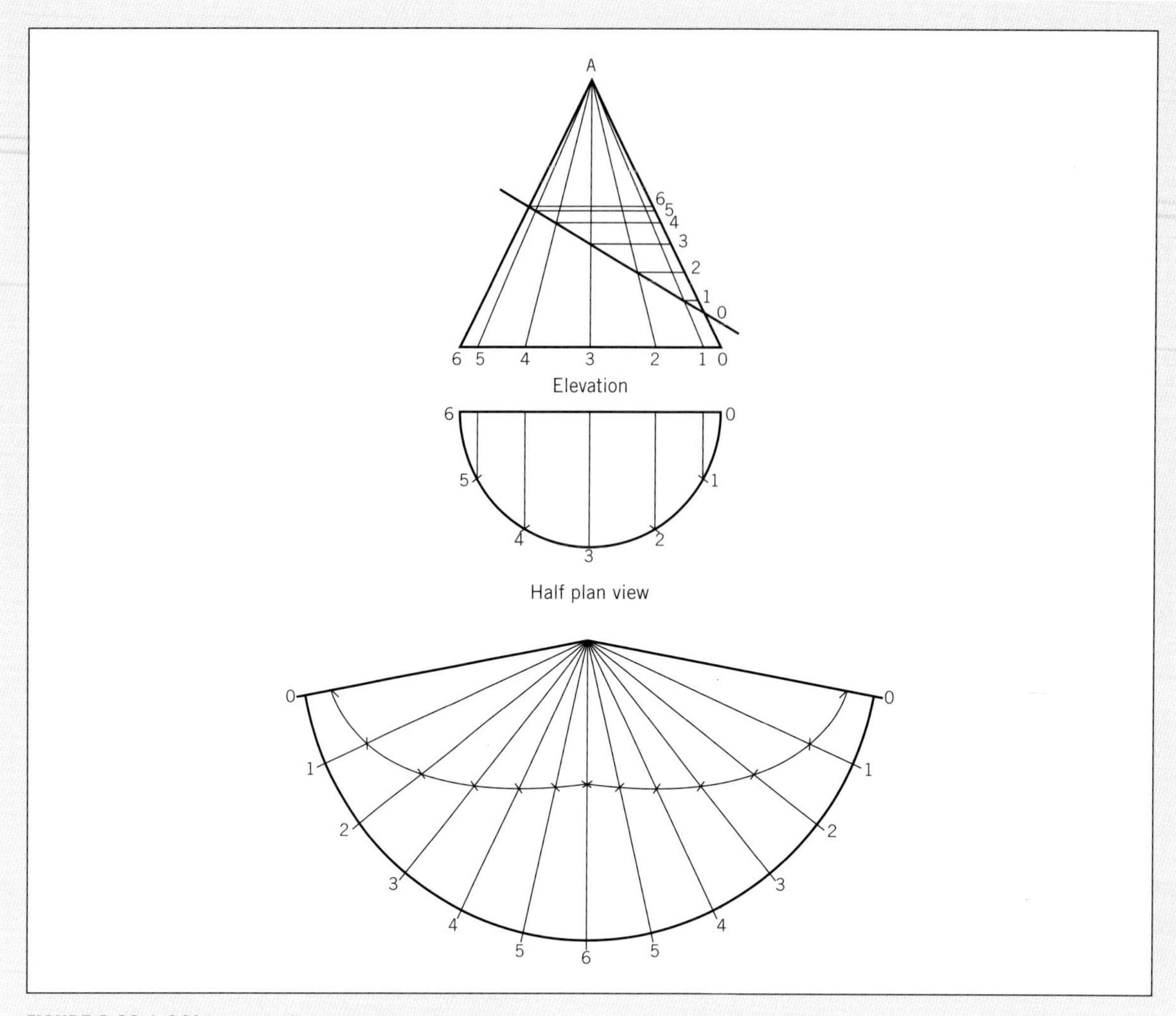

FIGURE 9.23 A 30° truncated cone pattern development

$C = D \times \pi$ (i.e. diameter × 3.1416), then divide the circumference length by the number of intervals that the plan view has been divided into. As a general guide, the greater the diameter of the plan view, the greater the number of intervals will be used.

10 Start where the arc has cut through the vertical line and mark off six intervals either side of this point. Referring to the base plan, identify your points around the arc and then draw in the radial lines from these points back to the apex.

11 To draw the line of truncation on the development we must take the distances off the elevation on a true length line. Take one more look at the elevation and then close your eyes and visualise the points on the base line converging to the apex. Notice that the points 1, 2, 3, 4 and 5 are not true length, only points 0 and 6.

Relating this back to our drawing, to change points 1, 2, 3, 4 and 5 to true length we project (draw a horizontal line) across from where lines 1, 2, 3, 4 and 5 are cut by the line of truncation to line 0 or 6, which is true length, and identify with corresponding numbers.

12 Take your compass or dividers and mark off each distance along line 0 from the apex to each number. Transfer this distance to the same number on the development. Now, with a pencil and flexible curve or freehand, join all the points.

Note: no lap allowance has been made for jointing. Should this be required the lap is normally placed on the shortest edge.

This completes the development of a 30° truncated right cone or frustum using the radial line method (see Figure 9.23).

Note: Another way of looking at true length in this problem is to think of a ladder leaning against a wall: the feet of the ladder are sitting on the base line and the top of the ladder is the apex of the cone. If you were directly behind the ladder and looked up towards the top of the ladder it would not appear to be the true length. The same would be said if you walked up the ladder and looked down. But if you walked to the side of the ladder you would get a true indication of how long the ladder really is!

HOW TO

PARALLEL LINE DEVELOPMENT OF A SQUARE VENT PASSING THROUGH A PITCHED ROOF

Figure 9.24 indicates a typical square penetration through a metal roof and some of the specific details associated with it. The information below explains the steps involved in developing the marking out and cutting process.

In this drawing we are simulating a square vent duct passing through the roof of a building (see Figure 9.24). You are required to develop and mark out in sheet metal material an appropriate flashing that could be used as part of the weatherproofing between the pipe and the building. To assist the tradespeople on site we will also develop a true shape of the section as a template to cut the hole in the roof.

1 About midway down your A3 piece of paper in the landscape orientation, draw a horizontal line across. We will refer to this as our *base line*.
2 Come in 90 mm from the left-hand side of your page or border line; put a mark on the base line and label this point 'P'. Now from point P draw a line up at 30° inclined to the left. This will simulate the pitch of the roof.
3 From point P along the base line, come in to the left 45 mm and draw a perpendicular line about 60 mm below the base line and draw a horizontal line. This vertical line will become our centre line.
4 On the perpendicular line and below the base line, measure down a further 50 mm and draw an additional horizontal line. Using the horizontal and perpendicular lines as the centre, measure 25 mm each side horizontally of the lines to draw a 50 mm square. This is the plan view of the duct. Now project the outside edges of the square up 100 mm above the base line and draw a horizontal line across to join the two outside edges, forming a side elevation. Now is a good time to identify the plan view with numbers or letters.
5 Set your compass at point P and where the outside lines cut through the 30° line, using point P as

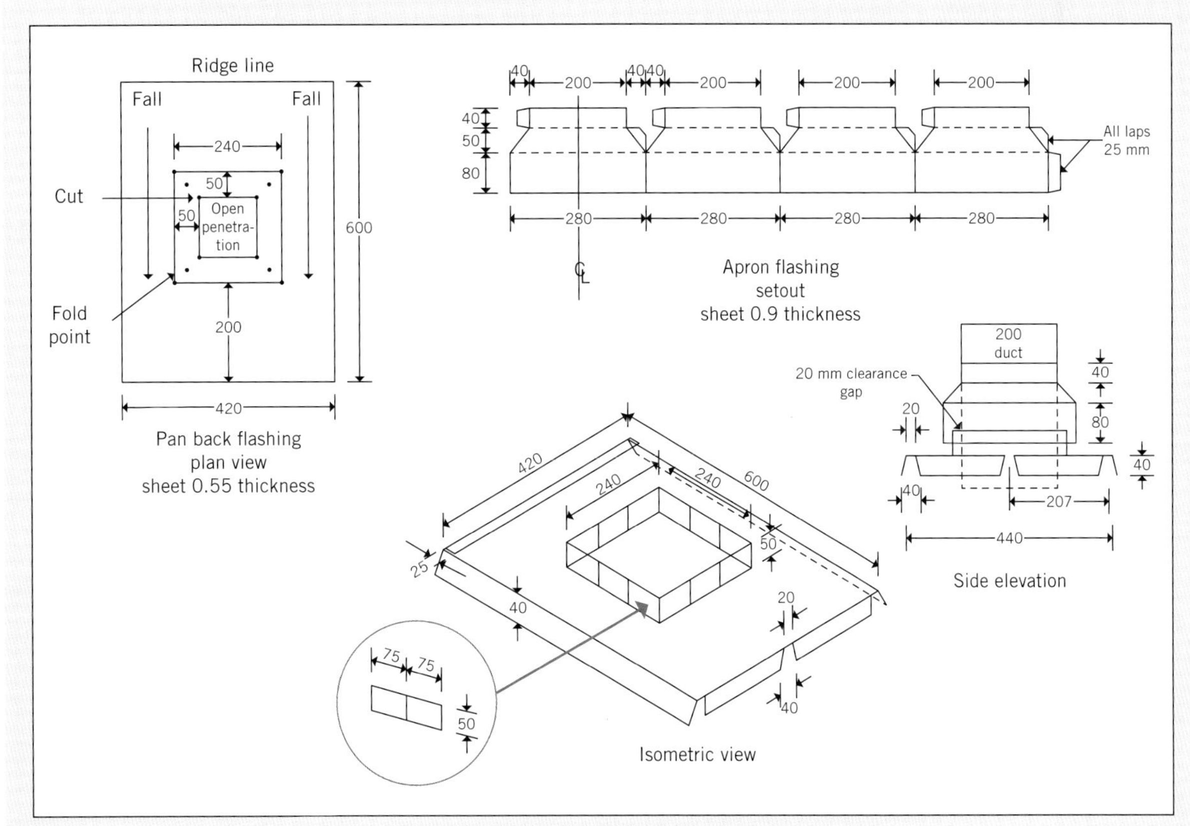

FIGURE 9.24 Weatherproofing a large square or rectangular penetration through a sloping roof

>>

your pivot point, swing the points across to the right and down to the base line. Identify these points along the base line for future reference.

6 Where the points are touching the base line, project these points down vertically approximately 110 mm. Now, from the plan view, project the points across to the right where the corresponding numbers or letters meet. This is the true shape of the section that would be required to be cut into the roof.

7 Measure in 180 mm along the base line from the left side as a development start point, then an additional 200 mm for the four 50 mm sides of the duct. Now draw the two points up perpendicular from the base line to a height equal to the top of the elevation. Now draw a line parallel to the base line joining the two vertical lines to form a rectangle.

Note: if there is insufficient room for this development along the base line, find an area large enough to fit the above dimensions; alternatively, use masking tape to add an additional piece of paper to accommodate the design.

8 In most developments a joint or lap is usually placed along the shortest side, in this case between points A and B. We will therefore start and finish between these two points and label them S to represent a seam or joint. Now measure half the distance on the plan view between A and B (in this case 25 mm from each end to S) and label points B and A respectively on the development view. Mark off the remaining three intervals (B-C, C-D, D-A) on the top horizontal line of the development at 50 mm each side and then project all points down vertically to the base line and complete the identification of the points.

9 Project the two lines horizontally across to the end of development, from the elevation at points A1, B1 and C1, D1 and mark where the respective lines intersect.

Note: If your development is not on the same base line, use the compass or dividers and transfer the measurements to the development.

10 Highlight around the outline to show where you will be required to cut out the development and label as required.

Note: This development would form the square upstand section that protrudes from the roof. A base plate, apron flashing and additional laps would also be required to be installed to complete this job. Allowance would also be required to ensure that an annular clearance is left between the square vent and the upstand to meet the requirements of SA HB 39:2015.

The same concept can be used on a flat roof as shown in Figure 9.25. An upstand is fabricated and attached to the back flashing, which will accommodate the square ducting. An apron flashing is fabricated to fit the size of the square duct and sits over the upstand and allows for expansion and contraction of the material while providing sufficient weatherproofing.

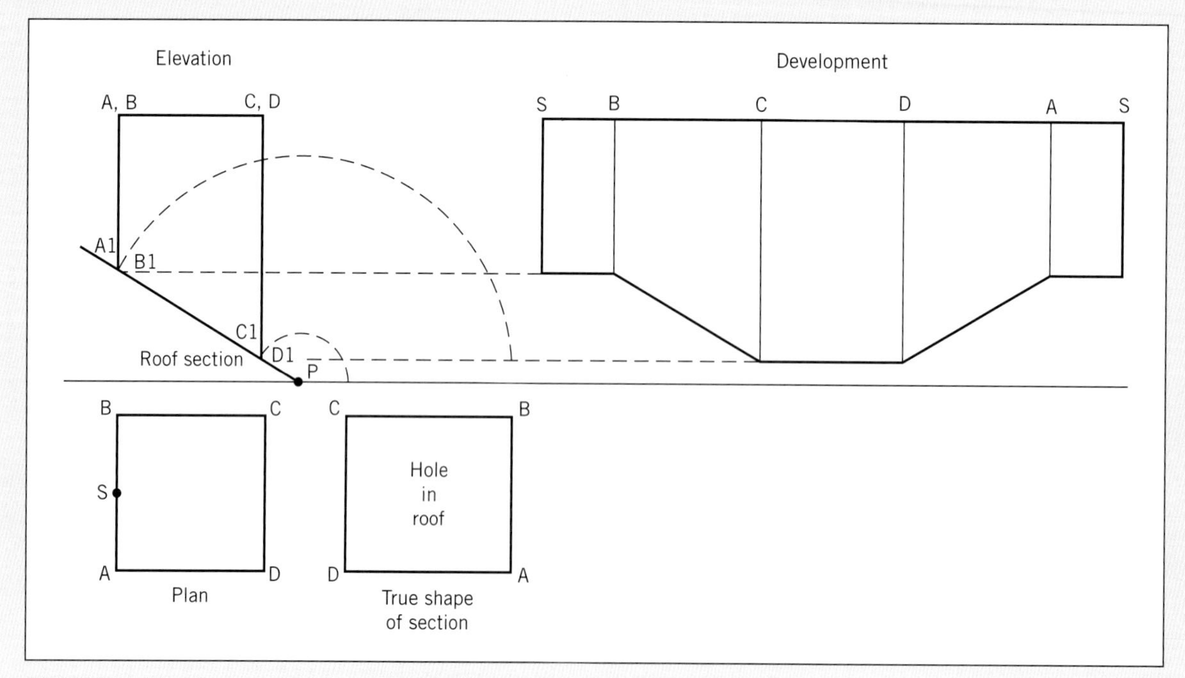

FIGURE 9.25 The development of a square duct penetrating through a pitched roof

HOW TO

PARALLEL LINE DEVELOPMENT OF A ROUND VENT OR PIPE PASSING THROUGH A ROOF

In this drawing we are simulating a pipe or vent duct passing through the roof of a building. You are required to develop and mark out in sheet metal material an appropriate flashing that could be used to weatherproof between the pipe and the building. To assist the tradespeople on site, we will also develop a true shape of the section as a template to cut the hole in the roof.

1 About midway down your A3 piece of paper in the landscape orientation, draw a horizontal line across. We will refer to this as our *base line*.
2 Come in 110 mm from the left-hand side of your page or border line and put a mark on the base line, labelling this point 'P'. Now from point P draw a line up at 30° inclined to the left. This will simulate the pitch of the roof.
3 From point P along the base line, come in to the left 45 mm and draw a perpendicular line about 100 mm above and 60 mm below the base line.
4 On the perpendicular line and below the base line, measure down 35 mm and draw a circle with a radius of 25 mm. This is the plan view of the pipe. Now project the outside edges of the circle up 100 mm above the base line and draw a horizontal line across to join the two outside edges, forming a side elevation.
5 Divide the plan view into 12 equal segments using the 60-30 set square. Nominate the points in the plan view and then project these points up into the elevation using light lines.
6 Set your compass at point P and where the light lines cut through the 30° line, using point P as your pivot point, swing the points across to the right and down to the base line. Identify these points along the base line for future reference.
7 Where the points are touching the base line, project these points down vertically approximately 65 mm. Now, from the plan view, project the points across to the right and put a mark where the corresponding numbers meet. Join the 12 points (freehand or with the aid of a flexible curve) to complete the true shape of the section. This true shape would be the hole required to be cut into the roof.
8 In order to develop a cylindrical section that protrudes past the 30° line in the elevation, we must work out its circumference. One method is to step off along the base line or, where room permits, the same number of intervals that we divided the plan view into (in this case 12 sections) or calculate the distance mathematically using the circumference of a circle formula, $C = D \times \pi$ (i.e. diameter × 3.1416), then divide the circumference length by the number of intervals the plan view has been divided into.
9 From these points draw perpendicular lines up from the base line and identify the 12 segments. *Note:* In most developments a joint or lap is usually placed on the shortest side, in this case point 6. Therefore, we will start and finish at this point.
10 Now either project the points where the light lines cut the 30° line across to the corresponding number in the development, or step off the distance with a compass or dividers from the top of the elevation down the light lines to where it is cut by the 30° line and then mark this distance off on the matching line in the development.
11 Join the 13 points (freehand or with the aid of a flexible curve) to complete the development (see Figure 9.26).
Note: This development would form the cylindrical upstand (flanged sleeve) section that protrudes beyond the roof (see Figure 9.25). A base plate fitted with a tapered skirt flashing would also be required to be installed to complete this job.

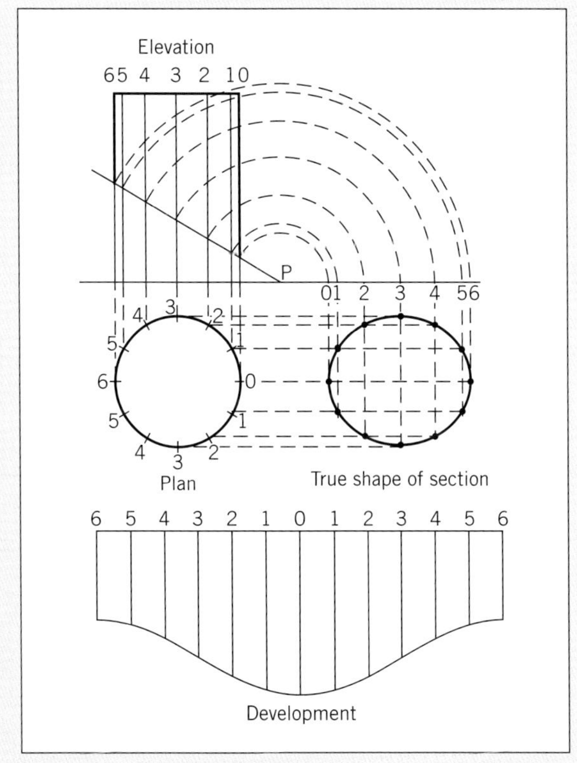

FIGURE 9.26 The development of a circular pipe upstand penetrating through a pitched roof

HOW TO

TRIANGULATION DEVELOPMENT OF A SQUARE TO ROUND TRANSITION PIECE

This development is an introduction into development by triangulation. The project is a square flat base tapering in to a centrally located circular top.

1 On the left-hand side of your A3 drawing paper in landscape orientation, about midway down, draw a horizontal line across the page. We will call this horizontal line our *base line*.

Plan view

2 From the left-hand side, come in 100 mm and draw a line down approximately 80 mm above the base line and 80 mm below down the perpendicular. Now come down the perpendicular line 45 mm below the base line and draw a circle with a radius of 20 mm at this point; around this circle draw a square with side length of 60 mm.

3 Divide the circle into 12 equal parts or 30° using your 60-30 set square and label the circle, starting with the number 0 at the top of the circle. Now identify around the outside of the square using letters as indicated. Join the points C-3, C-2, C-1, C-0, S-0, D-0, D-1, D-2, D-3, A-3, A-2, A-1, A-0, B-0, B-1, B-2, B-3. The letter S in between letters C and D will be the seam or joint.

Note: Because the circle is in the centre of this exercise and symmetrical in two directions, we really only need to mark out a quarter of the plan, which is the reason for numbering in this way. We will, however, draw the complete plan and elevation for clarity.

Elevation

4 Project up to the base line from the plan view points BC, 3, 2, 1, OS, AD, 3, 2, 1.

5 Come up the perpendicular line 75 mm above the base line and draw a horizontal line. From the base line, project the points 3, 2, 1, 0, 1, 2, 3 to the top of the elevation and join to the base line points as indicated (see Figure 9.27). You can see that we have now broken the plan and elevation into a series of triangles; however, neither view will indicate true length lines.

Finding true length

6 To find true length we must compare the plan view line length against the vertical height in the elevation. This measurement could be calculated mathematically; however, when marking out manually it is common practice to draw a true length key or diagram. To do this, come in along the base line 50 mm and call this point 'P', then draw a vertical line up at this point to the same height as the elevation and call this point '0'. With a compass or dividers measure out from point S to 0 in the plan view; transfer this measurement from point P on the true length key and mark this point along the base line. Identify this as point 0-S and draw a line from this mark up to the top of the vertical line at point 0. This line will be a true length. Repeat this

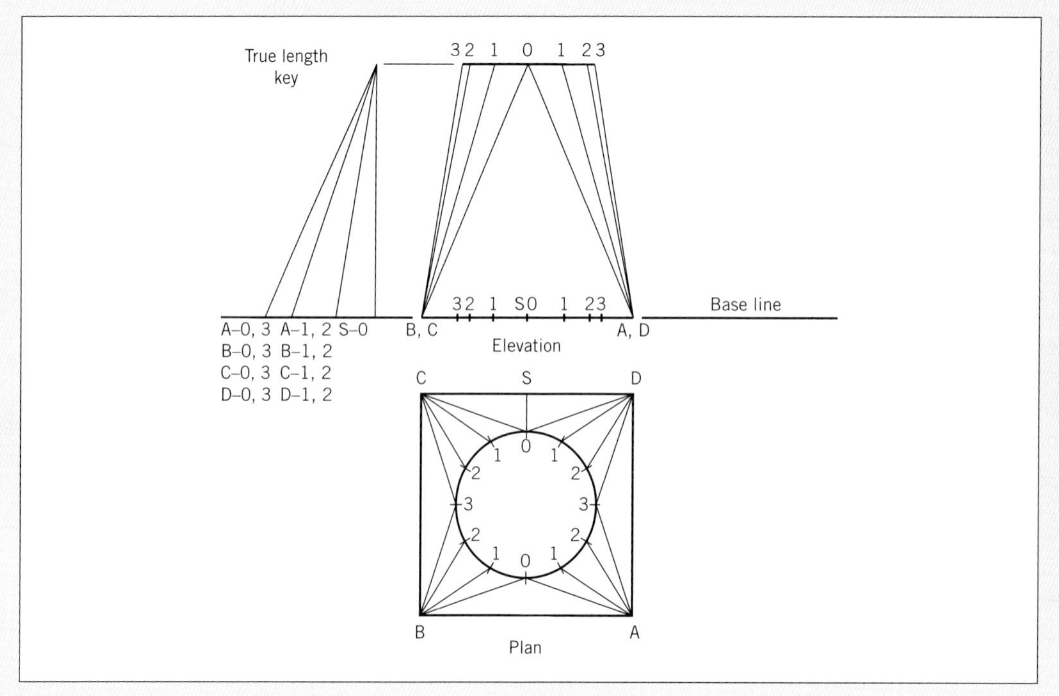

FIGURE 9.27 The plan and elevation of a square to round transition piece using triangulation

>>

>> exercise for points A-0 or 3 as well as A-1 or 2. By now you will have realised that, due to the symmetry of the plan view, all of the other points from B, C or D will be the same length as A-0 or A-1.

Label the points under their respective true length lines as indicated in Figure 9.27.

The development (Figure 9.28)

7 To start the development, draw a horizontal line at approximately the same height as points A and B in the plan view on the right side of the page. From the left side of the page or border, come in approximately 265 mm and draw a line vertically. To find points A and B on the development using your compass find a midpoint between A and B on the plan view (this will be the true length). Now mark this distance either side of the vertical line along the horizontal line in the development. Identify these points as A and B.

8 To locate point 0 for the development, set your compass to point A-0 or B-0 on the true length key. Now strike an arc from points A and B on the development vertical line. Where the arcs bisect this line, identify this as point 0. Now draw lines from point A to B, B-0, A-0.

9 To find points A-1 and B-1, locate the true length on the true length key and strike an arc from points A and B on the development. With your dividers or a second set of compasses, set a distance from point 0 to 1 in the plan view (which is true length) and transfer this measurement to the development. From point 0, swing an arc either side to bisect the arcs from A-1, B-1 and identify these as point 1. Draw a line from point A-1 and B-1.

10 To find points A-2 and B-2, locate the true length on the true length key and strike an arc from points A and B in the development. With your dividers or a second set of compasses, set a distance from point 0 to 1 in the plan view (which is true length) and transfer this measurement to the development. From point 1, swing an arc to bisect the arcs from A-2, B-2 and identify these as point 2. Now draw lines back from the points 2 to A and B.

11 To find points A-3 and B-3, locate the true length on the true length key and strike an arc from points A and B in the development. With your dividers or a second set of compasses, set a distance from point 0 to 1 in the plan view (which is true length) and transfer this measurement to the development. From point 2, swing an arc to bisect the arcs from A-3, B-3 and identify these as point 3. Now draw lines back from the points 3 to A and B.

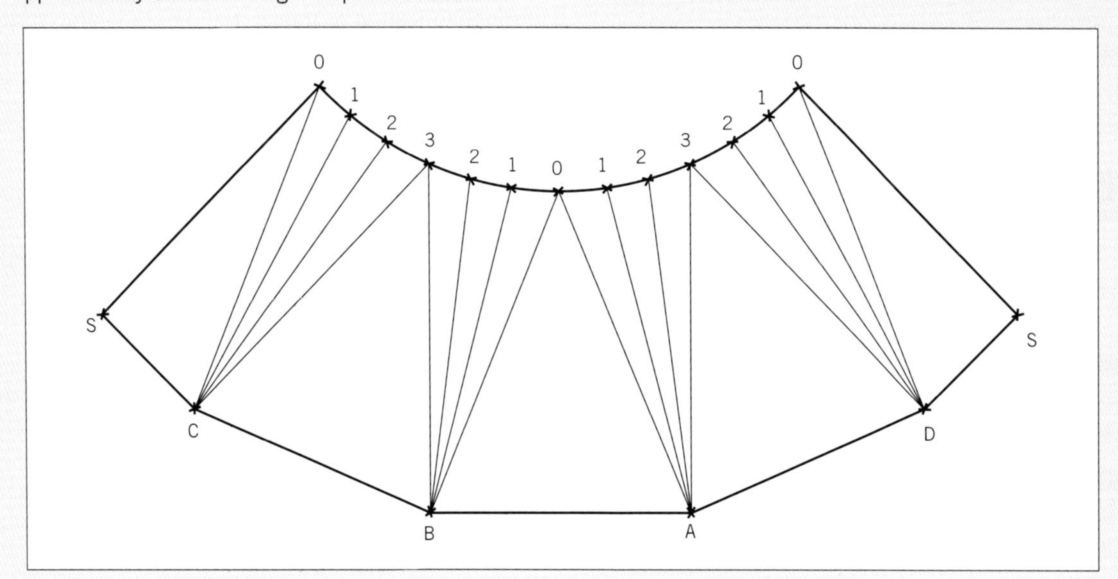

FIGURE 9.28 The development of a square to round transition piece using triangulation

Locating points C and D

12 To find points C and D, set the compass to either points A-D or B-C in the plan view and transfer this measurement to the development and strike arcs from points A and D. Reset compass to points B-3 or A-3 on the true length key. Swing arcs from point 3 to bisect the corresponding arcs from B-3 and A-3. Identify the new points as D and C. Draw lines from A-D-3 and B-C-3 then repeat steps 8, 9, and 10 to locate the new points for C-2, 1, 0 and B-2, 1, 0.

Locating point S

13 To find point S on the development, set your compass to either points C-S or D-S on the plan view. Using this length, strike an arc from points C and D. Then set the compass from the true length line 0-S and bisect the arc from the points C and D at the points 0 along the lines 0-C and 0-D. Draw in the lines C-S-0 and D-S-0. To complete the development, use a flexible curve or draw freehand to connect the points from C-0 to D-0. *Note:* No lap allowance has been made for jointing. Should this be required the lap is normally placed on the shortest edges.

LEARNING TASK 9.2

DETERMINE JOB REQUIREMENTS

Using the example in Figure 9.25 as a guide, draw a parallel line development for an upstand of a square vent passing through a pitched roof.

The dimensions for the upstand of the square duct are 100 mm × 100 mm and the angle of the pitched roof is 30°.

An allowance for the annular clearance between the upstand and square duct is required. Refer to *SA HB 39:2015 Installation code for metal roof and wall cladding* and provide a clause reference with your answer.

- SA HB 39:2015 Installation code for metal roof and wall cladding

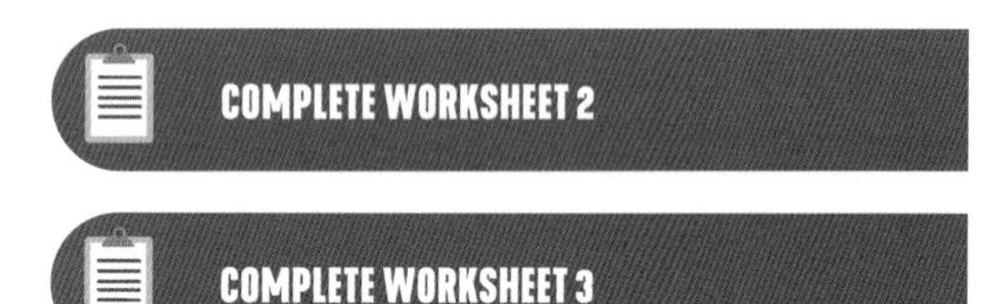

Mark out the job

When determining dimensions in the construction industry, correct measurement and tool choice is essential. Calculations may be required prior and during the marking out process, so it is important to have an understanding of what requirements need to be met. It is vital for a tradesperson to calculate measurements and transfer dimensions from the plans to the job to ensure that fabrication and assembly of plumbing work is undertaken correctly.

Marking out pipework

When marking out pipework, both the process and allowance considerations are similar, regardless of the material type. Where a job requires numerous layouts of the same pipework, real cost savings in both labour and materials can be achieved. An example of this can be seen in high-rise apartment blocks. The pipework and appliances connected in these types of buildings are often the same on each floor level. This repetition allows the person marking out the pipework to carry out this task very quickly, either at the job site or the plumbing workshop.

The process

Marking out pipework for fabrication involves taking as many dimensions as possible from either the job site or from plans and specifications. These measurements must be clearly drawn on a plan or pipe layout chart. The plan or drawing must show all correct drawing terminology and symbols, and be referenced to current codes and standards. This plan should then be identified and cross-referenced to the building's architectural plans, so that tradespeople fabricating or installing the pipe sections on site are able to clearly identify what section of pipe goes where.

The next step is to break the pipework into sections of a manageable size. This is to reduce the possibility of damage to the pipe sections or injury to workers during fabrication or transportation of the pipework, or during the installation process. By strategically placing the pipe section joints in a position where short lengths of pipework can be cut to length, any site variations can be accommodated.

Pipework allowances

When marking out pipework, consideration for accumulated gains or losses that will change the overall measured cutting dimension must be taken into account. Due to the possibility of confusion when measuring pipework with a change of direction, measurements will normally be made along or between the centre line of the pipe rather than back to back, face to face, or back to face. The following three allowances are typical regardless of the material used:

- *Fitting allowance* – unless forming a branch into the pipework or mechanically pulling a bend, when using prefabricated fittings such as Tees or junctions, a reduction must be made to the initial measurement for the overall distance that is gained by using the fitting (see Figure 9.29).

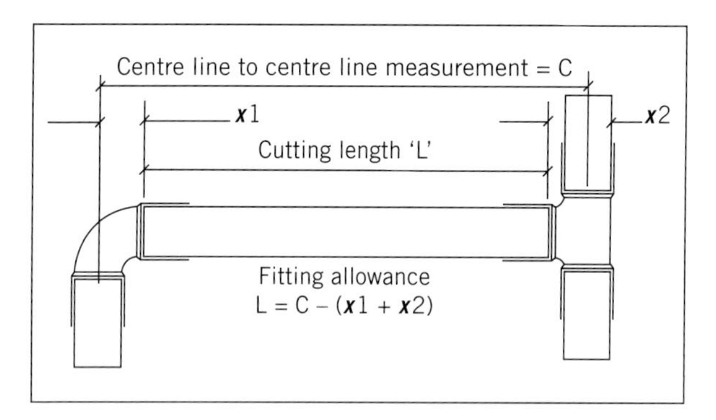

FIGURE 9.29 Fitting allowances

- *Jointing allowance* – the system used to join the pipework will determine whether the allowance will be a gain or loss. An expanded joint in copper pipework for silver brazing will require an addition to the overall measured distance due to the expanded copper socket. A solvent-welded PVC pipe and socket fitting would require a small reduction in measurement.
- *Bracket support allowance* – this allowance takes into account the distance that the bracketing support holds the pipework off the wall (see Figure 9.30). When following the contour of the wall,

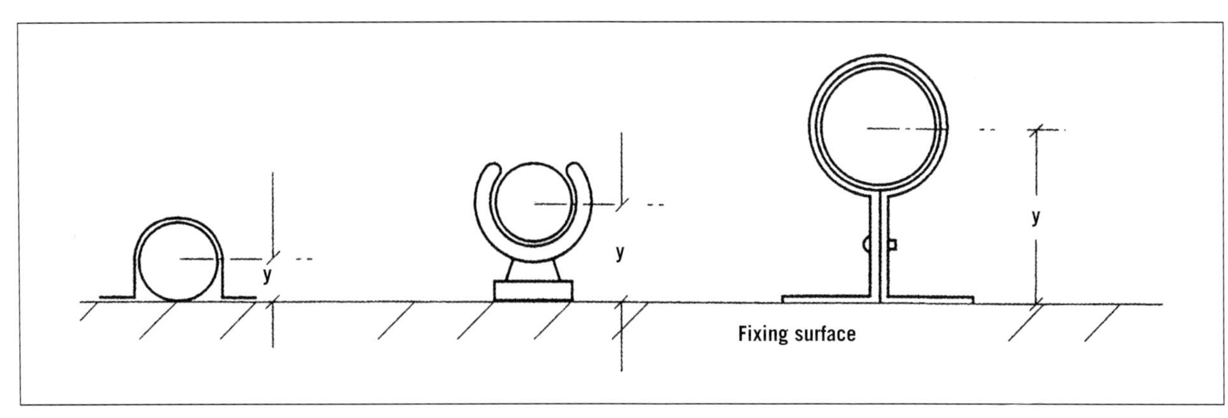

FIGURE 9.30 Pipework support stand-off allowance

a net gain in distance should be allowed for when the pipe goes around a pier or column, whereas a reduction in the dimension should be made when the pipework passes between an obstruction such as a recess in a wall (see Figures 9.31–33).

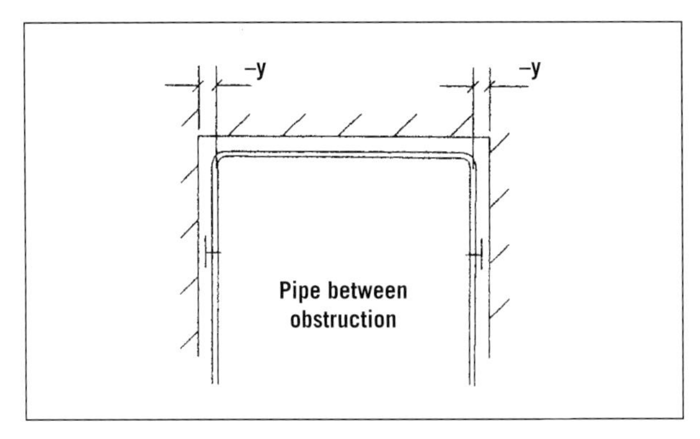

FIGURE 9.31 When the pipe passes between obstructions, the 'y' dimension must be *subtracted* from both ends

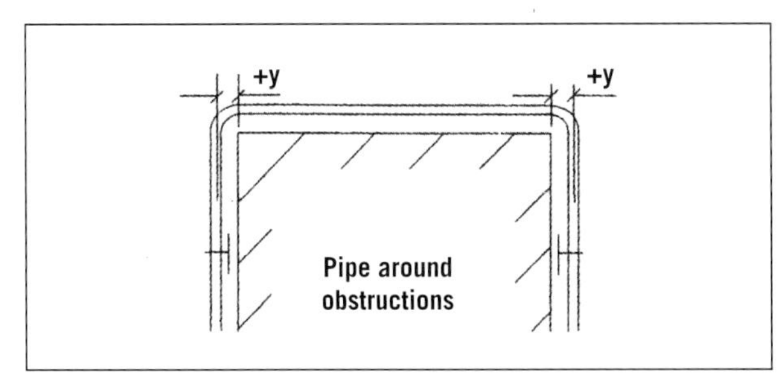

FIGURE 9.32 *When* the pipe passes around an obstruction, the 'y' dimension must be *added* to both ends

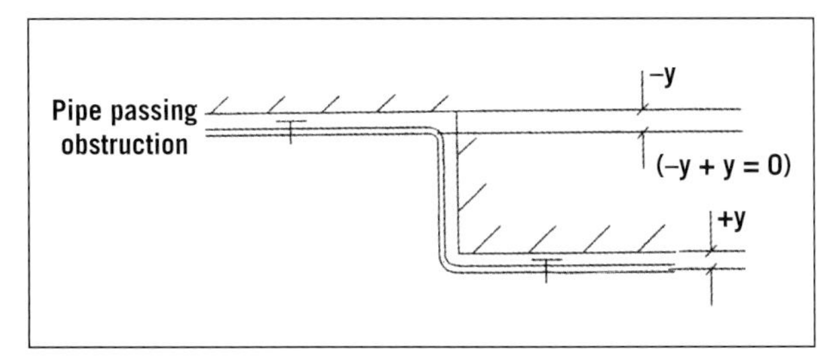

FIGURE 9.33 *When* the pipe goes past an obstruction, the 'y' dimension can be *ignored* because it is cancelled out

Marking out sheet metal roofing

When marking out roof sheets or cladding on a new job, it is common practice to mark the roof sheet or cladding on the underside. This allows for a smoother cutting action when using a cold metal cutting saw on deep-profile roof sheets, and any hot particles that are given off are less likely to burn the protective coating. Further, any surplus marks left from marking out the sheet will not be seen when the sheets are installed, as they will be on the underside.

When marking out on the underside of roofing and cladding, it is important to recognise that your left and right side of the sheet will be in the opposite position before the sheet is turned back over to install.

Sheet metal development

Within the plumbing trade, there is often a need to fabricate or reproduce gutters, downpipes, caps, trays and flashings out of thin-gauge sheet metal. When installing components for roof plumbing, the tradesperson must make sure that pipes or ducts that pass through a roof have been made weathertight. A competent tradesperson requires good problem-solving skills, a sound understanding of geometric shapes and the ability to be able to use their initiative.

Marking out and cutting for downpipe angles

Plumbers are often required to manufacture rainwater downpipe angles, particularly when working on domestic housing. There are several different methods used to mark out and cut the material.

Two methods are shown below: fabricating a downpipe (spouting) offset (see Figure 9.34) and fabricating a downpipe shoe (see Figure 9.35).

The position of laps plays an important role, as it provides a surface to allow a soldered joint for galvanised mild steel or copper, or a join using silicone and pop rivets for pre-painted material. The laps are positioned to minimise obstruction of flow within the downpipe, and allow full flow through to the stormwater drainage system.

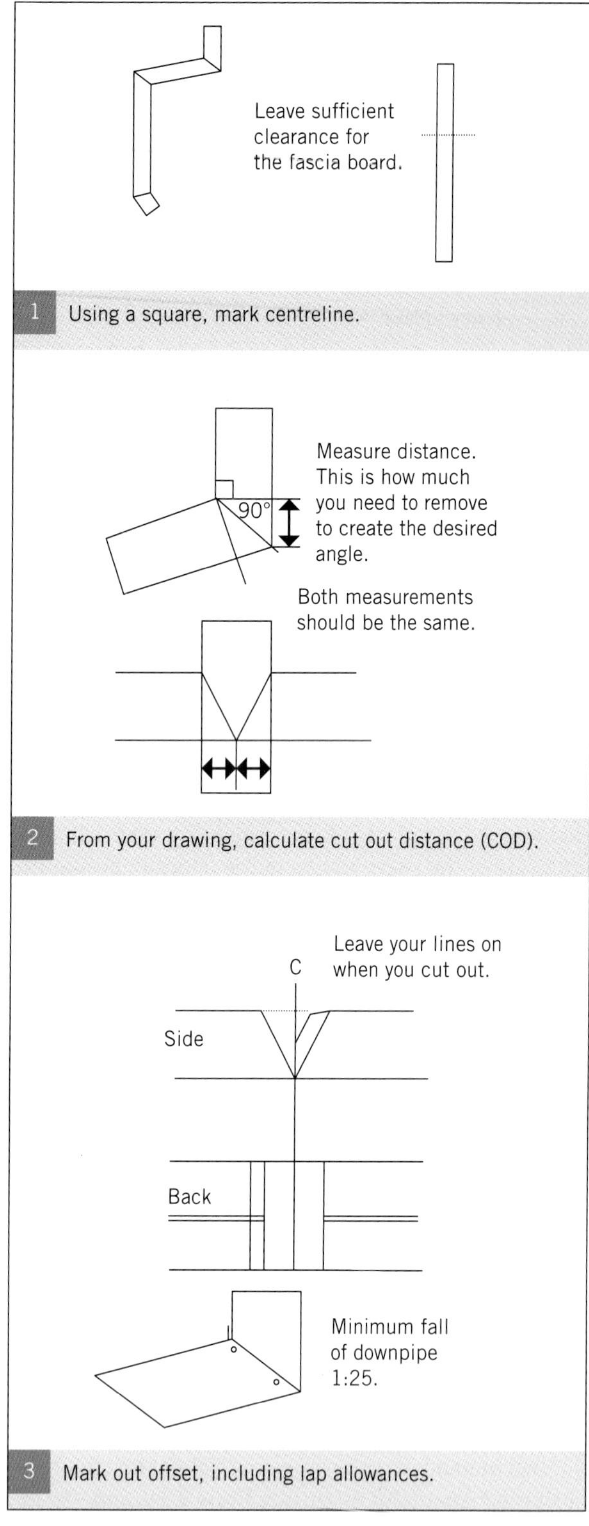

FIGURE 9.34 Fabricating a rectangular downpipe offset

The cut-out distance (COD) either side of the centre line must be an equal distance to ensure that when the shoe is folded into position, the sides and face of the downpipe all meet, as shown in Figure 9.34.

This is a critical part of the process, particularly with predetermined angles.

When marking the material to be cut, it is also important to use a scriber or fine-point pen, and to practise cutting on the 'waste' side of the cut to become more proficient with using the tools.

Box gutters and sumps

Plumbers are required to fabricate and install box gutters and sumps. Figure 9.36 shows an isometric view of a sump located in a box gutter. This isometric drawing shows a three-dimensional view of the installation, which provides more detail for fabrication purposes.

A sump is located within a box gutter and is required to collect rainwater and direct it into a downpipe located in the base of the sump. The position and size of a sump is determined by the amount of roof area and potential rainfall that would be collected and directed to the box gutter. An allowance for laps must be made for securing the sump to the box gutter.

The overflow is required to operate in case the downpipe becomes blocked and to help avoid flooding and possible damage to the building and its contents.

The box gutters and sumps are normally prefabricated off site then delivered to site for installation. In some cases where the sump has to be fitted on site, the plumber must ensure that it is installed correctly and is watertight.

There is a great need for accuracy by the person taking measurements on site in order to give the fabricator the necessary details for manufacture, and by the person installing the material. Being able to draw and interpret isometric drawings enables a great deal of information to be shown on a single drawing.

LEARNING TASK 9.3

MARK OUT JOB – PIPEWORK ALLOWANCE

You are required to compile a materials list for the room you are currently in:

1 Calculate the amount of 20 mm copper tubing in metres that would be required to run horizontally, starting from a chosen location and finishing back at its original location.
2 Select a suitable and approved bracketing system and make allowance for fittings, jointing and bracket support.
3 Provide a materials quantities list based on copper tubing lengths at 6 m with 20 mm copper elbows, and brackets installed 1.5 m apart.

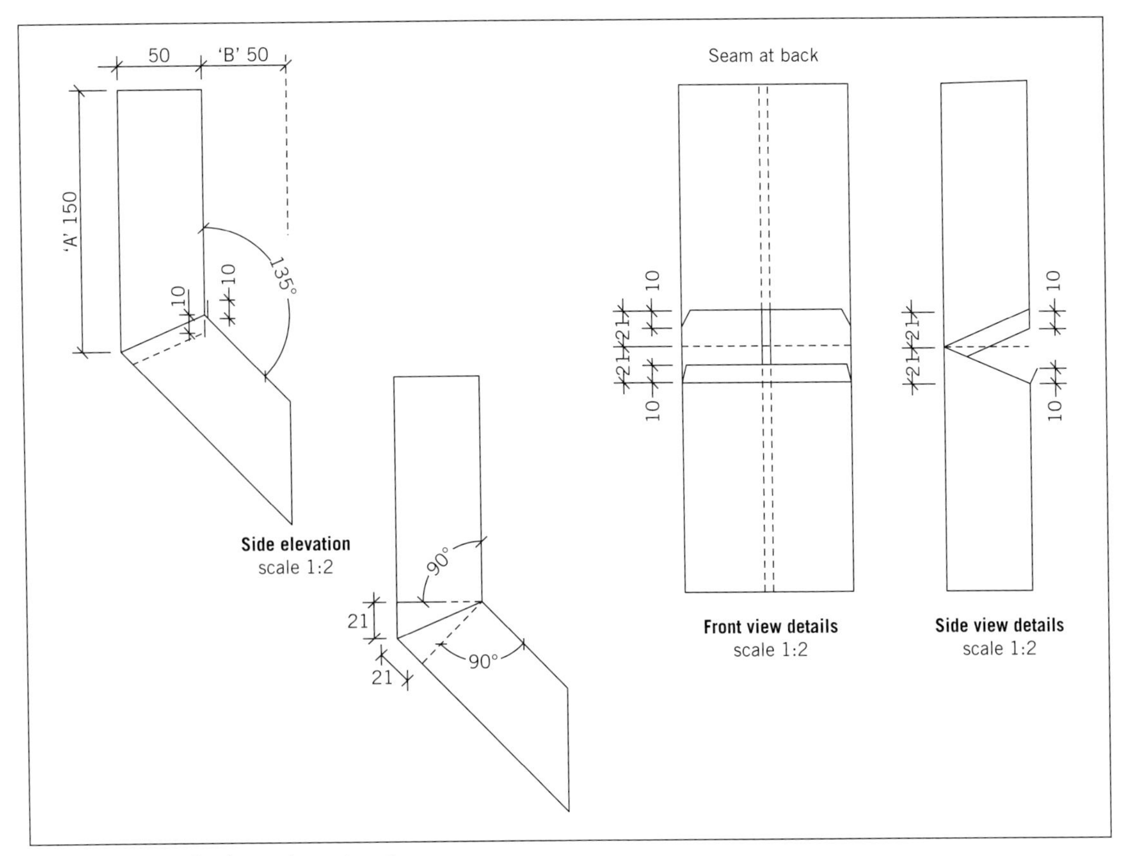

FIGURE 9.35 Fabricating a downpipe shoe

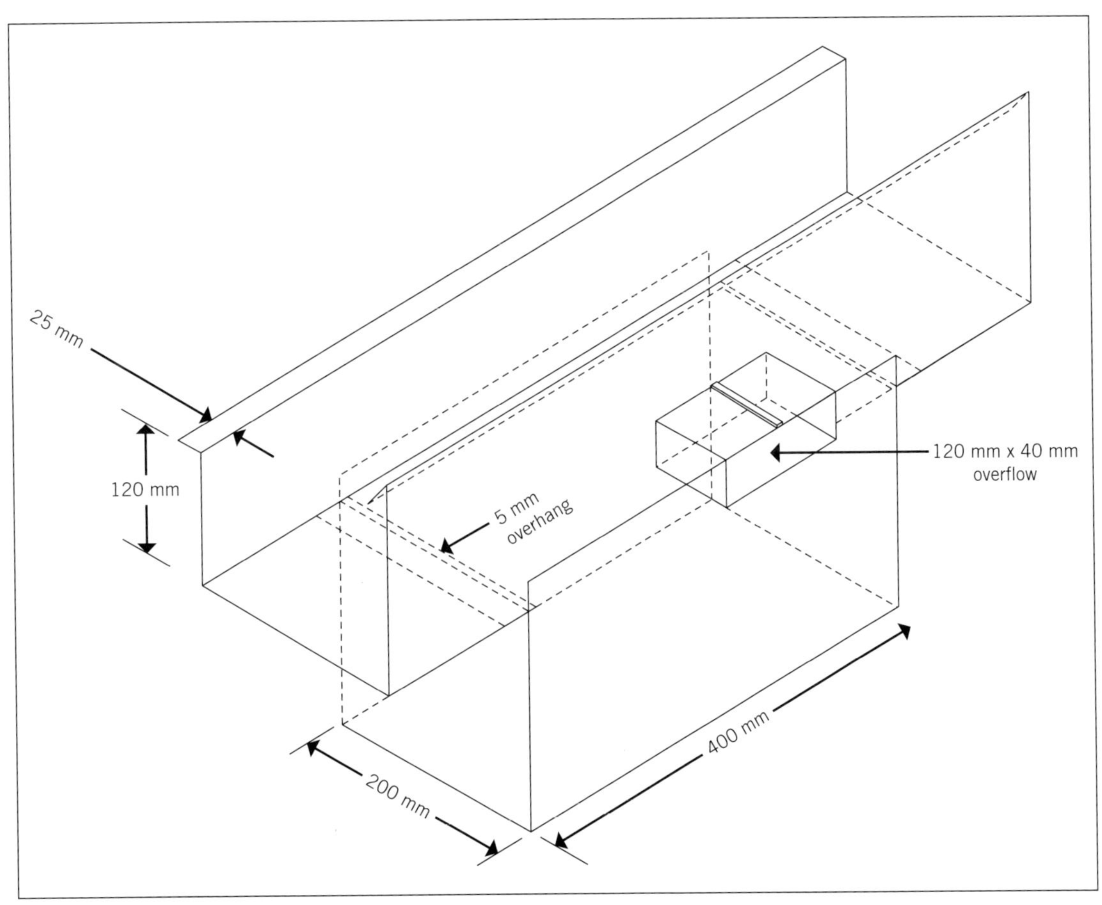

FIGURE 9.36 Isometric view of a box gutter, sump and overflow

Clean up

By keeping the work area clean and tidy and disposing of any resulting waste and correctly storing offcuts, the work area will remain safe, uncluttered and ready for the next job.

During marking out and when packing the tools or equipment away, you should quickly inspect the equipment for any faults. Should any tool or equipment require attention, it should be tagged out with a 'Do not use' safety tag immediately. Don't forget to identify the problem either on the tag or in a maintenance log and always inform a supervisor or follow workplace procedures to have the item repaired. If you continually maintain your work tools and equipment as required, they will give years of reliable service.

GREEN TIP

Always dispose of waste correctly and thoughtfully, and reuse or recycle unwanted waste.

Documentation completion

Once marking out is finished, all the tradespeople involved must document the work they have completed in accordance with workplace requirements. This paperwork will differ from one employer to another; however, typical information could include the following:

- job number
- job description
- materials used
- name of the tradesperson (or people) working on the project
- labour hours on the job
- quantity of jobs completed
- general comments or sketches.

Once gathered, the information needs to be accessible for those required in the organisation, and job sheets need to be accurate so that office staff can use the data for billing purposes, quality assurance, scheduling and historical records.

COMPLETE WORKSHEET 4

REFERENCES AND FURTHER READING

Acknowledgements

Many thanks to Gary Cook and Rodney Brunt for their assistance in providing Figures 9.24, 9.35, 9.36.

Reproduction of the following resource list references from DET, TAFE NSW C&T Division (Karl Dunkel, Program Manager, Housing and Furniture) and the Product Advisory Committee is acknowledged and appreciated.

Texts

Liebing, R.W. (1990), *Architectural working drawings*, Wiley, New York

Styles, K. (1986), *Working drawings handbook*, 3rd edn, Architectural Press, UK

Web-based resources

Resource tools and VET links

Training.gov.au: http://training.gov.au

NSW Education Standards Authority – Construction: **https://educationstandards.nsw.edu.au/wps/portal/nesa/11-12/stage-6-learning-areas/vet/construction-syllabus**

GET IT RIGHT

The photo below shows an incorrect practice that can be performed when marking out sheet metal material.

Identify the incorrect method, and provide reasons for your answer.

WORKSHEET 1

Student name: ______________________

Enrolment year: ______________________

Class code: ______________________

Competency name/Number: ______________________

To be completed by teachers	
Satisfactory	☐
Not satisfactory	☐

WORKSHEETS 9

Task: Review the section 'Prepare to work' in this chapter and answer the following questions.

1. List the three common sizes of the graduated steel rule.

2. How is a centre punch used to mark out materials?

3. State one advantage of a laser distance measure.

4. How can metal snips be used to mark materials?

5. What are the two types of set squares and what angles are contained within the set squares?

6. How can angles be determined without the use of a protractor?

7. What is the purpose of a plumb-bob?

8. What problems may arise from using a marker that is incompatible with the material being marked?

9. List a suitable marker for the following materials:
   i. Polished stainless steel
   ii. Sheet metal roofing profiles
   iii. Copper tube

WORKSHEET 2

To be completed by teachers	
Satisfactory	☐
Not satisfactory	☐

Student name: ______________________

Enrolment year: ______________________

Class code: ______________________

Competency name/Number: ______________________

Task: Review the section 'Pattern development' up to and including 'Development of a 30° truncated right cone' in this chapter and answer the following questions.

1. What are the three pattern development methods?

____

____

____

2. Using a 40 mm radius, draw a right cone with a 70 mm perpendicular height using the radial line development technique.

3. Calculate the circumference of a circle with a diameter of 650 mm.

____

4. Using a 40 mm radius, draw a truncated right cone with a 70 mm perpendicular height using the radial line development technique.

WORKSHEET 3

To be completed by teachers	
Satisfactory	☐
Not satisfactory	☐

Student name: ______________________

Enrolment year: ______________________

Class code: ______________________

Competency name/Number: ______________________

Task: Review the section 'Triangulation development of a square to round transition piece' in this chapter and answer the following question.

1. Using the notes for triangulation development as a guide, draw a plan view of the square to round transition development to show a circle with a 30 mm radius. Around this circle draw a square with side lengths of 70 mm and a perpendicular height of 85 mm. Develop the square to round transition using steps 1 to 12 (as per the notes).

WORKSHEET 4

To be completed by teachers

Satisfactory ☐

Not satisfactory ☐

WORKSHEETS 9

Student name: ______________________

Enrolment year: ______________________

Class code: ______________________

Competency name/Number: ______________________

Task: Review the sections 'Mark out the job' and 'Clean up' in this chapter and answer the following questions.

1. List the three allowance methods that must be taken into consideration when marking out pipework.

____

____

____

2. When marking out roof sheets, which side of the roof sheet is marked out on and why?

____

____

____

____

3. Using the information from Figure 9.34 and Figure 9.35 complete the following:

   i. Construct a full-size drawing of an angle to calculate the cut-out distance (COD) required to manufacture a 75 mm × 50 mm rectangular downpipe offset at an angle of 120°.

   ii. Record the COD required.

   ____

4. When cleaning up, what should be done if tools and equipment are found to be damaged or faulty?

   ____

   ____

   ____

   ____

WELD USING OXYACETYLENE EQUIPMENT

10A

This chapter addresses the following key elements for the competency 'Weld using oxyacetylene equipment':

- Prepare for work.
- Prepare materials and welding equipment.
- Perform welding.
- Clean up.

It introduces the oxyacetylene welding processes that a plumber may be required to undertake in their everyday work, including correct selection of personal protective equipment and safe handling and operating procedures. It also addresses the types of flames that can be obtained when using oxyacetylene welding equipment and their applications, and determination of appropriate action for reporting and rectifying defects in oxyacetylene welding.

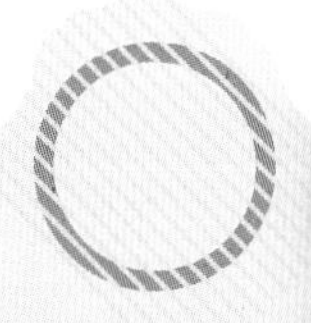

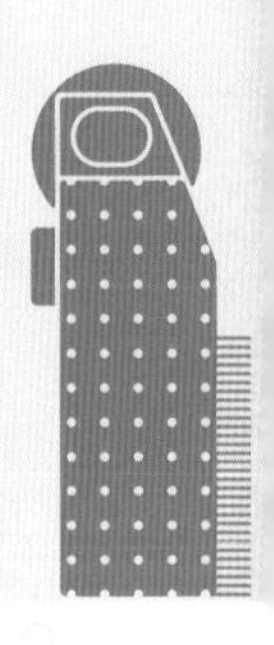

Prepare for work

Oxyacetylene welding is a process that combines oxygen and acetylene at a predetermined pressure, with combustion of oxygen and acetylene generating heat. This process has many uses in modern industry as an extensive range of welding processes can be performed on light- or heavy-gauge metals as well as ferrous and non-ferrous metals.

Oxyacetylene welding processes are highly hazardous and should be only be undertaken in environments that have been risk-assessed in accordance with safe work practices and company guidelines. Plans and specifications should be obtained from the builder or supervisor to plan and prepare for the work to be undertaken and to comply with associated work health and safety (WHS) and environmental requirements. Consideration must also be given to identifying and following quality assurance requirements in accordance with workplace procedures. Once the requirements have been determined from the plans and specifications, tasks relating to the job can be sequenced in conjunction with other trades or people that may be affected or required to complete the work. As oxyacetylene welding poses a number of risks, safe work practices and personal protective equipment (PPE) requirements must be determined and followed.

WHS requirements

WHS requirements must be observed in accordance with relevant state and territory legislation. Risk assessments and Safe Work Method Statements (SWMS) should be completed before commencing any welding using oxyacetylene equipment. A risk assessment should be developed for all welding jobs or hot works, which should include provision for:

- work to be carried out under hot work permit systems where required (see Figure 10A.65 at the end of this chapter)
- control of risk from fire or explosions
- control of inadequate ventilation of the work area.

Confined spaces present several hazards, so work permits and accreditation may be required. Precautions for working in confined spaces should also be considered. When working in confined spaces, the following precautions must be adhered to:

1. Gas cylinders must not be taken into the confined space.
2. Gas equipment should be removed, or gases turned off at the cylinder, once work has finished or ceased for a short time.
3. Gas hoses should be purged in a ventilated area away from the confined space.
4. Adequate ventilation or appropriate respiratory protection must be provided as there is a high risk of asphyxiation.

Also refer to work health and safety information in Chapter 1 of this text.

Personal protective equipment

In most cases, PPE is mandatory when undertaking oxyacetylene welding tasks. When used correctly, PPE will reduce the risks and hazards associated with oxyacetylene welding on the worksite. PPE should be provided to all workers by their employer to ensure that all tasks are carried out safely. All workers should receive appropriate instruction and training on the correct use and fitting of PPE to ensure it is used and protects as it was designed. PPE that does not fit or has not been fitted correctly may result in illness or injury and defeat the purpose of wearing the PPE in the first place.

Oxyacetylene welding requires the plumber to be protected from the radiant heat and rays associated with the process. Plumbers should wear long-sleeved cotton-combination overalls (see Figure 10A.1), firm-fitting leather safety boots, long-sleeved gauntlet leather gloves and Australian Standard class-appropriate tinted welding goggles (see Figure 10A.2) at a minimum. When welding in a confined space or carrying out vertical or overhead welding, a leather

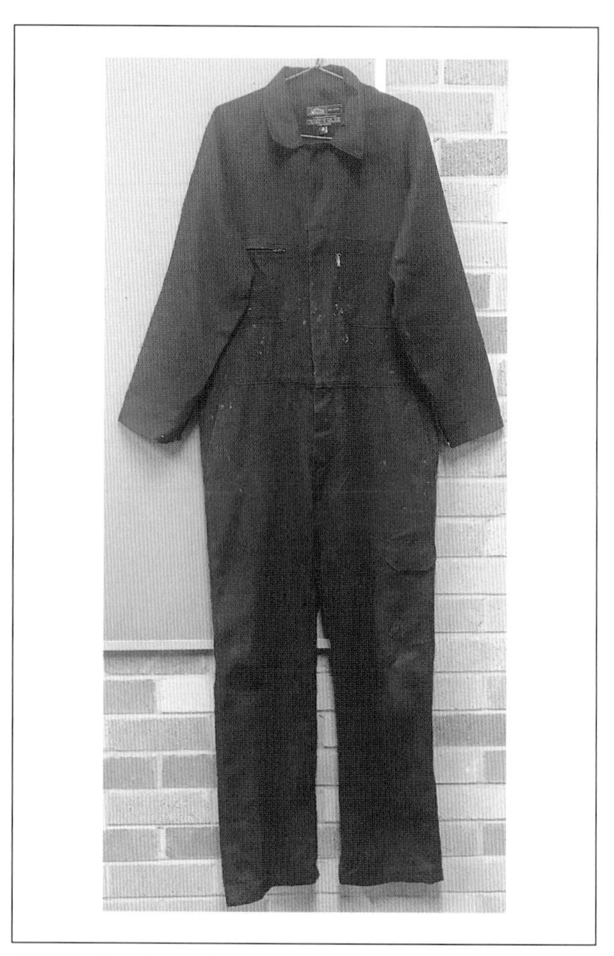

FIGURE 10A.1 Long-sleeved cotton-combination overalls

FIGURE 10A.2 Tinted welding goggles manufactured to Australian Standards

apron, spats and a cotton cap to protect your head should be worn. Nylon or similar fabrics should not be worn at any time when welding using oxyacetylene equipment, as they are flammable.

Tinted welding goggles

Sufficient lighting is necessary to see the work at hand, but an excessive amount of light is given off when welding. This means that light must be reduced considerably for the person welding, or they will suffer a great amount of discomfort and not be able to see properly.

The primary use of tinted welding goggles is to reduce the exposure of intense light emitted from the welding process to a satisfactory level for eye comfort. An equally important function is to protect the eyes from injury from sparks and weld spatter that may occur during the welding process.

Tinted lenses are graded in accordance with their visual density. *AS/NZS 1338 Filters for eye protectors, Part 1: Filters for protection against radiation generated in welding and allied operations*, provides a table of the suitable filter shades for specific welding applications.

10A.1 STANDARDS

- AS/NZS 1338 Filters for eye protectors, Part 1: Filters for protection against radiation generated in welding and allied operations

There are three commonly used shades of tinted lenses that will sufficiently filter ultraviolet light during welding operations. The list below is the recommended protection filters for gas welding processes:

- *Shade 4 lenses* – for silver brazing, fusion welding of zinc-based die casting, and braze welding of light-gauge copper pipe and light-gauge steel
- *Shade 5 lenses* – for oxygen cutting, general welding and for braze welding of heavy gauge steel and cast iron
- *Shade 6 lenses* – for fusion welding of heavy steel, heavy cast iron and steel castings, and for heavy cutting over 300 mm.

Protective leather gloves

Leather gloves or gauntlets are part of the PPE required by a plumber when welding or handling hot materials while using oxyacetylene equipment. They protect the hands from the heat and sparks generated by the welding process (see Figure 10A.3).

FIGURE 10A.3 Cotton-lined leather gloves (gauntlets)

Preparation of work area

Ensure that any flammable material and rubbish is well clear of the work area, as sparks will fly during fusion welding and can travel a long distance, staying hot for some time. Always alert those around you when undertaking welding tasks to prevent injuries and scheduling issues. The use of screens or barricading areas may be required to prevent injury or unauthorised access. The work area should be prepared to sufficiently support materials during the welding process. Sustainability principles and concepts should be observed by reducing the use of materials where possible, and reusing and recycling materials in accordance with state and territory guidelines

LEARNING TASK 10A.1

PREPARE FOR WORK

Provide short answers to the following questions:

1 What are three requirements relating to welding that must be considered once the plans and specifications are obtained?
2 A risk assessment must be completed before undertaking welding using oxyacetylene. What precautions must be considered?
3 What are the minimum PPE requirements for oxyacetylene welding?
4 What must be cleared in the work area before welding can commence?

Prepare materials and welding equipment

Plans and specifications should always be sought when preparing materials and welding equipment for the workplace. The materials and weld requirements must be identified and selected in accordance with workplace procedures. All welding equipment must be assembled and set up to meet manufacturers' and state or territory regulatory requirements.

Oxyacetylene welding equipment

Oxyacetylene equipment should be maintained and stored in an appropriate location when not in use. Due to the high risks associated with performing oxyacetylene welding, workplace procedures must be followed and adhered to. Listed below is some of the equipment that you will be required to use and maintain when performing oxyacetylene welding.

Oxygen cylinder

Oxygen cylinders are manufactured from steel or aluminium and must be of sufficient strength to hold the 15 000 kPa pressure to which they are filled. They are always painted *black* and have a *right-hand* thread on the regulator connecting outlet (see Figure 10A.4).

All cylinders have a safety bursting disc fitted that is designed to blow out at a much lower pressure than it would take to rupture the cylinder (see Figure 10A.5). Because of this feature there is less chance of cylinders exploding if they are heated due to an external fire or overfilling.

Oxygen cylinders come in several different sizes, and can be obtained from different suppliers. Volume sizes may vary between suppliers and are hired under the name of industrial grade oxygen. The three oxygen cylinder sizes that are commonly used in plumbing are:

- D – holds approximately 2 m^3 of oxygen
- E – holds approximately 4 m^3 of oxygen
- G – holds approximately 8 m^3 of oxygen.

Acetylene cylinder

Acetylene cylinders (see Figure 10A.6) are manufactured using a drawn seamless steel construction process.

FIGURE 10A.4 Size D oxygen cylinder

They are fitted with a valve that has a *left-hand* thread, which is indicated by a notch or groove (see Figure 10A.7), and to allow for attachment of a compatible fuel gas regulator. Their shoulders are squarer than those of oxygen cylinders, and are painted crimson, which is often referred to as maroon. These cylinders are also fitted with fusible plugs made of low-melting alloys and located in the bottom and top of the cylinder or in the outlet valve. These help to prevent explosions if the cylinder is overheated (see Figures 10A.7 and 10A.8). If acetylene was pressurised as a gas, there would be the danger of explosion, as there is no free gas present, so ordinary methods of storage in cylinders are not applicable. Consequently, acetylene is dissolved in a liquid called acetone while under an effective pressure

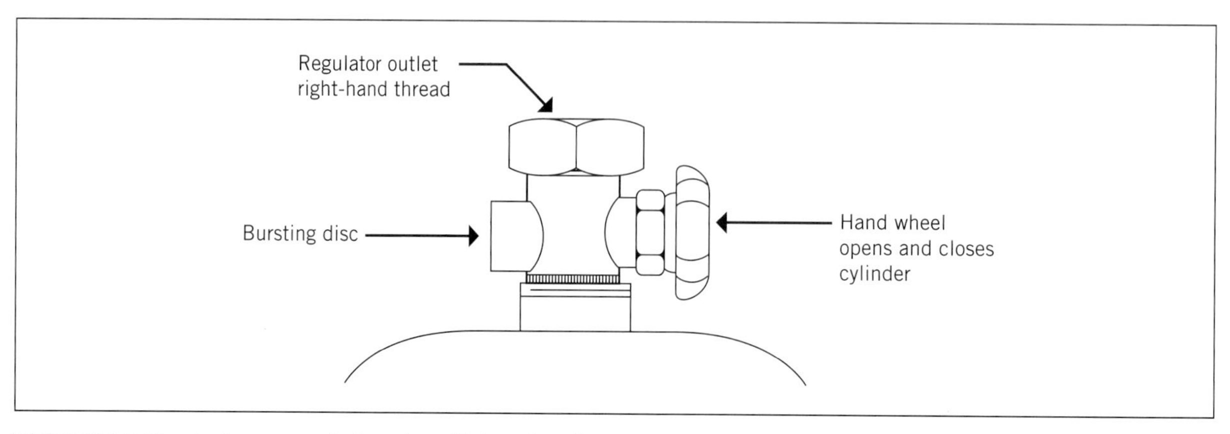

FIGURE 10A.5 Standard oxygen cylinder valve with bursting disc

FIGURE 10A.6 Size D acetylene cylinder

FIGURE 10A.7 Fusible plugs fitted at the top of an acetylene cylinder

of 1600 kPa. The acetone reduces the pressure and helps to stabilise the acetylene to avoid a reaction with the cylinder, thus allowing safe storage and usage. Due to the unstable nature of acetylene, it is recommended that the draw-off pressure does not exceed 100 kpa.

Like oxygen cylinders, there are three acetylene cylinder sizes that are commonly used in plumbing:

- D – holds approximately 1 m^3 of acetylene
- E – holds approximately 3.2 m^3 of acetylene
- G – holds approximately 7 m^3 of acetylene.

When stored on site, acetylene cylinders – and any other fuel gas cylinders stored in designated areas – must be separated from oxygen cylinders by not less than 3 m. If they are in a mixed cylinder storage shed, they must be separated from oxygen cylinders by not less than 5 m, or by a firewall in accordance with *AS 4289 Oxygen and acetylene gas reticulation systems*.

10A.2 STANDARDS

- AS 4289 Oxygen and acetylene gas reticulation systems

Caution: Acetylene cylinders should always be stored, transported and used in the upright position because of the liquid acetone in the cylinder. When using the acetylene, the draw-off rate should not be in excess of 20% of the cylinder capacity in an hour, which means that no cylinder should be emptied in less than five hours, in order to prevent acetone being drawn off.

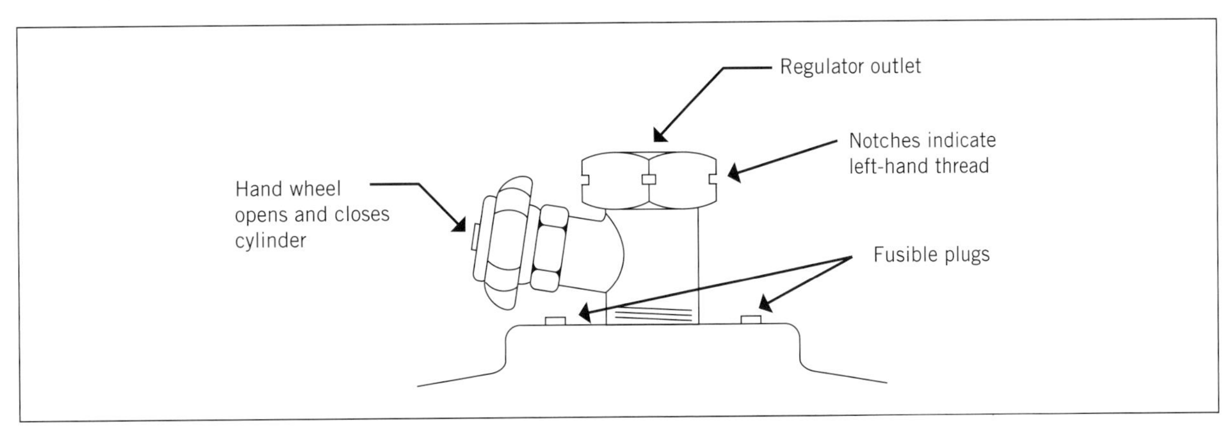

FIGURE 10A.8 Acetylene cylinder with fusible plugs and notches indicating left-hand thread

Always ensure that cylinders have been turned off correctly to avoid the escape of gases into the atmosphere.

Cylinders should always be transported and used in the upright position. This applies to acetylene cylinders in particular to prevent acetone from separating from the acetylene. Acetylene cylinders should be allowed to rest for one to two hours after transportation.

Twin oxygen/acetylene hose (tubing) assembly with fittings

A gas hose (tubing hose) is an essential yet vulnerable link in the system that gives flexibility of movement for the operator. Hoses are possibly the weakest point of the oxygen/acetylene system and potentially the greatest source of leaks. Only quality hoses that meet the requirements of *AS/NZS 1335: Gas welding equipment – Rubber hoses for welding, cutting and allied processes* should be used to maximise safety. While new hoses are colour coded to meet Australian Standards, and should be red for acetylene and blue for oxygen, some older hoses may be black. Hose fittings are designed to provide high mechanical strength and resistance, while being non-interchangeable by using right-hand threads for oxygen and left-hand threads for acetylene. Left-hand threads are notched for easy identification between the two gas types (see Figure 10A.9). All new hoses should be blown out before welding equipment is connected to them in order to remove talc, dust or water. They should also be checked regularly for damage and replaced immediately if perished, cut or burnt.

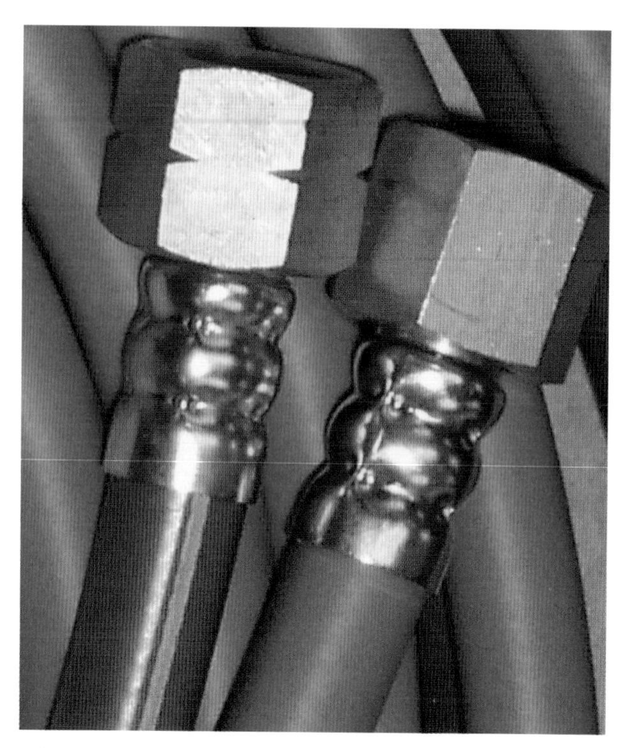

FIGURE 10A.9 Blue (oxygen) right-hand-fitting and red (acetylene) left-hand-fitting hoses

10A.3 STANDARDS

- AS/NZS 1335: Gas welding equipment – Rubber hoses for welding, cutting and allied processes

Oxygen regulator

An oxygen regulator is an apparatus for controlling the delivery of oxygen at a chosen constant pressure, regardless of variation in cylinder pressure.

The regulator is connected to the cylinder with a right-hand screwing thread that is sealed by an O-ring (see Figure 10A.10). There are two gauges on the regulator, with the gauge on the right indicating the pressure in the cylinder and the gauge on the left indicating the adjustable working pressure, which is adjusted by turning the control knob in or out until the desired safe working pressure is obtained.

FIGURE 10A.10 Oxygen regulator with right-hand connections

The oxygen hose is connected to the right-hand thread on the side of the regulator body. All fittings are tightened using the open-ended spanner that comes with the equipment.

Regulators are colour coded for the cylinders that they are to be connected to and cannot be interchanged due to the different thread type. They are precision instruments and should be treated with care. Do not work with damaged equipment and ensure that any leaking or damaged equipment is repaired only by an authorised repair person.

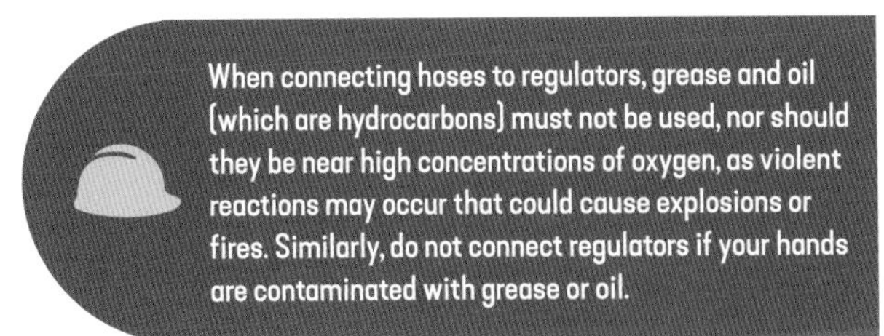
When connecting hoses to regulators, grease and oil (which are hydrocarbons) must not be used, nor should they be near high concentrations of oxygen, as violent reactions may occur that could cause explosions or fires. Similarly, do not connect regulators if your hands are contaminated with grease or oil.

Acetylene regulator

An acetylene regulator is an apparatus for controlling the delivery of acetylene at a constant pressure, regardless of variation in cylinder pressure.

The regulator is connected to the cylinder with a left-hand screwing thread that is sealed with an O-ring (see Figure 10A.11). There are two gauges on the regulator. The gauge on the right indicates the pressure in the cylinder and the gauge on the left indicates the adjustable working pressure, which is adjusted by turning the control knob in or out until the desired safe working pressure is obtained.

The acetylene hose is connected to the left-hand thread on the side of the regulator body. All fittings are tightened using the open-ended spanner that comes with the equipment.

FIGURE 10A.11 Acetylene regulator with LH connections

Open-ended spanner

Open-ended spanners are supplied with the welding equipment when it is purchased and should be used to tighten and undo all hexagon fittings on hoses, regulators and other equipment (see Figure 10A.12). The use of multi-grips and serrated wrenches should be avoided to prevent damage of hexagon fittings.

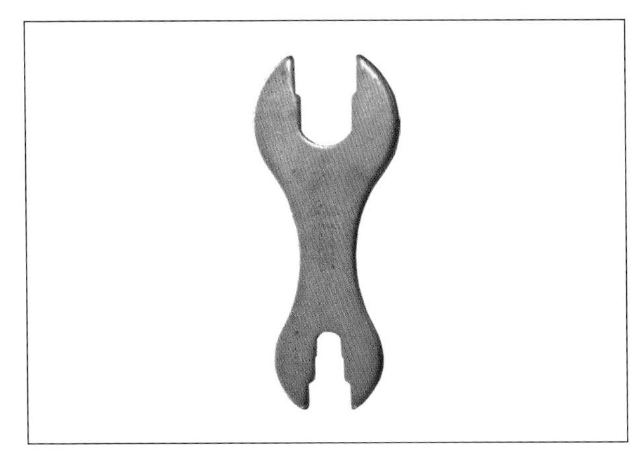

FIGURE 10A.12 Open-ended spanner

Oxygen and acetylene flashback arrestor

Flashbacks occur when mixed gases within a cutting or welding blowpipe system ignite and burn back against the flow of gas. This may lead to sooting in the hoses, serious equipment damage, fire or even explosions. Some common causes of flashback are:

- blockage in the welding or cutting tip
- faulty or damaged equipment
- incorrect operating pressures.

When used correctly, flashback arrestors virtually eliminate the possibility of a flashback causing any of these harmful and dangerous situations by rapidly closing off the flow of gas from the cylinders.

There are two types of flashback arrestors: regulator mounted (see Figures 10A.13–10A.16) and blowpipe (torch) mounted (see Figure 10A.17).

FIGURE 10A.13 Oxygen-regulator-mounted flashback arrestor

FIGURE 10A.14 Acetylene-regulator-mounted flashback arrestor

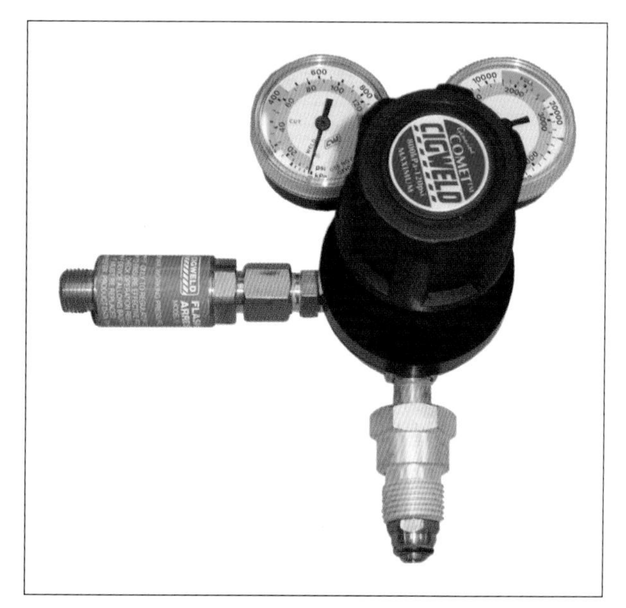

FIGURE 10A.15 Oxygen flashback arrestor mounted on regulator

FIGURE 10A.16 Acetylene flashback arrestor mounted on regulator

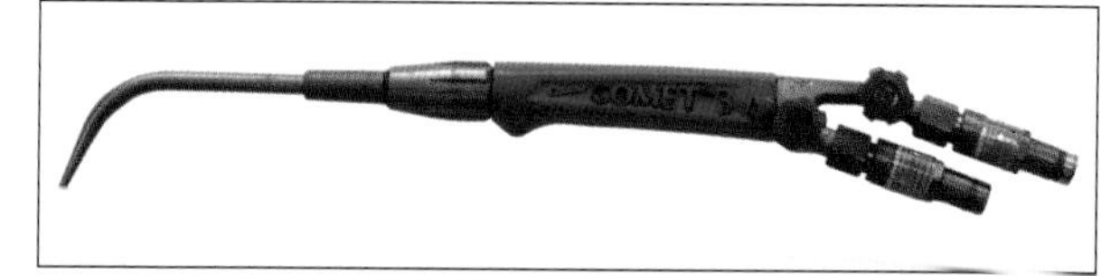

FIGURE 10A.17 Blowpipe (torch) with mixing chamber and flashback arrestors

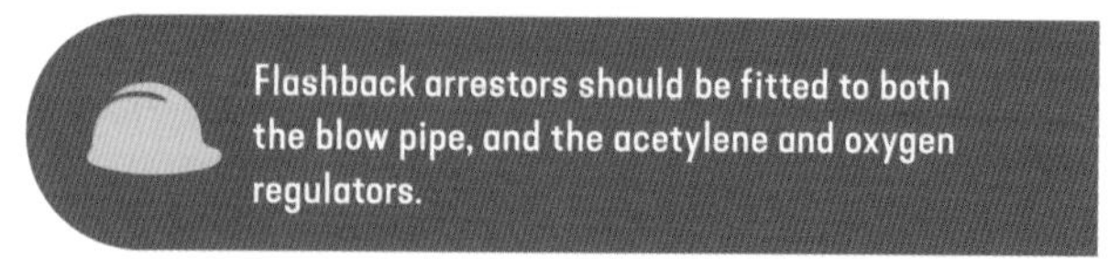

AS 4267 Pressure regulators for use with industrial gas cylinders recommends that flashback arrestors be used in oxyacetylene applications. In addition, AS *4289 Oxygen and acetylene gas reticulation systems* specifies the use of flashback arrestors in oxyacetylene applications. *AS 4603 Flashback arrestors – Safety devices for use with fuel gases and oxygen or compressed air* states that flashback arrestors must be tested once a year or if they have been subjected to backfiring or flashbacks.

10A.4 STANDARDS

- AS 4267 Pressure regulators for use with industrial gas cylinders
- AS 4289 Oxygen and acetylene gas reticulation systems
- AS 4603 Flashback arrestors – Safety devices for use with fuel gases and oxygen or compressed air

When accidents happen in welding it is generally due to the operator becoming over-confident and perhaps a little careless. It does not matter how safe the equipment is, if incorrectly handled it will become *unsafe*. Two examples of this are backfire and flashback:

- **Backfire**, which is the popping out of the flame, can be caused by touching the welding tip onto the work, particles blocking the flow of gas from the welding tip, the welding tip overheating due to incorrect angles, or an incorrect welding tip size being used.
- **Flashback** is the burning back of the flame inside the torch or hoses with a loud hissing sound. If this happens, turn off the blowpipe and, if in the hoses, turn off the cylinder control valves, acetylene first, and bleed the gases from the hoses. Flashbacks can be caused by having the wrong working pressures, loose joints, a blocked welding tip or an overheated tip due to incorrect angles, dirty tips or wrong tip size.

Blowpipe (welding torch)

The blowpipe and mixing chamber or torch (see Figures 10A.17–10A.19) is designed to mix oxygen and acetylene in the correct proportions while controlling the volume of gas combusted at the welding tip. Two hose connections are provided: one is for oxygen (RH) and the other for acetylene (LH). The torch has needle valves to control the flow of both the oxygen and acetylene. From the needle valves the gas is conveyed in separate tubes to the mixing chamber, from where the mixture of oxygen and acetylene passes to the tip for combustion.

FIGURE 10A.18 Blowpipe fitted to equipment ready for use

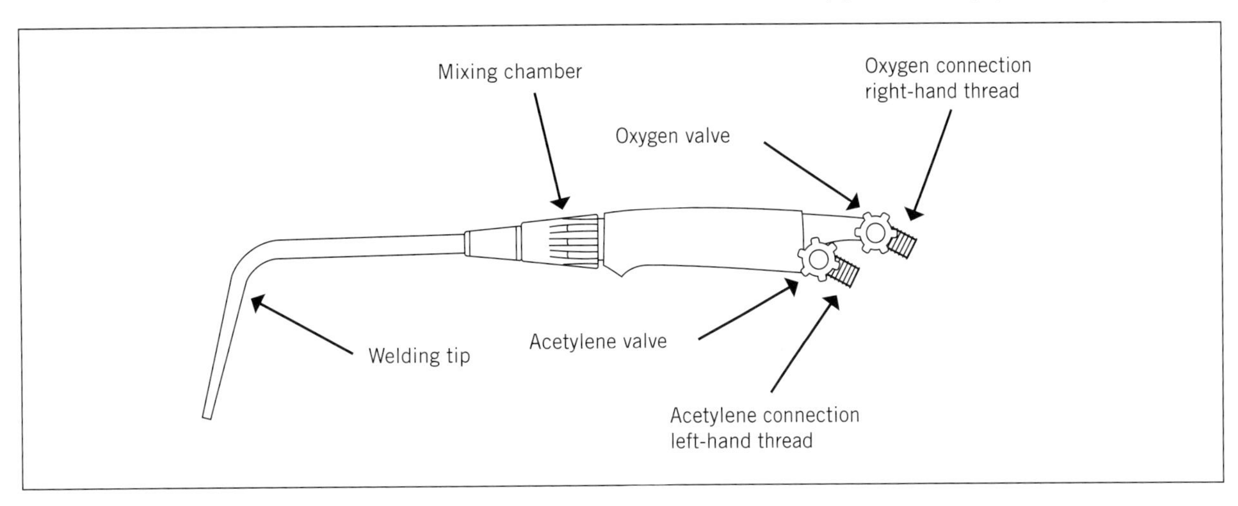

FIGURE 10A.19 The equipment that makes up the blowpipe

Oxyacetylene cutting attachment

The cutting attachment can be fitted to the welding blowpipe (handpiece) by removing the mixing chamber from the blowpipe and screwing on the attachment hand-tight (see Figure 10A.20).

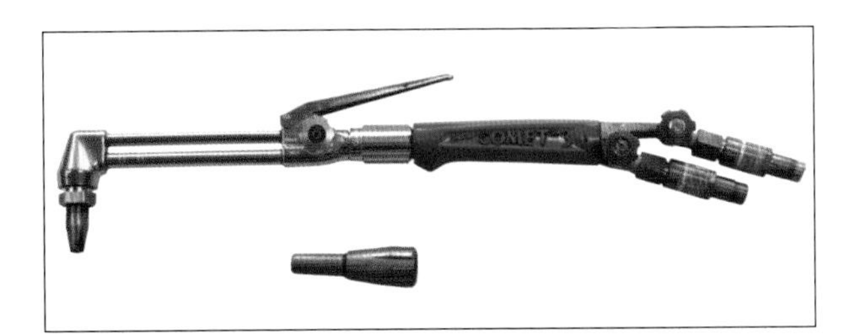

FIGURE 10A.20 Blowpipe with mixer removed and cutting attachment fitted

A range of cutting tips and accessories can be fitted when cutting circles and different thicknesses of ferrous material such as mild steel pipe and plate.

Welding tips

Welding tips are designed to screw into the mixing chamber on the end of the blowpipe (see Figure 10A.21). Tips direct the flame to the weld area, and the size of

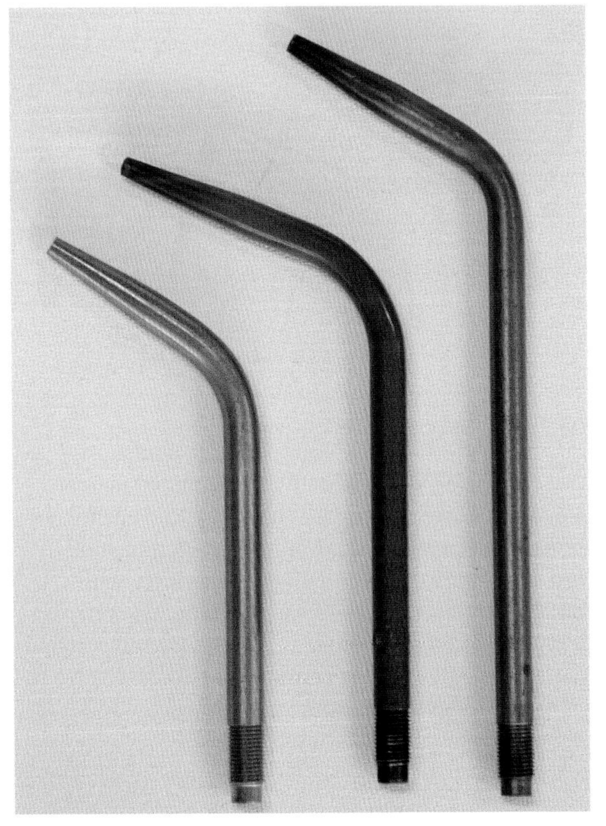

FIGURE 10A.21 A range of welding tips used for various applications

TABLE 10A.1 Selection of tip size and working pressures

Tip size	Orifice size (mm)	Steel thickness (mm)	Oxygen pressure (kPa)	Acetylene pressure (kPa)	Oxygen (L/min)	Acetylene (L/min)
8	0.8	1.0–1.6	50	50	2	2
10	1.0	1.6–2.4	50	50	3	3
12	1.2	2.4–3.0	50	50	4	4
15	1.5	3.0–5.0	50	50	6.5	6.5
20	2.0	5.0–8.0	50	50	12	12
26	2.6	8.0–10	50	50	22	22
32	3.2	10+	100	100	38	38

the tip dictates how much gas is supplied and therefore how much heat is directed at the weld area. The thicker the material, the larger the tip size. Table 10A.1 shows tip sizes used on different thicknesses of low-carbon steel, pressure settings, and gas consumption in litres per minute.

Tip cleaners

After prolonged use, welding tips can become dirty, due to not enough heat getting to the weld area. This can cause a number of problems including a lack of fusion, backfiring and even flashback. To prevent these problems from occurring, tip cleaners are used to clean the small orifice at the end of the tip.

They come in two different styles – reamers and drills (see Figure 10A.22) – and can be purchased from equipment suppliers.

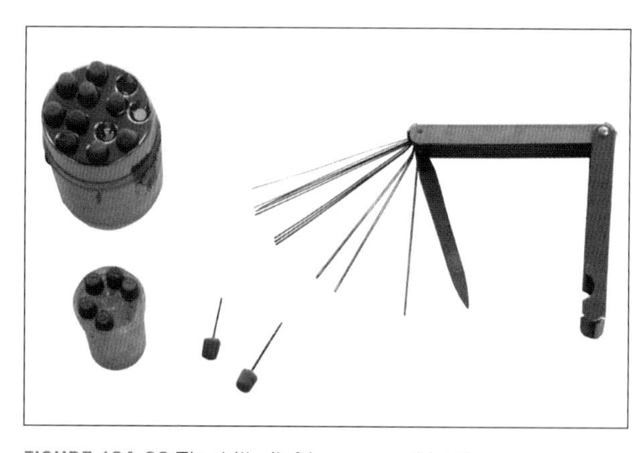

FIGURE 10A.22 Tip drills (left); reamers (right)

Adopting a simple process of moving the reamer in and out of the orifice (see Figure 10A.23) is generally sufficient to clean out the welding tip. At this point, *care must be taken* not to enlarge the orifice, as this can change the flow rate of gas to the tip. It is also recommended that, in order to prevent foreign material from entering the mixing chamber, the oxygen valve be left opened slightly. Alternatively, remove the welding tip from the mixing chamber.

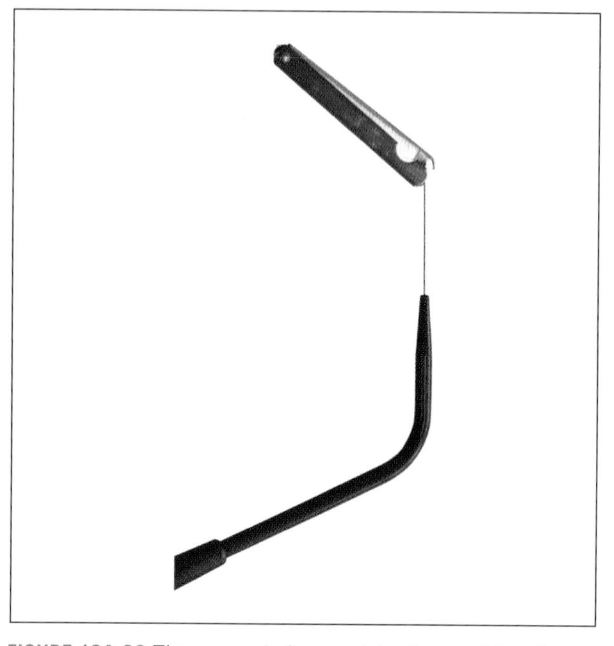

FIGURE 10A.23 Tip reamer being used to clean welding tip

Flint lighter (friction lighter)

A flint lighter (see Figure 10A.24) uses a replaceable flint that gives off a spark well away from the user's hand and safely ignites the oxyacetylene gas mixture. The hand is quickly and easily removed from the ignition area. A flint lighter or pilot light should be used to light the flame (see Figure 10A.25).

Never use cigarette lighters to ignite oxyacetylene equipment, as the hand is too close to the ignition area, and contains its own fuel source (butane). An unsafe situation such as burns or an explosion may occur.

Transport of oxygen and acetylene cylinders

Both oxygen and acetylene cylinders should be transported in the upright position and secured in purpose-built gas storage cabinets that are ventilated to the atmosphere and away from potential ignition sources or in an *open* or *well-ventilated* vehicle. In addition, if they are being transported from site to site, the cylinders must be secured to prevent them or their valves from being damaged.

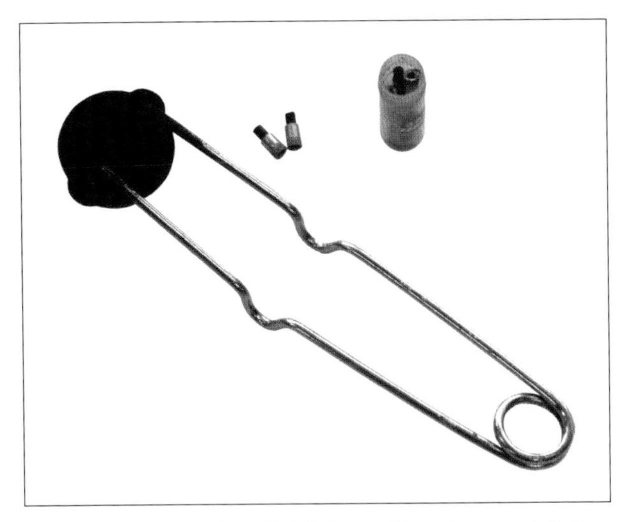

FIGURE 10A.24 Standard flint lighter with replacement flints

FIGURE 10A.25 Lighting the blowpipe with flint lighter

Recent explosions in plumbers' vans (see Figure 10A.26) have highlighted the need for tradespeople and apprentices to exercise extreme caution and adopt adequate safety measures when transporting or storing oxygen and acetylene cylinders. Overnight leakage and build-up of acetylene gas could ignite and cause an explosion within a van, which could be potentially

FIGURE 10A.26 Van explosion as a result of leaking acetylene cylinder

Source: Newspix/Trevor Pinder.

triggered by the electronic car-opening device. Before transportation or storage, check that cylinders are turned off at the control valve. Vehicles should not be used to store cylinders unless they are vented. Gauges should be removed in case cylinders are knocked over, causing further damage. When determining the safest way to transport and store cylinders within any vehicle used for plumbing purposes, tradespeople should check their relevant state and territory safety regulatory authorities' requirements and gas cylinder suppliers' information regarding the safe transport of cylinders.

Oxygen and acetylene cylinders should be secured upright in a trolley designed for the cylinder size being used, and an appropriate class fire extinguisher should be located with the equipment. For safety and transporting of welding equipment on site, the following extra equipment is recommended, and in some states and territories is required under WHS Acts, Codes and Regulations:

- a trolley that is designed for the cylinder in use, and is safely balanced to reduce the carry load, engages fully with the bases of the cylinders, and has a safety chain that securely positions and holds the cylinders
- an appropriate class fire extinguisher.

These should always be available next to welding and cutting equipment.

Figure 10A.27 shows a trolley designed for size D cylinders, while Figure 10A.28 shows a trolley safely loaded with size E cylinders and accompanied by an appropriate class fire extinguisher.

FIGURE 10A.27 Trolley to suit a size D cylinder, with safety chain

FIGURE 10A.28 Oxygen/acetylene size E cylinders securely held by safety chain, with CO_2 fire extinguisher (note yellow date test tag)

Cylinder maintenance and safety

It is essential that all safety procedures are followed before and after the welding process has taken place. All equipment must be checked for any damage and parts replaced if worn or damaged. When setting up cylinders it is important to follow the below points:

- Cylinders should be set up in a cool location to prevent them from overheating.
- Never use cylinders as support structures.
- Never roll cylinders along the ground.
- Never lubricate cylinder valves and fittings.
- Never apply sealants (liquid or tape form) or lubricants to any cylinder valves or connecting fittings.
- Never let oil or grease come into contact with the cylinder or its valve and fittings, and only use approved sealants or lubricants when connecting gas fittings or equipment.

High-pressure oxygen will react violently with oils and grease and can cause an explosion or localised ignition, leading to injury of the user and damage to equipment. Oxygen equipment is at most risk from oil and grease, so keep greasy hands, rags and gloves away from any part of the cylinder and fittings. Hands must be wiped clean and contact minimised with surfaces that might be subject to oxygen under pressure.

Valves will show test date tags that indicate when testing is required (see Figure 10A.29). In addition, test date tags can indicate if a cylinder has been heat affected as the tag will melt or distort at

FIGURE 10A.29 Top outlet valve showing a test date tag

Source: Supplied by BOC Limited, a Member of the Linde Group, http://www.boc.com.au.

a predetermined temperature when subjected to excessive heat.

Keeping cylinder valves clean

Cylinders are supplied with their cylinder-valve outlets capped or plugged, and in some cases PVC shrink-wrapped. The purpose of this is to indicate that the cylinder is full and to keep the outlet clean and contaminant-free. Top outlet valves are particularly prone to debris getting in the outlet.

If grit, debris, oil or dirty water enters the cylinder valve outlet, this may cause damage to the valve internals and result in leakage.

Before assembling regulators and fittings, make sure there are no particles or debris in the cylinder valve outlet. If a supply of clean, compressed, oil-free air or nitrogen is available, then, while wearing appropriate eye and ear protection, use this to blow any loose particles of debris out of the valve outlet.

If a supply of clean, compressed, oil-free air or nitrogen is unavailable, then use a clean, lint-free rag to clean the cylinder valve outlet, in particular the sealing surfaces.

Valve key

Valve keys come with the equipment when it is purchased and are needed to open and close cylinder valves on older-style oxygen and acetylene cylinders, though it is highly unlikely that you will see one in use. They should be left in the acetylene valve while welding so that if any fire or other dangerous situation occurs the acetylene cylinder valve can be turned off quickly. This key is not required on most modern cylinders as they have a cylinder-valve control knob fitted for ease of control and safety.

Never open a cylinder valve to clear the outlet of flammable or oxidising gases as this can lead to the ignition of escaping gas. Ejected particles and the resultant noise can also injure nearby personnel.

HOW TO

SET UP CYLINDERS READY FOR WELDING

Check cylinder valves then place the cylinders onto the trolley and secure into position with a safety chain.

1 Screw the regulators into the appropriate cylinder:
   a oxygen into the *black* cylinder, which has right-hand thread and needs to be turned clockwise to tighten
   b acetylene into the *crimson (maroon)* cylinder, which has left-hand thread and needs to be turned anti-clockwise to tighten. This should be done using the spanner provided or another suitable spanner (e.g. an adjustable spanner). Do *not* use any tools with teeth or serrated edges as they may damage the equipment and may also cause injury.
2 Connect the flashback arrestors to the regulators and then the hose assembly (tubing) to the flashback arrestors:
   a blue to the oxygen, which has a right-hand thread and needs to be turned clockwise to tighten
   b red to the acetylene, which has a left-hand thread and needs to be turned anti-clockwise to tighten.
3 Connect the blowpipe to the hose assembly (tubing):
   a the blue hose to the right-hand threaded oxygen, which needs to be turned clockwise to tighten
   b the red hose to the left-hand threaded acetylene, which needs to be turned anti-clockwise to tighten.

   Make sure that both oxygen and acetylene needle valves are closed.

   Note: Blowpipe flashback arrestors should be fitted to the blowpipe as well as to the regulators.
4 Screw the welding mixer onto the blowpipe until it is hand-tight.
5 Select the size welding tip required, check the seating for damage and screw the tip into the mixer. Loosen the joint made between the blowpipe and welding mixer, aligning the tip with the blowpipe to gain the best working position, and then re-tighten the welding mixer to hand-tight.
6 Check for leaks by slightly turning the oxygen needle on and spraying soapy water on all connections, including where the tip is inserted into the mixer. Observe if any bubbles occur.

   Note: If the tip seating is damaged, replace tip.

Lighting up procedure

1 Release the regulator, adjusting the knobs on both oxygen and acetylene regulators.
2 Open both the oxygen and acetylene cylinder valves, a maximum of one turn for oxygen and half turn for acetylene by slowly turning the cylinder valve control knob. This *must be done slowly* to prevent possible damage to the regulators. Cylinder pressure for both the oxygen and acetylene will be shown on the right gauge dial.
3 Open the acetylene gas valve on the blowpipe and adjust the acetylene regulator by rotating in a clockwise direction to increase and obtain the correct working pressure required. This will be indicated on the left gauge dial (see Figure 10A.30). When this is done, close the acetylene gas valve on the blowpipe.

FIGURE 10A.30 Acetylene regulator with cylinder pressure indicated

4 Open the oxygen gas valve on the blowpipe and adjust the oxygen regulator by rotating in a clockwise direction to increase and obtain the correct working pressure required. This will be indicated on the left gauge dial (see Figure 10A.31). When this is done, close the oxygen gas valve on the blowpipe. Refer to Table 10A.3 later in the chapter for oxygen and acetylene pressures required based on the weld to be undertaken.

>>

>>

FIGURE 10A.31 Oxygen regulator with cylinder pressure indicated

5. *Important* – check for leaks on all connections by spraying with a soapy water mixture and observe if any bubbles occur. Fix any leaks before welding.
6. Slightly open the acetylene gas valve on the blowpipe, pause a few seconds, and then light it using a flint lighter or pilot light.
7. Adjust the acetylene gas valve on the blowpipe until the flame just ceases to smoke. When this is done, open the oxygen gas valve and adjust the oxygen to obtain the desired neutral, oxidising or carburising flame, depending on the welding application to be carried out.

Welding gases and flames

With the mixture of acetylene and pure oxygen, a flame temperature of about 3100°C can be obtained. Oxygen can be mixed with other fuel gases, but no other fuel gas–oxygen mix can reach this temperature (see Figure 10A.32).

Oxygen

Oxygen is an odourless, colourless oxidising gas that supports combustion (burning).

Acetylene

Acetylene (C_2H_2) is the most suitable fuel gas for fusion welding of low-carbon steel. This is because it burns up to 3100°C and, when mixed with oxygen, produces a flame that is least likely to cause any contamination to the weld. Acetylene is a highly flammable, colourless gas with a strong, unpleasant odour.

If acetylene is released above 100 kPa as a free gas, it can become very unstable and may explode.

Flame types

There are three flame types used when oxyacetylene welding, and correct selection will depend on the material and welding process to be undertaken.

Neutral flame

A neutral flame has equal amounts of oxygen and acetylene supplied to the welding tip. This produces a cone that is characterised by an almost colourless outer envelope with a sharply defined inner cone and without a feather or secondary flame. It produces an inactive weld pool with very few sparks and is used for fusion welding, silver brazing and lead burning (see Figure 10A.33).

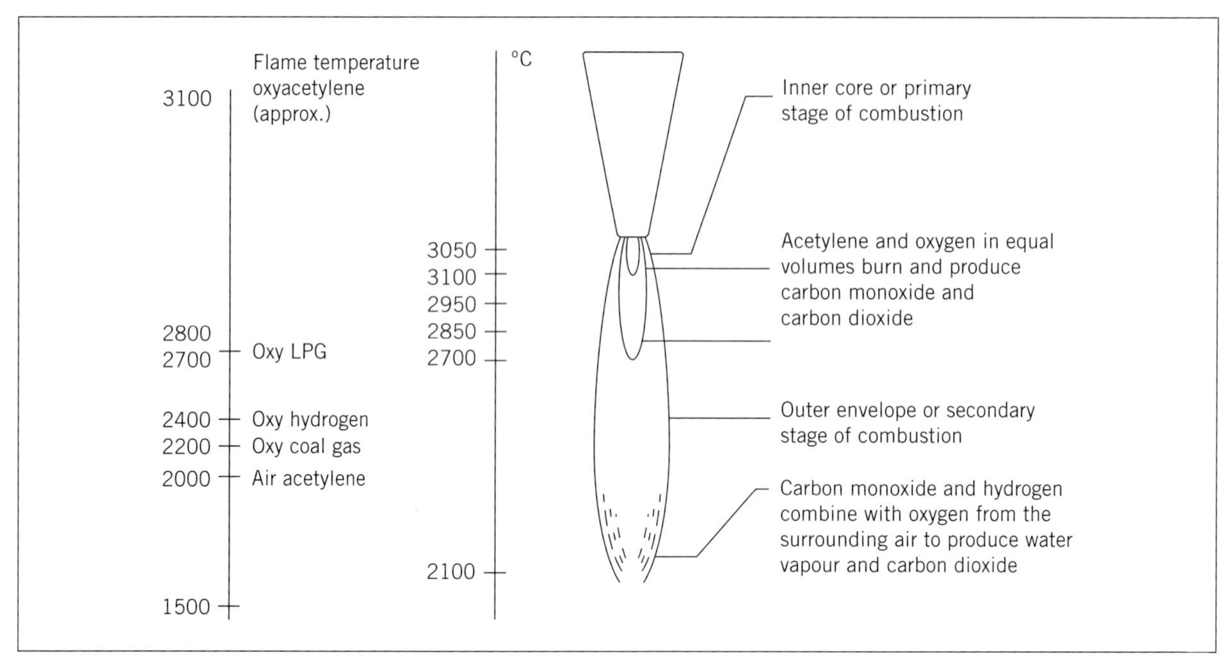

FIGURE 10A.32 Temperatures of oxygen/fuel flame

FIGURE 10A.33 Neutral oxyacetylene flame

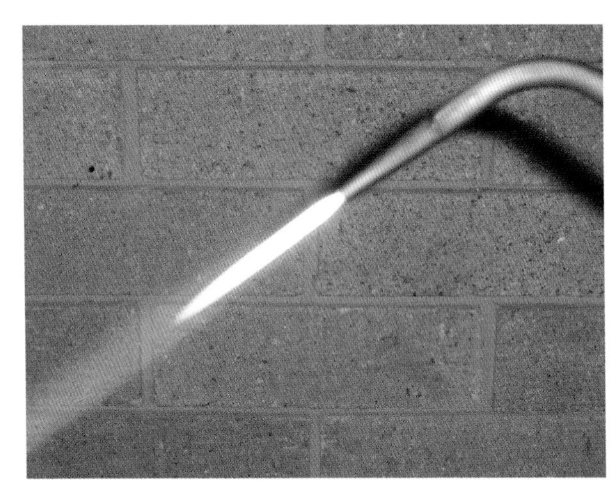

FIGURE 10A.34 Carburising oxyacetylene flame

Carburising flame (reducing flame)

A carburising flame has more acetylene than oxygen supplied to the welding tip. The inner cone is surrounded by a longer, whiter feather that would normally damage the weld area during fusion welding. If a highly carburising flame (see Figure 10A.34) is to be used, then carbon is added to the weld area in order to slow down the welding and cause sparks. This flame can be used to flame-harden a surface or a weld once it has been completed.

Oxidising flame

An oxidising flame has more oxygen supplied to the welding tip than acetylene. The cone becomes shorter and more pointed (see Figure 10A.35), the secondary stage of combustion is shorter, and a hissing sound is produced. This flame is used for braze welding (bronze welding) and for welding brass. If this flame were to be used for fusion welding of steel, the weld pool would boil and oxidise, destroying the properties of the steel in the weld.

Materials

To determine what preparation and jointing types are required, it must be established what material is to be joined. A range of materials can be used for oxyacetylene welding processes, with ferrous low-carbon steel the most frequently used. In addition to oxyacetylene welding, brazing may also be used on non-ferrous materials such as brass and copper.

Material preparation

Preparing material for welding will include cleaning surfaces, bevelling materials over 3 mm thick, and selecting the appropriate jointing technique. All surfaces must be free from contaminants such as rust, mill scale dirt, paint, galvanised coatings, oil and grease before welding, otherwise the weld quality will be unacceptable and dangerous fumes may be given off. These contaminants may be removed by grinding, filing, wire brushing or degreasing.

FIGURE 10A.35 Oxidising oxyacetylene flame

As the thickness of weld metal increases, it becomes harder to maintain full fusion (penetration) through the full thickness of the metal. This means that some form of edge preparation is required to ensure full fusion of the joint. Bevelling is used to prepare the edge of thicker metal and this can be done by flame cutting, filing or grinding.

Weld joint types

The selection of the type of joint that is required depends on the thickness of the material, where the weld is, how much stress the weld will be under and, in the case of pipework, what the service is to be used for.

Butt joints

Butt joints are commonly used for most jointing techniques, as they are easy to prepare and, for the skilled welder, easy to apply.

Flanged butt joints

The flanged butt joint is for materials with a thickness ranging from 0.8 mm to 1.6 mm (see Figure 10A.37). It is prepared by folding the edge up at right angles. The fold (flange) should be twice the thickness of the

HOW TO

SAFELY CLOSE DOWN OXYACETYLENE EQUIPMENT

When oxyacetylene equipment is to be closed down, the following procedure is to be followed:

1 Close the acetylene blowpipe valve:
   a Close the oxygen blowpipe valve.
   b Turn off the cylinder valve on the acetylene cylinder by using the valve control knob.
   c Turn off the cylinder valve on the oxygen cylinder by using the valve control knob.
   d Open the acetylene blowpipe valve and relieve the pressure in the acetylene hose and regulator, and then close the blowpipe valve.
   e Open the oxygen blowpipe valve and relieve the pressure in the oxygen hose and regulator, and then close the blowpipe valve.
   f Release the acetylene regulator control knob by screwing it out.
   g Release the oxygen regulator control knob by screwing it out. Ensure the control knobs are not completely wound out to it maximum to prevent damage.
   h Make sure the welding tip is cool and remove it from the blowpipe. Store the tip with other welding tips where they will not be damaged.
   i Roll up the oxygen and acetylene hoses and hang them on hose hooks or loosely on the trolley ready for use (see Figure 10A.36). *Do not* wrap the hoses around the regulators.
2 Check the gauges to ensure that the cylinders are correctly closed off. No pressure reading should be evident on the cylinder contents gauges.

If cylinders need to be removed from the cylinder trolley, both the oxygen and acetylene hoses should be disconnected from the regulators, the regulators unscrewed from each cylinder and dust plugs fitted into each cylinder valve. This should be done with the aid of the open-ended spanner. Both the oxygen and acetylene regulators should be stored in a safe place to protect them from damage.

Once the equipment set-up is complete, we must determine the material and preparation requirements for undertaking oxyacetylene welding tasks.

FIGURE 10A.36 Oxygen and acetylene hoses rolled up and safely supported on the cylinder trolley

metal. The weld is formed by fusing the upturned edges to a flat, neat weld using a neutral flame. Filler rod is not required.

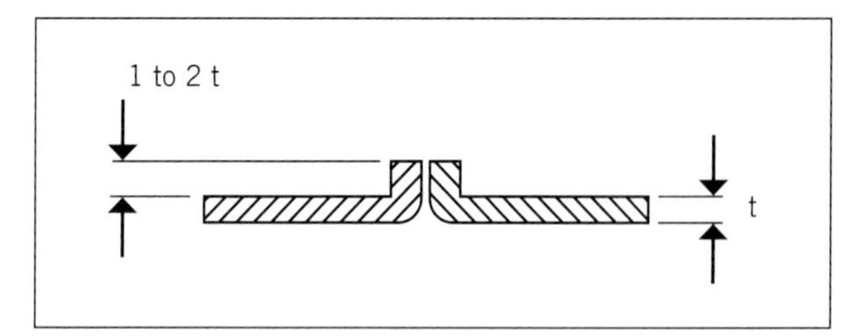

FIGURE 10A.37 Flanged butt joint for 0.8 mm to 1.6 mm low-carbon steel

Close butt joint

This type of joint is for materials with a thickness ranging from 1.2 mm to 2 mm (see Figure 10A.38). The edges are butted in a straight line, then fused together with or without the addition of filler rod for the first run (root weld). A capping run is then necessary to give the required build-up and strength to the joint.

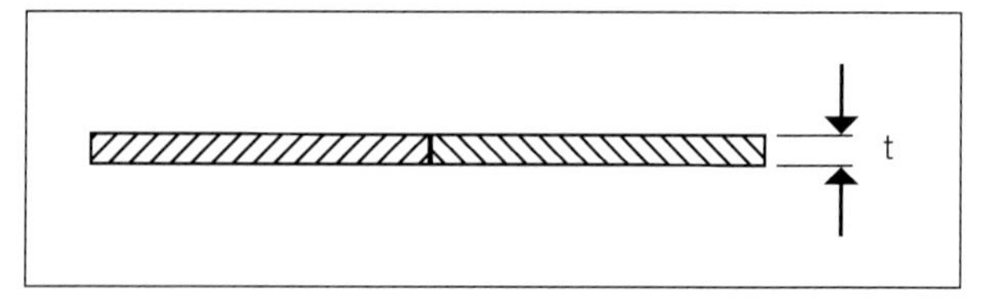

FIGURE 10A.38 Closed butt joint for 1.2 mm to 2 mm low-carbon steel

Open square butt joint

For this joint the clean, square edges are spaced half the thickness of the metal to be joined. It is generally used with metal with a thickness ranging between 1.6 mm and 3 mm. Fusion is undertaken using the forehand technique with the addition of filler rod. This may also require a capping run, as well as tacking the joint at about 50–75 mm intervals to minimise joint distortion (see Figure 10A.39).

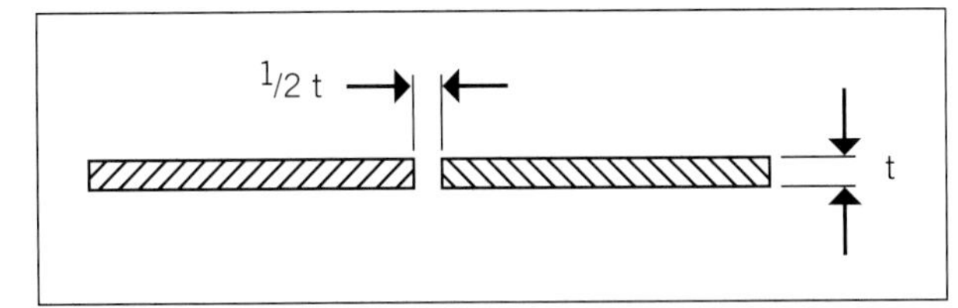

FIGURE 10A.39 Open square butt joint for 1.6 mm to 3 mm low-carbon steel

Single-V butt joint

This is a joint that requires the edges of the parent metal to be bevelled. Each edge must be bevelled by filing, grinding or using an oxyacetylene cutting torch to form a 60° included angle when assembled. Penetration is assisted by leaving a gap between the two pieces of metal, which is dependent upon the thickness of the metal to be welded (see Figure 10A.40). For metal that is between 3 mm and 5 mm thick, the gap is half the thickness of the metal; for metal that is greater than 5 mm thick, a 3 mm gap is required. The weld is formed by fusing the added filler rod on the bevelled edges and building it up to a desired contour. Vertical welds of this type require an 80° included angle. This may also require a capping run to give build-up and strength.

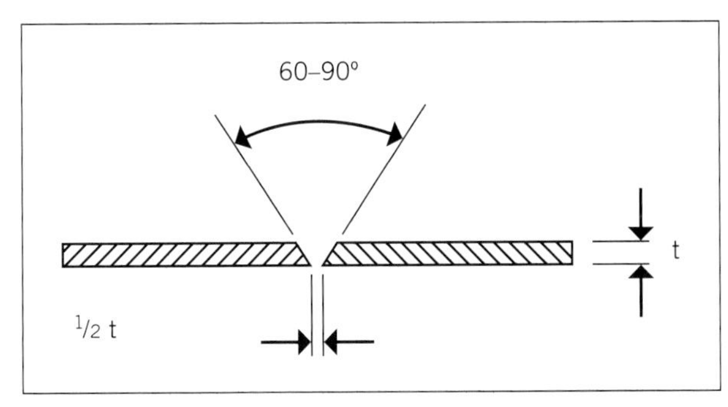

FIGURE 10A.40 Single-V butt joint for 3 mm to 15 mm low-carbon steel

Double-V butt joint

The double-V butt joint is for metal with a thickness of 16 mm and over, and where possible the weld is applied from both sides of the joint (see Figure 10A.41). The material must be prepared so that one side of the joint forms a 60° included angle and the other side an 80° included angle. The weld is started on the 60° angle side and completed by welding the 80° angle side. A gap of 3 mm is left between the joint.

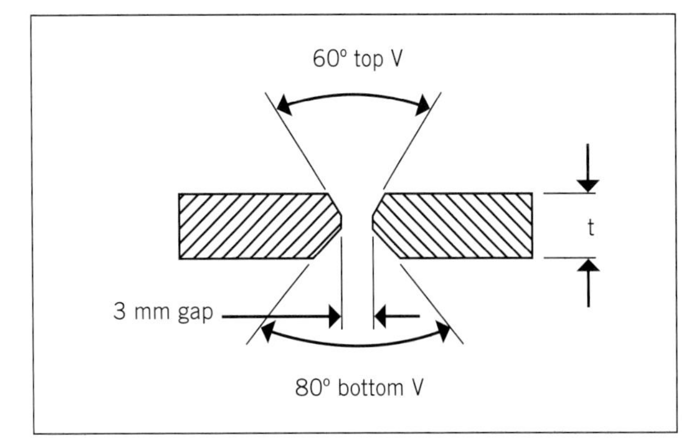

FIGURE 10A.41 Double-V butt joint for 16 mm and over low-carbon steel

Fillet welding

The fillet weld is one of the most common types of welds as it is strong and requires minimal preparation.

T-fillet joint

This is a common type of fillet joint that requires no special preparation except for cleaning. The two pieces of material must sit firmly against each other at 90° (see Figure 10A.42), with welds applied on both sides of the upright where it meets the base.

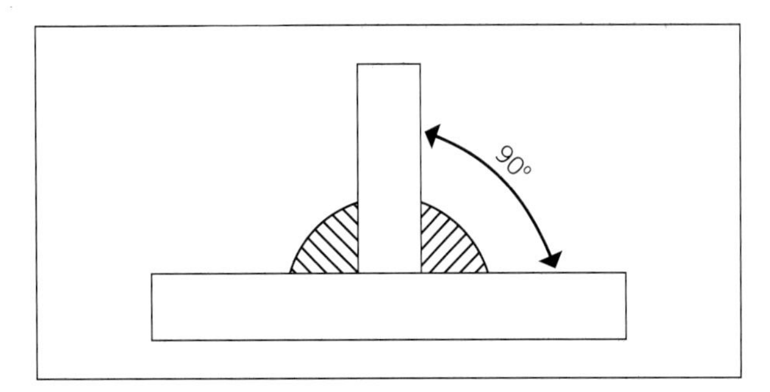

FIGURE 10A.42 T-fillet joint

Lap fillet joint

This joint is when materials are lapped, and the edges welded (see Figure 10A.43). Where possible this joint should be welded on each edge, as the strength of the weld is halved if both are not welded.

FIGURE 10A.43 Lap fillet joint

Outside-corner fillet joint

This is sometimes called a butt joint, although the weld that is applied is a fillet weld and is used on corner joints (see Figure 10A.44). This requires a minimum of preparation as the angle is generally 90° and the metal requires cleaning before welding. In this case, the weld fills the corner between two flat-edged pieces of metal.

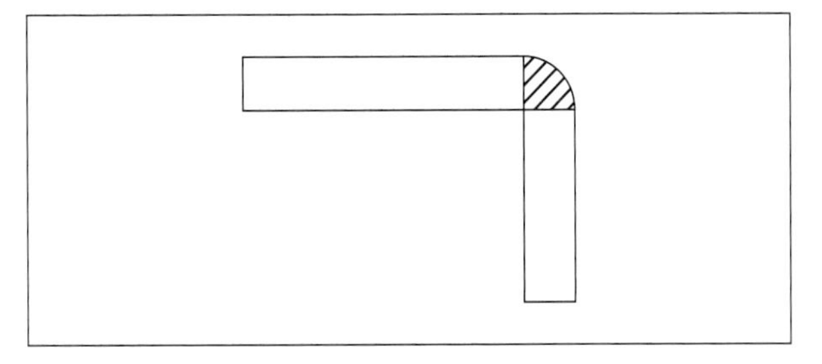

FIGURE 10A.44 Outside corner fillet joint

LEARNING TASK 10A.2

PREPARE MATERIALS AND WELDING EQUIPMENT

Provide short answers to the following questions:

1 How are oxygen and acetylene cylinders protected from explosion if overheated?
2 How long should an acetylene cylinder be allowed to rest after transportation before it can be used?
3 What tool is used to tighten hexagon fittings on regulators and hosing?
4 What dictates the size of welding tip to be used?
5 Why should only a flint lighter or pilot light be used to ignite an oxyacetylene torch?
6 Why are gauges removed from oxygen and acetylene cylinders when they are being transported?
7 What should be done once the oxyacetylene torch has been set up?
8 Why must contaminants be removed from surfaces before welding takes place?
9 Which weld joint type would be selected for joining 6 mm steel plate?
10 Why are fillet welds commonly used in welding processes?

COMPLETE WORKSHEET 1

COMPLETE WORKSHEET 2

Perform welding

Fusion welding is a technique that is predominantly used in, but not limited to, mechanical plumbing. The welding process can be used for a range of purposes, including pipework and bracketing systems.

Oxyacetylene fusion welding of low-carbon steel

This type of welding uses a high-temperature oxyacetylene flame for heat, in which the edges of two pieces of ferrous metal being joined are melted and completely fused together without pressure. If a filler rod is required, it must be a similar composition to the parent metal. Fusion relies on the thickness of the metal for strength, and the weld metal should have a build-up of approximately half the thickness of the parent metal to guarantee its strength.

The advantages of fusion welding over other welding techniques are that no electrical supply is required, and that the equipment is portable and can be taken to a job. However, fusion welding does have limitations, which include the following:

1 It requires a lot of skill and there are some processes that only a certified welder can undertake.
2 It is sometimes very slow.
3 Because of the high heat input, distortion will occur in the parent metal.
4 The equipment required is bulky and at times difficult to transport.

Consumables for fusion welding

Filler rods for low-carbon steel are classified in *AS/NZS 1167 Welding and brazing – Filler metals – Filler metal for welding*. There are two common filler rods used for fusion:

1 *RG low-carbon steel filler rod* – for general purpose welding; it has very low carbon content and produces ductile weld deposits.
2 *R1 low-alloy steel filler rod* – it has a much higher tensile strength and is used on pressure pipes for higher-strength welds. The rods are copper coated and no flux is required as the oxides are prevented from forming by deoxidisers in the filler rod.

10A.5 STANDARDS

- AS/NZS 1167 Welding and brazing – Filler metals – Filler metal for welding

Fusion welding of steel

The two main forms for fusion welding are the forehand and backhand welding techniques. Material thickness and the location of where the welding is to take place will determine what technique is to be used.

Forehand welding technique

This is the most commonly used welding technique for plumbers when fusion welding with oxyacetylene equipment. The filler rod leads the welding tip along the prepared joint, with the neutral flame pointing along the joint away from the weld puddle, and the filler rod is added by dipping or rotating in the molten puddle. For right-handed welders the travel is from right to left (see Figure 10A.45) and the reverse for left-handed welders. This technique is generally used on material up to and including 5 mm thick. If the angle of the welding tip is not correct, the tip will overheat and backfiring will occur, which is where mixed gases explode in the overheated tip.

Forehand welding is used on such joints as flanged butt, closed square butt (up to 3 mm), single-V butt (up to 5 mm) and fillet welds, and gives a neat, clean finished weld.

Forehand welding can also be used when welding vertical joints using open square butt joints up to 3 mm thick and single-V butt joints between 3 mm and 5 mm thick, but the included angle needs to be 80° minimum, not 60° to 70°. The weld is started from

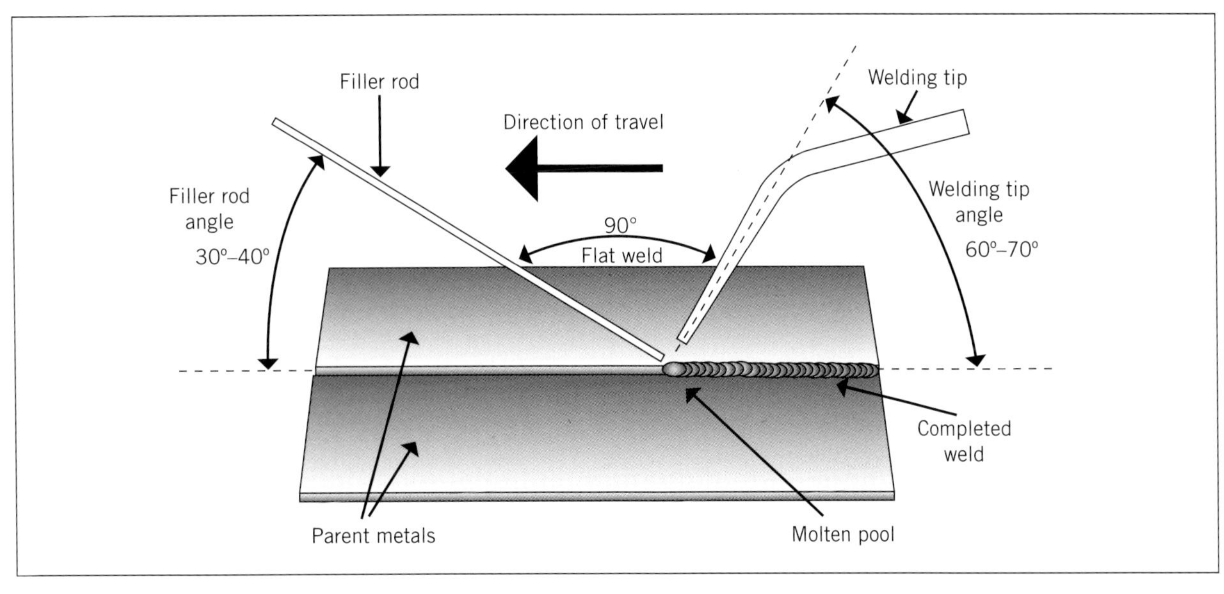

FIGURE 10A.45 Forehand welding technique for right-hand welders

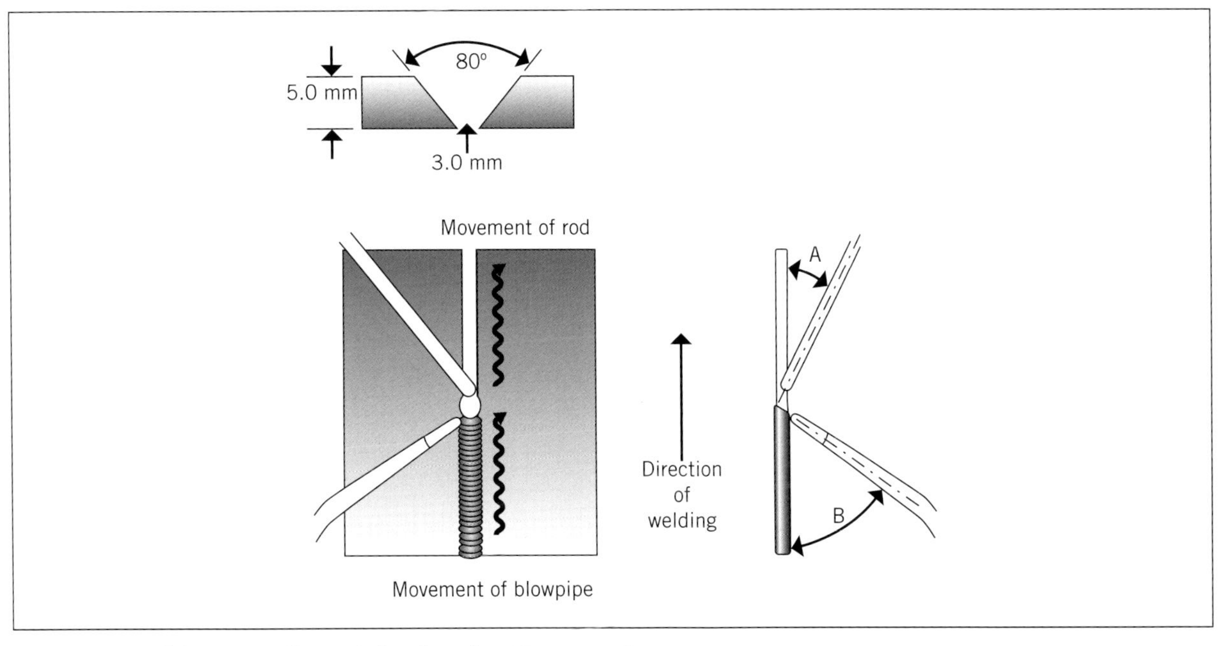

FIGURE 10A.46 Joint preparation and direction of weld movement

the bottom and proceeds vertically to the top of the joint (see Table 10A.2 and Figure 10A.46).

TABLE 10A.2 Welding tip and filler rod angles required for vertical welding

Steel thickness	Angle of filler rod A	Angle of blowpipe B	Distance of cone from metal
1.5 mm	30°	25°	1.5–3.0 mm
2.5 mm	30°	35°	1.5–3.0 mm
3.0 mm	30°	50°	1.5–3.0 mm
5.0 mm	30°	90°	1.5–3.0 mm

Backhand welding technique

For this welding technique the welding tip leads the filler rod along the prepared joint, with the neutral flame pointing back into the weld puddle, and the filler rod being added by dipping or rotating in the molten puddle. For a right-hand welder, the travel is from left to right (see Figure 10A.47). This technique is used on metal thicker than 5 mm. For a left-hand welder, the travel is from right to left along the weld joint.

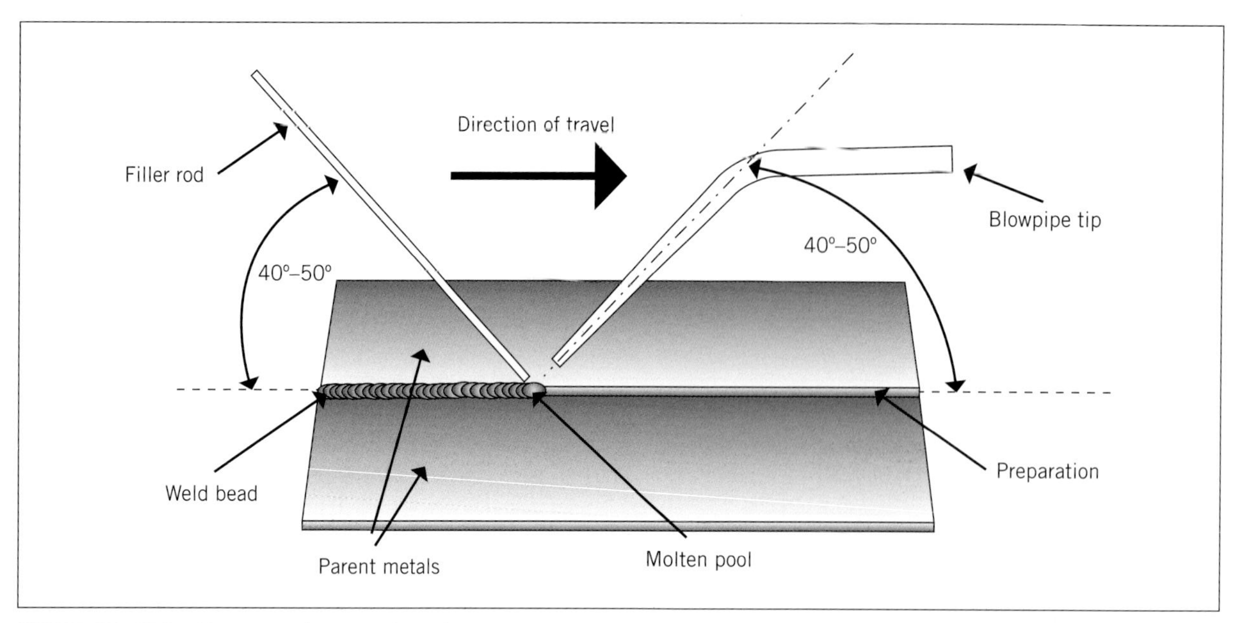

FIGURE 10A.47 Backhand welding technique for right-hand welders

Backhand welding can also be used when welding vertically, upright horizontally and overhead on steel plate thicker than 5 mm.

Before doing any fusion welding with oxyacetylene, the following should be used as a guide:

- *Flame setting* – neutral.
- *Tip size* – this is determined by the thickness of the steel (see Table 10A.3).
- *Tacking sequence* (see Figure 10A.48) – for material under 1.6 mm, tack every 50 mm. For material between 1.6 mm and 2.5 mm, tack every 75 mm. For material over 2.5 mm, tack every 100 mm. Tacks must fuse for the full thickness of material or root preparation. Tacks are used to hold the joint together and stop distortion/contraction (see Figure 10A.49) from taking place while welding.

TABLE 10A.3 Filler rod and tip size for forehand and backhand welding techniques in relation to the type of joint being welded

Joint preparation	Metal thickness	Filler rod diameter	Tip size	Oxygen pressure	Acetylene pressure
Forehand weld					
Flange butt and open square butt joints	0.8 mm	1.6 mm	8–10	50 kPa	50 kPa
	1.5 mm	1.6 mm	10–12	50 kPa	50 kPa
	2.5 mm	1.6 mm	10–12	50 kPa	50 kPa
	3.0 mm	2.4 mm	12–15	50 kPa	50 kPa
Single-V butt joint	3.0 mm	2.4 mm	12–15	50 kPa	50 kPa
	4.0 mm	2.4 mm	12–15	50 kPa	50 kPa
	5.0 mm	3.2 mm	15–20	50 kpa	50 kPa
	6.0 mm	5.0 mm	15–20	50 kPa	50 kPa
Backhand weld					
Open square butt joint	5.0 mm	3.2 mm	20–26	50 kPa	50 kPa
	6.0 mm	5.0 mm	20–26	50 kPa	50 kPa
	8.0 mm	5.0 mm	26	50 kPa	50 kPa
Single-V butt joint	6.0 mm	5.0 mm	20–26	50 kPa	50 kPa
	8.0 mm	5.0 mm	26	50 kPa	50 kPa
	10.0 mm	5.0 mm	26–32	100 kPa	100 kPa
	>13 mm	6.0 mm	26–32	100 kPa	100 kPa

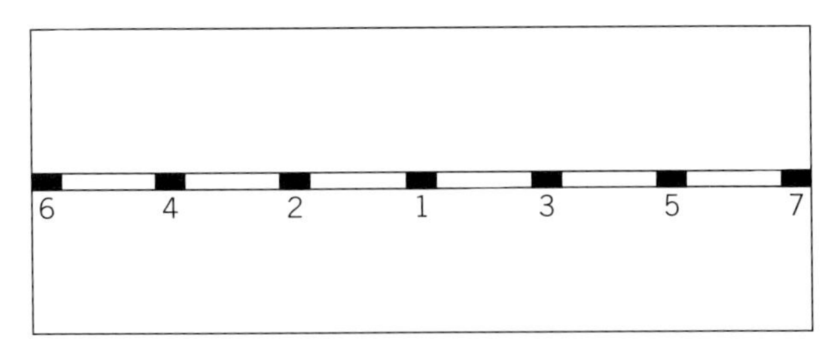

FIGURE 10A.48 A recommended tacking sequence for mild steel plate

- *Regulator pressure* – Refer to Table 10A.1 and Table 10A.3.
- *Filler rod* – Refer to job specifications and *AS/NZS 1167 Welding and brazing – Filler metals – Filler metal for welding.*
- *Joint preparation* – Refer to the job specifications. The type of joint specified will determine what preparation is required (see Figures 10A.37–44).
- *Welding technique* – Forehand weld or backhand weld is determined by the thickness of material being welded.
- *Distortion control* – Due to the high heat input, distortion will occur with thin metal. Tacking, clamping and backstep welding should be used.
- *Post weld* – To prevent large grain size, post-weld heating, which is called normalising, is sometimes required on critical or thick joints.

10A.6 STANDARDS

- AS/NZS 1167 Welding and brazin – Filler metals – Filler metal for welding

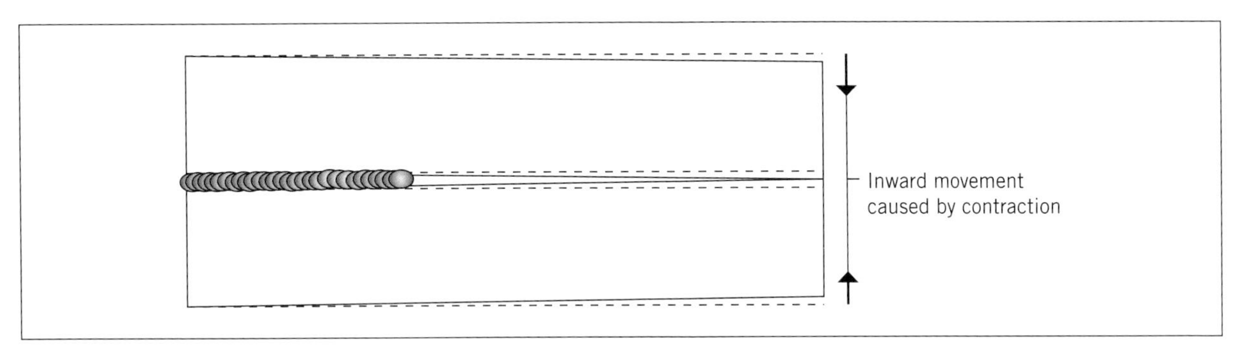

FIGURE 10A.49 Example of contraction occurring on a steel plate that was not tacked before welding took place

Fusion welding of pipe

The forehand method is the most common for welding steel pipe as its application requires only a slight adjustment. Both flame and filler rod selections stay the same.

There are two ways that pipe may be welded: pipe rotation and fixed position pipe.

Pipe rotation

Preparation of pipe is conducted in a similar manner to steel plate, and is based on the pipe wall thickness. The pipe is securely tack-welded on opposing sides of the pipe (see Figure 10A.50) and laid on some form of rollers if possible. Welding is then begun, as shown in Figure 10A.51. Although it looks like the starting point of the weld is vertical, it is described as inclined. Application of the filler rod in short runs, with the welder stopping and rotating the pipe, allows for excellent control of the weld metal and penetration. A continuous filler rod deposit can be obtained by having an assistant carefully rotate the pipe on the rollers while the welder carries on welding continuously until the joint is completed.

Fixed position pipe

Because the pipe is fixed and cannot be rotated during welding, the process is made more difficult for the welder, as techniques designed for welding flat, inclined, vertical and overhead positions must be used. While the techniques are varied, preparation of the joint remains the same, as in open square butt or single-V butt joints. Variations in filler rod and blowpipe positions must occur evenly to prevent undercut and other weld faults (see Figure 10A.52 for the procedure required). The weld is started at the lowest

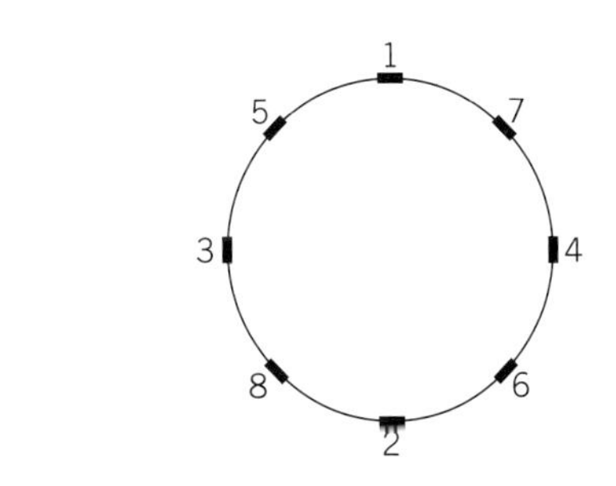

Tacks are placed so that the contraction of the first is balanced by the contraction of the second, which is located on the directly opposite section of the butt joint. The number of tacks required is governed by the diameter of the pipe, the larger the diameter the more tacks. Tacks should be spaced no greater than 50 mm apart.

FIGURE 10A.50 Recommended tacking sequence when two lengths of pipe are to be butt welded

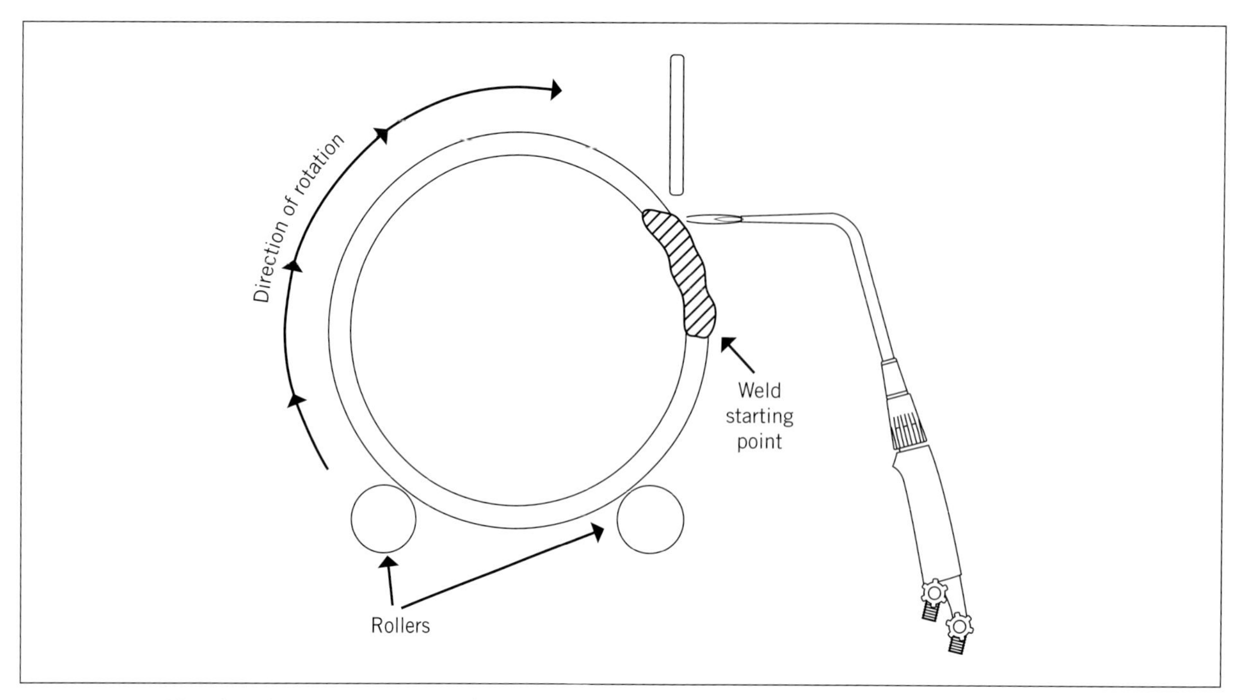

FIGURE 10A.51 Direction of weld travel and rotation of pipe

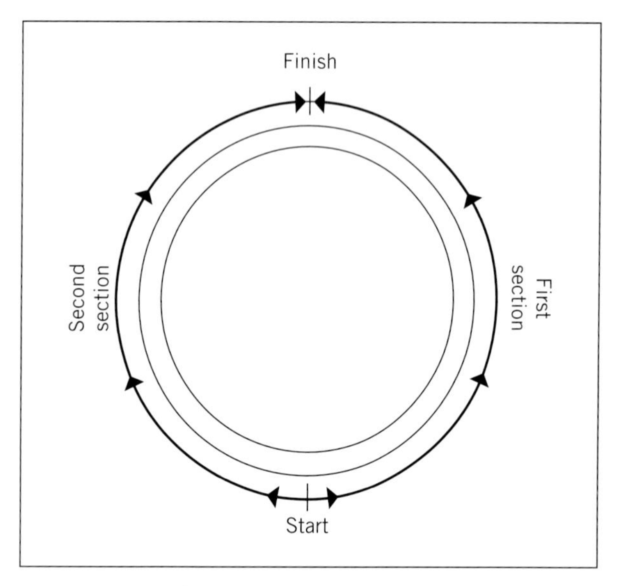

FIGURE 10A.52 Direction of welding when pipe is in a fixed position

point of the joint and progresses upwards to the highest point. Welding is then recommenced at the lowest point, making sure that complete fusion takes place between the initial run and the new one, and progresses upwards until the weld is completed by full fusion taking place between the end of the first run and the second.

GREEN TIP

Ferrous metals such as steel are recyclable. If materials are no longer required, ensure they are recycled to prevent them from going to land fill.

Weld defects

There are many weld defects that can occur within the weld or in the parent metal, and these can be detected by a close visual inspection and, where the weld is critical, by X-ray.

Undercut

Undercut is defined as a groove or channel in the parent metal along the toe of the weld caused by melting away the parent metal (see Figure 10A.53). It is

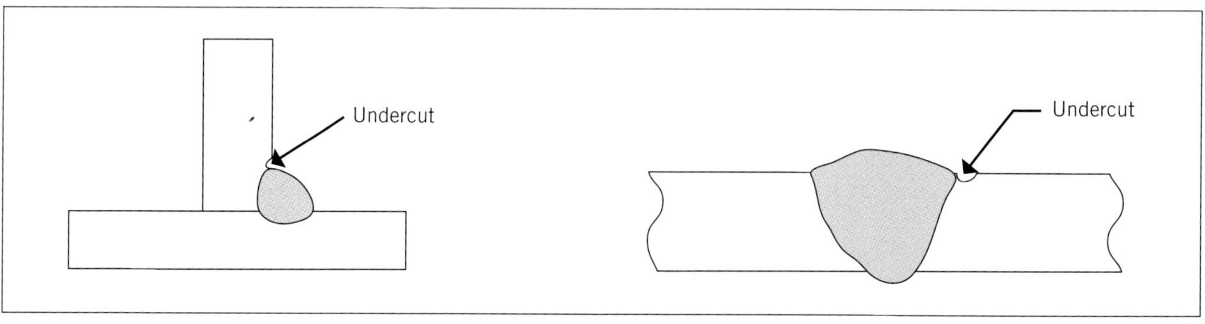

FIGURE 10A.53 Examples of undercut in a T-fillet and a butt weld

generally a sign of lack of skill or carelessness, and may be caused by the following faults in technique:

- too large a flame
- wrong angle of the welding tip
- poor filler rod manipulation
- excessive movement or weaving of the flame
- too fast a rate of travel.

Undercut is a serious weld defect that can lead to a weld failure, such as stress or fatigue failure, when in service as it reduces the wall thickness of the parent metal.

Overroll

Overroll (also called overlap) is a fault in the weld where molten weld metal overflows on the unmelted parent metal, leaving an unfused area (see Figure 10A.54). This may result from faults in welding techniques, such as:

- use of too large a flame, causing too large a weld pool
- lack of skill of the welder
- using the wrong size filler rod.

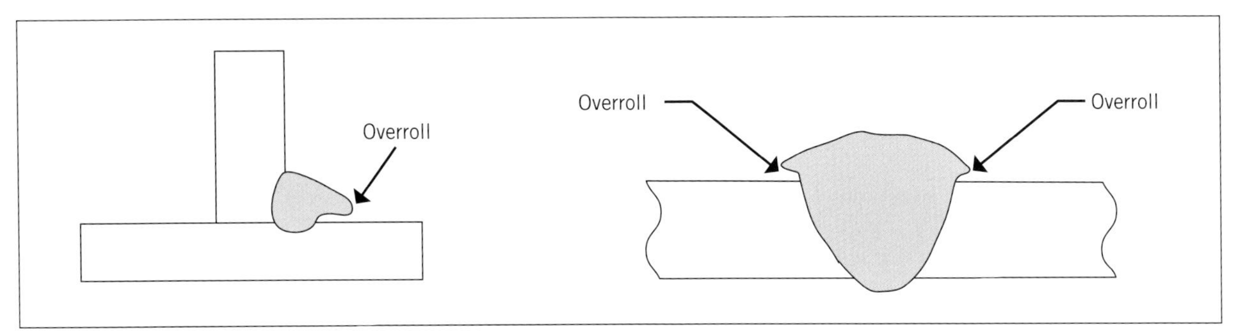

FIGURE 10A.54 Examples of overroll in a fillet weld and a butt weld

Overroll can affect the weld strength in the same way that undercut can, because of the notch effect of the weld metal overlapping the parent metal, with no fusion taking place.

Misalignment

Misalignment (see Figure 10A.55) is a result of poor preparation of the parent metals before the welding or movement happening during the welding process. This may happen due to:

- the job not being assembled or aligned correctly
- the job not being clamped correctly to prevent movement
- the tacks being too small and not fused correctly, causing cracking and movement during the welding process.

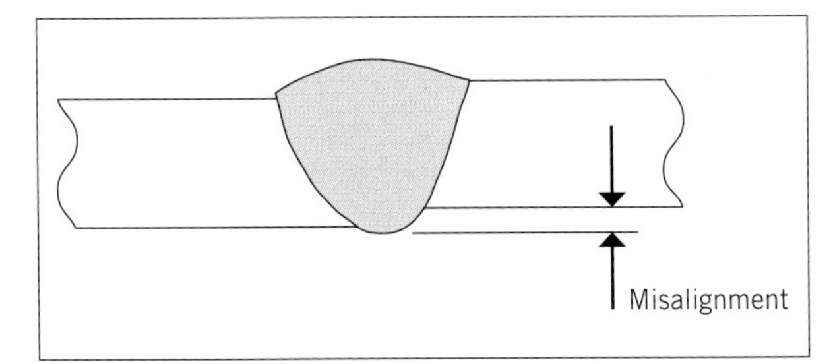

FIGURE 10A.55 Typical example of misalignment

Misalignment can cause serious weakness and a lack of strength and serviceability in the welded joint, and in the case of service pipes can restrict flow and pressure.

Excessive penetration

Excessive penetration (see Figure 10A.56) can occur on butt welds, with the main cause being a lack of control of the weld pool, which causes unwanted metal to protrude through to the underside of the joint. This can cause:

- stress
- joint rigidity
- flow restriction in pipes.

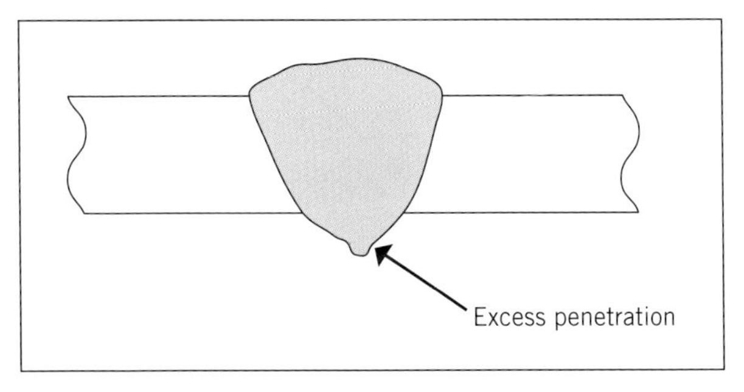

FIGURE 10A.56 Butt weld with excessive penetration

Incomplete penetration

This is the failure of the weld metal to fill the root of the weld. As the weld has not obtained full fusion, this means that it is not as thick as the parent metal and is therefore a source of weakness. Incomplete penetration (see Figure 10A.57) can be caused by:

- poor preparation
- insufficient heat, through the use of too small a welding tip

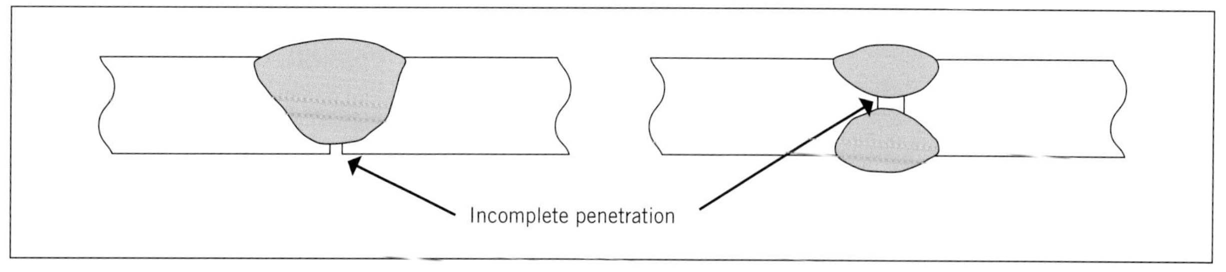

FIGURE 10A.57 Examples of incomplete penetration

- a lack of skill or carelessness on behalf of the welder
- a lack of fusion because of the failure to melt the underlying metal when depositing the filler rod
- using the wrong size filler rod.

Porosity

Porosity generally refers to small round holes in the weld metal (see Figure 10A.58) caused by gas trapped in the weld metal. The holes, which may occur in clusters or be scattered, may be caused by:

- gases given off by chemical actions within the molten metal
- gases given off by paint or chemicals on the parent metal or filler rod
- incorrect flame adjustment – either not enough or too much heat.

Cracking

When cracking (see Figure 10A.59) occurs within the weld, it may be caused by:

- contaminants on or in the parent metal
- the use of an unsuitable filler rod
- clamping the joint too rigidly
- the weld beads being too small.

Distortion

Distortion (see Figure 10A.60), as applied to welding, includes any departure from alignment or measurements resulting from the welding process. It is caused by:

- uneven expansion and contraction from localised heating of the parent metal when welding
- residual stresses that are present in the parent metal.

Lack of fusion

Lack of fusion (see Figure 10A.61) is the incomplete fusion between the added weld metal from the filler rod and the parent metal during the welding operation. It can be caused by:

- using too small a welding tip
- using an incorrect weld angle
- weld travel that is too rapid

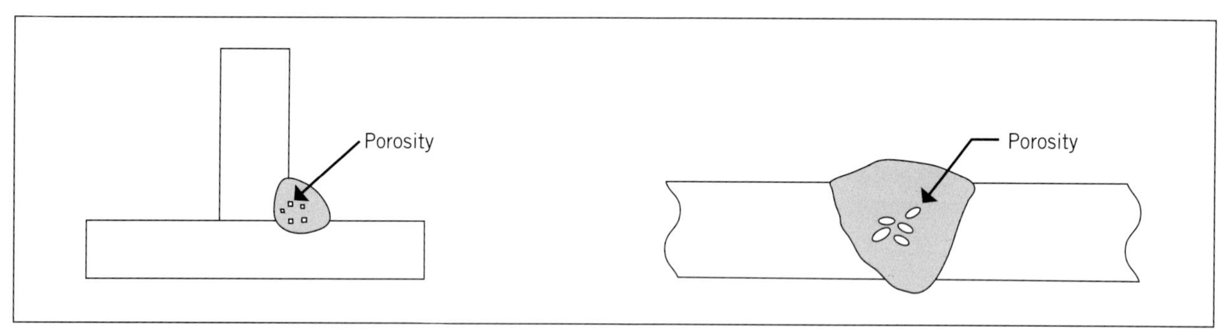

FIGURE 10A.58 Examples of porosity

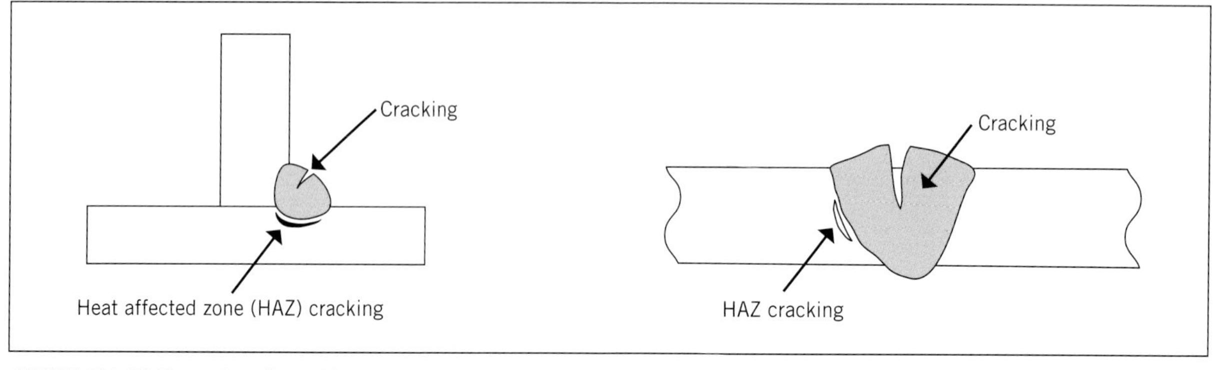

FIGURE 10A.59 Examples of cracking

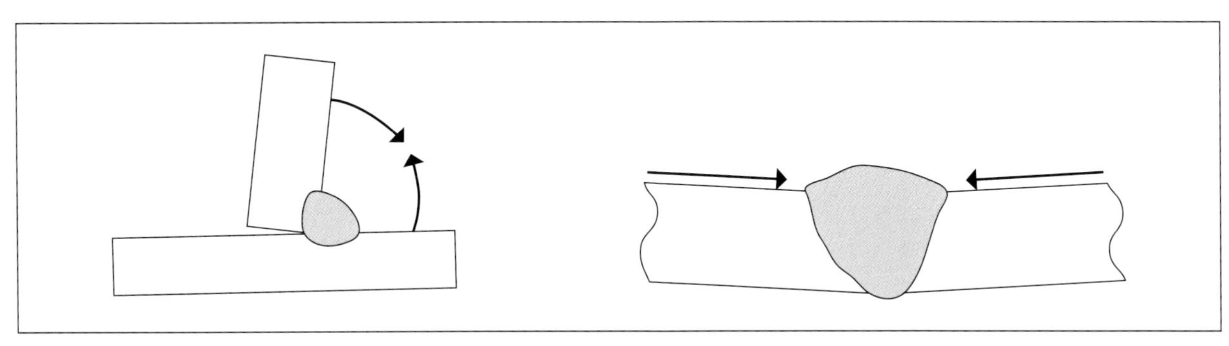

FIGURE 10A.60 Typical distortion in butt and fillet welds

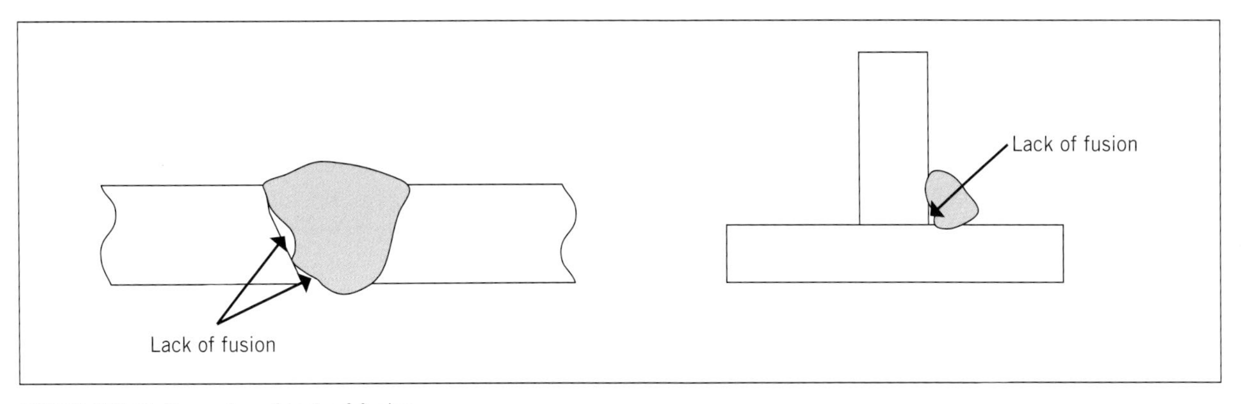

FIGURE 10A.61 Examples of lack of fusion

- using the wrong weld technique for the thickness of metal (e.g. forehand welding when backhand is required).

Inspection and cleaning of welds

Once all welding has taken place, completed welds must be cleaned and inspected and any defects rectified in accordance with workplace requirements. Services may require testing, and in some circumstances, X-rays may need to be provided on large joints.

Brazing processes used by plumbers

Other than fusion welding, there is another joining process that a plumber will use in day-to-day work. This common process is known as brazing.

Brazing

Brazing differs from welding as it joins two pieces of metal together with the use of a filler rod without melting the metal. The filler rod has a lower melting point than the metal to be joined, which means that only the filler rod melts. A flux may be required during the brazing process, depending on the materials that are required to be joined.

There are two common brazing processes used in plumbing: braze welding and silver brazing (silver soldering).

Braze welding

Braze welding, also commonly known as bronze welding (see Figure 10A.62), uses a process that relies on intergranular penetration to join two or more metals. When the metal is heated, the grains expand. A non-ferrous alloy molten filler flows between them and becomes entrapped as the parent metal cools. The strength of the weld relies on the surface area of the weld.

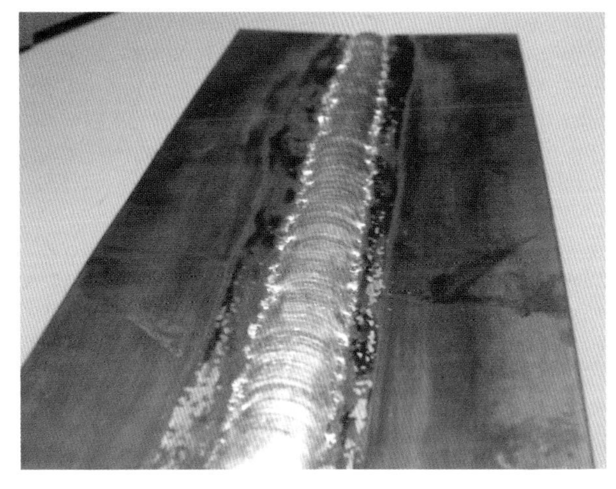

FIGURE 10A.62 Typical braze weld

The molten filler rod should penetrate the full depth of the joint as well as cover about 1.5 times the thickness of the parent metal on either side of the joint, while having a build-up of weld material of about 0.66 (two-thirds) of the diameter of the filler rod for reinforcement (see Figure 10A.63). Because these are bronze, and oxides can be formed when welding,

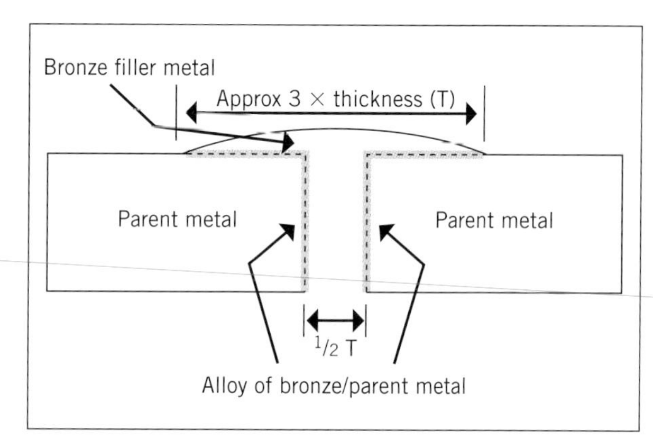

FIGURE 10A.63 Section of a typical braze weld joint

a copper and brass welding flux must be used. Braze welding is often used to join two dissimilar metals. There are two types of bronze filler rods used:

- *manganese bronze* – for ferrous metals
- *Tobin bronze* – for non-ferrous metals.

Silver brazing (silver soldering)

Silver brazing (silver soldering), is similar to bronze welding but the filler rod used has a much lower melting point than the bronze filler rod and is therefore not as strong, which means the joint design and fabrication must be strengthened.

The joint must be made with a lap that is close fitting so that the filler rod will be drawn into the space between the lap by capillary action. The metals are then bonded by intergranular penetration of the molten filler rod and hot expanded grains of the metal. The laps need to be at least three times the thickness of the parent metal being joined (see Figure 10A.64).

Plumbers use this brazing method for joining copper tube used in a number of different services. Silver brazing is used to join:

- copper to copper (tubing)
- copper to copper alloy (brass fittings).

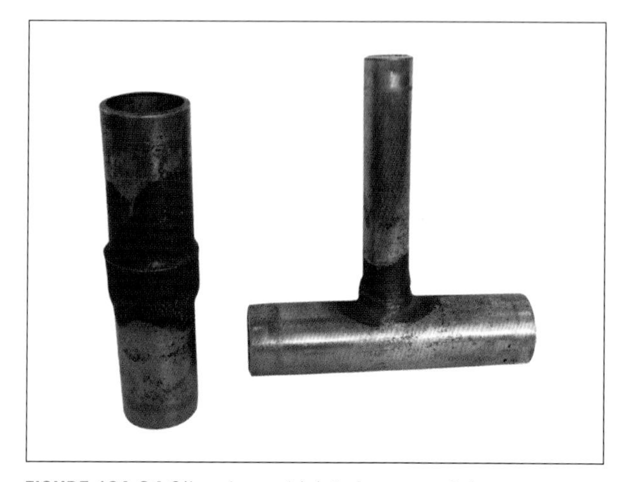

FIGURE 10A.64 Silver brazed joints in copper tube

Copper to copper

When joining copper to copper, there are two basic processes used:

- slipped and brazed
- branch-formed.

Slipped and brazed

This process is used to join copper tube and involves annealing and expanding the end of a copper tube with a set of tube expanders designed for this task. The spigot end of another copper tube is then placed (slipped) into the expanded end. The joint is then silver brazed (soldered) (see Figure 10A.64).

Branch-formed

This process is used for making branches (Tees or Ts) in copper tube. It involves drilling a hole into the tube, where the T section is to be formed, then using a special forming tool to draw out the branch. Before placing the spigot end of the tube into the formed branch, it is advisable to use dimple pliers on the end of the tube to help prevent it from penetrating too far into the main tube and restricting the flow. Note that Australian Standards do not permit the use of equal-sized Ts with this process. If an equal T is required, then a fitting should be used, whether it is a copper or copper alloy fitting.

Copper to copper alloy

Fittings used for copper tubing are generally:

- copper fittings such as tees, elbows or bends
- a copper alloy such as brass, which is generally but not limited to threaded fittings.

While the process of silver brazing itself is relatively simple, certain conditions apply.

When silver brazing copper to copper, it is advisable to ensure that the tubing (as well as fittings when used) are free from dirt and grime. This is achieved simply by cleaning the spigot end of the tube and the inside of the tube end or fitting with steel wool or an approved abrasive cloth before the process of silver brazing is started. This will also ensure that the thin oxide layer on the copper tube has been removed. If the surfaces are clean, then no flux is required.

When silver brazing copper tube to copper alloy fittings, it is essential to use a proprietary flux in the process. This is normally called silver-brazing flux or general-purpose brazing flux. It is extremely important to clean both the tube and fitting, generally with steel wool or an approved abrasive cloth, before applying the flux to both surfaces. It is also recommended to use a brush to apply the flux and check the Safety Data Sheet (SDS) for any flux that is used.

Silver brazing rod

Silver brazing rod, often referred to as silver solder, is used for the process of silver brazing and is available with many different percentages of silver content. The percentage content will depend on the requirements of the medium being conveyed through the pipe and what material is to be joined.

The most common types used for plumbing projects are either 2% or 5% silver solder. This is governed by plumbing regulations that specify the grade or percentage to be used. The percentage of silver influences the flow of the solder in the jointing process, with the higher percentages flowing more easily. *AS/NZS 3500.1 Plumbing and drainage, Part 1: Water services* specifies 1.8% minimum silver content for potable water applications.

- AS/NZS 3500.1 Plumbing and drainage, Part 1: Water services

The silver brazing rod is colour coded for easy identification. The three most commonly used brazing rods are:

- yellow tip – 2% silver
- silver tip – 5% silver
- brown tip – 15% silver.

LEARNING TASK 10A.3

PERFORM WELDING

Determine the requirements for joining two steel plates using oxyacetylene welding processes. The material is 3 mm thick and has the dimensions 150 mm × 50 mm.

1. Which weld joint type would be selected based on the material thickness?
2. Is joint preparation required?
3. Is a filler rod required? If so, which type would be used?
4. Which welding technique would be used?
5. For a right-handed welder, what would be the direction of travel for the welding technique used?
6. What is the recommended flame setting?
7. What pressure setting is required on the oxygen and acetylene regulators?
8. What is the recommended tip size?

Clean up

Good housekeeping and clearing of the work area must be undertaken at the end of each workday. All debris and offcuts should be disposed of appropriately, and unused materials should be stacked and stored for reuse or recycled in accordance with state and territory requirements.

Maintenance of tools and equipment

All tools and equipment must be checked to ensure they are in working order and cleaned before appropriate storage in accordance with workplace procedures. Maintenance of tools and equipment must be regularly carried out to manufacturers' recommendations to ensure they are in working order and safe to use every time they are required. Always inform your supervisor if tools are in need of repair.

Accessing and completing documentation

Whenever plumbers are required to carry out any work, they must complete a risk assessment and develop a SWMS for each job and site. In addition, on many large construction sites or maintenance contracts, plumbers will also be required by the company to which they are contracting to fill out a hot work permit whenever they are using oxyacetylene welding equipment or LPG heating equipment. This is not only for the company's own risk assessment but also may be an insurance requirement.

Hot work permits must be completed in accordance with *AS 1674.1 Safety in welding and allied processes – Fire precautions*. It should have a precautions checklist that needs to be checked off for each permit; it should be checked before, during and after any hot works (see Table 10A.4 as an example). A typical hot work permit (see Figure 10A.65) should include such information as company name, contact person, description of work, equipment being used, start and finish dates, emergency procedures to follow if something should go wrong, and a list of all actions that must occur before work is begun.

- AS 1674.1 Safety in welding and allied processes – Fire precautions

TABLE 10A.4 Precautions checklist for hot works

Precautions checklist			
General precautions	**Yes**	**No**	**N/A**
Have flammable and combustible materials been removed or protected?			
Are sprinklers, hose reels and extinguishers available and operating?			
Has the floor been swept clean and wetted down where required?			
Is there adequate ventilation?			
Is all welding/heating equipment well maintained?			
Is a fire watcher required (an extra person on watch)?			
Has the fire panel been isolated (building security)?			
Have the smoke/thermal detectors been isolated (building security)?			
Has management and/or maintenance been informed of details of the hot work?			
Precautions within 15 m of hot work			
Have all combustible liquids, vapour and gases been removed or protected?			
Have combustible floors been protected?			
Has all flammable dust and lint been removed or protected?			
Has any explosive atmosphere in the area been eliminated?			
Have all wall and floor openings been covered?			
Work on walls or ceilings			
Are walls/ceilings constructed of non-combustibles and without any combustible coverings or insulation?			
Have all combustible materials on the other side of the wall/ceiling been moved?			
Are there fire-resistant coverings under the work area to collect sparks?			
Work on enclosed equipment			
Has the enclosed equipment been cleaned of all combustibles?			
Have all containers been purged of flammable liquids/vapours?			
Fire watcher			
Is a fire watcher required?			
Has a fire watcher been organised (if required)?			
Has the fire watcher been trained in the use of equipment and sounding the fire alarm?			
Are the appropriate fire extinguishers full and ready for use?			
Has the fire panel been reinstated on completion of work?			
Other precautions that may be highlighted from the risk assessment			

Permit No.	This permit must be completed for all cutting, welding and other hot work performed outside a dedicated workshop area. The permit must be displayed at the work site and returned on completion of work.

Application for hot work				
Company/Dept Performing Work for			:	
Contact Name:			**Tel: ( )**	
Location of Work:				
Description of Work:				
Equipment to be used:				

Permit begins		**Permit expires**	
Date: / / **Time:**	**am/pm**	**Date: / /** **Time:**	**am/pm**

Emergency information	
If a fire occurs, call	**Tel: ( )**
Nearest fire alarm:	

Authorisation by company representative		
The above work is authorised to proceed subject to the following action being taken prior to work starting and procedures being maintained for the duration of the work. Each item is to be checked by the Authorised Company Representatives prior to work starting for each period (delete and initial if and where not applicable).		
Authorised by:	**Signed:**	**Date: / /**

1	Fire sprinklers and/or thermal detectors must be confirmed as operational (where installed).	❑	6	Combustible materials located within 10 m must be removed or protected with non-combustible curtains, metal guards or flame-proof covers (not ordinary tarpaulins). In a retail/office environment, if 15 m clearance is not practical then the largest distance possible (minimum of 3 m) is acceptable.	❑
2	Smoke detectors must be isolated in the work area and impairment procedures followed.	❑	7	All floor and wall openings within 15 m must be covered to prevent transmission of sparks.	❑
3	Fire equipment to be provided as follows: • Fire hose reel • Fire extinguisher Mandatory fire watcher present	 ❑ ❑ ❑	8	The hot work area and any adjoining areas must be patrolled from the start of work until 30 minutes after the work is completed (including break periods).	❑
4	Barricades, warning signs and spark/flash screens must be provided.	❑	9	Special conditions. **(Please detail)**	❑
5	Work area, trenches, pits, etc. must be clear of flammable liquids, gases or vapours.	❑			

Work completed and area safe		
The work area has been inspected by the Authorised Company Representative 30 minutes after completion of work.		
Signed:	**Date: / /**	**Time:** **am/pm**

This permit is only valid for 24 hours. Ensure contractor returns this form.

FIGURE 10A.65 Typical hot work permit

REFERENCES AND FURTHER READING

Acknowledgements

Information on cylinder safety supplied by BOC Limited, A Member of the Linde Group: **http://www.boc.com.au**. Reproduction of the following resource list references from DET, TAFE NSW C&T Division (Karl Dunkel – Program Manager – Housing and Furniture) and the Product Advisory Committee is acknowledged and appreciated.

Texts

The following resources are suggested as extra information and you should try to read or view them to help you in this unit of competency:

manufacturers' instructions and guidelines
job plans, drawings and specifications
copies of the relevant work health and safety and WorkCover Code/Regulations
basic training manuals.

TAFE NSW (1985), *Theory of oxyacetylene welding for trade students*

Web-based resources

BOC – Guidelines for cylinder safety: **http://www.boc.com.au**
ESAB – Links to CIGWELD and other welding suppliers: **http://apac.thermadyne.com**
Standards Australia: **http://www.standards.org.au**
Standards New Zealand: **http://www.standards.govt.nz**

Audiovisual resources

Short videos covering topics such as oxyacetylene safety, welding hazards, and prevention of eye damage are available from the following organisations:

Safetycare: http://www.safetycare.com.au
TAFE NSW: **https://www.tafensw.edu.au**
Vocam Australia: **http://www.vocam.com.au.**

GET IT RIGHT

The photo below shows two incorrect practices that can be performed when oxyacetylene welding of steel. Identify the incorrect methods and provide reasoning for your answer.

To be completed by teachers

Satisfactory ☐

Not satisfactory ☐

Student name: ______________________

Enrolment year: ______________________

Class code: ______________________

Competency name/Number: ______________________

Task: Review the section 'Prepare for work' up to and including 'Tip cleaners' in this chapter and answer the following questions.

1. What two WHS documents should be completed before commencing oxyacetylene welding processes?

2. When a plumber is welding using fluxes or in a confined area, there can be problems with fumes given off. What safety precautions need to be taken?

3. Name the three shade lenses used for oxyacetylene welding and describe what they are used for.

4. What is the pressure in a full oxygen cylinder and on which gauge of the oxygen regulator is this pressure indicated?

5. What are the colours of the oxygen and acetylene cylinders?

6. At what distance should oxygen and acetylene cylinders be kept apart when stored separately on site?

7. Threads on the oxygen equipment are right-handed and those on the acetylene equipment are left-handed. Why is this so?

8. How can a right-hand thread be distinguished from a left-hand thread?

9. State the function of oxygen and acetylene regulators.

10. State the function of a flashback arrestor.

11. List the two types of flashback arrestors.

12. If, while welding, a flashback occurs in the welding blowpipe, what action should be taken?

13. What is the purpose of tip cleaners?

WORKSHEET 2

To be completed by teachers	
Satisfactory	☐
Not satisfactory	☐

Student name: ______________________

Enrolment year: ______________________

Class code: ______________________

Competency name/Number: ______________________

Task: Review the section 'Transport of oxygen and acetylene cylinders' up to and including 'Weld joint types' in this chapter and answer the following questions.

____

____

____

1. Describe how oxygen and acetylene cylinders should be transported in work vehicles.

____

____

____

____

____

2. Before fitting a regulator to an oxygen or acetylene cylinder valve, the valve seat must be cleaned. How and why is this done?

____

____

____

____

____

3. Oil and grease must be kept clear of oxyacetylene equipment. Explain why this is so important.

____

____

____

____

4. Once the oxyacetylene equipment has been assembled, a plumber must check the equipment for leaks. Explain how this test is carried out.

5. When carrying out fusion welding on mild steel, acetylene is the preferred gas to be mixed with oxygen. Why is this?

6. Three different flames can be obtained when using oxyacetylene equipment for welding. Name and describe these three flames.

7. Neatly sketch and label the various parts of the three different flames (mentioned above) for oxyacetylene welding.

8. Draw two different types of butt joints that a plumber might need to prepare before fusion welding 3 mm mild steel plate. Show any angles or dimensions required.

WORKSHEET 3

To be completed by teachers

Satisfactory ☐

Not satisfactory ☐

Student name: ______________________

Enrolment year: ______________________

Class code: ______________________

Competency name/Number: ______________________

Task: Review the sections 'Perform welding' and 'Clean up' in this chapter and answer the following questions.

1. List the two different filler rods that are used for fusion welding of mild steel and state the differences between the two.

____

____

____

____

____

____

2. What is the most common fusion welding technique used by plumbers?

____

3. In which direction would a right-handed welder travel when using the forehand welding technique?

____

4. When setting up for a forehand fusion weld on 3 mm mild steel plate using an open square butt joint, what are the working pressures required for both oxygen and acetylene and the recommended tip size for the weld?

____

____

____

____

5. What will happen to the welding tip if the weld angle is not correct?

____

____

____

____

6. List the two angles that the filler rod and the welding tip are to be held at when using a forehand welding technique.

7. What is the purpose of tacking the joint together before welding?

8. Draw the tacking sequence that should be used to prevent any distortion from occurring when setting up an open square butt joint on 2 mm mild steel plate.

9. What are the two welding techniques for fusion welding of steel pipe?

10. Describe the following weld defects and what may be the cause of each.

Undercut

Misalignment

Porosity

Distortion

Cracking

11. Neatly sketch the filler rod and welding tip angle and direction of travel required to carry out a forehand fusion weld on 3 mm steel plate.

12. Using a table, list the tip sizes, diameter of filler rod, and oxygen and acetylene pressure settings required to carry out a forehand fusion weld on a single-V joint for the following thicknesses of mild steel plate: 2 mm, 4 mm and 5 mm.

13. With the aid of a neat sketch, show the tack weld sequence required for tacking a butt joint on 50 mm mild steel pipe.

14. Explain the term 'slipped and brazed' when joining copper tube to copper tube.

____

____

____

____

15. Describe how copper tube is branch-formed.

____

____

____

16. What are the three most commonly used silver brazing rods?

____

17. What are their silver content percentages?

____

____

18. What document should be used in conjunction with a hot work permit, that needs to be checked off?

____

WELD USING MANUAL METAL ARC WELDING EQUIPMENT

10B

This chapter addresses the following key elements for the competency 'Weld using manual metal arc welding equipment':

- Prepare for work.
- Identify welding requirements.
- Prepare materials and equipment for welding.
- Weld items.
- Clean up.

It introduces the welding processes that a plumber may be required to undertake in their everyday work, including the identification of **weld** requirements from plans and specifications. It also addresses the safe handling and operating procedures involved with various arc welding processes and gives basic examples of how each is approached and carried out.

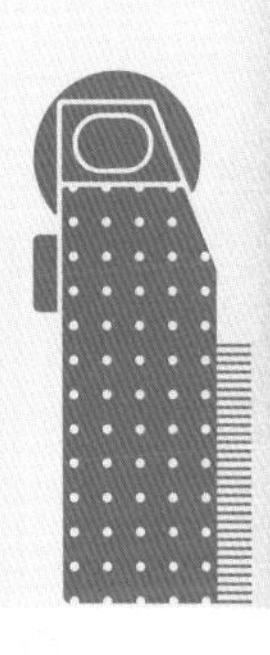

Prepare for work

Plumbers on occasion are required to carry out basic arc-welding applications for the manufacture of brackets on piping installations, general repairs with equipment, and the installation of mechanical services on commercial properties. However, for most mechanical services, a certified welder will be required to undertake this activity.

It is important that anyone operating an arc welder is properly instructed on its safe use by a competent person or welder. Should you wish to have further training with arc welding, it is suggested that you complete a specific course that specialises in this field of work.

When undertaking arc welding processes, there is a high risk for injury to many parts of the body, so strict observation of and adherence to work health and safety (WHS) requirements must be identified in all parts of the process (see Figure 10B.1). Plans and specifications should be obtained from the builder or supervisor to plan and prepare for the work to be undertaken and to comply with associated WHS and environmental requirements. Consideration must also be given to identifying and following quality assurance requirements, in accordance with workplace procedures. Tasks that need to be undertaken must be identified and sequenced in conjunction with other trades or people that may be affected or required to complete the work.

FIGURE 10B.1 A typical arc welding process

It is also important to remember that an emphasis on the correct preparation of materials, tools and equipment will help to ensure that a job is performed safely and effectively.

WHS requirements

WHS requirements must be observed in accordance with relevant state and territory legislation. Risk assessments and Safe Work Method Statements (SWMS) should be completed before commencing arc welding processes. A risk assessment should be developed for all welding jobs or hot works, which should include provision for:

- work to be carried out under hot work permit systems, where required (see Figure 10B.36 at the end of this chapter)
- control of risk from fire or explosions
- control of inadequate ventilation of the work area
- fume extraction
- working in confined spaces.

Confined spaces present several potential hazards, and work permits and accreditation may be required. Precautions for working in confined spaces should also be considered, including adequate ventilation and fume extraction. Appropriate respiratory protection and apparatus should be provided as required as there is a high risk of asphyxiation.

Also refer to work health and safety information in Chapter 1 of this text.

Safe practices, precautions and hazards when arc welding

Arc welding can be a dangerous and unhealthy practice without the proper precautions. However, using new technology (such as fume extraction systems), personal protective equipment (PPE) and safe work practices means the risk of injury or death associated with welding can be greatly reduced. The most common types of injuries that occur during the arc welding process are burns and eye injuries. It is extremely important to protect not only yourself but also those around you (see Figure 10B.2).

Always take the following precautions:

- Wear sufficient closely woven clothing to protect all parts of the body from ultraviolet (UV) light and infrared radiation. This is also to protect you from being burnt by hot metal.
- Use an approved welding helmet (face shield) to protect your eyes and face.
- When changing electrodes, keep yourself insulated from the ground or metal objects nearby, through appropriate footwear and approved insulated mats.
- Whenever possible, or required, use a welder's curtain to protect everyone nearby and the general public from the UV light and infrared radiation. Warn those around you and your fellow workers not to look at the arc. It is a good practice to call out a warning such as 'Watch your eyes' or simply 'Eyes' prior to striking an arc.
- If others must work nearby and you are unable to shield them from the UV light, advise them to obtain and wear appropriate PPE and maintain a safe distance from the area.
- Be careful when making welding machine workpiece connections. Make these direct to the workpiece whenever possible. When connecting to metal structures, make sure you will not cause a fire at a distant point.

Safe Work Procedure
MANUAL METAL ARC WELDER

DO NOT use this equipment unless you have been instructed in its safe use and operation, and have been given permission

PERSONAL PROTECTIVE EQUIPMENT

Safety glasses must be worn at all times in addition to welding mask.

A welding mask with correct grade lens for GMAW must be worn.

Oil free leather gloves and spats must be worn.

Sturdy footwear with rubber soles must be worn.

Long and loose hair must be contained.

Rings and jewellery must not be worn.

Respiratory protection devices may be required.

Close fitting/protective clothing to cover arms and legs must be worn.

PRE-OPERATIONAL SAFETY CHECKS

- ✓ Locate and ensure you are familiar with all machine operations and controls.
- ✓ Check workspaces and walkways to ensure no slip/trip hazards are present.
- ✓ Ensure the work area is clean and clear of grease, oil and any flammable materials.
- ✓ Keep the welding equipment, work area and your gloves dry to avoid electric shocks.
- ✓ Ensure electrode holder and work leads are in good condition.
- ✓ Start the fume extraction unit before beginning to weld.
- ✓ Ensure other people are protected from flashes by closing the curtain to the welding bay or by erecting screens.

OPERATIONAL SAFETY CHECKS

- ✓ Keep welding leads as short as possible and coil them to minimise inductance.
- ✓ Ensure work return earth cables make firm contact to provide a good electrical connection.
- ✓ Ensure the electrode holder has no electrode in it before turning on the welding machine.
- ✓ Ensure current is correctly set according to electrode selection.

ENDING OPERATIONS AND CLEANING UP

- ✓ Switch off the machine and fume extraction unit when work is completed.
- ✓ Remove electrode stub from holder and switch off power source.
- ✓ Hang up electrode holder and welding cables. Leave the work area in a safe, clean and tidy state.

POTENTIAL HAZARDS

- ⓘ Electric shock
- ⓘ Fumes
- ⓘ Radiation burns to eyes or body
- ⓘ Body burns due to hot or molten materials
- ⓘ Flying sparks
- ⓘ Fire

DON'T

- ✗ Do not use faulty equipment. Immediately report suspect equipment.
- ✗ Do not use bare hands and never wrap electrode leads around yourself.

FIGURE 10B.2 Safety requirements for arc welding

- Never use combustible materials to support your work.
- Always wear clear safety glasses or goggles to protect your eyes when cleaning scale and removing slag.
- Keep the electric supply cable to the welding machine safely overhead and out of the reach of anyone standing on the ground.
- Never weld any pipeline, tank or portable container without first obtaining clear proof that it is free from an explosive mixture of vapours.
- When working near an area where there is the chance of causing a fire, keep an appropriate class fire extinguisher within reach.
- Ensure that there is adequate ventilation or fume extraction to help remove harmful fumes.

Do not work with damaged equipment. If equipment needs to be repaired, ensure the work is done by an authorised person.

Personal protective equipment (PPE)

PPE is mandatory when undertaking arc welding tasks, and when used correctly will reduce the risks and hazards associated with arc welding on the worksite. PPE should be provided to all workers from their employer to ensure that all tasks are carried out safely. All workers should receive appropriate instruction and training on the correct use and fitting of PPE to ensure it is used and protects as it was designed. PPE that does not fit or has not been fitted correctly may result in illness or injury and defeat the purpose of wearing the PPE in the first place.

Arc welding requires the plumber to be protected from UV light and infrared radiation associated with the process. At a minimum, plumbers should wear long-sleeved cotton-combination overalls, firm-fitting leather safety boots, long-sleeved gauntlet leather gloves and Australian Standards–approved welding helmets. However, if you are welding in a confined space, or carrying out vertical or overhead welding, then leather apron and spats should be used, along with a cotton cap to protect your head. Nylon or similar fabrics should not be worn at any time when welding as they are flammable.

Listed below are some required items of PPE when undertaking arc welding processes.

Protective leather gloves

Leather gloves, or gauntlets, are required when welding or handling hot materials while using arc welding equipment. They protect the hands from heat, sparks and rays generated by the welding process (see Figure 10B.3).

FIGURE 10B.3 Cotton-lined leather gloves (gauntlets)

Welding helmet

Welding helmets are mandatory PPE items when arc welding and are used to reduce the UV light and infrared radiation produced from the welding process to a satisfactory intensity for eye comfort. If exposed to radiation during the welding process, the skin can be damaged in a manner similar to severe sun burn, and may cause blisters and sores. If the eyes are exposed to radiation, a condition known as welder's flash, or arc eye, can occur, in which the cornea becomes inflamed. At first the eyes will feel itchy, followed by a throbbing pain that may feel like there is sand in the eye. Exposure to UV light for only a few seconds can cause these conditions and symptoms may not be seen until hours after exposure.

An equally important function is to protect the eyes from injury from sparks and weld spatter during welding processes. It is advisable to wear safety glasses underneath the welding helmet to prevent hot slag or sparks coming underneath or over the top of the helmet and causing eye injuries.

Welding helmets range in style and price, with the most common being front-lift and auto-darkening helmets (see Figure 10B.4). Auto-darkening helmets are battery- or solar-operated and have a darkening response time of less than a millisecond. Front-lift helmets require the operator to flip the filtered lenses up when chipping slag or observing the weld, or to flip the filtered lenses down just before welding takes place.

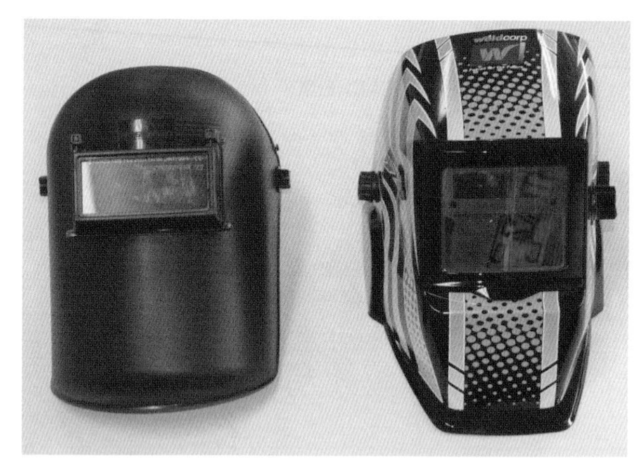

FIGURE 10B.4 Welding helmets: front-flip (left); auto-darkening (right)

Filter lenses are specially designed to filter out harmful radiation and are graded in accordance with *AS/NZS 1338 Filters for eye protectors, Part 1: Filters for protection against radiation generated in welding and allied operations*.

10B.1 STANDARDS

- AS/NZS 1338 Filters for eye protectors, Part 1: Filters for protection against radiation generated in welding and allied operations

Table 10B.1 provides the recommended filter lenses for arc welding processes.

TABLE 10B.1 Recommended protective filters for manual metal arc welding (MMAW)

Recommnended Protective filters for electric welding		
Description of process	**Approximate welding current range**	**Minimum shade number of filter(s)**
Manual metal arc welding (MMAW)	≤ 100	8
	100 to 200	10
	200 to 300	11
	300 to 400	12
	>400	13

Foot protection

It is mandatory to wear protective footwear when undertaking welding practices at all times. Footwear should conform to *AS/NZS 2210.1: Safety, protective and occupational footwear, Part 1: Guide to selection, care and use.*

- AS/NZS 2210.1: Safety, protective and occupational footwear, Part 1: Guide to selection, care and use

All safety footwear must have:

- stout soles or steel midsoles to protect against sharp objects and hot metals
- good uppers to protect against sharp tools and materials
- reinforced toecaps to protect against heavy falling objects (see Figure 10B.5).

Preparation of work area

The use of screens in the work area is required to prevent injury and the possibility of welder's flash affecting people working near or passing by the area. Barricading areas may be required to ensure that there is no unauthorised access, and welding fume extraction and filtration systems may be required in confined areas. Extraction units can be set up as a centralised system or an individual unit, depending on the location and requirements of the welder. In addition to extraction systems, well-ventilated areas are essential, and permanent or temporary mechanical ventilation may be required.

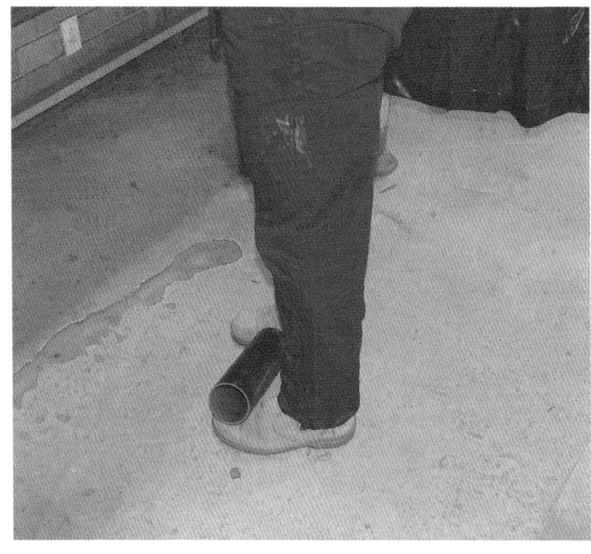

FIGURE 10B.5 Steel falling on boot

The work area should be prepared to sufficiently support materials during the welding process. Ensure that any flammable material and rubbish is well clear of the work area, as sparks will fly during the arc welding process and can travel a long distance, staying hot for some time. Always alert those around you when undertaking welding tasks to prevent injuries and avoid scheduling issues.

LEARNING TASK 10B.1

PREPARE FOR WORK

Provide short answers to the following questions:

1. What are three requirements that must be considered once plans and specifications are obtained?
2. A risk assessment must be completed before undertaking arc welding. What precautions must be considered?
3. What is the importance of fume extraction and ventilation?
4. What are the minimum PPE requirements for arc welding?
5. What is the purpose of using screens during arc welding?

Identify welding requirements

Plans and specifications should always be sought when preparing materials and welding equipment for the workplace. The materials and weld requirements must be identified and selected in accordance with workplace procedures.

Arc welding process

Manual metal arc welding (MMAW), or stick welding as it is informally known, is a type of welding that uses a power supply to join materials together by creating an electric arc between an electrode and the parent metals (see Figure 10B.6). Arc welders can use either direct current (DC) or alternating current (AC) along with consumable electrodes. During the welding process the weld area is protected by an inert or semi-inert gas, known as a shielding gas, which is produced from the flux on an electrode.

The welding current on a machine is adjustable to suit the type and thickness of a material and is measured in amps. The temperature produced is approximately 6000°C, which is almost twice that generated by oxyacetylene welding. Although its proportion of the total welding market decreases yearly, it is still a widely used and popular welding process because of its low running costs.

Before arc welding can begin, it is important to identify and select the material to be welded, in accordance with workplace procedures. This information can be obtained through plans and specifications, and can be used to establish materials, weld requirements and the location of welds.

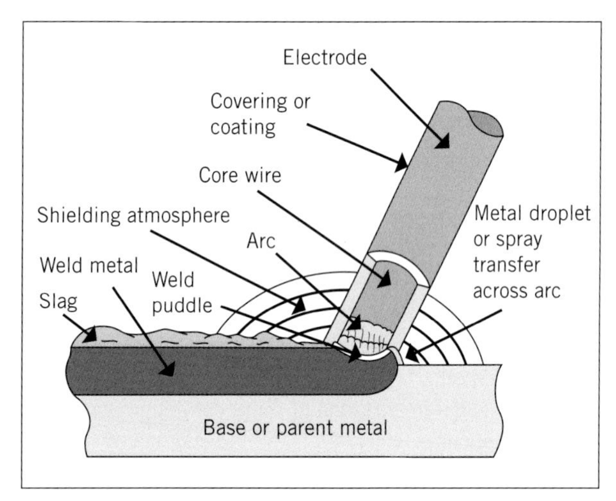

FIGURE 10B.6 Metallic welding arc

Welding symbols

There are many terms and symbols related to welding that tradespeople need to be able to interpret in order to be able to read plans and specifications and be proficient in the skill of welding.

AS 1101.3 Graphical symbols for general engineering, Part 3: Welding and non-destructive examination lists the approved welding symbols to be used on all plans relating to steel plate and pipe fabrication. A plumber may be required to interpret and apply these symbols when fabricating or installing work of this nature. Figure 10B.7 gives examples of welding symbols and how they are applied.

10B.3 STANDARDS

- AS 1101.3 Graphical symbols for general engineering, Part 3: Welding and non-destructive examination

Weld joint types

The selection of the type of joint that is required depends on the thickness of the material, where the weld is, how much stress the weld will be under, and, in the case of pipework, what the service is to be used for.

The thickness of the material will also determine how many weld runs will be required to make a sound joint. There are several different types of joints that can be undertaken – as the welder gains more experience, they may attempt them and become more skilled.

Below is a list of commonly used joints when undertaking the arc welding process.

Butt joints

Butt joints are commonly used for most jointing techniques, as they are easy to prepare and, for the skilled welder, easy to apply.

Open square butt joint

For this joint, the clean, square edges are spaced by half the thickness of the metal to be joined. It is generally used for metal with a thickness of between 1.6 mm and 3 mm. This may also require a capping run, as well as tacking the joint at about 50–75 mm intervals to minimise joint distortion (see Figure 10B.8). On thinner materials below 2 mm in thickness, arc welding can be difficult and, in some circumstances, unsuitable.

Single-V butt joint

The single-V butt joint is a joint that requires the edges of the parent metal to be bevelled. Each edge requires bevelling by filing, grinding or using an oxyacetylene cutting torch to form a 60° included angle when assembled. Penetration is assisted by leaving a gap between the two pieces of metal, which is dependent upon the thickness of the metal to be welded (see Figure 10B.9). For metal that is between 3 mm and 5 mm thick, the gap is half the thickness of the metal; for metal that is greater than 5 mm thick, a 3 mm gap is required. The weld is formed by welding on the bevelled edges and building it up to a desired contour. Vertical welds of this type require an 80° included angle. This may also require a capping run to give build-up and strength.

Double-V butt joint

This type of joint is for metal 16 mm thick and over, and where possible the weld is applied from both sides of the joint (see Figure 10B.10). The material must be prepared so that one side of the joint forms a 60° included angle and the other side an 80° included

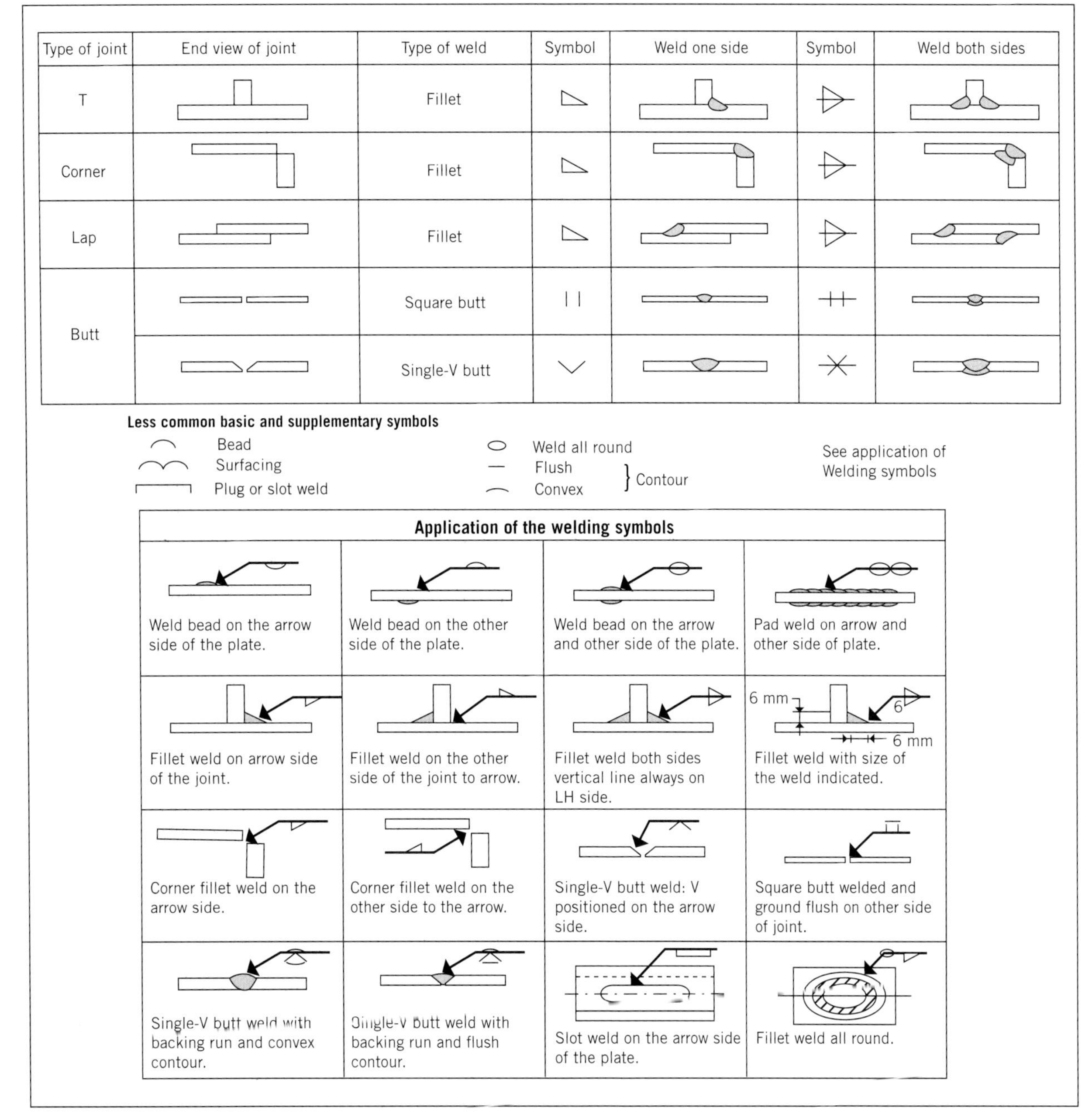

FIGURE 10B.7 Welding symbols and their applications

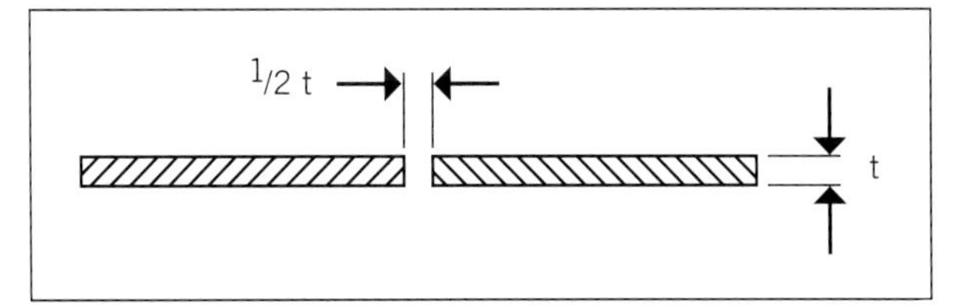

FIGURE 10B.8 Open square butt joint for 1.6 mm to 3 mm thick low-carbon steel

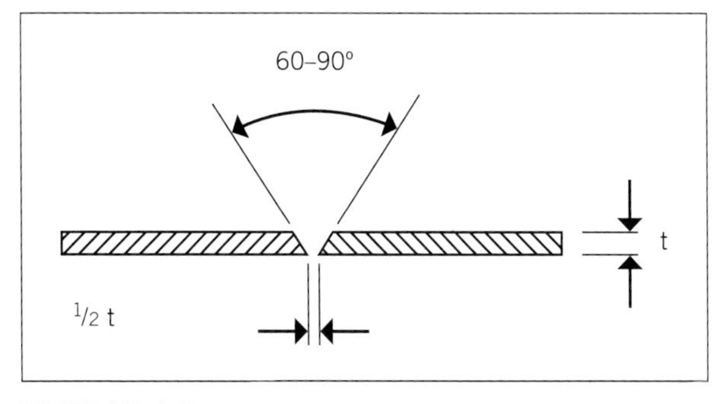

FIGURE 10B.9 Single-V butt joint for 3 mm to 15 mm low-carbon steel

angle. The weld is started on the 60° angle side and completed by welding the 80° angle side. A gap of 3 mm is left between the joint.

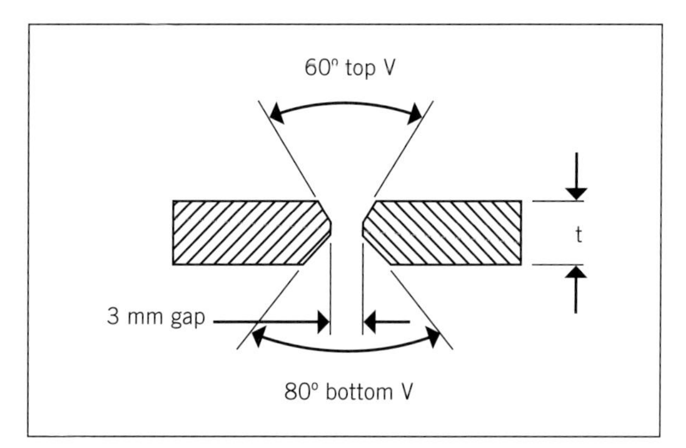

FIGURE 10B.10 Double-V butt joint for 16 mm and over low-carbon steel

Fillet joints

The fillet weld is one of the most common types of welds as it is strong and requires minimal preparation.

T-fillet joint

A T-fillet joint is a common type of fillet joint that requires no special preparation except for cleaning. The two pieces of material must sit firmly against each other at 90° (see Figure 10B.11), with welds applied on both sides of the upright where it meets the base.

Lap fillet joint

This joint is when materials are lapped, and the edges welded (see Figure 10B.12). Where possible this joint should be welded on each edge, as the strength of the weld is halved if both are not welded.

Outside-corner fillet joint

This is sometimes called a butt joint, although the weld that is applied is a fillet weld and is used on corner joints (see Figure 10B.13). This requires minimal preparation as the angle is generally 90° and the metal requires cleaning before welding. In this case, the weld fills the corner between two flat-edged pieces of metal.

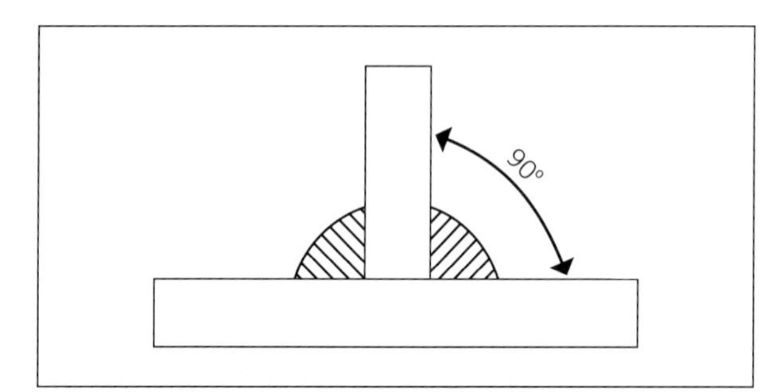

FIGURE 10B.11 T-fillet

FIGURE 10B.12 Lap fillet

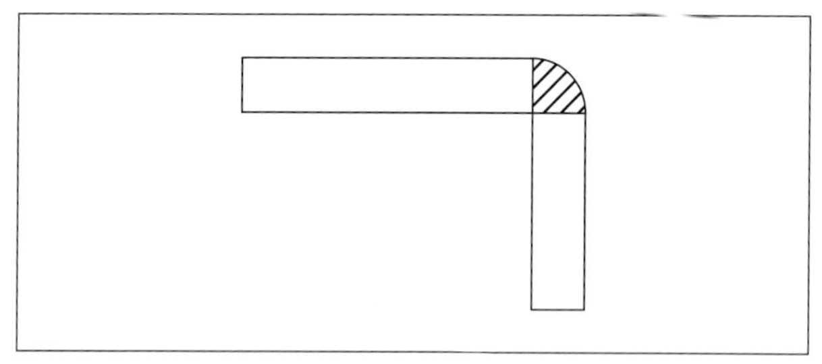

FIGURE 10B.13 Corner fillet

LEARNING TASK 10B.2

IDENTIFY WELDING REQUIREMENTS

You are required to determine the type of joint required and preparation necessary to join two 150 mm × 50 mm steel plates with a thickness of 6 mm. Answer the following questions:

1 Select the most suitable joint to complete the task.
2 What preparation is required before welding is to commence?
3 What angle of bevel is required?
4 How would the bevelling be undertaken?
5 What gap is required between the plates before welding?

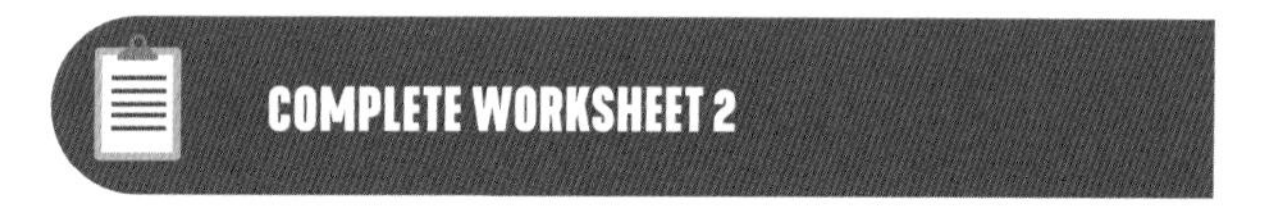

Prepare materials and equipment for welding

Once the welding requirements of the job have been determined through plans and specifications, the welding equipment can be assembled and set up in accordance with manufacturers' guidelines and regulatory requirements in preparation for the job.

Material preparation

Preparation of material will be determined from joint selection and is based on information that has been obtained from job specifications and workplace procedures. This will include cleaning surfaces and bevelling materials over 3 mm thick. All surfaces must be free from contaminants such as rust, mill scale dirt, paint, galvanised coatings, and oil and grease before welding, or the weld quality will be unacceptable and dangerous fumes may be given off. These contaminants may be removed by grinding, filing, wire brushing or degreasing.

As the thickness of weld metal increases, it is harder to maintain full fusion (penetration) through the full thickness of the metal. This means that a form of edge preparation is required to ensure full fusion of the joint. Bevelling is used to prepare the edge of thicker metal and this can be done by flame cutting, filing or grinding. Joint preparation is very important in gaining a high-quality, strong weld and applies to all arc welding processes. A joint that has an overly wide gap, or is dirty, rusty or greasy, will not give a good result. The design of the work piece will influence the structural integrity of the type of joint, which will then determine what preparation is required.

Fundamentals of arc welding

An arc welding machine requires a constant supply of electricity, usually from mains power at 415V or 240V. The actual welding process uses lower voltages for safety reasons (below 80V). The welding machine reduces the mains power voltage to 45–80V for striking the arc, and once welding begins the voltage drops again to 20–35V (see Figure 10B.14).

The process is very versatile, requiring little operator training and inexpensive equipment. However, weld times are rather slow, since the consumable electrodes must be frequently replaced and because slag, the residue from the flux, must be chipped away after welding. The process is generally limited to welding ferrous materials, although specialty electrodes have made possible the welding of cast iron, nickel, aluminium, copper and other metals. The versatility of arc welding makes it popular in several applications, including repair work and construction.

FIGURE 10B.14 Voltage drop during the arc welding process

Source: © Copyright Commonwealth of Australia, 1978.

Set up welding equipment

After the welding machine has been connected to the power source, the electrode cable should be connected to the terminal labelled 'Electrode', and the work cable should be connected to the terminal labelled 'Work' or 'Work piece'.

The work cable, which is supplied with a solid clamp, must make good, clean contact with the workbench or job to be welded (see Figure 10B.15). As all arc welding machines vary, it is essential that the manufacturer's specifications are consulted to ensure that there is correct connection of arc welding cables to the arc welding machine.

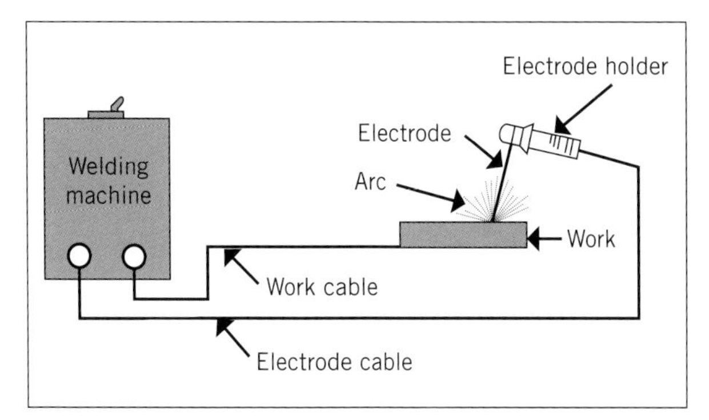

FIGURE 10B.15 Diagram showing how the circuit is made for arc welding

Source: © Copyright Commonwealth of Australia, 1978.

Note: Cables may often be referred to as leads.

Welding set-up considerations

When setting up for arc welding, the correct electrode (size and type) and correct welding current (measured in amps) need to be selected.

Electrode selection

Electrodes are manufactured from a metal wire coated in flux (see Figures 10B.16 and 10B.17). The composition of flux coating varies depending on the material and type of weld to be undertaken. Different electrodes are made for welding all types of metals, including mild steel, high-tensile steel, high-carbon steel, stainless steel, aluminium, copper and copper alloys.

Electrodes are manufactured in a range of diameters from 1.6 mm up to 11 mm, and lengths of between 250 mm and 450 mm. The diameter of an electrode refers to the diameter of the wire, not the overall diameter including the flux (see Figure 10B.18).

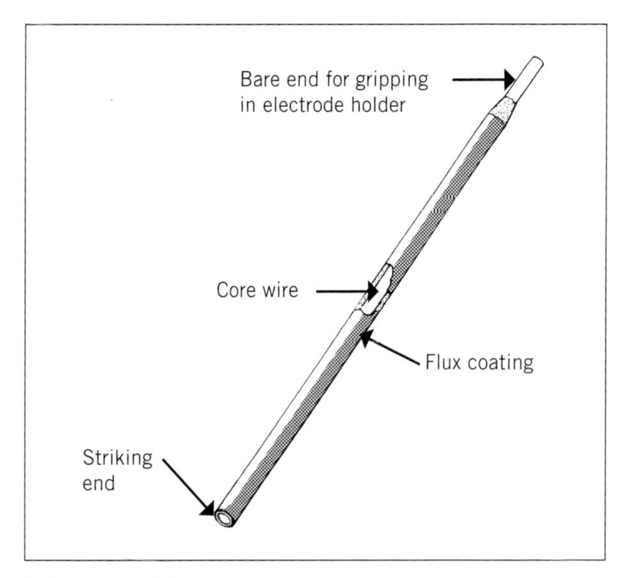

FIGURE 10B.16 An electrode

FIGURE 10B.17 Typical electrodes

Source: Shutterstock.com/Rus S.

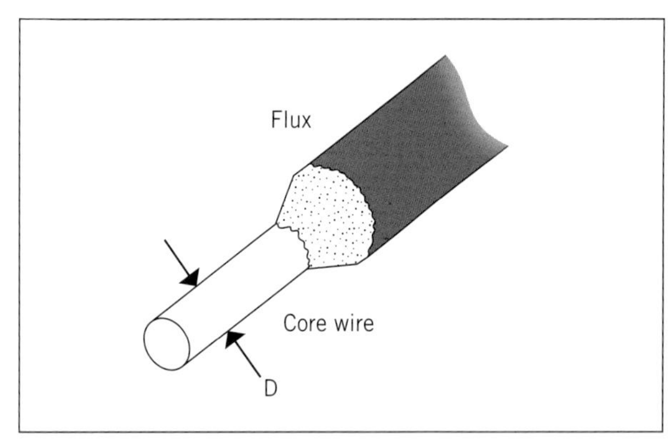

FIGURE 10B.18 Size of electrode

Source: © Copyright Commonwealth of Australia, 1978.

The flux coatings on electrodes serve different purposes during the welding process. The common functions of flux are to:

- provide a gas shield in order to prevent contamination of the weld metal by the atmosphere
- stabilise the arc by controlling ionisation around the arc
- control the fluidity of the weld metal and the penetration and shape of the weld bead
- form a protective layer of slag over the weld metal in order to prevent oxides from forming during solidification
- add alloying elements to the weld that are not in the core wire, or to add deoxidants to the molten pool.

The size and type of electrode depends on the thickness of the material to be welded and the position of the weld. Plumbers would most commonly use a general-purpose steel electrode either 2.5 mm or 3.2 mm in diameter, which is suitable for most steel arc-welding techniques.

Welding current (amps) selection

The selection of the correct welding current for a job is extremely important (see Table 10B.2). The recommended amps range for an electrode is printed on the packet that houses the electrodes, and it is best to try to weld near the high end of the recommended range if possible. The following illustrates the importance of correct current selection:

- *Too low an amperage* will cause difficulty in striking the arc, and penetration will be poor.
- *Too high an amperage* will make the job overheated, and cause undercut, burning through of material and excessive spatter.
- *Normal amperage* is considered the maximum without burning through the work, overheating the electrode or producing a rough, spattered surface.

TABLE 10B.2 A guide for general arc welding of mild steel

Material thickness (mm)	Electrode size (mm)	Current range (amps)
Up to 3	2.5	55–80
3–5	3.25	95–130
5–8	4.0	130–180
Over 8	5.0	170–230

Consult electrode suppliers or manufacturers for information on the correct electrode for a specific purpose. Most manufacturers provide information booklets.

The electrode is clamped in a hand-held electrode holder by the operator (see Figure 10B.19), and is manually guided along the joint as the weld is made.

The electrode coating burns in the intense heat of the arc and forms a blanket of gas and slag that completely shields the arc and weld puddle from the atmosphere and provides the filler metal for the weld.

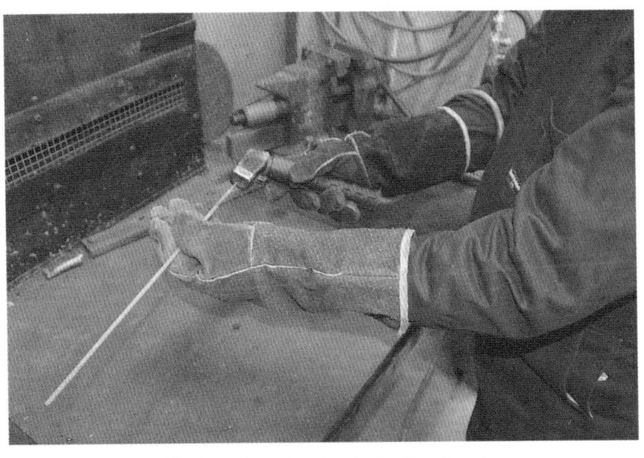

FIGURE 10B.19 Fitting the electrode to the holder

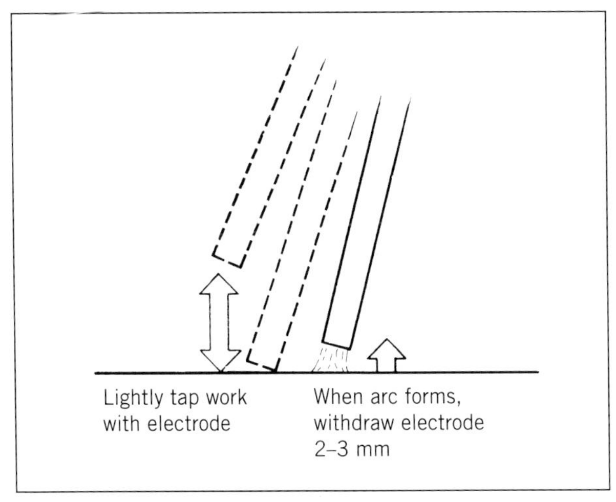

FIGURE 10B.20 Striking the arc

Source: © Copyright Commonwealth of Australia, 1978.

LEARNING TASK 10B.3

PREPARE MATERIALS AND EQUIPMENT FOR WELDING

Provide short answers to the following questions:

1 How would you select the correct electrode before arc welding?
2 What is the purpose of the flux coating on an electrode?
3 When arc welding 6 mm steel plate, what size electrode and amps range would be selected?
4 What is the purpose of slag during the arc welding process?

COMPLETE WORKSHEET 3

Weld items

Once materials are cleaned, prepared and all necessary equipment has been set up, the items can now be arc welded in accordance with plans, specifications and job requirements.

Welding technique

Apart from the selection of electrode and current, the following factors are critical to producing a sound weld:

- striking the arc
- electrode angle
- correct arc length and speed of travel.

Striking the arc

One of the hardest aspects of arc welding for a beginner is striking the arc. The arc must be struck like a match is lit. Scraping the end of the electrode across the work, or lightly tapping the electrode (see Figure 10B.20) and then rapidly lifting it slightly are two ways in which to strike the arc. However, it's often not as easy as it seems. It's a good idea to practise striking the arc on a piece of scrap, and adjusting the amps and seeing the effect this has on the arc striking. The arc should be the shortest that will produce a good weld (e.g. 1.5 mm), so be careful not to lift it too far after the arc has started.

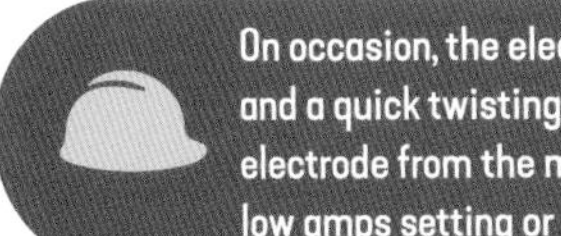

On occasion, the electrode may stick to the material and a quick twisting action will generally release the electrode from the material. This is often caused by a low amps setting or incorrect technique when welding.

It is good practice to call out the warning 'Watch your eyes' or 'Eyes' before you strike the arc and ensure approved welding screens are drawn and in good condition. This helps to protect those in the vicinity of the welding area from getting 'welder's flash' and possibly damaging their eyes.

Electrode angle

The angle at which the electrode comes into contact with the job dictates whether there is a smooth, even transfer to the metal. For most arc welding applications, the electrode should be pointing back towards the weld pool against the direction of travel at an angle of 70° to the joint and should bisect the angle formed by the parent metals to be joined (see Figures 10B.21 and 10B.22).

Correct arc length and speed of travel

The arc length is the distance between the electrode and the material that is to be welded. Achieving the correct arc length and speed is probably the most difficult part of the process to learn, yet it is essential to produce a neat weld run. This takes considerable practice to achieve.

The simple rule for proper arc length is to maintain the shortest arc, which gives a good weld appearance. If the arc is too long, penetration is reduced, and spatter and a rough weld surface are produced. If the arc is too short, the electrode could stick to the job and a rough weld with slag inclusion may result.

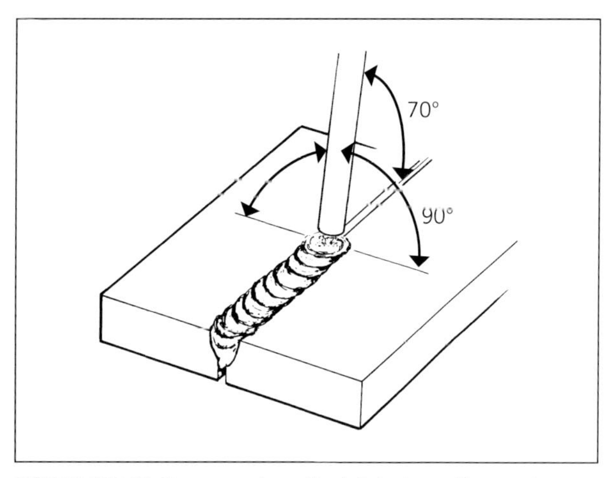

FIGURE 10B.21 Progress along the joint at a uniform rate

Source: © Copyright Commonwealth of Australia, 1978.

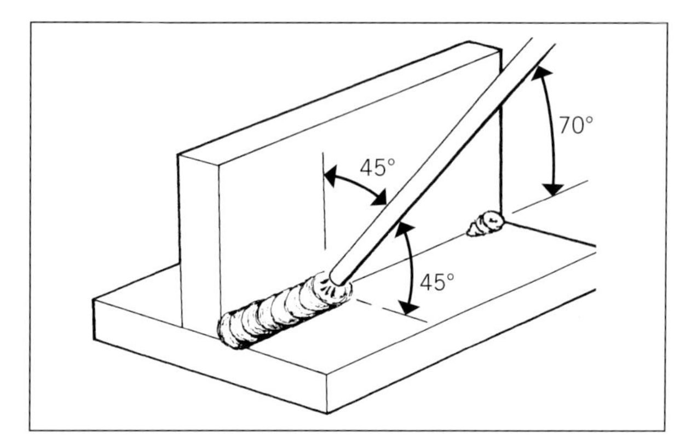

FIGURE 10B.22 Single-pass fillet weld T-joint

Source: © Copyright Commonwealth of Australia, 1978.

One trick is to learn to look beyond the arc to the pool of molten metal. It will tell you a lot more about what is going on than the appearance of the arc itself.

Travel speed is also critical:

- Travel speed that is too fast produces poor fusion and lack of penetration.
- Travel speed that is too slow produces arc instability, slag inclusion and overheated metal.

It is also important to know what good and bad welds look like and the reasons why bad ones appear the way they do. Figure 10B.23 provides both plan and cross-sectional views of finished welds and shows where errors may occur. Cutting a practice piece in half and assessing the penetration of the bead is worthwhile to see how the welds are progressing.

Another way to see what is happening is to deliberately weld in error. What does a weld look like when the current is way too high, or far too low? What is the bead's appearance when welding too slowly, or far too rapidly? If you deliberately create the error, you'll learn to recognise when a bad weld occurs.

Running multiple beads, one over the top of the other, and weaving from side to side are often suggested for the experienced welder. If you're a competent welder and apply multiple beads, make sure that the slag is removed after running each bead (see Figure 10B.24). However, if you're a beginner, using multiple beads and weaving will potentially create slag inclusions and voids in the weld.

Tacking materials

Metals expand when heated and contract when cooled. Tack welds are used to hold the material prior to the welding process and will stop distortion and contraction (see Figure 10B.25) from taking place while welding. Distortion can occur with high heat applied to the weld area, but can be reduced by clamping the job while welding. Tacks must fuse for the full thickness of material or root preparation.

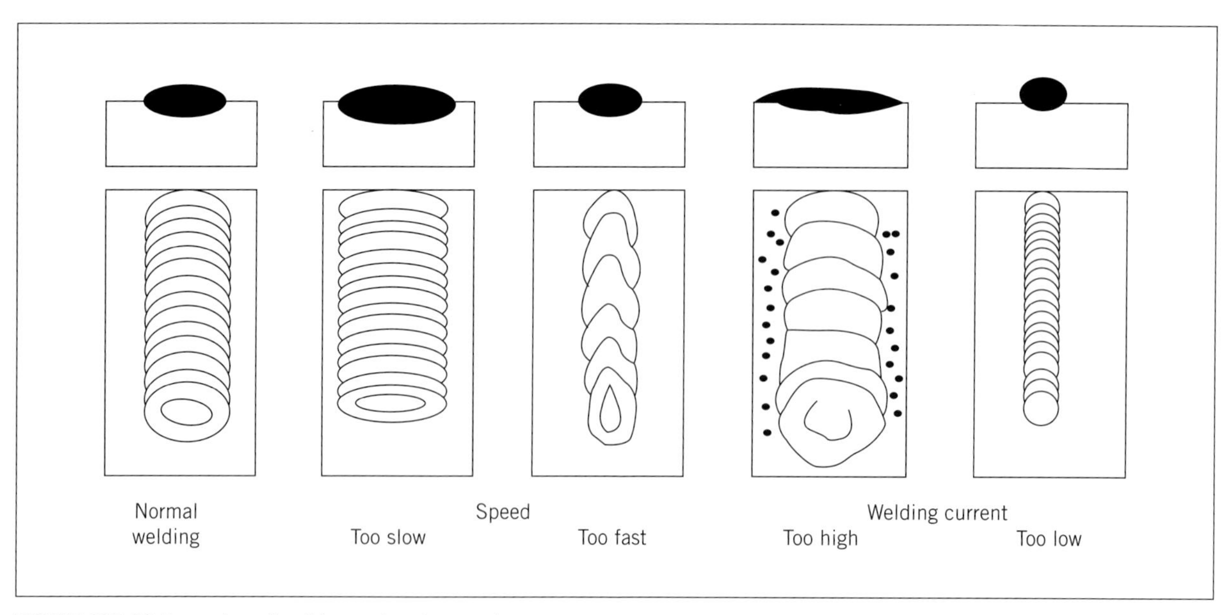

FIGURE 10B.23 Examples of weld speed and current

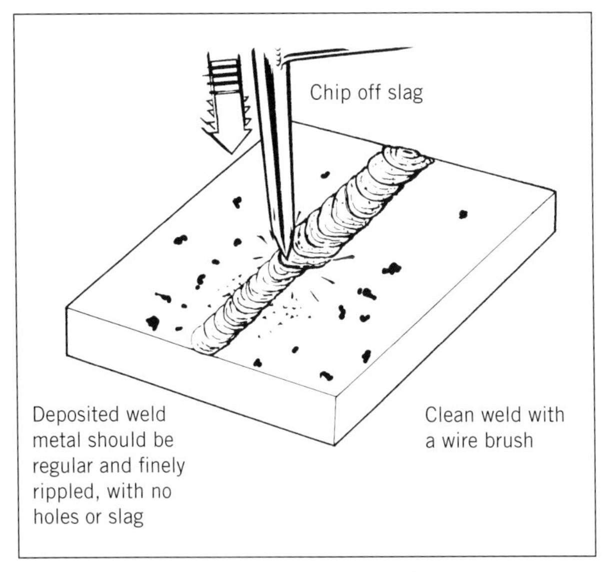

FIGURE 10B.24 Clean and inspect the weld

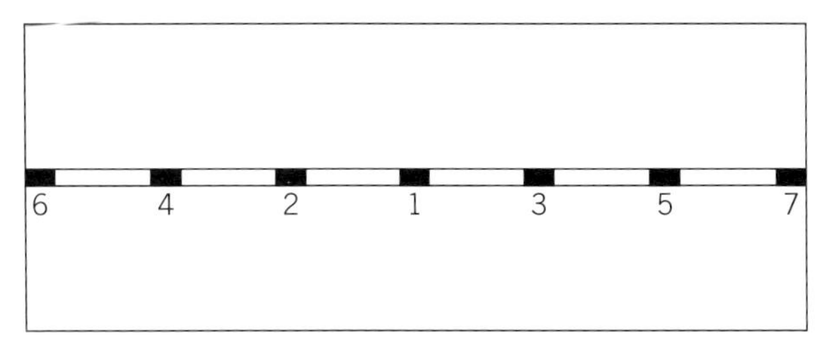

FIGURE 10B.26 A recommended tacking sequence for mild steel plate

A tacking sequence (see Figure 10B.26) is recommended when applying tacks to minimise distortion. For material between 1.6 mm and 2.5 mm, a tack is recommended every 75 mm, and for material over 2.5 mm, a tack is recommended every 100 mm.

Weld defects

There are many weld defects that can occur within the weld or in the parent metal, and these can be detected by a close visual inspection and, where the weld is critical, by X-ray.

Undercut

Undercut is defined as a groove or channel in the parent metal along the toe of the weld caused by melting away the parent metal (see Figure 10B.27). It is generally a sign of lack of skill or carelessness, and may be caused by the following faults in technique:

- poor electrode manipulation
- too fast a rate of travel.

Undercut is a serious weld defect that can lead to a weld failure, such as stress or fatigue failure, when in service as it reduces the wall thickness of the parent metal.

Overroll

Overroll (also called overlap) is a fault in the weld where molten weld metal overflows on the unmelted parent metal, leaving an unfused area (see Figure 10B.28). This may result from faults in welding techniques, such as:

- lack of skill of the welder
- using the wrong size electrode.

Overroll can affect the weld strength in the same way that undercut can, because of the notch effect of the weld metal overlapping the parent metal, with no fusion taking place.

Misalignment

Misalignment (see Figure 10B.29) is a result of either poor preparation of the parent metals before the welding or movement occurring during the welding process. This may happen due to:

- the job not being assembled or aligned correctly
- the job has not been clamped correctly to prevent movement
- the tacks being too small and not fused correctly, causing cracking during the welding process by allowing the material to move before the weld is completed.

Misalignment can cause serious weakness and a lack of strength and serviceability in the welded joint, and in the case of service pipes can restrict flow and pressure.

Excessive penetration

Excessive penetration (see Figure 10B.30) can occur on butt welds, with the main cause being a lack of control

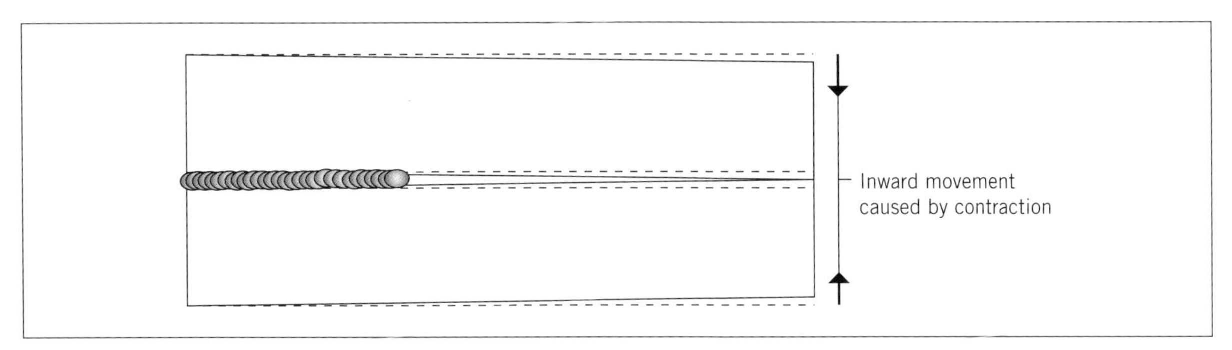

FIGURE 10B.25 Example of contraction occurring on a steel plate that was not tacked before welding took place

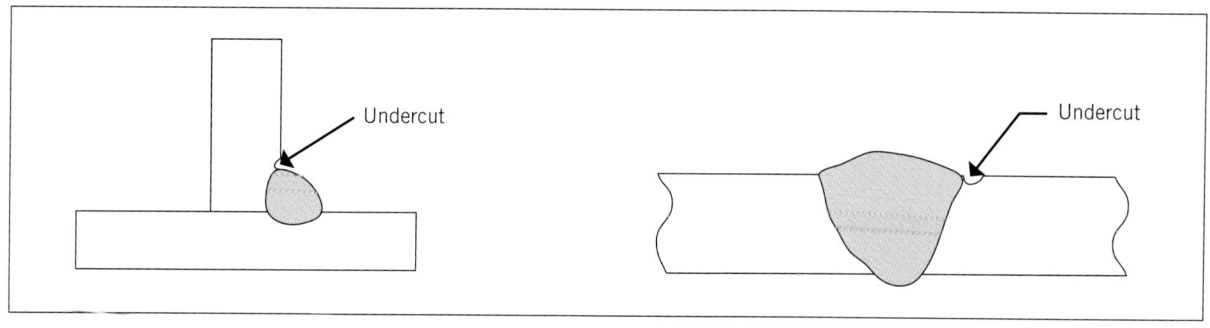

FIGURE 10B.27 Examples of undercut in a T-fillet and a butt weld

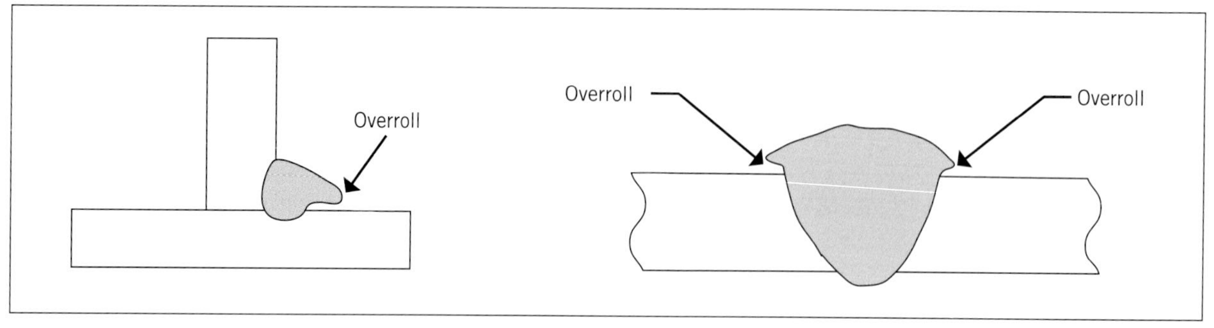

FIGURE 10B.28 Examples of overroll in a fillet weld and a butt weld

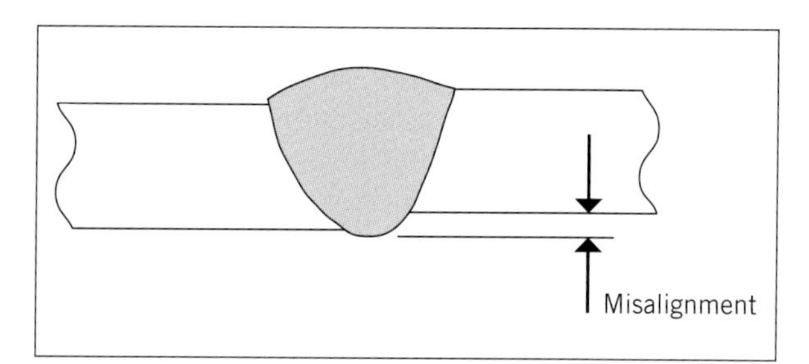

FIGURE 10B.29 Typical example of misalignment

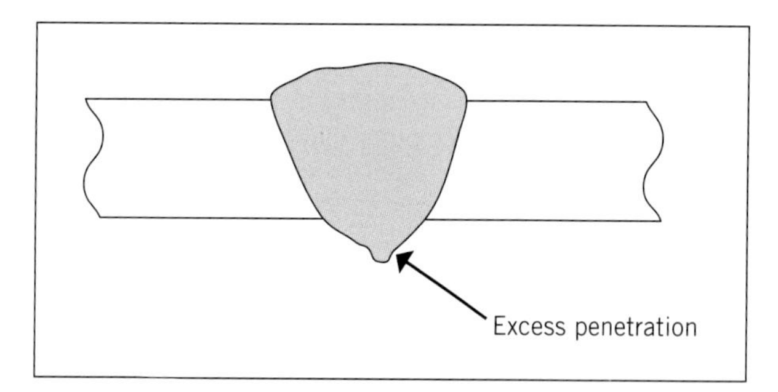

FIGURE 10B.30 Butt weld with excess penetrationc

of the weld pool, which causes unwanted metal to protrude through to the underside of the joint. This can cause:

- stress
- joint rigidity
- flow restriction in pipes.

Incomplete penetration

This is the failure of the weld metal to fill the root of the weld. As the weld has not obtained full fusion, this means that it is not as thick as the parent metal and is therefore a source of weakness. Incomplete penetration (see Figure 10B.31) can be caused by:

- poor preparation
- too little current
- a lack of skill or carelessness on behalf of the welder
- a lack of fusion because of the failure to melt the underlying metal when depositing the weld metal
- using the wrong size electrode.

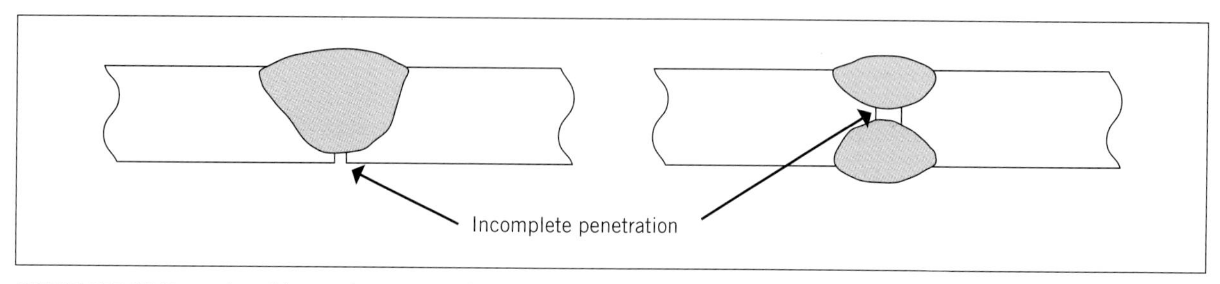

FIGURE 10B.31 Examples of incomplete penetration – the failure of the weld metal to extend into the root of the joint

Porosity

Porosity generally refers to small round holes in the weld metal (see Figure 10B.32) caused by gas trapped in the weld metal. These holes, which can occur in clusters or be scattered, may be caused by:

- gases given off by chemical actions within the molten metal
- gases given off by paint or chemicals on the parent metal
- incorrect use of electrode causing slag inclusion.

Cracking

When cracking (see Figure 10B.33) occurs within the weld, it may be caused by:

- contaminants on or in the parent metal
- the use of an unsuitable electrode
- clamping the joint too rigidly
- the weld beads being too small.

Distortion

Distortion (see Figure 10B.34), as applied to welding, includes any departure from alignment or measurements resulting from the welding process. It is caused by:

- uneven expansion and contraction from localised heating of the parent metal when welding
- residual stresses that are present in the parent metal.

Lack of fusion

Lack of fusion (see Figure 10B.35) is the incomplete fusion between the added weld metal from the electrode and the parent metal during the welding operation. It can be caused by:

- using an incorrect weld angle
- weld travel that is too rapid
- using the wrong weld technique for the thickness of metal.

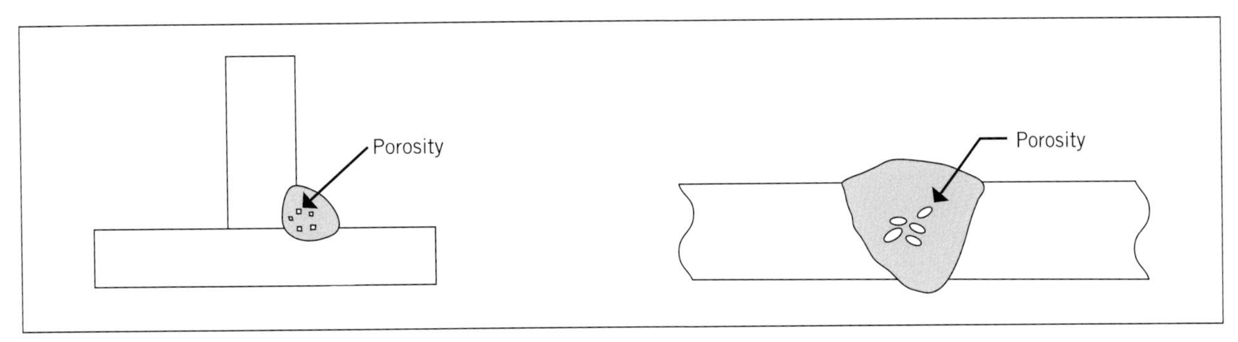

FIGURE 10B.32 Examples of porosity

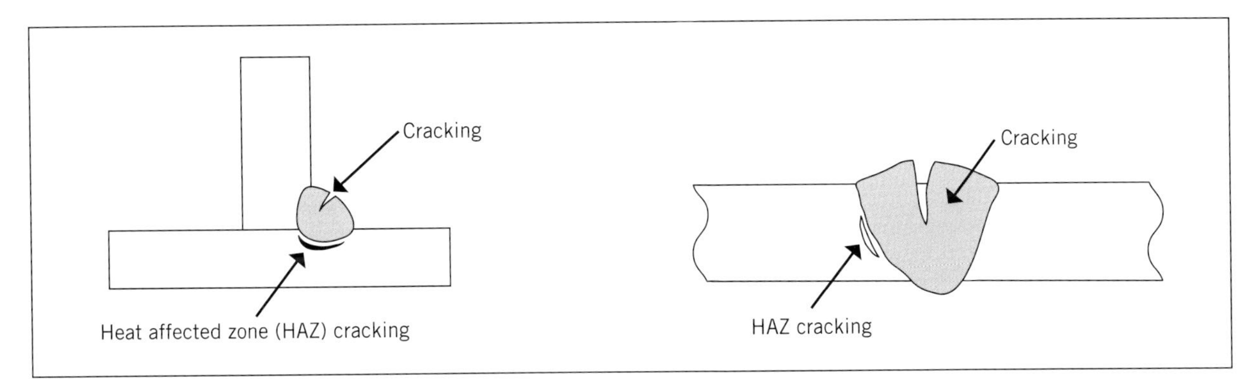

FIGURE 10B.33 Examples of cracking

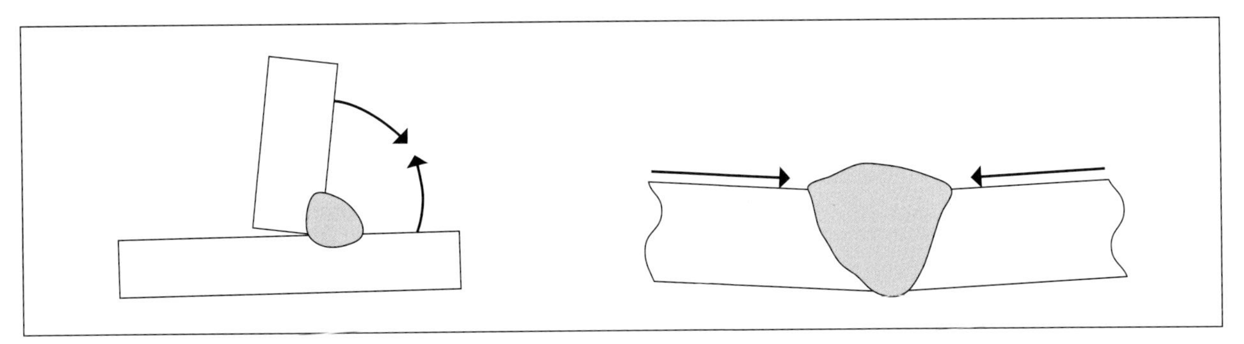

FIGURE 10B.34 Typical distortion in butt and fillet welds

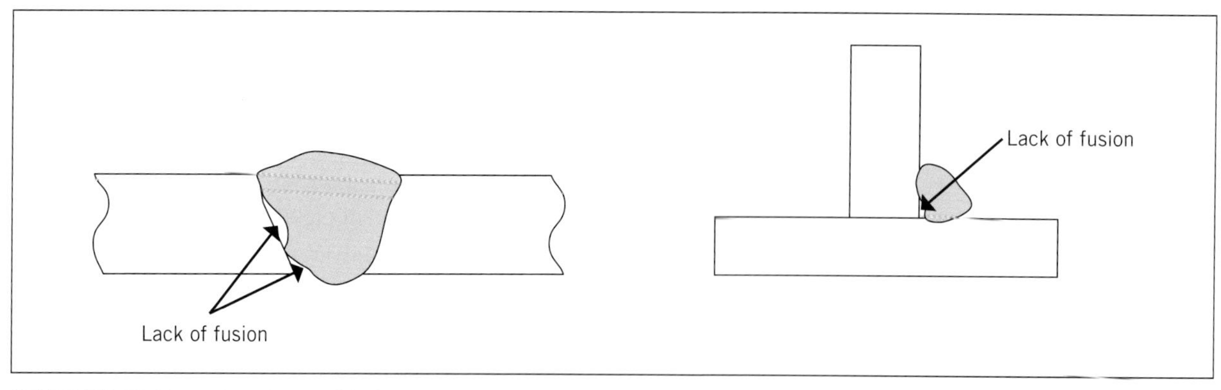

FIGURE 10B.35 Examples of lack of fusion

When chipping slag from completed welds, goggles that cover the eyes entirely must be worn as hot, hardened slag may come over the top of ordinary safety glasses and cause potential burns and injuries to the eyes.

LEARNING TASK 10B.4

WELD ITEMS

Provide short answers to the following questions:

1 When you are about to weld, what should you do to alert others working around you?
2 At what angle pointing backwards should the electrode be travelling at when welding?
3 When practising arc welding, what is the purpose of welding in error?
4 Is running multiple beads suggested for the beginner welder?
5 Why does distortion occur when welding?
6 How are critical welds checked for defects?

Clean up

Cleaning up at the end of each job is essential to ensure WHS, workplace and regulatory requirements are met. When the job is completed, it is important to turn off all power, clean up the work area and return all tools and equipment. It is equally important to remove all the slag from finished joints, clean them thoroughly with a wire brush or similar, then treat the joints to help prevent them from rusting. Any materials that are no longer required should then be reused or stored appropriately and safely in an area that can be located easily. Sustainability principles and concepts must be observed, where applicable, and materials that can be recycled should be disposed of in accordance with state and territory requirements.

GREEN TIP

Ferrous metals such as steel can be recycled, which reduces potential waste that may otherwise go to landfill.

Maintenance of tools and equipment

Maintenance of tools and arc welding equipment is essential to ensure that all equipment is in good working order and is safe before every use. Before carrying out any maintenance, all electrical equipment must be isolated or switched off.

The welding machine must be kept dry and dust-free, and machine terminals should be clean and tight. Cables and clamps should be checked for loose joints to prevent overheating and unstable arcing. Electrode and return cables should be checked for damage prior to and after use, and rolled up neatly and stored appropriately when not in use. If any equipment is found to be faulty, your job supervisor should be notified immediately.

Accessing and completing documentation

When plumbers are required to carry out any work, they must complete a risk assessment and develop a SWMS for each job and site. A hot work permit may also be required before using electric arc welding equipment on large construction sites or maintenance contracts. This is not only for the company's own risk assessment but also may be an insurance requirement.

Hot work permits must be completed in accordance with *AS 1674.1 Safety in welding and allied processes – Fire precautions*. It should have a precaution checklist that needs to be checked off for each permit; it should be checked before, during and after any hot works (see Table 10B.3 as an example). A typical hot work permit (see Figure 10B.36) should include such information as company name, contact person, description of work, equipment being used, start and finish dates, emergency procedures to follow if something should go wrong, and a list of all actions that must occur before work is begun.

10B.4 STANDARDS

- AS 1674.1 Safety in welding and allied processes – Fire precautions

TABLE 10B.3 Precautions checklist for hot works

Precautions Checklist			
General precautions	**Yes**	**No**	**N/A**
Have flammable and combustible materials been removed or protected?			
Are sprinklers, hose reels and extinguishers available and operating?			
Has the floor been swept clean and wetted down where required?			
Are arc welding curtains required and in place?			
Is there adequate ventilation and fume extraction?			
Is all welding/heating equipment well maintained?			
Is a fire watcher required (an extra person on watch)?			
Has the fire panel been isolated (building security)?			
Have the smoke/thermal detectors been isolated (building security)?			
Have management and/or maintenance been informed of details of the hot work?			
Precautions within 15 m of hot work			
Have all combustible liquids, vapour and gases been removed or protected?			
Have combustible floors been protected?			
Has all flammable dust and lint been removed or protected?			
Has any explosive atmosphere in the area been eliminated?			
Have all wall and floor openings been covered?			
Work on walls or ceilings			
Are walls/ceilings constructed of non-combustibles and without any combustible coverings or insulation?			
Have all combustible materials on the other side of the wall/ceiling been moved?			
Are there fire-resistant coverings under the work area to collect sparks?			
Work on enclosed equipment			
Has the enclosed equipment been cleaned of all combustibles?			
Have all containers been purged of flammable liquids/vapours?			
Fire watcher			
Is a fire watcher required?			
Has a fire watcher been organised (if required)?			
Has the fire watcher been trained in the use of equipment and sounding the fire alarm?			
Are the appropriate fire extinguishers full and ready for use?			
Has the fire panel been reinstated on completion of work?			
Other precautions that may be highlighted from the risk assessment			

Permit No.	This permit must be completed for all cutting, welding and other hot work performed outside a dedicated workshop area. The permit must be displayed at the work site and returned on completion of work.

Application for hot work

Company/Dept Performing Work for :

Contact Name: **Tel: ( )**

Location of Work:

Description of Work:

Equipment to be used:

Permit begins	**Permit expires**
Date: / / Time: am/pm	**Date: / / Time: am/pm**

Emergency information

If a fire occurs, call **Tel: ( )**

Nearest fire alarm:

Authorisation by company representative

The above work is authorised to proceed subject to the following action being taken prior to work starting and procedures being maintained for the duration of the work. Each item is to be checked by the Authorised Company Representatives prior to work starting for each period (delete and initial if and where not applicable).

Authorised by: **Signed:** **Date: / /**

1	Fire sprinklers and/or thermal detectors must be confirmed as operational (where installed).	❑	6	Combustible materials located within 10 m must be removed or protected with non-combustible curtains, metal guards or flame-proof covers (not ordinary tarpaulins). In a retail/office environment, if 15 m clearance is not practical then the largest distance possible (minimum of 3 m) is acceptable.	❑
2	Smoke detectors must be isolated in the work area and impairment procedures followed.	❑	7	All floor and wall openings within 15 m must be covered to prevent transmission of sparks.	❑
3	Fire equipment to be provided as follows: • Fire hose reel • Fire extinguisher Mandatory fire watcher present	 ❑ ❑ ❑	8	The hot work area and any adjoining areas must be patrolled from the start of work until 30 minutes after the work is completed (including break periods).	❑
4	Barricades, warning signs and spark/flash screens must be provided.	❑	9	Special conditions. **(Please detail)**	❑
5	Work area, trenches, pits, etc. must be clear of flammable liquids, gases or vapours.	❑			

Work completed and area safe

The work area has been inspected by the Authorised Company Representative 30 minutes after completion of work.

Signed: **Date: / /** **Time: am/pm**

This permit is only valid for 24 hours. Ensure contractor returns this form.

FIGURE 10B.36 Typical hot work permit that may be required prior to commencement

REFERENCES AND FURTHER READING

Acknowledgements

Reproduction of the following resource list references from DET, TAFE NSW C&T Division (Karl Dunkel – Program Manager – Housing and Furniture) and the Product Advisory Committee is acknowledged and appreciated.

Texts

The following resources are suggested as extra information and you should try to read or view them to help you in this unit of competency:

- manufacturers' instructions and guidelines
- job plans, drawings and specifications
- copies of the relevant WH&S and WorkCover Code/Regulations
- basic training manuals.

TAFE NSW (1985), *Theory of oxyacetylene welding for trade students*

Web-based resources

BOC – Guidelines for cylinder safety: **http://www.boc.com.au**
ESAB – Links to CIGWELD and other welding suppliers: **http://apac.thermadyne.com**
Standards Australia: **http://www.standards.org.au**
Standards New Zealand: **http://www.standards.govt.nz**

Audiovisual resources

Short videos covering topics such as oxyacetylene safety, welding hazards, and prevention of eye damage are available from the following organisations:

- Safetycare: **http://www.safetycare.com.au**
- TAFE NSW: **https://www.tafensw.edu.au**
- Vocam Australia: **http://www.vocam.com.au**.

GET IT RIGHT

The photo below shows an incorrect practice that can be performed when arc welding. Identify the incorrect method and provide reasoning for your answer

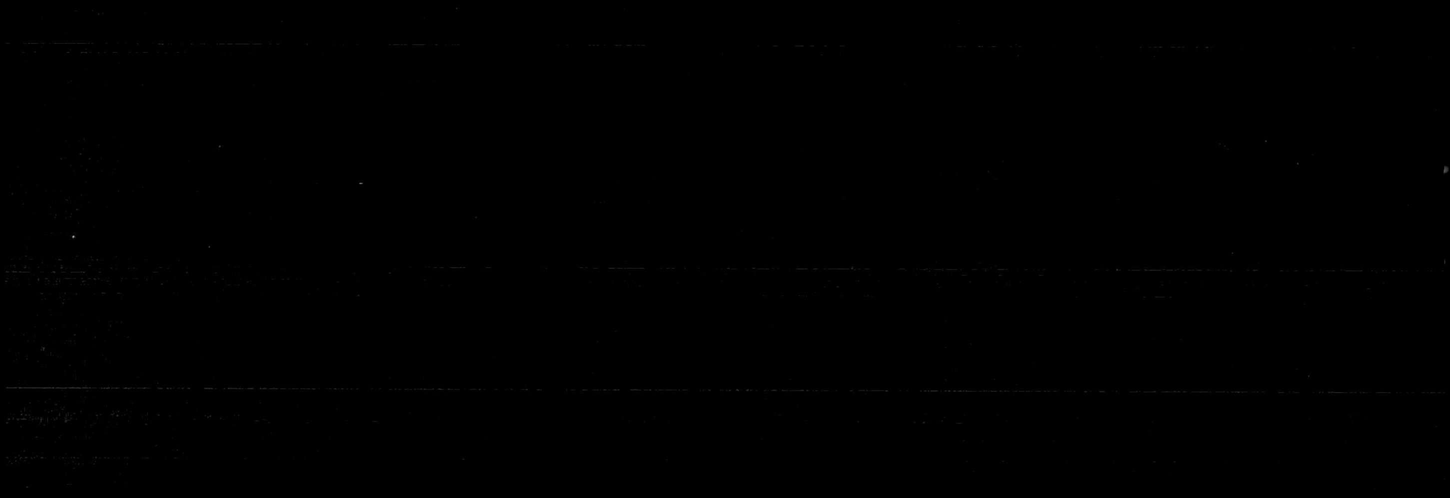

WORKSHEET 1

To be completed by teachers	
Satisfactory	☐
Not satisfactory	☐

Student name: ______________________

Enrolment year: ______________________

Class code: ______________________

Competency name/Number: ______________________

Task: Review the section 'Prepare for work' in this chapter and answer the following questions:

1. What two WHS documents should be completed before commencing arc welding processes?

____

____

2. What hazard is considered to be a high risk when working in a confined space?

3. How should dangerous fumes created by the welding process be removed?

4. To what area of the body is the most common injury sustained during arc welding?

5. How would you protect other workers on site when arc welding?

6. How would you keep yourself insulated from the ground when arc welding?

7. List four different safe practices and precautions that should be followed while arc welding.

____

____

____

____

8. What is the minimum recommended shade filter when the arc welding machine is set at 120 amps?

____

9. What is the purpose of a barricade when preparing the work area?

____

WORKSHEET 2

To be completed by teachers	
Satisfactory	☐
Not satisfactory	☐

Student name: ______________________

Enrolment year: ______________________

Class code: ______________________

Competency name/Number: ______________________

Task: Review the section 'Identify welding requirements' in this chapter and answer the following questions:

1. What is the approximate temperature that is produced by the arc welding process?

____

2. Explain the meaning of the following welding symbols that would be found on a typical welding drawing.

i.

____

____

ii.

____

____

iii.

____

____

iv.

____

____

3. List the three common butt joints used in arc welding processes.

____

____

____

4. Describe a T-fillet joint.

____

____

5. When arc welding steel plate, why would a single-V butt joint be selected?

____

WORKSHEET 3

To be completed by teachers	
Satisfactory	☐
Not satisfactory	☐

WORKSHEETS

10B

Student name: ______________________

Enrolment year: ______________________

Class code: ______________________

Competency name/Number: ______________________

Task: Review the section 'Prepare materials and equipment for welding' in this chapter and answer the following questions.

1. When preparing material for welding, why should the material be free from contaminants?

____

____

2. Why is bevelling used and how is it performed?

____

____

3. Describe, with the aid of a sketch, how the circuit is made for arc welding.

4. List five factors that contribute to successful manual arc welding.

____

____

____

____

____

5. List the functions of the flux coating found on an electrode.

6. How is welding current measured?

7. How is the size and type of electrode selected when arc welding?

WORKSHEET 4

To be completed by teachers

Satisfactory ☐

Not satisfactory ☐

Student name: ______________________

Enrolment year: ______________________

Class code: ______________________

Competency name/Number: ______________________

Task: Review the sections 'Weld items' and 'Clean up' in this chapter and answer the following questions.

1. List the three points that must be taken into consideration to develop correct welding technique.

____

____

____

____

2. Describe why you should warn others around you before striking an arc.

____

____

3. List the two ways how to strike an arc on the parent metal.

____

____

4. What is the purpose of a tacking sequence?

____

____

5. List two weld defects that could occur due to an incorrectly sized electrode.

____

____

6. Describe the following weld defects and what may be the cause of each.

Undercut

____

____

Misalignment

____

____

Porosity

Distortion

Cracking

7. What is a hot work permit and when may it be required?

8. Refer to the glossary and define the following arc welding terms.

Current

Polarity

Slag

Slag inclusion

Tack weld

Voltage

GLOSSARY

A

acetone A flammable and volatile liquid used as a solvent in acetylene cylinders to dissolve and stabilise acetylene under pressure

acetylene (C_2H_2) A highly combustible gas that is used as a fuel gas in oxyacetylene welding; when burned with oxygen it produces a flame temperature of around 3100°C

acute health effect An adverse health effect from the short-term exposure to a chemical

alloy A mixture of two or more metals mixed together to form another metal with characteristics different from those of the parent metal

alternating current An electrical current in which magnitude and direction change in cycles (as opposed to direct current, in which direction is constant)

annealing A process of gradually heating a metal to make it soft for mechanical working; it will relieve any stresses that may result from a welding operation

arc welding (AW) A welding power supply creates an electric arc between the base material and an electrode to melt the metals at the welding point

arc welding cables The lines that carry the current from the machine to the welding materials; they are made up of lots of copper wires woven into one in order to conduct electricity, and covered with a non-conductive rubber or plastic wrap

availability The system/goods work instantly when required

award The law that establishes minimum wages and conditions of employment for specific occupations

B

backfire A loud snapping or popping noise caused when the blowpipe flame goes out suddenly and then re-ignites

backhand weld Welding with the flame pointing in the direction opposite to the weld progress (towards the finished portion of the weld)

bevel A sloping edge on the metal prepared before welding in order to obtain a single butt joint

blowpipe An instrument designed to bring together and mix acetylene and oxygen in such a manner that the mixture, when ignited, will produce a controlled flame

bonding The union of a surface that in braze welding is achieved by intergranular penetration

braze welding This does not rely on capillary action and the parent metal is not melted but the joint design is similar to that used for a fusion weld; the filler rod is generally a non-ferrous alloy with a melting point above 500°C

brazing A jointing process in which a molten filler rod is drawn between two closely fitting surfaces by capillary action, the filler rod being a non-ferrous alloy with a melting point above 500°C but lower than the melting point of the metal being joined

brittleness The ease with which a metal can be fractured if subjected to forces of bending or impact

bronze weld A term that has been used to describe a braze weld in which a copper-rich filler rod is used

Building Code of Australia (BCA) A uniform set of technical provisions for the design and construction of buildings and other structures throughout Australia

butt joint A weld in which the two edges of metal are abutted together

C

chronic health effect An adverse effect from the long-term exposure to a chemical

Codes of Practice A practical guide that provides guidance to achieve the standards of health and safety required under relevant Work Health and Safety (WHS) Acts and model WHS regulations

coefficient of linear expansion The amount of change in length of a material in relation to every degree Celsius of temperature change

coefficient of volumetric expansion The amount of change in volume of a material in relation to every degree Celsius of temperature change

cone The part of the oxyacetylene flame that is conical in shape at the end of the welding tip

current The flow of electricity in the electric circuit; what you are welding on resists the flow, and that forms heat; amps are the measurement of your current

customer A person who pays a professional to undertake work on their behalf

cylinder A steel or aluminium container for storing and transporting industrial gases such as compressed oxygen

D

datum point Any known point, line or level from which a level line may be transferred to another position

direct current The constant flow of electrons in the single direction from low to high potential; also known as continuous current

double-V butt joint A butt joint in which the edges of the weld metal are double bevelled so the fusion faces form two opposite Vs

ductile or malleable Materials that can be drawn into lengths or flattened out without breaking

durable Refers to a material that will last for a suitable length of time

duty of care A legal obligation that requires a worker/employee to adhere to a standard of reasonable care in the prevention of foreseeable harm to others

E

earth leakage circuit breaker (ELCB) A safety device that is used to prevent someone from being electrocuted if they come into contact with a bare wire by immediately cutting off electricity when a leakage of current is detected

elastic Refers to the ability of a deformed material to return to its original shape

electrode A rod made up of filler metal with a flux coating on it that is designed to aid and protect the bead during the welding process

electrode holder The 'handle' portion of the arc welder that holds the electrode in place

expansion coefficients The different rates at which materials expand and contract when they are heated or cooled

expansion fracture Cracking that occurs when a welded joint stresses due to expansion and contraction

F

filler rod A metal rod that is melted with the oxyacetylene flame and deposited in the molten weld to supply additional metal

fillet weld A weld joint made in a corner, as in a lap or a T-joint

fire protection equipment Equipment such as fire hoses and fire extinguishers used to prevent, control and extinguish fires

first aid and safety equipment Equipment such as emergency showers, emergency eye washes and emergency exits used in first aid situations

flanged butt joint A welded joint in which the two edges to be welded are turned up and fused together with or without the use of filler rod

flashback The burning back of a flame into the tip or blowpipe and hoses

flashback arrestor A safety device that is installed on oxygen and acetylene welding equipment and is designed to stop the reversal of flow or a flame towards the regulators and cylinders

flux A powder or chemical paste used to clean the metal and remove oxides while preventing oxidisation when welding or brazing and to assist in the flow of the filler rod; when burned, flux makes a shielding gas that protects the weld pool or puddle from atmospheric contaminants that cause defects

forehand welding Oxyacetylene welding with the flame pointing in the direction in which the weld progresses

fusible plug A safety plug made of a low-melting alloy, which is fitted in an acetylene cylinder to help prevent explosions if the cylinder is subjected to heat

fusion weld The welding technique in which the edges of the metals being joined are melted and completely fused together without pressure and with or without the use of a filler rod

G

Globally Harmonised System of Classification and Labelling of Chemicals (GHS) A single internationally agreed system of chemical classification and hazard communication through labelling and Safety Data Sheets (SDS)

H

heat of fusion When heat is added to a material to change it from a solid to a liquid

heat of vaporisation When heat is added to a material to change it from a liquid to a gas

helmet (face shield) A safety device worn over the face to protect the eyes and face from the arc, sparks and molten metal; arc welding without proper eye protection can lead (quickly) to permanent damage to the eyes

hot work permit A procedure that is followed to monitor hot work activities, which identify the hazards, risks, control measures and safe practices of work with a potential to create a source of ignition

I

incident report form Records serious work-related illnesses, injuries or dangerous occurrences as required by the regulatory authorities of the various states and territories

inner cone The bright, short part of the oxyacetylene flame that comes from and is in contact with the orifice of the welding tip

invert level (IL) A vertical height taken from the bottom internal surface of drainage pipework

L

lap joint A type of joint formed by two overlapping pieces of metal with the edges of the metals welded to the face of the other

level line Any horizontal line that is parallel to the surface of still water

levelling The determination and representation of the elevation of points on the surface of the earth from a known datum, using a surveyor's level to measure the differences in elevation by direct or trigonometric methods

M

maintainability Parts and service are readily repaired if necessary

mixing chamber The part of the oxyacetylene welding blowpipe in which the gases are mixed, ready for combustion

N

neutral flame An oxyacetylene flame in which the inner cone is neither oxidising nor carburising, which means there are equal parts of oxygen and acetylene

Noble Scale The degree to which metals and semi-metals are resistant to corrosion or oxidation

O

oxyacetylene welding A welding process that depends on the combustion of oxygen and acetylene to produce the heat for welding

oxygen A colourless and odourless gas that supports combustion; when the correct mixture of oxygen and acetylene is burnt, a flame of approximately 3100°C is obtained

P

parent metal The metal on which the weld is deposited

PCBU (Person Conducting a Business or Undertaking) Can range from a sole trader or a small partnership through to a large company. Commonly referred to as an employer

penetration A dimensional expression of the depth of the fusion below the original surface of the parent metal in a weld

personal protective equipment (PPE) Designed and manufactured to provide protection from a specific hazard to a particular part of the body; it's the last line of defence to protect your health and safety from workplace hazards. It must fit properly, and must be properly cleaned and maintained

plumb any vertical line or surface that is exactly vertical

polarity The direction of current flow; polarity can be obtained only on a direct current machine

Q

quality The level of excellence that goes into a product or service

quantity A defined quantity of materials that is calculated and required for a particular task

R

recognition of prior learning (RPL) A process that allows learners to obtain accreditation for skills, knowledge and experience gained previously through learning and working

reduced level (RL) A vertical height or elevation taken in relation to a datum point

regulator A device used to control the delivery of a gas at a pre-set constant pressure regardless of the cylinder pressure

reinforcement Where another weld (or welds) is added to the initial weld and overlaps the toes of the initial weld

reliability A product will continue to work for its guaranteed life

reportable illness One where a worker has a medical certificate stating that they are suffering from a work-related illness that stops the worker from carrying out their usual duties for a continuous period that is specified by the relevant WHS state or territory regulator

reportable injury One occurring from a workplace incident where a fatality occurs or a person cannot carry out their usual duties for a continuous period that is specified by the relevant WHS state or territory regulator

residual current device (RCD) A safety device that is used to prevent someone from being electrocuted if they come into contact with a bare wire by immediately cutting off electricity when a leakage of current is detected

root The zone at the bottom or inmost part of the space for, or occupied by, a fusion weld

S

Safe Work Australia An agency working with members that assists with the implementation of model work health and safety legislation and the coordination and development of national policy and strategies

Safe Work Method Statement A document that sets out the high-risk construction work activities to be carried out at a workplace.

Safety Data Sheet (SDS) Previously called a Material Safety Data Sheet (MSDS); a document that provides information on the properties of hazardous chemicals and how they affect health and safety in the workplace

silver brazing (silver soldering) A low-temperature brazing process in which a silver alloy is used as filler metal

single-V butt joint A joint used for welding in which the two edges to be joined are each bevelled at one side only

slag Where the flux on a welding rod melts it produces a shielding gas to protect the weld, which hardens to a protective coating over the weld; this slag has to be chipped off and thoroughly cleaned, usually by wire brushing

slag inclusion One of the main weld defects: a chunk of slag left in the bead; if you don't properly clean the slag from a bead, you run the risk of it becoming part of the weld when you run the next bead

specific gravity (relative density) A measurement that is used to compare the densities of solids, liquids and gases

specific heat capacity The amount of heat required to raise a substance by 1°C

specification A precise description of written instructions that are used in conjunction with plans that list the materials, colours, finishes, style and workmanship of the job to be completed

supplier A person giving customers goods and materials with which to work

swarf Potentially hazardous metal, plastic or wood debris resulting from machining, woodworking and/or manufacturing

T

T-fillet joint A joint where two plates or pipes are located at 90° to each other

tack weld A temporary weld used to hold two edges of metal in a proper alignment while the full weld is carried out; used extensively in fabricating, tack welds can be easily broken off if a change needs to be made

tip The removable end of the welding blowpipe; it has an orifice from which the gas issues before ignition

toe The point at which the contour of the weld metal joins the parent metal

tubing hose Reinforced rubber hose built to resist the pressure of the gases and to withstand constant bending and twisting

V

vertical weld A weld in a position where the axis of the weld is vertical

voltage The pressure required to move the electric current

W

'Watch your eyes' or 'Eyes' A warning that lets people know you are about to strike an arc; you should always call out this warning before you strike an arc, so that people don't get flash burn

weld A union between two or more pieces of metal where the faces are rendered plastic or liquid by heat and filler metal may or may not be used

weld defect Defects such as undercut, porosity, slag inclusion, and underfill, all of which can adversely affect a weld (usually causing a crack which weakens it); weak welds can damage equipment or materials, injure or even kill

weld metal The metal that has been melted in making the weld, including the filler rod and the parent metals, which is different from the unmelted parent metal

welding rod A means of delivering filler metal to the weld; in arc welding, the rod is called an electrode

worker Someone who is assigned to work by another party (the PCBU), whether they are in paid employment or working as a volunteer

INDEX

Note: Bold page numbers refer to definitions; italic page numbers to figures.

T

U